Natural Product Chemistry for Drug Discovery

RSC Biomolecular Sciences

Titles in the Series:

1: Biophysical and Structural Aspects of Bioenergetics
2: Exploiting Chemical Diversity for Drug Discovery
3: Structure-based Drug Discovery: An Overview
4: Structural Biology of Membrane Proteins
5: Protein–Carbohydrate Interactions in Infectious Disease
6: Sequence-specific DNA Binding Agents
7: Quadruplex Nucleic Acids
8: Computational and Structural Approaches to Drug Discovery: Ligand–Protein Interactions
9: Metabolomics, Metabonomics and Metabolite Profiling
10: Ribozymes and RNA Catalysis
11: Protein–Nucleic Acid Interactions: Structural Biology
12: Therapeutic Oligonucleotides
13: Protein Folding, Misfolding and Aggregation: Classical Themes and Novel Approaches
14: Nucleic Acid–Metal Ion Interactions
15: Oxidative Folding of Peptides and Proteins
16: RNA Polymerases as Molecular Motors
17: Quantum Tunnelling in Enzyme-Catalysed Reactions
18: Natural Product Chemistry for Drug Discovery

Preface

Natural products hold a special place in drug discovery having provided and inspired numerous life saving medicines and medical breakthroughs, particularly in the treatment of infectious diseases, cancer, hypercholesterolemia and immunological disorders. Twenty one drugs approved for marketing between 2003 and 2008 owe their existence to natural product leads discovered from mainly actinomycete, bacteria and fungal sources.[1] It has been our intention with this book to not only provide insights into the likely sources and methodologies that may be used to discover new natural product based drugs in the future, but also to stress the utility and importance of this approach to drug discovery in terms of new clinical candidates and recent commercial successes.

The final section of this book provides fascinating accounts of the twists, turns and pitfalls, as well as the role serendipity played, in the successful development and commercialisation of daptomycin and micafungin. Accounts of natural product derived drug candidates which are currently being evaluated in clinical trials may be found in Chapters 11–13, with salinosporamide A and bevirimat described in detail. The pipeline of 36 drug candidates which are in late stage clinical development may imply a continuing role for natural products in drug discovery, but we will return to this issue towards the end of this preface. Before then, let us look at the earlier chapters which follow in this book.

Well known for their thorough analyses of the sources of new and approved drugs, Newman and Cragg set the scene in Chapter 1 with a discussion on the historical influence natural products have had on the drug discovery process, with particular emphasis on antibacterial, antifungal and anticancer agents. The reason for the success of natural product chemistry in drug discovery is multifactorial, but certainly includes the unique "chemical space" that is occupied by such molecules. A particularly elegant account of how this applies to the required physicochemical property space for antibacterial compounds

RSC Biomolecular Sciences No. 18
Natural Product Chemistry for Drug Discovery
Edited by Antony D. Buss and Mark S. Butler

Published by the Royal Society of Chemistry, www.rsc.org

has been made recently by O'Shea and Moser.[2] In Chapter 2, Singh and Culberson expand on this theme with a comparison of the diversity of natural products with various synthetic compound libraries and their impact as drug leads in general.

La Clair, in Chapter 3, adopts a cinematic approach whilst delving into the mechanistic modes of action and the complex roles that natural products play. Included in this account are descriptions of how natural products have led to a better understanding of the regulation of tubulin and actin assembly in tumour cells and to the identification of an array of new, putative anticancer drug targets.

In Chapter 4, Cordell thoroughly evaluates the impact of the Convention on Biological Diversity (CBD) and other related agreements on academic and industrial natural product research. While the CBD has resulted in the development of laws and practices that have protected the sovereign rights of countries over their genetic resources, it has also led to natural product research programmes being compromised in scope and has perhaps contributed, at least in part, to many pharmaceutical companies terminating their natural product research activities.

Plants, microorganisms and, to a lesser extent, macromarines have been the main sources of natural product based drugs (produced as secondary metabolites). Reviews of these traditional sources of naturally occurring chemical compounds are found in Chapters 5–7, together with hints and suggestions as to how these sources may be better utilised to continue supplying new drug leads in the future.

Advances in high throughput screening technology, particularly with regard to detection methods and readouts, are reviewed in Chapter 8. These advances in biological screening, coupled with improvements in chromatographic and analytical techniques (highlighted in Chapter 9), have led to a significant reduction in the time required to purify active compounds from complex mixtures and to determine their chemical structures. In addition to conventional natural product discovery approaches, new versions of two major classes of natural products, the non-ribosomal peptides and polyketides, can now be engineered and produced using genetic manipulation techniques because of the ability to correlate gene sequence with amino acid sequence and thus, the chemical structure of the biosynthetic product. In Chapter 10, Udwary reviews the advances made in this field of combinatorial biosynthesis over the last 15 years, together with an account of some of the significant technical limitations that still need to be overcome before the rational engineering of biosynthetic pathways can be more readily harnessed for drug discovery.

We promised to return to the earlier statement that a healthy development pipeline of natural product derived candidates implies that natural products will still have a role to play in modern day drug discovery. In fact, this is far from reality. Firstly, these late-stage clinical candidates reflect the output from research activities undertaken at least 10 years ago and certainly not the current situation. Secondly, there is a lack of truly novel chemical templates in the pipeline and thirdly, it is clear that very few pharmaceutical companies remain engaged, at least internally, in natural product drug discovery activities.

In 2007, the US Food and Drug Administration approved only 16 new molecular entities, the lowest in a single year since 1983.[3] Despite a slight improvement in 2008, there remains a disturbing overall decline in pharmaceutical R&D productivity that is exacerbated by exponential rises in R&D costs, erosion of sales as many key products face patent expiration and increasing regulatory hurdles. With a burgeoning and aging population, the need for innovative new medicines throughout the world will not diminish. So is there a place for natural product based drug discovery in the future and, if so, where will new biologically active natural products come from?

In this book many of the significant technical advances which have accelerated the screening, purification and structural identification of bioactive natural products have been highlighted. As Bugni *et al.* remind us in Chapter 9, many of the previous bottlenecks that made natural products discovery a slow, laborious process have indeed been removed. However, for natural product based drug discovery to become cost effective and remain competitive, a number of key problems must be addressed, including the continual discovery of known compounds from existing natural product extract collections, the scarcity of novel bioactive chemical templates and the challenge of structurally modifying sometimes complex, often oxygen-rich, chiral natural product lead structures.

With the concept that secondary metabolites have evolved to specifically interact with protein targets and that these are not so different from human proteins, the construction of synthetic compound libraries inspired or based on natural product templates will continue to gain popularity and general acceptance as a valid drug discovery approach. Given that access to biologically relevant, drug-like chemical space is central to the drug discovery process and that natural products often occupy very different areas of this "space" compared to synthetic compounds,[4] then we believe that the search for drug leads from natural products offers a complimentary and much needed approach to other drug discovery strategies.

Antony D. Buss and Mark S. Butler
MerLion Pharmaceuticals, Singapore

References

1. M. S. Butler, *Nat. Prod. Rep.*, 2008, **25**, 475.
2. R. O'Shea and H. E. Moser, *J. Med. Chem.*, 2008, **51**, 2871.
3. B. Hughes, *Nat. Rev. Drug Discov.*, 2009, **8**, 93.
4. J. Rosen, J. Gottfries, S. Muresan, A. Backlund and T. I. Oprea, *J. Med. Chem.*, 2009, **52**, 1953.

Contents

RSC Biomolecular Sciences No. 18
Natural Product Chemistry for Drug Discovery
Edited by Antony D. Buss and Mark S. Butler

Published by the Royal Society of Chemistry, www.rsc.org

Section 2 Sources of Compounds

Section 5 Case Studies of Marketed Natural Product-derived Drugs

Section 1 Introduction to Natural Products for Drug Discovery

CHAPTER 1

Natural Products as Drugs and Leads to Drugs: The Historical Perspective†

DAVID J. NEWMAN AND GORDON M. CRAGG

Natural Products Branch, Developmental Therapeutics Program, NCI-Frederick, PO Box B, Frederick, Maryland, 21702, USA

1 Ancient History (>2900 BCE to 1800 CE)

It is always a little difficult to know where to start and when to stop in time when discussing the historical influence of natural products upon drug discovery because, even today, materials that were identified as late as the 1970s (though used for many centuries as a mixture) are still influencing chemists and biological scientists to use the "native product" and/or a modification as either probes for specific targets or as a treatment in its own right. Later in the chapter, we will demonstrate how natural product structures are still valid models upon which to base 21st century drugs.

Throughout the ages, humans have relied on Nature to cater for their basic needs—not the least of which are medicines for the treatment of a wide spectrum of diseases. Plants, in particular, have formed the basis of sophisticated traditional medicine systems, with the earliest records, dating from around 2900–2600 BCE,[1] documenting the uses of approximately 1000 plant-derived

† Note: The opinions in this chapter are those of the authors and not necessarily those of the US Government.

RSC Biomolecular Sciences No. 18
Natural Product Chemistry for Drug Discovery
Edited by Antony D. Buss and Mark S. Butler

Published by the Royal Society of Chemistry, www.rsc.org

substances in Mesopotamia[2] and the active transportation of medicinal plants and oils around what is now known as Southwest Asia. These include oils of *Cedrus* species (cedar) and *Cupressus sempervirens* (cypress), *Glycyrrhiza glabra* (liquorice), *Commiphora* species (myrrh) and *Papaver somniferum* (poppy juice), all of which are still used today for the treatment of ailments ranging from coughs and colds to parasitic infections and inflammation. In addition to plants, around 120 mineral substances were also listed as "medicinal in nature" including materials now identified as arsenic sulfide, sulfur, lime, potassium permanganate and even rock salt. In most cases, the materials were delivered as infusions (teas), ointments, medicated wines, enemas and even by fumigation—methods still in use in pharmaceutical delivery systems even today. By approximately 700 BCE, the concept of "contagion" was developing, though it would be millennia before the relationship of microbes to plagues, *etc.*, was formally established. Although what is interesting is a description of the use of "rotten grain" in treating wounds; it is tempting to speculate that this might have been a method of administering a crude antibiotic formulation to a patient.

Egyptian medicine dates from about 2900 BCE with the best known record being the "Ebers Papyrus" dating from 1500 BCE, documenting over 850 drugs, mostly of plant origin[3] including opium, cannabis, linseed, aloe and garlic. At around the same time, the Chinese Materia Medica was being extensively documented, with the first record dating from about 1100 BCE (Wu Shi Er Bing Fang, containing 52 prescriptions), although records from the Pent'sao are reputed to date from ~2700 BCE. These were followed by works such as the Shennong Herbal (~100 BCE, 365 drugs) and the Tang Herbal (659 CE, 850 drugs).[4,5] Likewise, documentation of the Indian Ayurvedic system dates from before 1000 BCE (Charaka; Sushruta and Samhitas with 341 and 516 drugs respectively).[6,7]

The Greeks and Romans contributed substantially to the rational development of the use of herbal drugs in the ancient Western world with Hippocrates (~460 to 377 BCE) being considered the father of medicine through an anonymous treatise known as *Corpus Hippocraticum*, which covered the usage of mainly plant-based mixtures but with an emphasis on the correct diet. Sources included extracts of poppy, henbane and mandrake, alongside juniper and saffron. Entertainingly, one might well argue that establishing a potential resistance to poisoning (assassination rather than happenstance) also contributed to the evolution of Greek pharmacy around 100 BCE, with the preparation of *Mithridaticum*, a combination of 54 ingredients, made for Mithridates, who was the King of Pontus at that time.[8] The mixture was "improved" by Andromachus, Nero's physician to contain 70 ingredients and was still available under the name *Theriac* in various European pharmacopoeias until the 19th century. Dioscorides, a Greek physician (100 CE), accurately recorded the collection, storage and use of medicinal herbs during his travels with Roman armies throughout the then "known world", publishing his famous five volume botanical work, *De Materia Medica* at that time; details are available[9] from the US National Library of Medicine (NLM).

The majority of his "drugs" (80%) were based on plant sources, with animal and mineral sources making up ~10% each. Almost at the same time, Galen (130–200 CE.), a practitioner and teacher of pharmacy and medicine in Rome, was well-known for his complex prescriptions and formulae used in compounding drugs, with details given in his herbal, *De Simplicibus*, of 473 drug entities.

During the Dark and Middle Ages (5th to 12th centuries), the Arabs (covering Southwest and Central Asia) preserved much of the Greco-Roman expertise and expanded it to include the use of their own resources, together with Chinese and Indian herbs unknown to the Greco-Roman world. Thus Rhazes, a Persian physician in the early 900s, gave the first accurate descriptions of measles and smallpox and Avicenna, an Arab physician of the late 900–early 1000 era, codified the then current knowledge with his epic and encyclopaedic work, *Canon Medicinae*, a book that influenced the practice of medicine for the next 600 plus years.[10] This was subsequently superseded by the work of Ibn al-Baitar (whose full name was Abu Muhammad Abdallah Ibn Ahmad Ibn al-Baitar Dhiya al-Din al-Malaqi) an Arab, born in Malaga towards the end of the 12th Century (died 1248 CE), but who travelled extensively over the Muslim world and produced two extremely well known treatises, one on botany, that described over 1400 plants, over 200 of which had never previously been recorded (*Kitab al-Jami fi al-Adwiya al-Mufrada*) and the other, a comprehensive compilation known as the *Corpus of Simples* in English (*Kitab al-Mlughni fi al-Adwiya al-Mufrada* in Arabic). Both books were translated into Western languages in later centuries. For those readers wanting to further investigate the influence of the Arabic schools, details can be found at the NLM.

From a Western perspective, following on from the ~1500 CE time frame, the person whose ideas permeated the West for the next two hundred or so years was Paracelsus, whose real name was Theophrastus Phillipus Aureolus Bombastus von Hohenheim, born in Switzerland in 1493. He attempted to "modernise" the then existing works of his forerunners by perhaps the first use of alchemy to separate "good from bad" effects of treatments. Although he was probably responsible for the derivation of what later became known as the "doctrine of signatures", he did place pharmacy on a relatively sound chemical footing and may best be known for the use of mercury as a treatment for syphilis and for the value of mineral waters, plus substituting more simple herbal remedies for some of the complex mixtures handed down from the time of Galen. However, pharmacy was still an empiric science, as shown by publication of the herbal *The London Pharmacopoeia* in 1618 and then in 1676, the book, *Observationes Medicae*, by the English physician Thomas Sydenham. This latter publication was used for close to two centuries as a standard textbook and contained such "observational remedies" as the use of laudanum (opium in alcohol), quinquina (Jesuit bark preparations for malaria) and iron for iron-deficiency anaemias.

2 The Initial Influence of Chemistry upon Drug Discovery

The subtitle for this section could easily be "The experimental chemist discovers the active principles of major drug preparations". Although it is not often realised, the initial discoveries that can be considered to have revolutionised drug discovery and development were made by European chemists (known as apothecaries at that time) in the 1803–1805 time frame, building upon the physico-chemical principles evolving in the recent past from the work of experimental and theoretical chemists such as Proust, Davy, Gay-Lussac, Berzelius, Dalton, *etc.* This body of theory and experiment led away from "polypharmacy" towards the "pharmacology" of single (pure) agents" which was probably first enunciated by Cadet de Gassicourt[11] in 1809.

2.1 Alkaloids

An excellent example of this change would be the story of morphine **1**. The initial report of the isolation of fractions from the opium poppy was reputedly made by Derosne[12] in 1803 at the Institute of France and then published in 1814.[13] However, this preparation had no narcotic properties whatsoever and was probably noscapine with a little meconic acid extracted by the ethanol–water system that he used. A controversy arose because the German pharmacist Seturner then published his work in 1805,[14] claiming that he had commenced work before Derosne; however inspection of this title implies investigation of the acidic and not the basic fractions of opium, probably meconic acid, as shown in a paper the following year.[15] In 1817, Seturner's use of a different extraction technique—hot water extraction followed by precipitation with ammonia—led to colourless crystals that had the narcotic properties of opium.[16] What surprised the scientists reading this publication at the time was that the material obtained was alkaline, not acidic; thus this was the first non-acidic compound with biological properties purified from a plant.

Subsequent conversion into heroin **2** was first reported in 1874 by Wright in the UK as a result of boiling morphine acetate; the process was commercialised by Bayer AG in 1898. The subsequent use and abuse of these compounds is much too complex to discuss here, but one major discovery came in the early 1970s when Pert and Synder reported the identification of opioid receptors in brain tissue.[17] This report was followed closely by the identification of "endogenous morphine-like substances" in 1975 by Kosterlitz and Hughes,[18] which over the next few years led to the identification of enkephalins, endorphins and dynorphins—all of which had the common N-terminal sequence of Tyr-Gly-Gly-Phe-(Met/Leu), leading to the concept that morphine actually mimics this sequence.[19]

Irrespective of the exact timing of the isolation of morphine, alkaloids were discovered at an ever increasing rate from plant sources over the next

50 or so years, confirming the influence of chemistry on pharmacology and drug development in its simplest form from 1817 to roughly the middle of the 19th century. Thus, emetine **3** was probably the first alkaloid to be purified and reported by Pelletier and Magendie[20] in 1817 from *Ipecacuanha*, closely followed the same year by the isolation of strychnine **4** from *Strychnos* by Pelletier, now working with Caventou. Then in 1820, the same workers reported the isolation of quinine **5** from *Cinchona* species.[21] They developed a commercial process for the preparation of quinine and, in addition to its subsequent use in the treatment of malaria, it was also used extensively as a "tonic" and an antifever drug. Though not documented specifically, the realisation of its use for malaria probably followed after the usage for other "illnesses" in northern Europe.

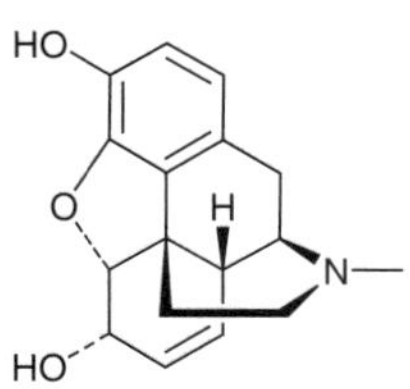

1

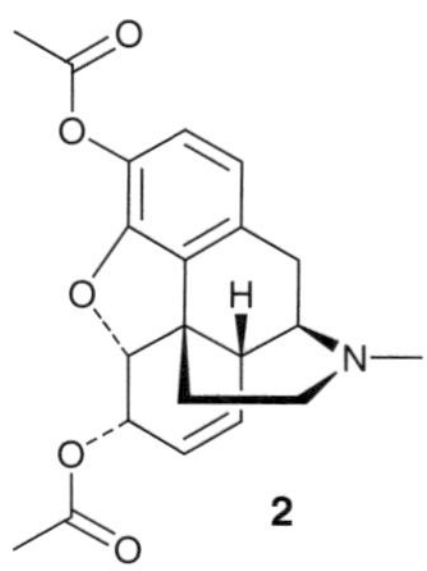

2

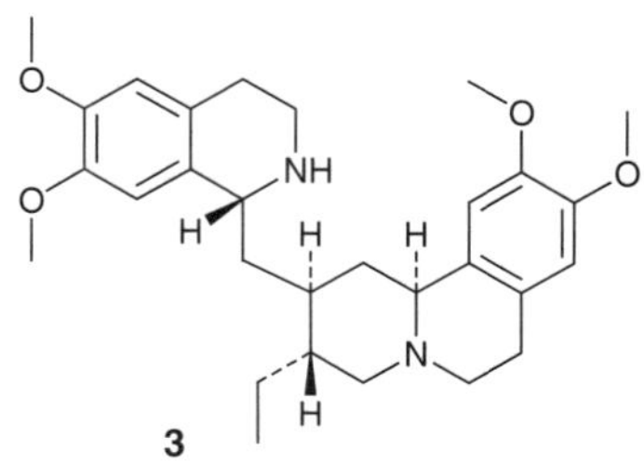

3

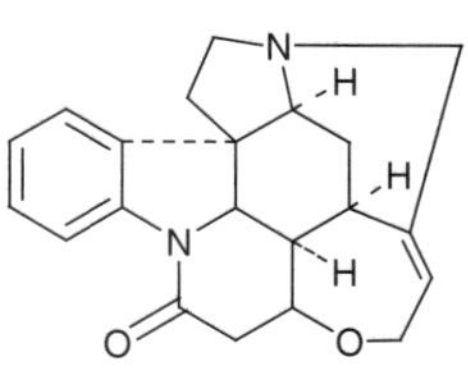

4

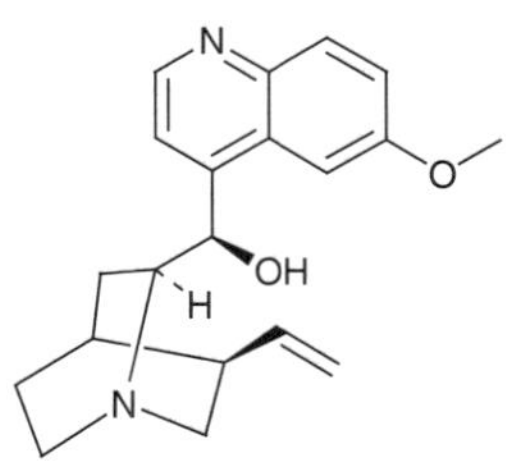

5

Over the next few years, a veritable "treasure trove" of potential drug structures was reported, though one must realise that the structures were not identified for many years if not many decades following their initial isolation. Thus in 1819, brucine **6** and caffeine **7** were purified, followed in 1820 by colchicine **8**, in 1820, codeine **9**,[22] in 1833, atropine **10**[23] and papaverine **11**[24] in 1848. During this time frame, the first plant-derived alkaloid to be purified, have its structure elucidated and finally synthesised was coniine **12**. The compound was extracted in 1826, followed by determination of its structure in 1870 and then synthesised by Ladenberg[25] in 1881. Even today, over 180 years since it initial isolation, the compound is still a candidate for drug discovery, this time being a model for induction of apoptosis in trypanosomal infections.[26]

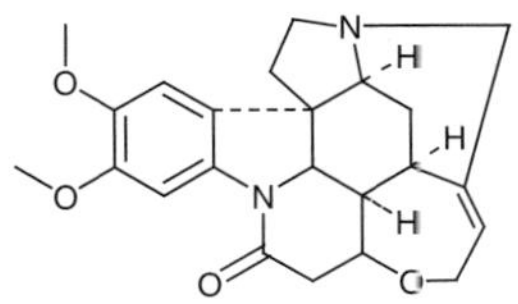

6

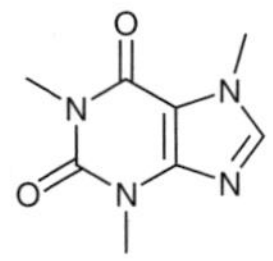

7

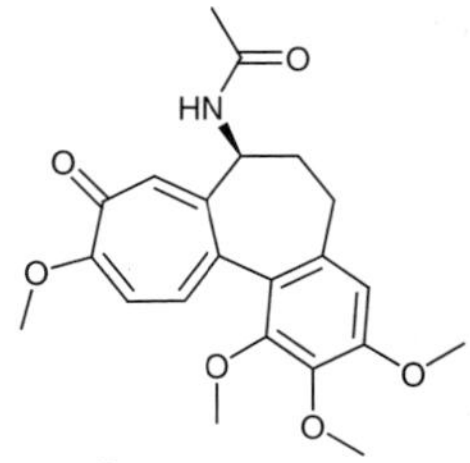

8

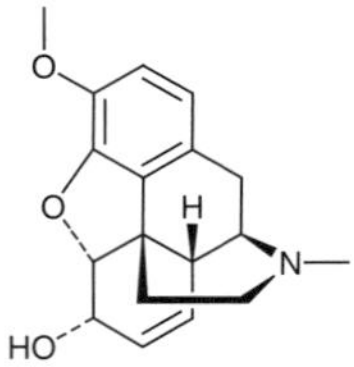

9

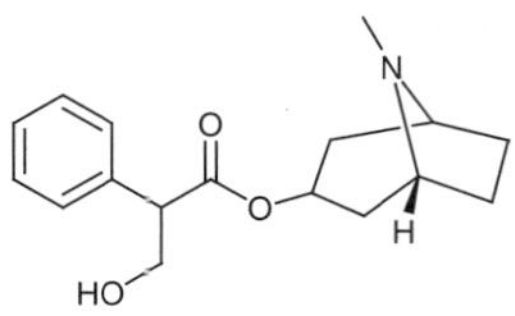

10

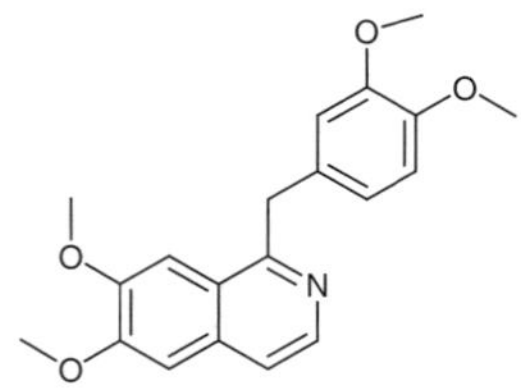

11

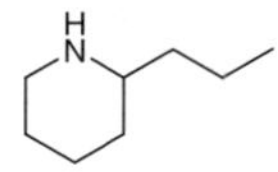

12

2.2 Aspirin

No historical perspective of natural product derived drugs would be complete without a discussion of aspirin (acetylsalicylic acid)—probably the most widely utilised drug of all time when the numbers of tablets consumed worldwide on an annual basis are considered. Even today, where presumably the major pharmacological effect is modulation of the cyclooxygenase isoforms, its full activity is still not fully defined.

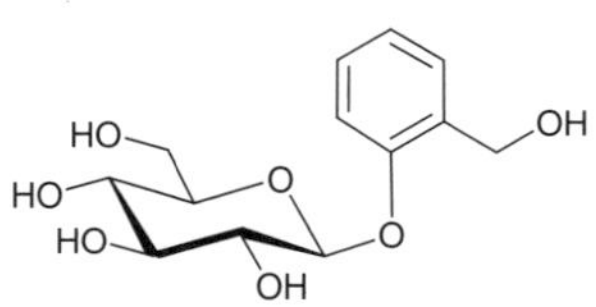

13

Salicin **13** was first introduced into medicinal use by Maclagan[27] in 1876 as the single agent, although as a part of "herbals" the use of extracts of *Salix* or *Spiraea ulmaria* (the source of "spirin" in "aspirin") for treatments of fevers and pain dated from the days of Hippocrates. It is probable that the formal identification of salicin in more "modern" medicine would be the letter from the Reverend Edmund Stone to the President of the Royal Society in 1763 covering the use of the compound for treatment of "fever". There are various "stories and/or anecdotes" over the transition from salicin to acetylsalicylic acid (aspirin), but the most probable steps were as follows.

Piria, working with *Spiraea* species, first isolated salicylaldehyde in 1839 while working in Dumas' laboratory and then prepared salicylic acid.[28] This was followed by the first synthesis of acetylsalicylic acid in 1853 by Gerhardt[29] but without any pharmacological investigation. The story that Hoffmann at Bayer synthesised acetylsalicylic acid to overcome the taste problems with salicylic acid in order to help his father "take his medicine" has been revised relatively recently by Sneader.[30] It would appear, in spite of some discussion on Sneader's paper in later issues of the *British Medical Journal*, that acetylsalicylic acid was synthesised by Hoffmann, but "under the direction of" Arthur Eichengrün at Bayer and that the compound was in fact under "impromptu clinical testing" before the 1898 time frame. Eichengrün left Bayer in 1908 and proceeded to develop materials and processes such as flame-retardant fibres and also injection moulding of plastics. However, when the Hoffmann story was published as a footnote in a book on chemical engineering in 1934, [31] Eichengrün, being of Jewish descent, could not comment due to the political climate of 1930s Germany and it was not until 1949 that he was able to publish his story of the discovery.[32]

2.3 Digitalis

In 1775, the English physician Withering reported his extensive work on a potential treatment for "dropsy" that he had developed as a result of studies on a plant decoction that the local inhabitants were using for their own treatment. He subsequently investigated the methods used and identified the foxglove, *Digitalis purpurea*, as the potential source and also demonstrated what today would be known as a "narrow therapeutic index" for the preparation. It was realised subsequently that purification would have to be performed, but in spite of significant efforts, it was not until 1867 that Nativelle[33] was able to produce an effective crystalline preparation that he named "crystallised digitalin". A few years later in 1875, the German pharmacologist, Schmiedeberg was able to produce digitoxin **14**.[34] Subsequently, other compounds with a similar pharmacology were discovered by a combination of what would now be known as ethnopharmacology/ethnobotany—yielding ouabain from *Acocanthera* bark and roots and strophantin from *Strophantus* species. Both of these agents were used as arrow poisons in Africa, albeit in very crude form.

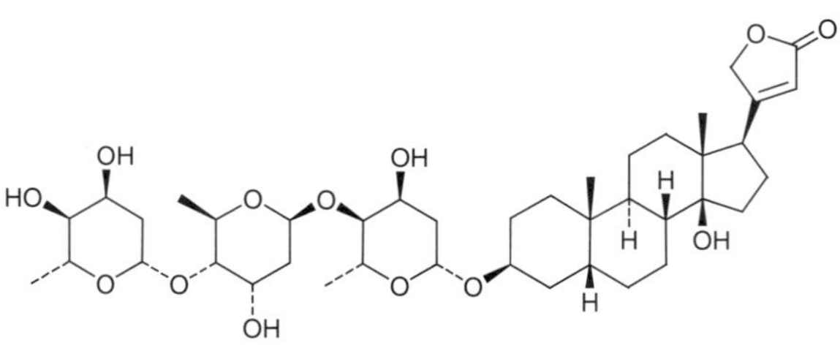

14

Inspection of any textbook of pharmacognosy, pharmacology or natural product chemistry in the middle of the 20th century would show a large number of very similar "cardiac glycosides" isolated from a multiplicity of plant sources and, even as the bufodienolides, from *Amphibia*. However, it was not for three-quarters of a century after the purification of digitalis that a mechanism of action of these agents was firmly established as inhibitors of the sodium/potassium ATPase pump in membranes. Even today, novel agents based upon variations of an unusual plant-sourced "cardiac glycoside" are in preclinical trials as antitumour drugs with an example being UBS-1450 **15**; oleandrin **16** is the subject of a 2008 PCT patent application for use as a "functional food" in cancer therapy by a Japanese company.

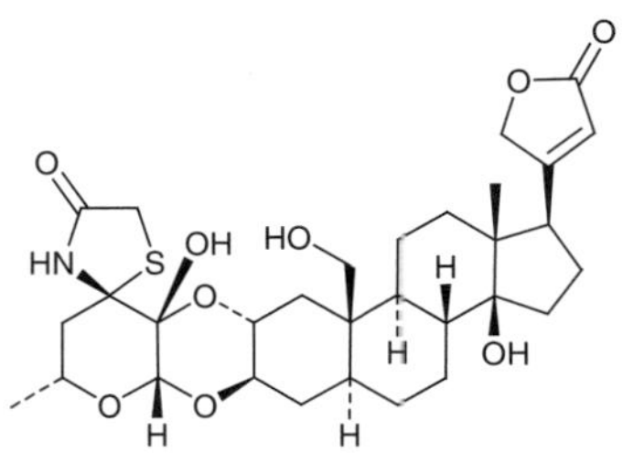

15

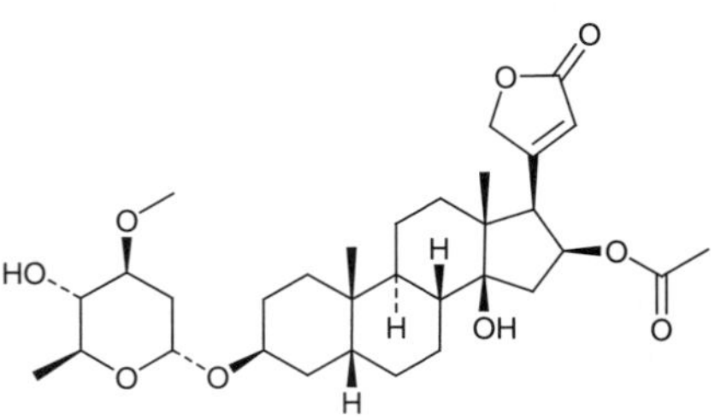

16

3 20th and 21st Century Drugs/Leads from Nature

The breadth of disease areas that could be covered from a historical perspective starting in the mid-1920s or so is staggering. Due to space limitations and because specific areas are covered by other authors, we have limited our discussions to early antibiotics (antibacterial, fungal and viral) and cancer, even though the initial work in this "series of diseases" (cancer) was performed using materials from war gases (the nitrogen mustards/alkylating agents) and then, in the early 1960s, workers using both microbes and plant sources proceeded to report and then to use multiple agents from these sources as treatments for cancers. For all these diseases, we give examples in each case of how the early historical structures have led to novel agents based upon them still being synthesised (or even biosynthesised) in the 2008 time frame.

3.1 Antibacterial and Antifungal Antibiotics

It is probably true that if one had to name the natural product that has saved the most lives, directly or indirectly since its original discovery, penicillin G **17** would be the molecule of choice. In this day and age, there are few people in the developed countries who can remember the pre-antibiotic age with any clarity. Some, over the age of 75, may have hazy memories of relatives dying at young ages due to bacterial infections, but that is not the norm. However, let us set the stage for the reader.

Antibacterials

The first usage of natural products as true antibacterials rather than as surface sterilants (*e.g.* use of thymol and other essential oils) can be fixed in time as the later stage of World War II, with the use of microbial derived secondary metabolites such as penicillin and streptomycin being the exemplars known in the West. This occurred as a result of the recognition by Fleming in the late 1920s of the activity of penicillin (though there were anecdotal reports of scientists such as Tyndall, Roberts and Pasteur in the 1870s recognising antagonism between various bacteria), leading ultimately to the well-known and documented use of penicillins G and V[35] and streptomycin (discovered by Waksman[36]) in the early 1940s. However, it now appears that the (then) USSR was using the antibiotic Gramicidin S (Soviet Gramicidin[37–39]) as a treatment for war wounds at the end of World War II.

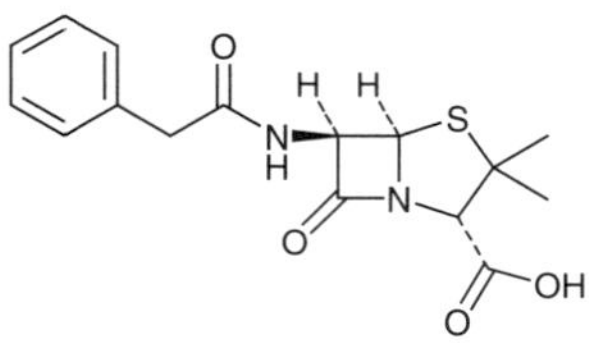

17

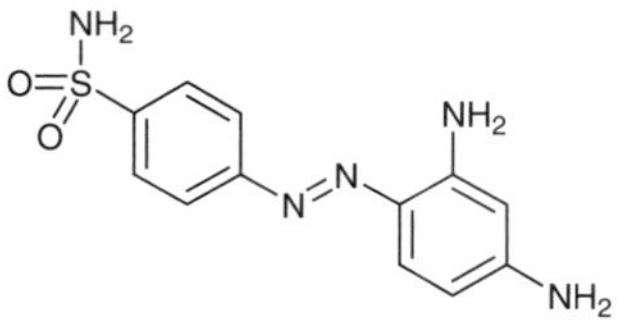

18

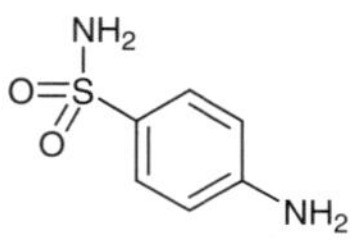

19

It is recognised, however, that the use of Prontosil® **18** pre-WWII led to the introduction of synthetic antibacterials, with the first clinical efficacy report in 1933 ultimately leading to the award of the Nobel Prize for Medicine in 1938 to Domagk. This could also be thought of as the first formal prodrug in the antibiotic field as the active principle, sulfanilamide **19** is a structural analogue of *para*-aminobenzoic acid (PABA) and an essential nutrient of many bacteria and in particular, the cocci. PABA competitively inhibits dihydropteroate synthase, thus leading to inhibition of folic acid and bacterial death. So although synthesised in the absence of such knowledge and for an entirely different purpose, it was in retrospect an isostere of a natural product.[40] Using the nomenclature of Newman *et al.*[41] this would now be classified as an "S/NM" or "synthetic but natural product mimic".

Other "Early Antibacterial Classes"

Although the aminoglycosides such as streptomycin, neomycin and the gentamicins have a long and storied history as treatments for antibacterial infections, particularly in the early days when streptomycin was a treatment for both infected wounds and also for tuberculosis, few modifications of the basic molecule(s) went into clinical use, mainly due to the complexity of chemical modification of saccharidic-based structures. Thus, we do not discuss this class further or molecules such as the rifamycins and their manifold derivatives. Instead, due to space constraints, we show how β-lactams, macrolides, tetracyclines, glycopeptides and pleuromutilins—all "ancient antibiotic structures"—are even today still being utilised as base scaffolds on which to build molecules.

β-Lactams of All Classes

To date, the number of penicillin and cephalosporin-based molecules produced by semi- and total synthesis is well in excess of 20 000. Most started with modification of the fermentation product, 6-amino-penicillanic acid **20** or the corresponding cephalosporin, 7-amino-cephalosporanic acid **21**, both of which can be produced by simple chemical or biochemical deacylation from penicillin or cephalosporin C. The number above is only approximate as a significant proportion of structures from industry were never formally published, or were only mentioned in the patent literature—particularly if they had marginal or no significant activity levels over those which had been reported previously.

In 1948, the ring-expanded version of penicillin, cephalosporin C **22**, was reported from *Cephalosporium* sp. by Brotzu; its structure was determined in 1961 by the Oxford group.[42,43] As with the penicillin nucleus, this ring-expanded molecule, 7-aminocephalosporanic acid **21**, also served as the building block for many thousands of cephalosporins with the first orally active molecule, cephalexin **23** being introduced in 1970. Since that time, a multitude of cephalosporins have been synthesised with the aim of producing molecules that are more resistant to β-lactamases.

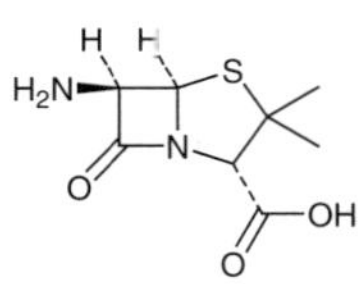

20

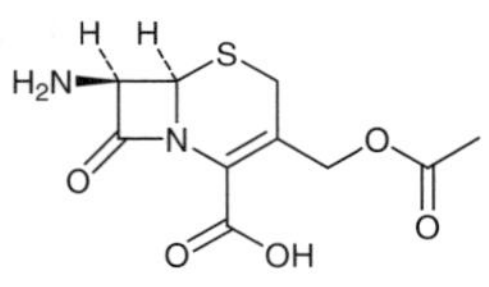

21

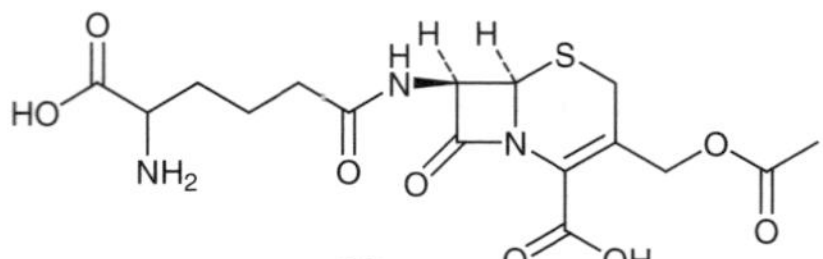

22

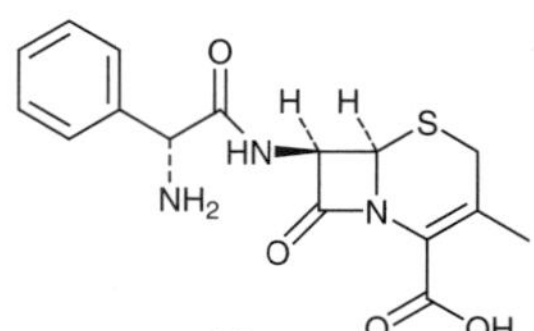

23

In order to give extra "medicinal life" to β-lactams that were no longer useful due to the presence of both constitutive and inducible β-lactamases, efforts were made in the late 1960s and early 1970s—particularly by Beecham (now part of GlaxoSmithKline) and Pfizer—to find molecules that would have similar pharmacokinetics to the β-lactams but would inhibit the "regular" β-lactamases that were part of the pathogenic microbe's defence systems. Beecham discovered the naturally occurring clavulanate family, with clavulanic acid **24** being incorporated into the combination known as Augmentin®

(a 1 : 1 mixture of amoxicillin and clavulanic acid launched in 1981), thus extending the franchise of this particular β-lactam well beyond its original patent date. The Pfizer entrant was basically des-amino penicillanic acid **25** with a sulfoxide in place of the sulfur; in tazobactam **26**, which was originally synthesised by Taiho and launched by Lederle (now Wyeth), one of the gem methyl groups was replaced by a 1,2,3-triazol-1-yl-methyl substituent. Even today, 17 years after the last introduction, no other β-lactamase inhibitors have made it to commercialisation. Currently, as mentioned earlier, amoxicillin is combined with clavulanate or ticarcillin, sulbactam with ampicillin, and tazobactam with piperacillin. All these inhibit only class A serine-based β-lactamases, leaving a vast number of other enzymes where inhibitors are required.[44]

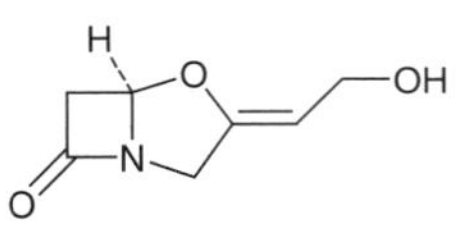

24

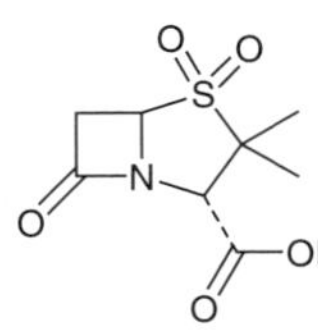

25

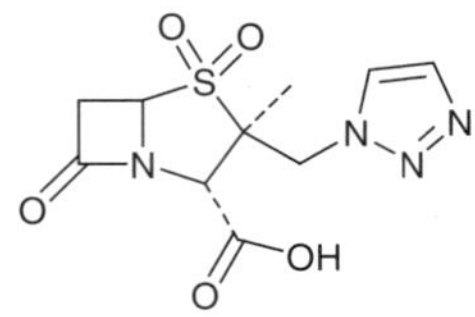

26

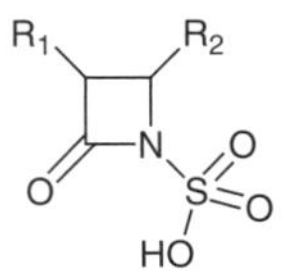

27

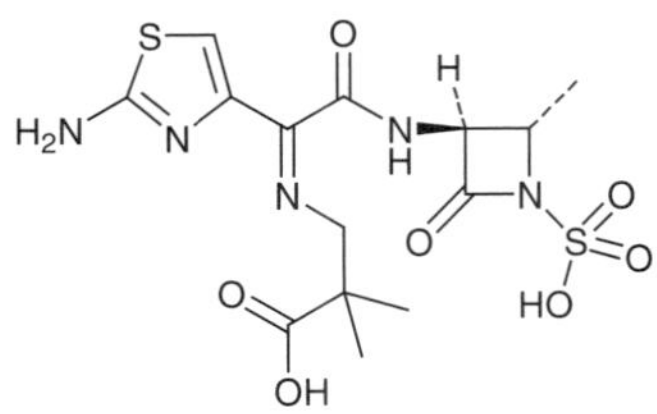

28

Along with the search for the β-lactamase inhibitors, efforts were underway to produce the simplest β-lactam, the monobactam. Following many years of unsuccessful research at major pharmaceutical houses, predominately in the synthetic chemistry areas, came reports from Imada *et al.* in 1981[45] and a Squibb group led by Sykes,[46] who both demonstrated the same basic monobactam nucleus **27**. What is important to realise is that no molecules synthesised before the discoveries of these natural products had a sulfonyl group attached to the lactam nitrogen, which is an excellent method for stabilising the single four-membered ring.

Since that time, a significant number of variations on that theme have been placed into clinical trials and, in some cases, such as Aztreonam® **28**, into commerce. Recently (late 2007, early 2008), this compound was submitted for approval in the EU and the USA as the lysinate salt for the inhalation treatment of *Pseudomonas aeruginosa* in cystic fibrosis under an Orphan drug category. As of late 2008, the Food and Drug Administration (FDA) was requiring further information and the status of the EU application was not yet known.

That these base structures and others discovered after the early 1940s are still valid as scaffolds upon which to base new drugs is shown by the following data. Since 2000, three penems (biapenem **29**, ertapenem **30** and doripenem **31**), which although produced synthetically were based upon the structure of thienamycin **32**, and two cephalosporins—cefovecin **33** (a veterinary drug) and ceftobiprole medocaril **34**—have been approved for marketing. Currently, there is one penem, tebipenem pivoxil **35**, which has been pre-registered in Japan with the aim of approval in early 2009.

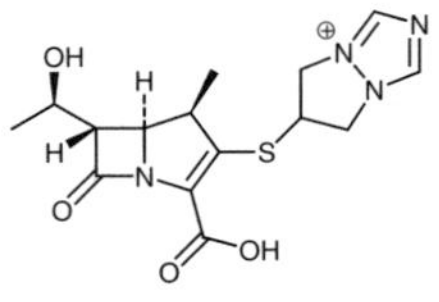

29

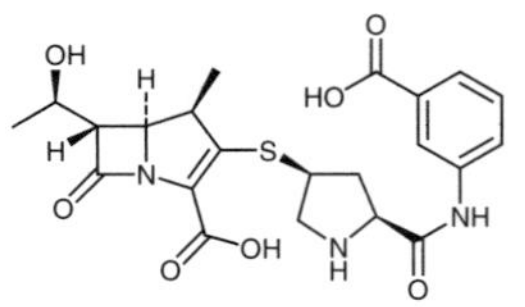

30

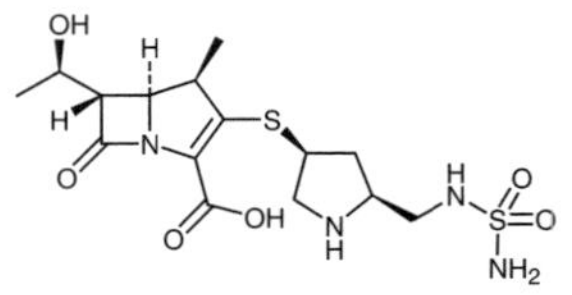

31

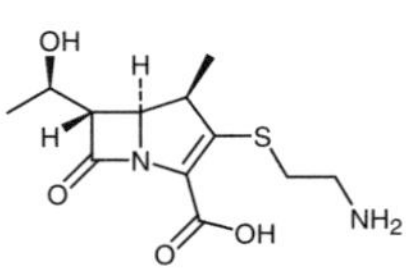

32

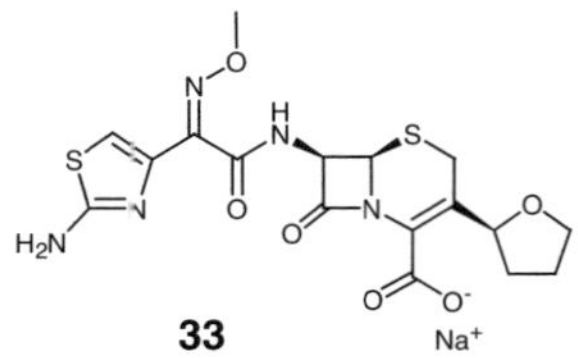

33

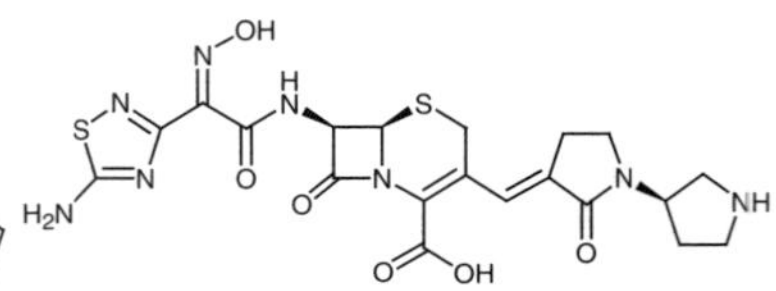

34

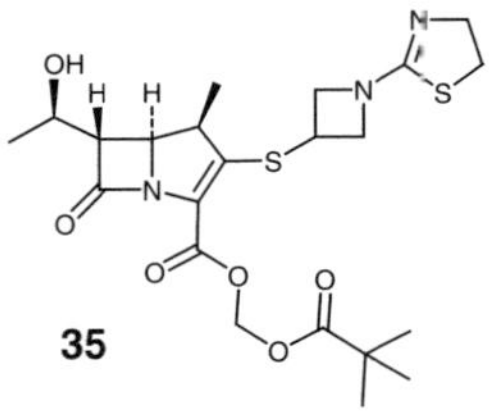

35

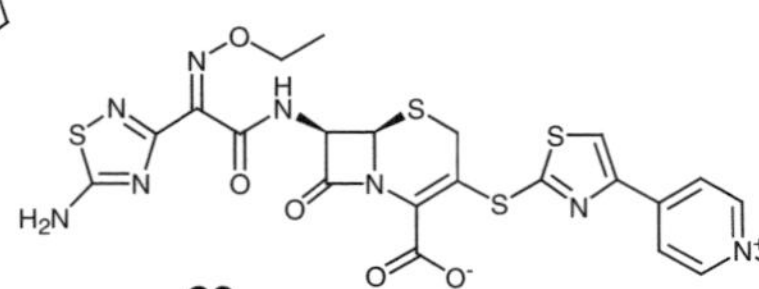

36

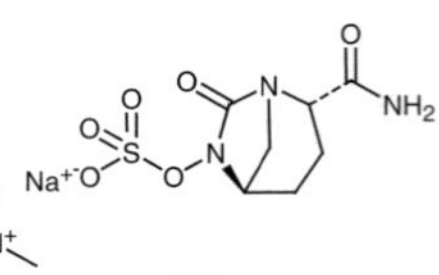

37

Although we mentioned earlier that only three β-lactamase inhibitors have been marketed, there is now a potential "cepahalosporinase inhibitor/cephalosporin combination" in trials. Forest Pharmaceuticals recently announced that the cephalosporin, ceftaroline fosamil acetate **36**, which is currently in Phase III clinical trials, has been combined with Novexel's synthetic β-lactamase inhibitor,[47] NXL-104 (AVE-1330A) **37** and has entered Phase I trials.

Tetracycline Derivatives

Even though the base molecule or its better known chloro-derivative, aureomycin and later the dimethyl amino derivative, doxycycline, have been stalwart members of the physician's armamentarium for 40–50 years, in 2005, Wyeth had the glycyl derivative of a modified doxycycline molecule, tigecycline **38**, approved for complicated skin and soft tissue infections. Tigecycline has broad-spectrum activity including both Gram-positive and Gram-negative bacteria and methicillin-resistant *Staphylococcus aureus* (MRSA). It shows that relatively simple chemical modifications can even give very old base structures a new lease on life and be effective against clinically important infections.

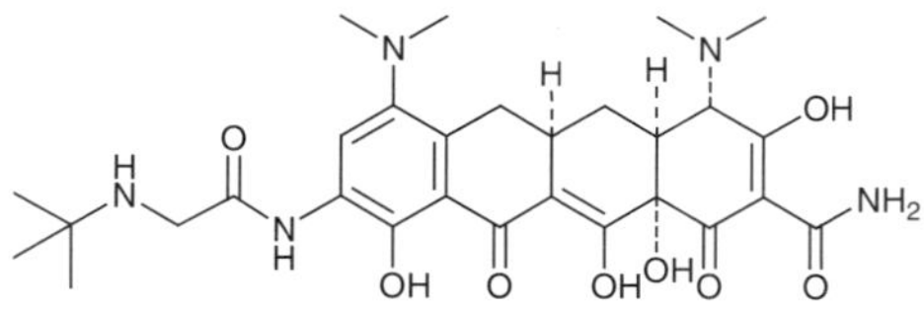

38

Glycopeptide Antibacterials

Vancomycin, a natural product that was first approved in 1955, is still the prototype for structural variations with the same mechanism of action: the binding to the terminal L-Lys-D-Ala-D-Ala tripeptide in Gram-positive cell wall biosynthesis. The compounds below are semi-synthetic modifications of the same basic structural class (glycopeptides) as the prototype vancomycin, thus following in the "chemical footsteps" of the β-lactams; currently, there are three semi-synthetic glycopeptides, oritavancin **39**, telavancin **40** and dalbavancin **41**, in late stage clinical development.

In all cases, their antibacterial mechanism involves inhibition of cell wall biosynthesis initially *via* the vancomycin target, although the exact mechanisms can vary with the individual agent. In the case of oritavancin, from very recent data it would appear that the agent is comparable with vancomycin in its inhibition of trans-glycosylation, but is more effective as a transpeptidation inhibitor.[48] As mentioned above, all are semi-synthetic derivatives of natural products, with oritavancin being a modified chloroeremomycin (a vancomycin analogue), dalbavancin being based on the teicoplanin relative, B0-A40926 and telavancin (TD-6424) being directly based on chemical modification of vancomycin.[49]

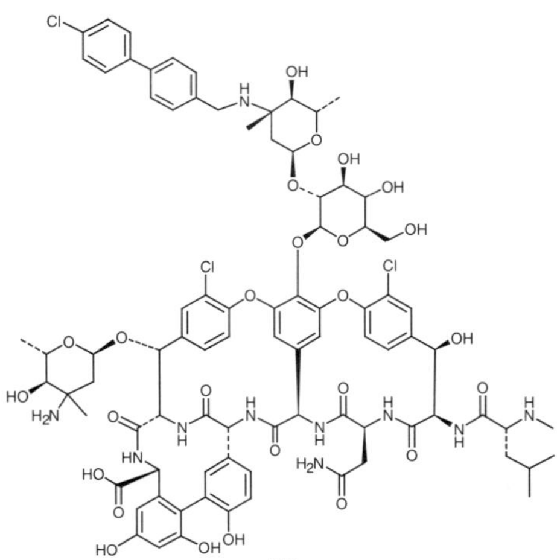

39

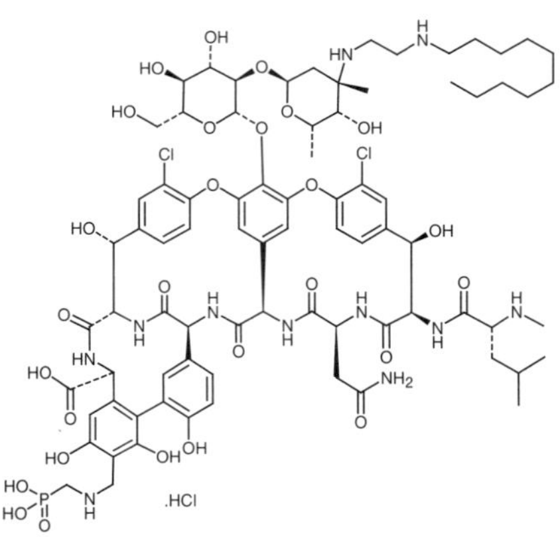

40

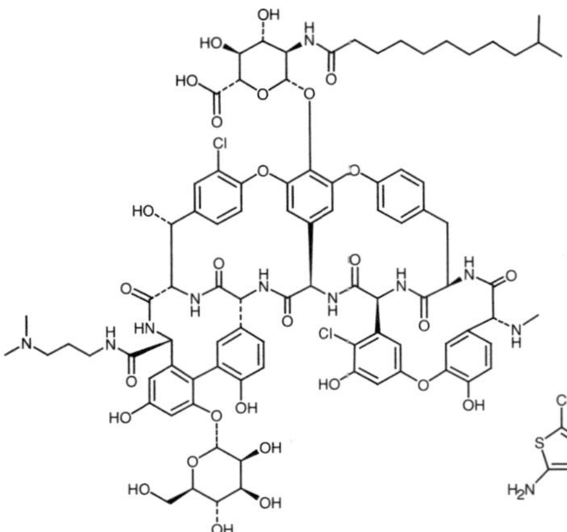

41

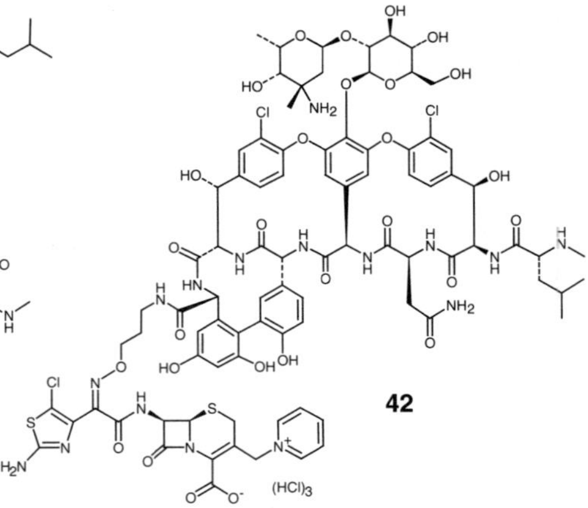

42

That one may combine the characteristics of two separate agents working at different targets within the same basic biological area is shown by the work of Theravance (also the originator of telavancin), which has combined a cephalosporin with vancomycin itself to produce the hybrid TD-1792 **42**, which is currently in Phase II trials against complicated skin and soft tissue infections. Thus, two old antibiotic classes can produce novel agents—again underscoring the possibilities of reworking older structures if one understands their history.

Macrolidic Antibiotics

Following on the track of novel modifications of old structures, since 2000 there have been four molecules formally based upon the erythromycin chemotype that have either been approved (telithromycin **43** in 2001) or entered clinical trials; cethromycin (ABT-773; Phase III; **44**), EDP-420 (EP-013420, S-013420; **45**) and the product of glyco-optimisation, CEM-101 (Phase I; **46**). Cethromycin **44** is currently in Phase III trials for use against community acquired pneumonia (CAP) and is being evaluated as an anti-anthrax agent (and against other biodefence targets) by the National Institute of Allergy and Infectious Diseases (NIAID) and the US Army. The interesting modification of the base erythromycin structure, the "bicyclolide" EDP-420 (**45**) a novel, bridged bicyclic derivative originally designed by Enanta Pharmaceuticals,[50,51] is currently in Phase II trials for treatment of CAP by both Enanta and Shionogi. Interestingly, this molecule is also active in a murine model of *Mycobacterium avium*, a common infection in immunosuppressed patients,[52] which may well expand its usage in the future.

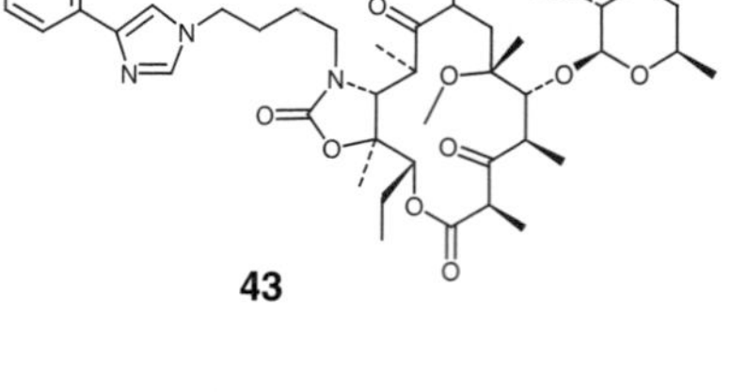

43

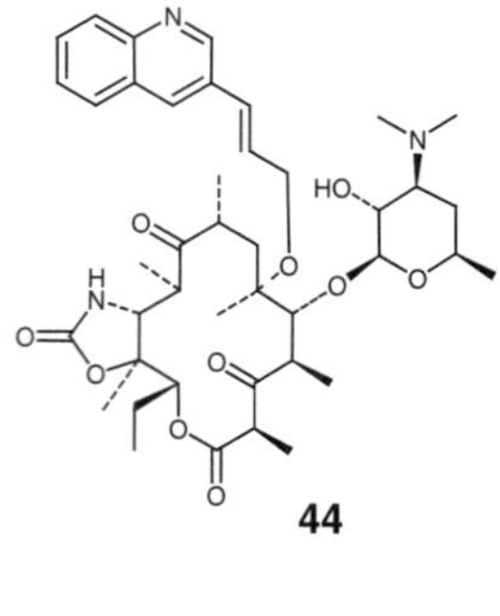

44

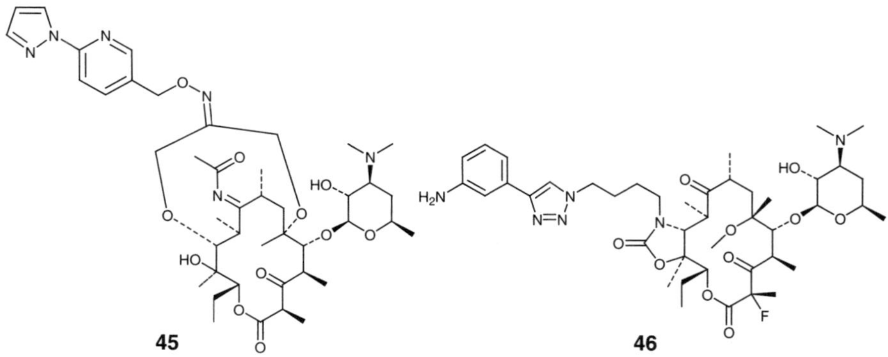

45 46

Pleuromutilin Derivatives

Demonstrating yet again that older antibiotic structures have significant validity for today's diseases, GlaxoSmithKline (GSK) received approval in 2007 for a modified pleuromutilin, retapamulin **47**, for the treatment of impetigo in paediatric patients. The base structure, pleuromutilin **48**, was first reported in 1951 from the basidiomycete *Pleurotus mutilis* (FR.) Sacc and *Pleurotus passeckerianus* Pilat.[53] In the mid-1970s, a significant amount of work was reported on the use of derivatives of pleuromutilin as veterinary antibiotics;[54] thus the subsequent utilisation of the base molecule as a source of human use antibiotics is very reminiscent of the work that led to the approval of Synercid® in the late 1990s, as the base molecules in that case were also used extensively in veterinary applications.

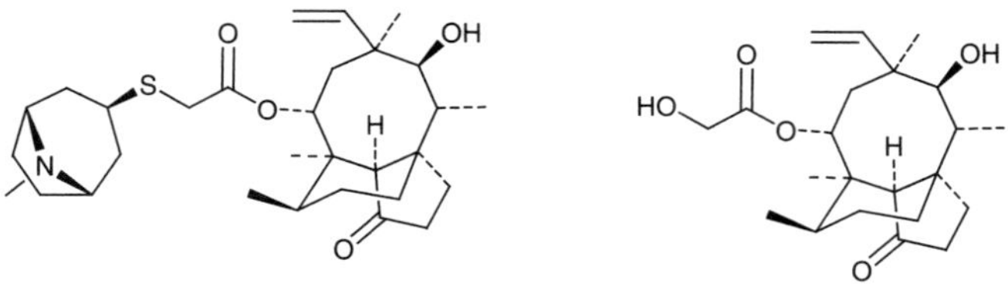

47 48

It is quite possible that a number of antibiotics based upon this elderly scaffold will enter late stage human trials as currently there are two "mutulins" in Phase I clinical trials for use against Gram-positive infections. Although structures have not yet been released, they are BC-3205 and BC-7013 from Nabriva in Vienna, with the former for oral use and the latter for topical use.

Antifungal Antibiotics

Although a very considerable amount of time and effort was expended in the early days of antibiotic discovery (by this we mean the mid to late 1940s), only three general use antifungal agents entered clinical practice as a result. Perhaps the first clinically used antifungal natural product (our information on

Russian efforts in this field under the old USSR is effectively nil), was griseofulvin **49**, which although launched in 1958, was originally reported in 1939. Its non-polyene structure was defined in a series of papers in 1952 using classical techniques[55] and, even today, close to 70 years after it was first described, is still in clinical use against dermatophytes—the only class of fungi that it is active against; long-term treatment is necessary due to its insolubility.

Perhaps the best known clinical agent is the heptaene polyene, amphotericin B **50**, isolated from *Streptomyces nodosus* and first reported in 1956. The full structure was not elucidated until 1970 when it was determined by X-ray crystallography,[56] closely followed by a description of the absolute configuration determined by utilising the iodo-derivative for X-ray and by mass spectroscopy.[57] Quite recently, 50 years after its initial discovery, a full review giving the highlights of the chemistry around the compound was published by Cereghetti and Carreira.[58]

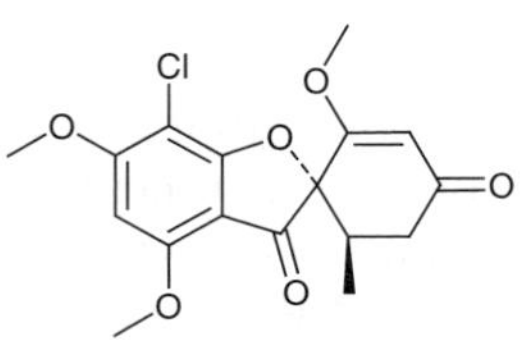

49

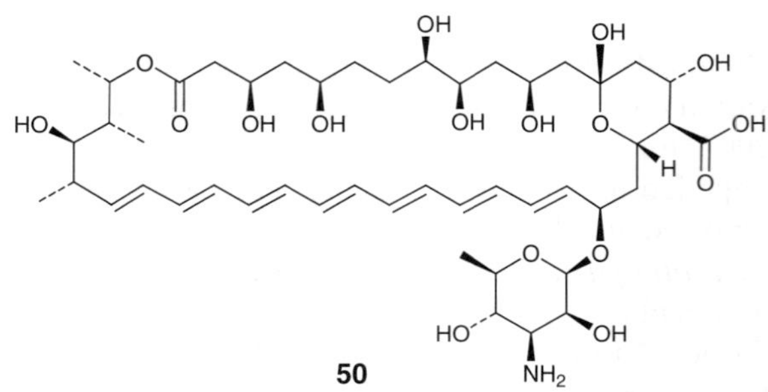

50

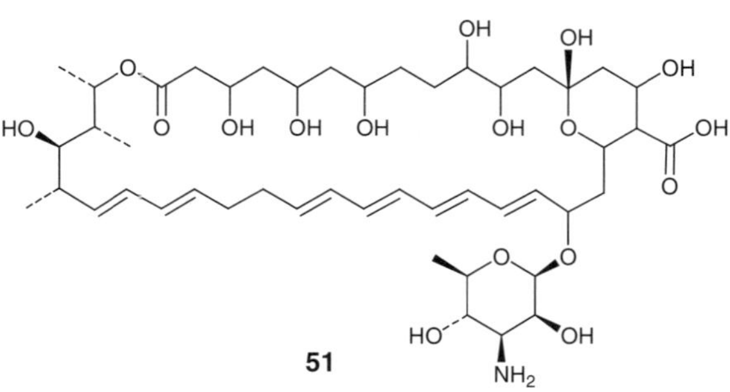

51

Although many polyenes with varying numbers of conjugated double bonds have been reported since those early days, only one other compound of this class (in fact the first identified in 1950 of this general structural class), the tetraene nystatin **51**, has gone into general clinical use and like amphotericin B its primary indication is for candidiasis. It was first reported from *Streptomyces noursei* and, as with amphotericin, its structure was reported in the 1970 time frame by two groups, one using classical chemical degradation plus proton NMR[59] and the other *via* mass spectroscopy.[60] Confirmation of the proposed hemiketal structures of both amphotericin B and nystatin was published subsequently by the Rinehart laboratory in 1976.[61]

Current Status of Natural Product-Derived Antifungal Antibiotics

Since 2000, three natural product-derived antifungal drugs from the echinocandin/pneumocandin class of glucan synthase inhibitors have been approved for human use.[62,63] In order, these were caspofungin **52** (2001, Merck) which recently has been shown to function successfully in both invasive candidiasis and in candidaemia,[64] micafungin **53** (2002, Astellas, see Chapter 15), which is currently in clinical trials for paediatric disease[65] and anidulafungin **54** (2006, Pfizer).[66,67] Another modification of the basic echinocandin structure, aminocandin **55** (HMR-3270), a semi-synthetic derivative of deoxymulundocandin, is currently in Phase I clinical trials with Phase II studies reported as being scheduled.[68]

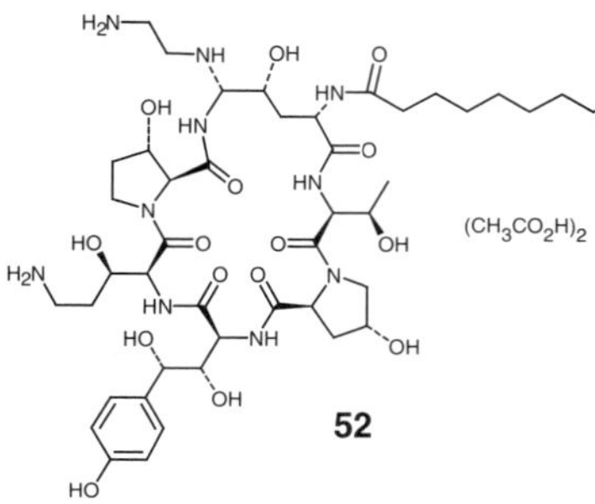

52

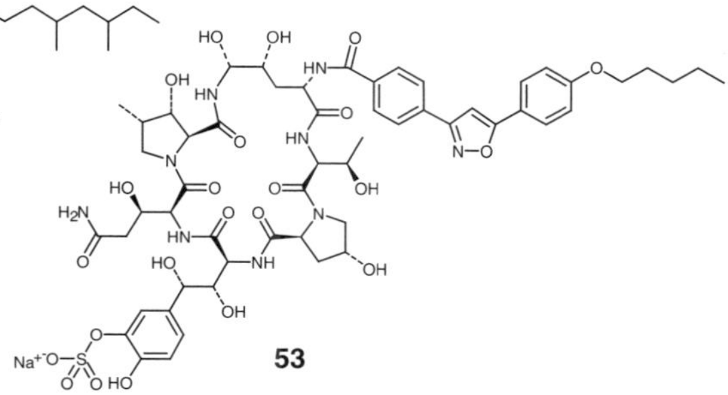

53

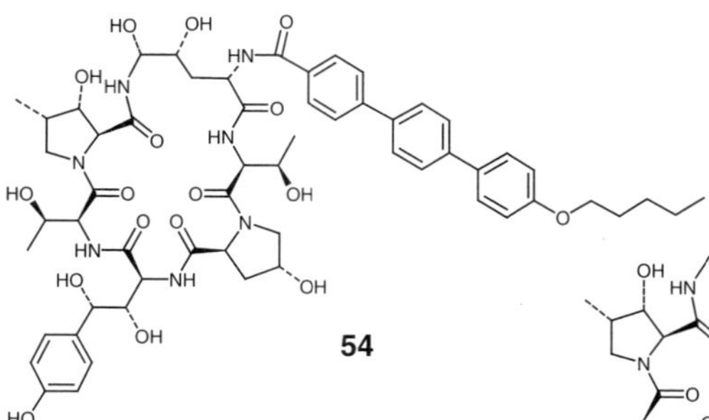

54

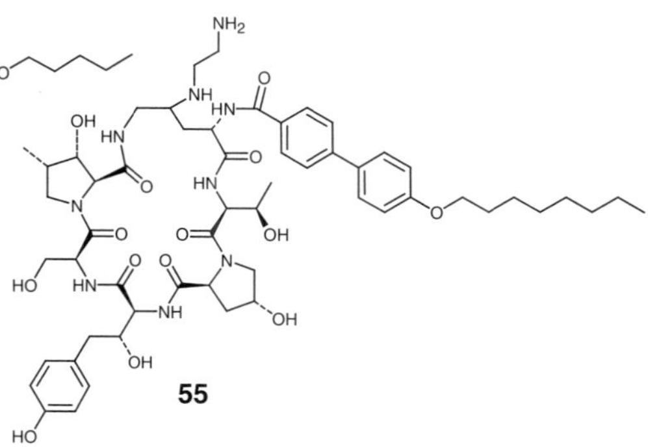

55

3.2 Antiviral Agents

It can be argued quite successfully (and has been a number of times) that the derivation of the nucleoside-based antiviral agents can be traced back to the time frame 1950 to 1956, when Bergmann *et al.* reported[69–71] on two compounds they had isolated from marine sponges, spongouridine **56** and spongothymidine **57**. What was significant about these materials was that they demonstrated, for the first time, that naturally occurring nucleosides could be found using sugars; importantly other than ribose or deoxyribose and having biological activity, as it had been "then current dogma" that one could change the "nucleoside bases" but ribose or deoxyribose had to be the sugar to maintain biological activity. These two compounds can be thought of as the prototypes of all of the modified nucleoside analogues made by chemists that have crossed the antiviral and antitumour stages since then.

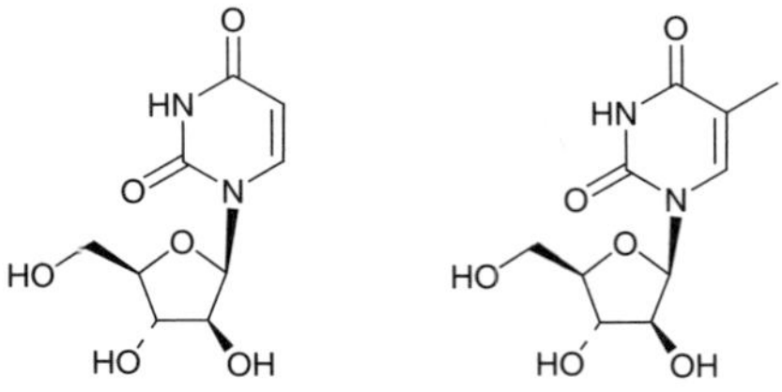

56 57

Once it was realised that biological systems would recognise the base and not pay too much attention to the sugar moiety, chemists began to substitute the "regular pentoses" with acyclic entities and with cyclic sugars with unusual substituents. These experiments led to a vast number of derivatives that were tested extensively as antiviral and antitumour agents over the next 30+ years. Suckling, in a 1991 review,[72] showed how such structures evolved in the (then) Wellcome laboratories, leading to AZT and, incidentally to Nobel Prizes for Hitchens and Elion, though no direct mention was made of the original arabinose-containing leads from natural sources.

Showing that "Mother Nature" may follow chemists rather than the reverse, or conversely that it was always there but the natural product chemists were "slow off the mark", arabinosyladenine **58** (Ara-A or Vidarabine®) was synthesised in 1960 as a potential antitumour agent,[73] with its antiviral activities reported by Schabel[74] in 1968 with production *via* fermentation of *Streptomyces antibioticus* NRRL3238 being reported in a British patent in 1969[75] and isolated, together with spongouridine, from a Mediterranean gorgonian (*Eunicella cavolini*) in 1984.[76]

Building on from these original discoveries, medicinal chemists over the next 40+ years made a very large number of "substituted nucleosides" varying the base and the sugar moieties (including molecules that were acyclic), leading to the very well-known antiviral agents, acyclovir **59** and its later prodrug derivatives and AZT **60**.

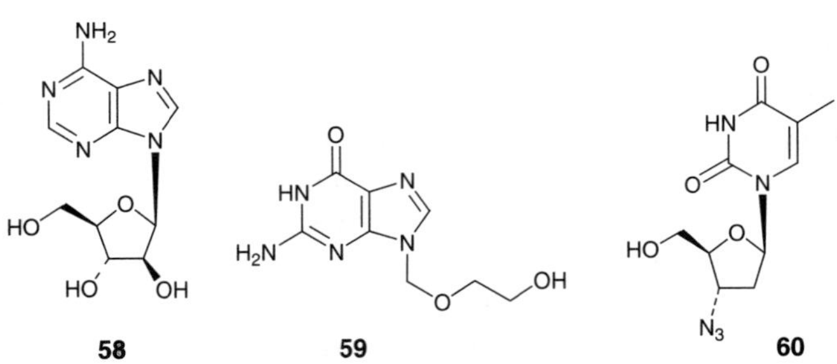

58 59 60

Although a significant number of antiviral vaccines have either been approved or are in clinical trials for a variety of viral diseases, small molecules based upon "modified nucleosides" are still being approved by either the FDA or the European Medicines Agency (EMEA). As in earlier days, agents originally approved as antiviral agents may later be shown to have potential utility as antitumour agents.

Since 2000, seven such agents have been approved for antiviral treatments covering anti-HIV, hepatitis B and cytomegalovirus (CMV). Rather than give details of each, we discuss below the importance of just two compounds of this class that would not have been synthesised without the historical perspective.

In 2001, tenofovir disoproxil fumarate **61**, a prodrug of tenofovir was approved for treatment of HIV, subsequently being preregistered in the USA for treatment of hepatitis B. Emtricitabine **62**, a reverse transcriptase inhibitor, was approved in 2003 for HIV. What is of import is that these compounds are now part of fixed dose combination therapies for treatment of HIV, either two drug (tenofovir disoproxil fumarate/emtricitabine) or three drug Atripla® (tenofovir disoproxil fumarate/emtricitabine/efavirenz) formulations. Thus, even 50+ years after Bergmann's discovery of bioactive arabinose nucleosides, small molecules synthesised as result of his discoveries are still in clinical use and others are in clinical trials for treatment of viral diseases.

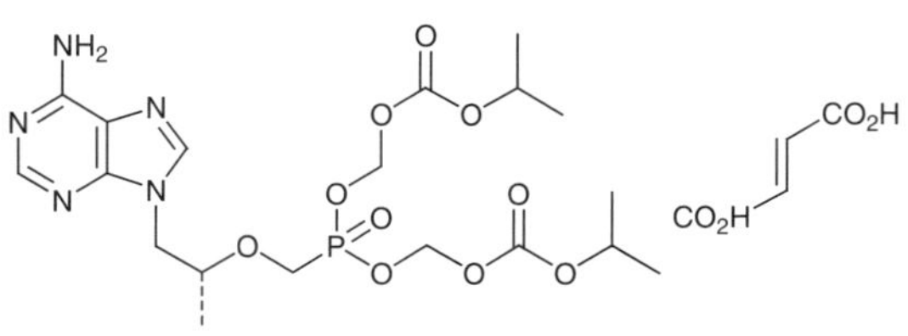

61

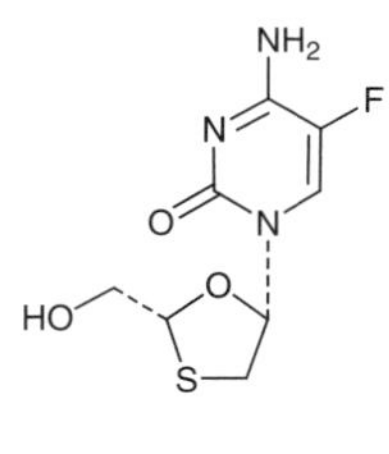

62

3.3 Natural Product Based Antitumour Agents

A recent book[77] details most of the agents from natural sources that have entered the oncologist's armamentarium over the last 50 or so years. However, it is indicative of the importance of natural product sources that 14 microbial-sourced pure compounds have been approved for use in various countries since 1954 (Table 1.1) and ten modified microbial-sourced natural products (Table 1.2) for a total of 24 compounds from this source.

In addition to these, there are 13 products from (nominal) plant sources that have entered clinical use (Table 1.3). Of these, only three are the natural products, the rest are derivatives. However, just as in the marine environment, where there is now significant evidence of the involvement of microbes/protists (single-celled organisms from all three domains of life) in the production of the secondary metabolites isolated from the host macroorganism, there have now been a significant number of recent publications[78–86] that demonstrate that endophytic fungi isolated from the plants thought to be the sources of the base

Table 1.1 Pure microbial products.

Name	*Year approved*
Carzinophilin	1954
Sarkomycin	1954
Mitomycin C	1956
Chromomycin A3	1961
Mithramycin	1961
Actinomycin D	1964
Bleomycin	1966
Doxorubicin	1966
Daunomycin	1967
Neocarzinostatin	1976
Aclarubicin	1981
Peplomycin	1981
Pentostatin	1992
Trabectedin[a]	2007

[a]Trabectedin is probably produced in Nature by an as yet uncultured microbe in the nominal producing tunicate *Ecteinascidia turbinata*.

Table 1.2 Modified microbial products.

Name	*Year approved*
Epirubicin HCl	1984
Pirarubicin	1988
Idarubicin HCl	1990
Zinostatin stimalamer	1994
Valrubicin	1999
Gemtuzumab ozogamicin	2000
Amrubicin HCl	2002
Hexyl aminolevulinate	2004
Ixabepilone	2007
Temsirolimus	2007

Table 1.3 Plant-sourced products.

Name	*Source*	*Year approved*
Vinblastine	*Catharanthus roseus*	1961
Vincristine	*Catharanthus roseus*	1963
Vindesine[a]		1979
Vinorelbine[a]		1989
Taxol®	*Taxus brevifolia*	1993
Docetaxel[a]		1995
Abraxane[a]		2005
Nanoxel[a]		2007
Irinotecan[a]		1994
Topotecan[a]		1996
Belotecan[a]		2004
Teniposide[a]		1967
Etoposide[a]		1980

[a]Modified

compounds in Table 1.3 can produce the same compound—albeit in very low yield—on fermentation of the purified microbes under conditions where "carryover" is not feasible. An argument that was used against such reports was based on the very low yields seen. However, as shown by Keller's group, the genetic control of secondary metabolic clusters in fungi (*Aspergillus nidulans*) is extremely complex[87] and it is possible that the very low yields are due to a lack of information as to the control systems involved.

4 Final Comments

From the history and examples presented above, it can be seen that natural products in the developed world have led to many different drug entities. It should be emphasised that, in the other ~80% of the world, combinations of

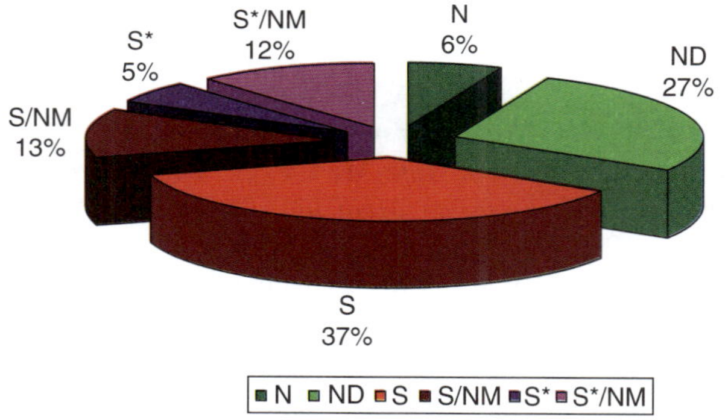

Figure 1.1 Sources of small molecule drugs, 1 January 1981–12 October 2008 ($n = 1024$). N Natural product; ND Nat. Prod. Derivative; S Synthetic; S/NM Synthetic/Nat. Prod. Mimic; S* Nat. Prod. Pharmacophore; S*/NM Nat. Prod. Pharmacophore Mimic.

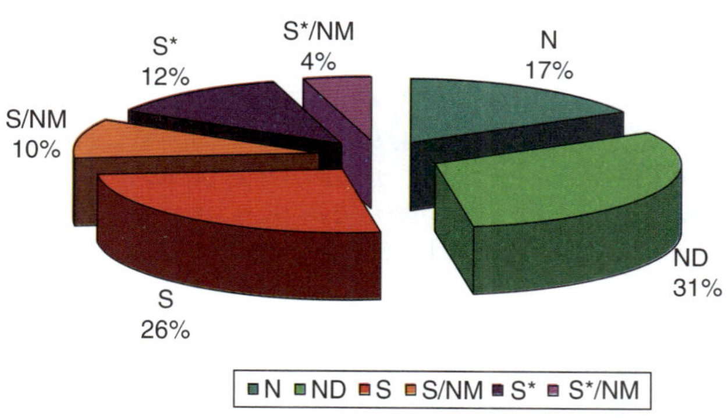

Figure 1.2 Sources of small molecule anti-tumor drugs as of 12 October 2008 ($n = 162$). N Natural product; ND Nat. Prod. Derivative; S Synthetic; S/NM Synthetic/Nat. Prod. Mimic; S* Nat. Prod. Pharmacophore; S*/NM Nat. Prod. Pharmacophore Mimic.

plants (and their associated microflora) are *still* the major source of medications for the manifold illnesses that afflict mankind.

Lest one might assume that natural products have had their day, we finish with two pie charts (codes from Newman *et al.*[41]) that demonstrate the continued involvement of Mother Nature's chemistry late in 2008 covering the sources of small molecule drugs, January 1981 to October 2008 (Figure 1.1), and sources of small molecule antitumour agents the 1930s to October 2008 (Figure 1.2).

In closing, we suggest that interested readers should consult the following three recent review papers that demonstrate how, even today, almost 20 years into the combinatorial chemistry era, chemists and biologists are still "learning chemical history from Nature"—the review by Kaiser *et al.* on biology-inspired compound libraries,[88] and the two reviews on natural products in the modern age by Ganesan[89] and Butler.[90]

References

1. J. K. Borchardt, *Drug News Perspect.*, 2002, **15**, 187.
2. O. Feenstra and I. Seybold, *Acta Med. Leg. Soc. (Liege)*, 1989, **39**, 335.
3. J. F. Nunn, *Ancient Egyptian Medicine*, University of Oklahoma Press, Norman, OK, 1996.
4. H. M. Chang and P. P. H. But, *Pharmacology and the Applications of Chinese Materia Medica*, World Scientific Publishing, Singapore, 1986.
5. K. C. Huang, *The Pharmacology of Chinese Herbs*, 2nd edn. CRC Press, Boca Raton, FL, 1999.
6. L. D. Kapoor, *CRC Handbook of Ayurvedic Medicinal Plants*, CRC Press, Boca Raton, FL, 1990.
7. S. Dev, *Environ. Health Perspect.*, 1999, **107**, 783.
8. J. P. Griffin, *Adverse Drug React. Toxicol. Rev.*, 1995, **14**, 6.
9. www.nlm.nih.gov/hmd/greek/greek_dioscorides.html.
10. M. A. Nurhussein, *Ann. Intern. Med.*, 1989, **111**, 691.
11. C. L. Cadet de Gassicourt, *Bull. Pharm.*, 1809, **1**, 5.
12. J. F. Derosne, *Ann. Chim.*, 1803, **45**, 257.
13. M. A. Seguin, *Ann. Chim.*, 1814, **92**, 225.
14. F. Seturner, *Journal der Pharmazie für Artze, Apotheke*, 1805, **13**, 29.
15. F. Seturner, *Journal der Pharmazie für Artze, Apotheke*, 1806, **14**, 47.
16. F. Seturner, *Ann. Physik.*, 1817, **55**, 56.
17. C. B. Pert and S. H. Synder, *Science*, 1973, **179**, 1011.
18. H. W. Kosterlitz and J. Hughes, *Life Sci.*, 1975, **17**, 91.
19. F. A. Gorin and G. R. Marshall, *Proc. Nat. Acad. Sci. USA*, 1977, **74**, 5179.
20. P. J. Pelletier and F. Magendie, *Ann. Chim. Phys.*, 1817, **4**, 172.
21. P. J. Pelletier and J. B. Caventou, *Ann. Chim. Phys.*, 1820, **15**, 289.
22. P. J. Robiquet, *Ann. Chim. Phys.*, 1832, **51**, 225.
23. H. F. Mein, *Ann. Chem. Pharm.*, 1833, **6**, 67.
24. G. F. Merck, *Ann. Phys. Chem.*, 1848, **66**, 125.

25. A. Ladenberg, *Ann. Chem. Pharm.*, 1881, **206**, 274.
26. V. Rosenkranz and M. Wink, *Molecules*, 2008, **13**, 2462.
27. T. Maclagan, *Lancet*, 1876, **107**, 342.
28. R. Piria, *Comptes Rendues de l'Académie des Sciences*, 1839, **8**, 479.
29. C. F. Gerhardt, *Ann. Chem. Pharm.*, 1853, **87**, 149.
30. W. Sneader, *Br. Med. J.*, 2000, **321**, 1591.
31. A. Schmidt, *Die industrielle Chemie in ihrer Bedeutung im Wetbild und Errinnerungen an ihren Aufban*, De Greuter, Berlin, 1934.
32. A. Eichengrun, *Pharmazie*, 1949, **4**, 582.
33. C. A. Nativelle, *Bull. Acad. Med.*, 1872, **2**, 201.
34. J. E. O. Schmiedeberg, *Arch. Exp. Path. Pharmacol.*, 1875, **3**, 16.
35. R. I. Mateles, *Penicillin: A Paradigm for Biotechnology*, Candida Corporation, Chicago, 1998.
36. A. Schatz, E. Bugie and S. A. Waksman, *Proc. Soc. Exp. Biol. Med.*, 1944, **55**, 66.
37. G. F. Gauze and M. G. Brazhnikova, *Am. Rev. Sov. Med.*, 1944, **2**, 134.
38. G. F. Gauze, M. G. Brazhnikova, A. N. Belozerskii and T. S. Paskhina, *Byull. Eksp. Biol. Med.*, 1944, **18**, 3.
39. Y. V. Dudnik and G. G. Gause, *SIM News*, 2002, **52**, 230.
40. T. Hager, *The Demon under the Microscope*, Harmony Books, New York, 2006.
41. D. J. Newman, G. M. Cragg and K. M. Snader, *J. Nat. Prod.*, 2003, **66**, 1022.
42. G. G. F. Newton and E. P. Abraham, *Biochem. J.*, 1956, **62**, 651.
43. E. P. Abraham and G. G. F. Newton, *Biochem. J.*, 1961, **79**, 377.
44. T. S. Mansour, P. A. Bradford and A. M. Venkatesan, in *Annual Reports in Medicinal Chemistry*, ed. J. E. Macor, Academic Press, San Diego, 2008, **vol. 43**, pp. 247–267.
45. A. Imada, K. Kitano, K. Kintaka, M. Muroi and M. Asai, *Nature*, 1981, **289**, 590.
46. R. B. Sykes, C. M. Cimarusti, D. P. Bonner, K. Bush, D. M. Floyd, N. H. Georgopapadakou, W. H. Koster, W. C. Liu, W. L. Parker, P. A. Principe, M. L. Rathnum, W. A. Slusarchyk, W. H. Trejo and J. S. Wells, *Nature*, 1981, **291**, 489.
47. A. Bonnefoy, C. Dupuis-Hamelin, V. Steier, C. Delachaume, C. Seys, T. Stchyra, M. Fairley, M. Guitton and M. Lampilas, *J. Antimicrob. Chemother.*, 2004, **54**, 410.
48. S. J. Kim, L. Cegelski, D. Stueber, M. Singh, E. Dietrich, K. S. Tanaka, T. R. Parr, Jr, A. R. Far and J. Schaefer, *J. Mol. Biol.*, 2008, **377**, 281.
49. M. R. Leadbetter, S. M. Adama, B. Bazzini, P. R. Fatheree, D. E. Karr, K. M. Krause, M. Lam, M. S. Linsell, M. B. Nodwell, J. L. Pace, K. Quast, J. P. Shaw, E. Soriano, S. G. Trapp, J. D. Villena, T. X. Wu, B. G. Christensen and J. K. Judice, *J. Antibiot. (Tokyo)*, 2004, **57**, 326.
50. G. Wang, D. Niu, Y. L. Qiu, L. T. Phan, Z. Chen, A. Polemeropoulos and Y. S. Or, *Org. Lett.*, 2004, **6**, 4455.
51. L. Xiong, Y. Korkhin and A. S. Mankin, *Antimicrob. Agents Chemother.*, 2005, **49**, 281.

52. L. E. Bermudez, N. Motamedi, C. Chee, G. Baimukanova, P. Kolonoski, C. Inderlied, P. Aralar, G. Wang, L. T. Phan and L. S. Young, *Antimicrob. Agents Chemother.*, 2007, **51**, 1666.
53. F. Kavanagh, A. Hervey and W. J. Robbins, *Proc. Nat. Acad. Sci. USA*, 1951, **37**, 570.
54. J. G. Meingassner, F. P. Schmook, R. Czok and H. Mieth, *Poultry Sci.*, 1979, **58**, 308.
55. J. F. Grove, J. MacMillan, T. P. C. Mulholland and M. A. T. Rogers, *J. Chem. Soc.*, 1952, 3977.
56. W. Mechlinski, C. P. Schaffner, P. Ganis and G. Avitabile, *Tetrahedron Lett.*, 1970, **11**, 3873.
57. E. Borowski, J. Zielinski, T. Ziminski, L. Falowski, P. Kolodziejczyk, J. Golik and E. Jereczek, *Tetrahedron Lett.*, 1970, **11**, 3909.
58. D. M. Cereghetti and E. M. Carreira, *Synthesis*, 2006, **6**, 914.
59. C. N. Chong and R. W. Rickards, *Tetrahedron Lett.*, 1970, **11**, 5145.
60. E. Borowski, J. Zielinski, L. Falowski, T. Ziminski, J. Golik, P. Kolodziejczyk, E. Jereczek, M. Gdulewicz, Y. Shenin and T. Kotienko, *Tetrahedron Lett.*, 1971, **12**, 685.
61. R. C. Pandey and K. L. Rinehart, *J. Antibiot. (Tokyo)*, 1976, **29**, 1035.
62. V. A. Morrison, *Expert Rev. Anti. Infect. Ther.*, 2006, **4**, 325.
63. M. S. Turner, R. H. Drew and J. R. Perfect, *Exp. Opin. Emerg. Drugs*, 2006, **11**, 231.
64. O. A. Cornely, M. Lasso, R. Betts, N. Klimko, J. Vazquez, G. Dobb, J. Velez, A. Williams-Diaz, J. Lipka, A. Taylor, C. Sable and N. Kartsonis, *J. Antimicrob. Chemother.*, 2007, **60**, 363.
65. W. W. Hope, D. Mickiene, V. Petraitis, R. Petraitiene, A. M. Kelaher, J. E. Hughes, M. P. Cotton, J. Bacher, J. J. Keirns, D. Buell, G. Heresi, D. K. Benjamin Jr, A. H. Groll, G. L. Drusano and T. J. Walsh, *J. Infect. Dis.*, 2008, **197**, 163.
66. G. Aperis, N. Myriounis, E. K. Spanakis and E. Mylonakis, *Exp. Opin. Invest. Drugs*, 2006, **15**, 1319.
67. D. Cappelletty and K. Eiselstein-McKitrick, *Pharmacotherapy*, 2007, **27**, 369.
68. A. C. Pasqualotto and D. W. Denning, *J. Antimicrob. Chemother.*, 2008, **61** (Suppl. 1), i19.
69. W. Bergmann and R. J. Feeney, *J. Am. Chem. Soc.*, 1950, **72**, 2809.
70. W. Bergmann and R. J. Feeney, *J. Org. Chem.*, 1951, **16**, 981.
71. W. Bergmann and D. C. Burke, *J. Org. Chem.*, 1956, **21**, 226.
72. C. J. Suckling, *Sci. Prog.*, 1991, **75**, 323.
73. W. W. Lee, A. Benitez, L. Goodman and B. R. Baker, *J. Am. Chem. Soc.*, 1960, **82**, 2648.
74. F. M. Schabel Jr, *Chemotherapy*, 1968, **13**, 321.
75. Parke Davis & Co, *GB Patent* 1159290, 1969.
76. G. Cimino, S. De Rosa and S. De Stefano, *Experientia*, 1984, **40**, 339.
77. G. M. Cragg, D. G. I. Kingston and D. J. Newman, *Anticancer Agents from Natural Products*, Taylor and Francis, Boca Raton, FL, 2005.

78. T. Amna, S. C. Puri, V. Verma, J. P. Sharma, R. K. Khajuria, J. Musarrat, M. Spiteller and G. N. Qazi, *Can. J. Microbiol.*, 2006, **52**, 189.
79. A. L. Eyberger, R. Dondapati and J. R. Porter, *J. Nat. Prod.*, 2006, **69**, 1121.
80. B. Guo, H. Li and L. Zhang, *J. Yunnan Univ.*, 1998, **20**, 214.
81. J.-Y. Li, R. S. Sidhu, A. Bollon and G. A. Strobel, *Mycolog. Res.*, 1998, **102**, 461.
82. S. C. Puri, A. Nazir, R. Chawla, R. Arora, S. Riyaz-ul-Hasan, T. Amna, B. Ahmed, V. Verma, S. Singh, R. Sagar, A. Sharma, R. Kumar, R. K. Sharma and G. N. Qazi, *J. Biotech.*, 2006, **122**, 494.
83. S. C. Puri, V. Verma, T. Amna, G. N. Qazi and M. Spiteller, *J. Nat. Prod.*, 2005, **68**, 1717.
84. A. Stierle, G. Strobel and D. Stierle, *Science*, 1993, **260**, 214.
85. X. Yang, L. Zhang, B. Guo and S. Guo, *Zhong Cao Yao* [Chinese Traditional and Herbal Drugs], 2004, **35**, 79.
86. L. Q. Zhang, B. Guo, H. Li, S. Zeng, H. Shao, S. Gu and R. Wei, *Zhong Cao Yao* [Chinese Traditional and Herbal Drugs], 2000, **31**, 805.
87. D. Hoffmeister and N. P. Keller, *Nat. Prod. Rep.*, 2007, **24**, 393.
88. M. Kaiser, S. Wetzel, K. Kumar and H. Waldmann, *Cell. Mol. Life Sci.*, 2008, **65**, 1186.
89. A. Ganesan, *Curr. Opin. Chem. Biol.*, 2008, **12**, 306.
90. M. S. Butler, *Nat. Prod. Rep.*, 2008, **25**, 475.

CHAPTER 2

Chemical Space and the Difference Between Natural Products and Synthetics

SHEO B. SINGH AND J. CHRIS CULBERSON

Merck Research Laboratories, Rahway, NJ and West Point, PA, USA

1 Introduction

Chemical space can be defined as a total descriptor space that encompasses all the small carbon-based molecules that could theoretically be created.[1] Chemical space is very vast if not infinite. The vastness of the chemical space can be appreciated by considering that the number of possible computed structures of compounds with molecular weight less than 500 Daltons consisting C, H, O, N, S and a few other atoms generally employed in drugs exceeds 10^{200}. Just a subset of molecules containing up to 30 C, N, O and S atoms may have more than 10^{60} possible structures.[2] The number of structures increases exponentially with increasing numbers of atoms. Approximately 10^{7} of these chemical structures have been reported thus far in SciFinder. The number of known compounds is certainly larger since most synthetic compounds and many natural products that reside in corporate libraries have not been reported publicly.

In the context of drug discovery, the fraction of chemical space that is relevant to biological space, which is called "biologically relevant chemical space", is of prime importance and is significantly smaller than chemical space. It appears that the simplest living organisms can function and survive with just a few hundred different types of molecules.

RSC Biomolecular Sciences No. 18
Natural Product Chemistry for Drug Discovery
Edited by Antony D. Buss and Mark S. Butler

Published by the Royal Society of Chemistry, www.rsc.org

The current chemical space is occupied by compounds isolated from nature, synthesised by conventional solution phase synthesis, solid phase combinatorial synthesis and by smaller fragment based libraries. These libraries vary significantly both in their size and complexity. The comparison of diversity of these libraries and their impact as drug leads is reviewed in this chapter.

2 Sources of Organic Compounds and Drug Leads

There are two major sources of organic compounds. They are either produced by living cells (being termed as natural products) or synthesised in the laboratory from smaller building blocks, which are predominantly obtained from petroleum products.

2.1 Natural Products

The natural products are produced by living cells. They are either produced as primary metabolites, which are used by the cells for their own function or biosynthesised as secondary metabolites for various purposes, most of which unknown to us.

Most of the interesting molecules that are considered as drug leads are secondary metabolites. These are generally produced by plants, microorganisms (fungi and bacteria) and marine organisms. These products are generally architecturally complex and highly functionalised.

The Dictionary of Natural Products version 16.2 (2008) has 265 123 entries as natural products and their derivatives.[3] Natural products remain a major source of drugs even today. In fact natural products, their derivatives and natural product mimics constitute over 50% of all drugs that are used clinically.[4]

2.2 Natural Product Derivatives

These are semi-synthetic compounds derived directly from natural products by chemical or biological (biotransformation) methods. Since organisms produce natural products for their own use and do not prepare them as drugs for human use, they can be deficient in certain pharmaceutical properties. Therefore, structural modifications are often needed before they can become drug candidates to improve physical properties and selected properties such as ADME (absorption, distribution, metabolism and excretion). For example, the conversion of the natural product lead pneumocandin Bo to the marketed drug caspofungin was accomplished in one or two transformational steps.[5] However, some compounds can be formulated directly as natural products into clinically used drugs (*e.g.* lovastatin and FK506).[5]

3 Synthetic Compounds

Synthetic compounds are synthesised *de novo* from small building blocks. This compound class can be divided in four segments depending on the method and purpose of synthesis.

3.1 Synthetic Compound Libraries

Many of these compounds are synthesised by chemists for structure–activity relationship purposes and deposited in sample collections for future high throughput screening against new targets/assays. Other classes of compounds are synthesised as a curiosity and to complement a structural void in the sample collection.

The sample collection of a pharmaceutical company is one of the largest assets of the corporation and its diversity, quality and quantity determines the future success of the organisation. These collections are widely used for high throughput screening campaigns against new assays and targets leading to the discovery of most of the drug leads, which then proceed to lead optimisation. The new compounds generated during the new lead optimisation phase are again deposited in the collection to increase size and diversity. Millions of these compounds exist in corporate and university laboratory collections, and their structures and associated biological activities are not known to the public at large.

3.2 Combinatorial Libraries

The foundation of combinatorial synthesis was laid with the development of solid phase peptide synthesis by Merrifield in the 1960s. This allowed a rapid and convenient way for the elimination of reagents from products by simple washing after each step of the synthesis followed by cleavage of the product from the solid support.

The outcome of this technology was the synthesis of many compounds in parallel by mixing and matching various intermediates. This technology certainly allowed for the preparation, with limited resources, of libraries of compounds consisting of many, many thousands to millions of compounds.

The quality of the product in the library depended on the efficiency of individual reaction in the sequence. Any inefficiency was magnified due to the iterative nature of reactions advancing the product to the next step without purification. Additionally, the number of reactions that was amenable for solid phase synthesis applied in high throughput mode was highly limited.

Unfortunately, these issues did not allow the technique to deliver new quality hits against a variety of biological targets, which was the original premise. While early libraries were prepared as mixtures, many newer libraries were prepared as a single compound contained in a unique well of a microtitre plate.

A number of multi-million member combinatorial libraries exist in many corporations. Screening cost plays a major role in decision-making as to which library or part of a library to screen.

3.3 Diversity-Oriented Synthetic (DOS) Libraries

After the advent of combinatorial chemistry and its lack of success, it was recognised that the combinatorial libraries lacked diversity. Hence, new approaches were applied to design compounds that allowed the introduction of many points of diversity, such as chiral centres to mimic natural products.[6] These libraries are also called natural product-like libraries.[7]

Since the goal of DOS libraries is to make compounds more like natural products, structural features of natural products play a bigger role in the design strategies of such libraries. The DOS libraries fall in three broad groups:

- Libraries based on the core scaffold of an individual natural product.
- Libraries based on specific structural motifs of a class of natural products.
- Libraries that mimic structural features of natural products in a more general sense.

The goal of library design for the most part was to address activity against a variety of drug targets and to discover tools to interrogate biology, in addition to finding lead compounds. This led to the identification of a number of natural product fragments as privileged structures that are used in synthetic drugs such as purines, indoles and benzopyrans.[7,8] Cascade and multi component reactions are employed routinely in DOS library design strategies.[7]

3.4 Fragment Libraries

This is an alternative strategy for the identification of drug leads. The idea behind this approach is to identify low molecular weight "fragments" of compounds with low affinity to a protein target and build onto it high affinity ligands either by rational modifications by addition of chemical groups or by joining two or more low affinity ligands into a single higher affinity ligand.

The definition of fragment varies, but usually refers to molecules with a molecular weight less than 200–300 Daltons and consisting of fewer than 15–20 heavy atoms.[9,10] A small ($\sim$50 000 member) diverse library is generally sufficient for high throughput screening to yield a low affinity hit.

In addition to wet screening (a high concentration is required), NMR, X-ray crystallographic, functional and structure-guided fragment screening, the fragment-based approach is highly amenable for virtual screening by the generation of all possible theoretical structures by computer algorithms.

Virtual screening requires knowledge of the three-dimensional (3D) structure of the protein targets. Fink *et al.*[11,12] designed and applied algorithms to generate possible structures of compounds containing up to 11 atoms of C, N, O, F and H. This led to the generation of 26.4 million molecules and 110.9 million structures (if one considers stereoisomers). Because of their small molecular mass they obey Lipinski's bioavailability rule (see Section 4).[7] Half of these molecules also follow Congreve's "rule of three" for lead-likeness.[13]

This library consisted of molecules with an average molecular weight of 157.3 Daltons.

Virtual screening of this virtual library was undertaken against three targets including a G-protein coupled receptor (GPCR), a kinase and an ion channel. For comparative purposes, a reference database of structures of known compounds consisting of the same 11 atoms was prepared and identical virtual screening was performed.[12] Fink *et al.* found that 90% of the chemical space of these virtual hits fell in the regions of chemical space covered by both databases and 10% of the hits were found only in the virtual database, but not in the reference database.[11,12] It appears that fragment-based screening is gaining popularity.

These investigators discovered that exhaustive enumeration of chemical space consisting of 11 atoms provided a rich source of information, but this approach may not be successful for molecules with more than 11 atoms at this moment of time.[12] In fact, the size of the virtual library generated from 11 atoms is larger than the entire collection of known compounds in the Chemical Abstract Index.[12] The virtual library grows exponentially with the addition of an extra atom. However, with more powerful computers and computer memory it is possible to generate structures with 12–13 atoms that would extend the database by an order of magnitude of 2–4. Based on these calculations, it can be extrapolated that the size of a drug-like library consisting of 25 atoms would be approximately 10^{27}.[12] It is not possible to exhaustively enumerate and generate a virtual library of all these structures with current computing power, let alone have the capacity to synthesise these compounds. Synthesis and wet screening of these compounds will be out of the question for a long time, if ever. A virtual library may be the only way to access such structural diversity when the computing power becomes available.

4 Lipinski's "Rule of Five" for Orally Active Drugs

The failure of drugs at later stages of development, particularly in clinical trials, is very expensive for drug developers and, more importantly, patients. To better understand the key reasons for these failures, Lipinski *et al.*[14] undertook an analysis of the properties of compounds that entered Phase II human clinical trials. They selected a subset of 2245 compounds from the World Drug Index (WDI) database of over 50 000 compounds after eliminating the majority of compounds for various well-reasoned criteria. This subset of compounds had assigned trade names and, as a result, were assumed to have entered Phase II oral efficacy studies and be expected to have superior physico-chemical properties since they would have passed most of the other earlier clinical trial hurdles.

Compound solubility and permeability were identified as key factors in advancing these 2245 compounds into human trials. The authors compared the calculated properties of these compounds with the corresponding properties of the compounds of the entire database. Four parameters appeared to be correlated with solubility and permeability, namely molecular weight (MW), log P, the number of H-bond donors (HBD) and the number of H-bond acceptors (HBA).[14] The investigators asked what were the approximate numerical values of these parameters that met the 90% confidence interval for better solubility and permeability.

The analysis of the numerical values of these four parameters led to cutoff numbers close to five or multiples of five. This led to the formulation of the rule now famously called "rule of five". The rule states that poor absorption or permeation are more likely to result when a compound has a molecular weight greater than 500 Daltons, has log P greater than five (or MlogP greater than 4.15), has more than five H-bond donors (expressed as the sum of OHs and NHs) and has more than ten H-bond acceptors (expressed as the sum of Ns and Os).

A number of companies, such as Pfizer, have instituted an organisation-wide rule in their registration systems in which compounds are flagged automatically if two or more of these parameters are out of range from the "rule of five", alerting them to the fact that poor absorption or permeability is possible with such compounds.[14]

The "rule of five" analysis focused on oral drugs and so is not applicable to all drug classes, *e.g.* compound classes that are substrates for biological transporters, parental drugs, and most drugs that are natural products.[15] This exclusion also applies to essentially all anti-infective drugs including synthetic and natural antibiotics, antifungals, antiparasitics, antivirals, vitamins and a large number of other critical life-saving parental drugs.

5 Assessment of Diversity of Libraries with Respect to Drugs

Assessing the diversity of various compound classes has been challenging. A number of descriptor-based approaches have been applied to assess the diversity of various libraries and compared with existing drugs.

In 1999, Henkel *et al.*[16] were first to statistically analyse structural parameters of organic compounds. They used the following sources:

- Natural product structures were obtained from the *Dictionary of Natural Products* (DNP, Chapman & Hall, n=78 318) and the biologically active Natural Products Database (BNPD, compiled by Berdy of Szenzor Management Consulting Company, Budapest, n=29 432). These two databases cover most of the natural products and their derivatives.
- Synthetic chemical structures were obtained from the Available Chemicals Directory (ACD, MDL Information Systems Inc., San Leardo, CA, n=182 822) and a representative pool of test substances from the Synthetics database (Bayer AG, n=not reported).
- The drug structures were obtained from the Drugs database (pharmaceutical products/compounds in development recorded in Pharmaprojects, RDFocus and in the active compounds pool of Bayer AG, n=14 596).

Henkel *et al.* did not differentiate the natural products from natural product derivatives. The descriptors they used for comparison were molecular weight, number and type of hetero atoms, pharmacophore groups (*e.g.* CO, CONH, OH, CN, NH, *etc*), bridgehead atoms, rotatable C–C bonds, rings per molecule, chiral centres per molecule and rotatable bonds per molecule.

In 2001, Lee and Schneider[17] compared the properties of trade drugs (taken from the Derwent World Drug Index, WDI, n=5757) and natural products (taken from the BioScreenNP database, n=10 495). These investigators described the comparison of parameters applicable only to the "rule of five" (molecular weight, log P, number of H donors per molecule, number of N per molecule, number of O per molecule and percentage of "rule of five" alerts).

In 2003, Feher and Schmidt[18] published a more comprehensive comparison of the structures, which included different databases and also for the first time used combinatorial libraries. For this comparison, drug molecules were taken from Chapman & Hall's *Dictionary of Drugs* (n=10 968). Combinatorial libraries (perhaps corresponding to any synthetic compounds, n=670 536) were sourced from the following databases: Maybridge HTS; ChemBridge EXPRESS-PICK; ComGenex; ChemDiv International Diversity; ChemDiv CombiLab™ Probe Libraries; and SPECS screening compounds. Natural products (n=3287) and their derivatives (n=27 338) were obtained from the following databases: BioSPECS natural products; ChemDiv natural products; and InterBioScreen IBS2001N and HTS-NC.

The parameters used to compare these data sets[18] were a combination of the two earlier investigations.

We have also made similar descriptor comparisons of compounds from the Merck chemical collection, the top 200 selling drugs in 2006, and Merck's natural product collection.

Although the three papers discussed above used data from different databases, the overall conclusions were essentially the same—as indeed were those from our analysis. The results from many of these descriptors are highlighted below.

5.1 Molecular Weight

In all the reports, the molecular weight distribution of drugs generally follows a Gaussian distribution with median value of 312.[18] Natural products generally peak at a similar position (mean value of 362) compared with drugs, but skewed towards higher molecular weight. This higher mean value is a reflection of the wider distribution of molecular weights of natural products. The molecular weight distribution of combinatorial libraries peaked at an even higher value (mean 389) and showed a narrower distribution than drugs and natural products.[18] The molecular weight distribution of the top 200 selling drugs followed a Gaussian distribution with flattening in the middle (Figure 2.1). Compounds in the Merck sample collection followed the molecular weight distribution pattern of the top 200 selling drugs. Similar to the observations made in other

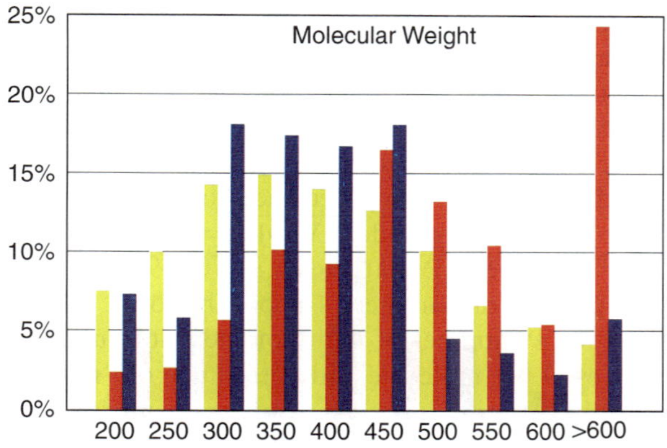

Figure 2.1 Distribution of molecular weight of the Merck sample collection (yellow), Merck natural products (red, $n=595$) and 137 of the top 200 drugs from 2006 (blue).

reports, the molecular weight distribution of Merck's natural products was also on the higher side than drugs.

5.2 Distribution of Atom Types: H-bond Donors and Acceptors

The combinatorial compounds, natural products and drugs differ drastically in their elemental composition. The natural products on average contain three times fewer nitrogen atoms per molecule than combinatorial compounds and twice as many oxygen atoms compared with the combinatorial compounds.[18] In general, natural products contain a higher number of H-bond acceptors and substantially higher numbers of H-bond donors per molecule than synthetic compounds. Merck's sample collection follows the same trend when compared with the top 200 selling drugs and the Merck natural products collection (Figures 2.2 and 2.3). Again H-bonding properties of drugs fall between the synthetics and natural products. These H-bond donors and acceptors are predominantly represented by a higher number of OH, ether, ester and lactone groups in natural products. Combinatorial compounds generally contain a higher number of sulfur atoms per molecule than natural products. The difference in the distribution of the hetero atoms is reflective of the precursor availability for chemical synthesis and biosynthesis of the combinatorial and natural products respectively. The description of H-bond donor and acceptor properties has been expanded by Feher and Schmidt.[18]

The therapeutic effects of drugs result from their interaction with enzymes and receptors. These proteins have evolved over time and receptor–ligand specificity is dependent on the matched interaction between protein and ligand. The binding potency and specificity originates from proper matching of complementary polar and non-polar interactions of the bound complex.

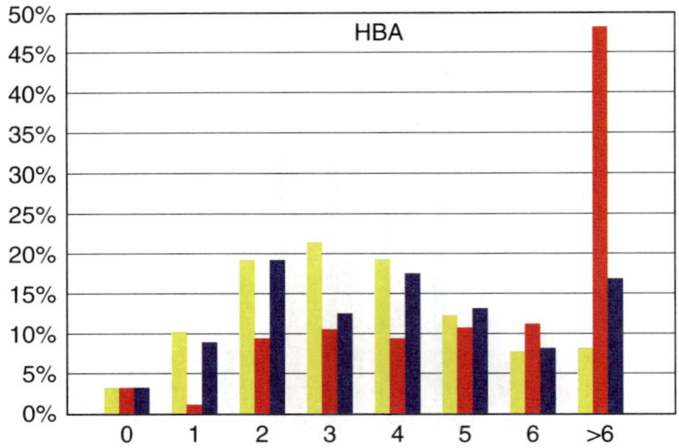

Figure 2.2 Distribution of hydrogen bond acceptors (HBA) per molecule of the Merck sample collection (yellow), Merck natural products (red, n=595) and 137 of the top 200 drugs from 2006 (blue).

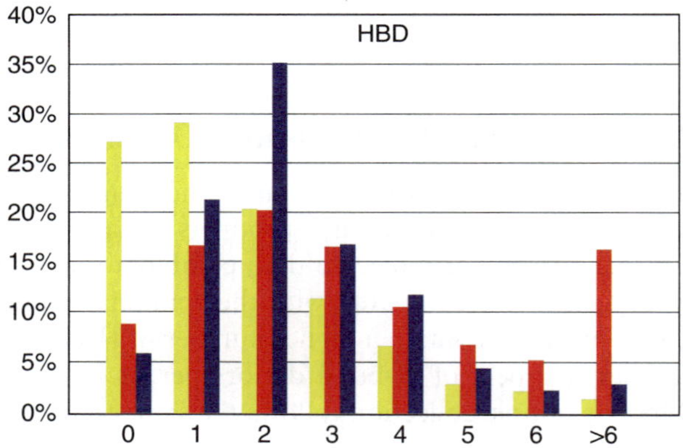

Figure 2.3 Distribution of hydrogen bond donors (HBD) per molecule of the Merck sample collection (yellow), Merck natural products (red, n=595) and 137 of the top 200 drugs from 2006 (blue).

It is obvious that the outcome would be poor when protein and ligand interactions are mismatched, *e.g.* polar groups of a ligand interacting with the non-polar region of the protein. Thus, differences in the distribution of the type and number of polar atoms in synthetic and natural product compounds could have significant consequences for their binding behaviours. Although it is not clear why natural products are produced, they are certainly not purposefully produced for human consumption; their binding advantage due to their co-evolution with enzymes and receptors should not be overlooked.

5.3 Lipophilicities (Log P)

The distribution of calculated Slog P (or clog P or Alog P) of natural products, drugs and combinatorial compounds suggest that the lipophilicities of natural products and drugs are much closer to each other and markedly differ from combinatorial compounds. The latter compounds are generally significantly more lipophilic.[18] Higher lipophilicity has a negative impact on the drug-like behaviour of the compounds, leading to poor adsorption and permeability of the combinatorial compounds. This is particularly important since most of the drug structures contain major portions of original lead structures.[19] The Merck sample collection fared much better in the distribution of clog P properties and was similar to the top 200 selling drugs. Merck natural products showed a much wider distribution of clog P properties (Figure 2.4).

5.4 Chiral Centres

The percentage distribution of the number of chiral centres in three classes of compounds (drugs, natural products and combinatorial) was markedly different.[18] The mean numbers of chiral centres were 6.2 for natural products, 2.3 for drugs and 0.4 for combinatorial molecules.[18] Since natural products are synthesised by enzymes the introduction of a chiral centre appears to be effortless, whereas it requires special attention to synthesise chiral compounds in the laboratory.

The presence of chiral centres in the natural products provides for higher affinity and target specificity, and clearly differentiates them from non-natural compounds. Henkel *et al.*[16] found a different numerical value for this parameter but the trend was identical.

The presence of larger numbers of chiral centres in natural products compared with *de novo* man-made synthetic compounds is the biggest and most

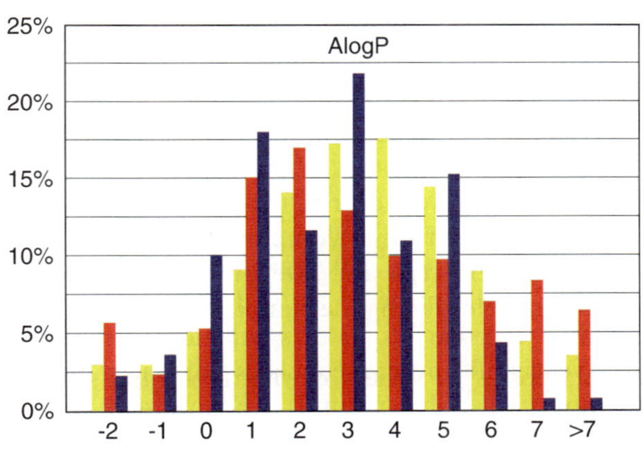

Figure 2.4 Distribution of AlogP of the Merck sample collection (yellow), Merck natural products (red, n=595) and 182 of the top 200 drugs from 2006 (blue).

important difference between the two groups of compounds. Since the biological targets for these drugs are chiral, ligands with correctly constructed chiral centres should provide increased target engagement. Attempts have been made during the synthesis of DOS libraries to synthesise compounds with more chiral centres, thus making them more like natural products. Despite these efforts, natural products still provide most structural and chiral complexity.

5.5 Rotatable Bonds, Unsaturations, Rings, Chains and Ring Topology

Comparisons of the databases of combinatorial, natural products and drugs by Feher and Schmidt[18] suggest that a higher percentage of natural products have no rotatable bonds and provide a wider, but non-Gaussian distribution of this property. Combinatorial compounds show a different distribution, with the highest percentages possessing a minimum of two rotatable bonds. The drugs again fall between the two groups.

It is notable that, due to higher entropic losses, the flexible ligands with identical H-bond and hydrophobic interactions are generally expected to show weaker binding affinity with a protein than the corresponding rigid ligand. Hence, the improvement of rigidity of the molecule is employed as one of the key factors for lead optimisation by medicinal chemists to gain additional ligand affinity with the protein.

The level of unsaturation is another property of the molecule that provides rigidity. Combinatorial compounds possess higher numbers of aromatic rings which provides for larger numbers of unsaturation per molecule, but if the aromatic ring is considered as a single degree of unsaturation, then natural products are, on average, more unsaturated (median six) than combinatorial compounds (median five). Surprisingly, drugs are more saturated (median unsaturation four) than both natural products and combinatorial compounds.

The rigidity of the natural products is also confirmed by the presence of larger numbers of rings (median four rings) than combinatorial compounds (median three rings). The median number of rings for drugs is only two, which is lower than both natural products and combinatorial compounds. The distribution of the number of rings varies in different combinatorial collections.[18] The numbers of rings themselves do not tell the complete story of the differences of natural products and other compounds. In fact, while the numbers of rings are important, the type and size of the rings and ring fusion are even more important in determining the complexity and overall topology of the molecule.

Natural products possess a variety of ring types and sizes including fused rings, bridged rings and spiro rings which are essentially absent in combinatorial libraries. Combinatorial libraries possess more aromatic rings than natural products, with drugs falling in between the two groups.[18] This high level of ring fusions also provides a higher degree of rigidity to natural products.

6 Principal Component Analysis (PCA)

While the individual property comparisons amply differentiate the natural products from combinatorial and drug libraries, the cumulative diversity can only be visualised by principal component analysis (PCA).

The ten molecular descriptor PCA analysis of a subset of three libraries performed by Feher and Schmidt[18] suggested that the combinatorial library clustered very tightly and represented significantly lower diversity than natural products or drugs. The drugs and natural products showed wider diversity and covered a diversity map that was not represented by the combinatorial library.

A similar PCA analysis of 595 natural products from the Merck sample collection, 137 small molecules from the top 200 best selling drugs in 2006 and the 65 000 sample combinatorial library from ChemBridge (one of the libraries used by Feher and Schmidt in their analysis) showed a similar plot and provided similar conclusions (Figure 2.5).

A more profound visual difference of the chemical space covered by natural products and synthetic drugs was presented by Derek Tan, in which he applied a similar PCA analysis of 20 synthetic drugs (including ten best sellers of 2004) and 20 natural products.[7] For this analysis, Tan used nine molecular descriptors—molecular weight, clog P, H-bond donors, H-bond acceptors, rotatable bonds, polar surface area (PSA), chiral centres, N and O atoms—and then applied PCA to reduce nine-dimensional vectors to two-dimensional vectors before re-plotting the data.

A similar nine molecular descriptor analysis of the 200 top selling drugs in 2006 (including natural product drugs) and 595 Merck natural products

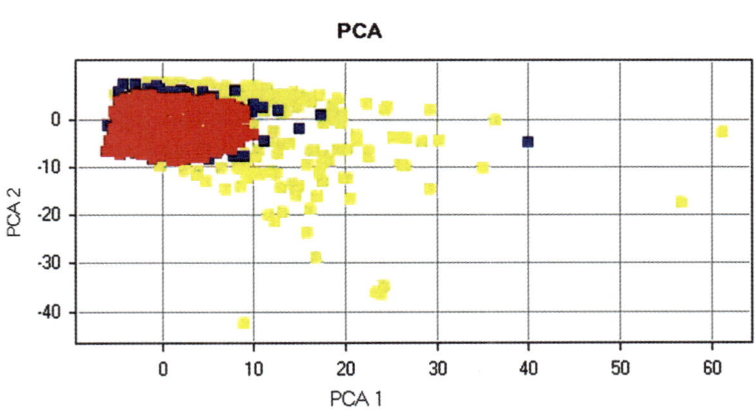

Figure 2.5 Plot of the first two principal components obtained from a database containing: (a) a random selection of combinatorial compounds (red, n=65 000); (b) Merck natural products (yellow, n=595); and (c) the top 200 drugs in 2006 (blue, n=137).

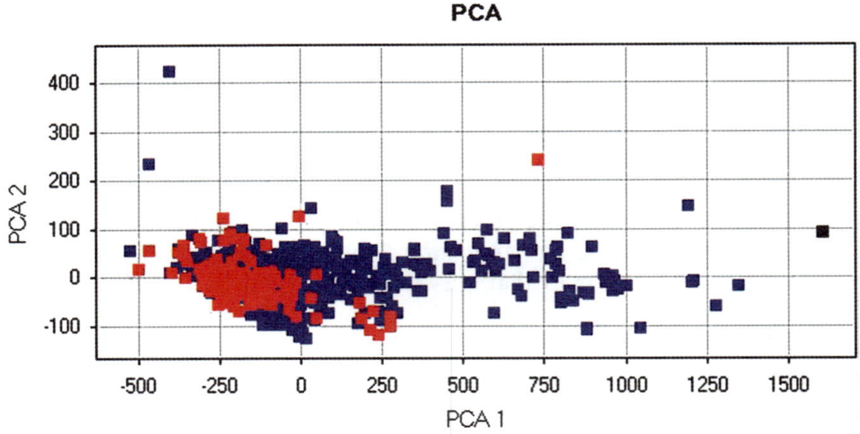

Figure 2.6 PCA of the top 200 small molecule drugs in 2006 including natural product drugs (red) and 595 Merck natural products (blue). The descriptors MW, HBA, HBD, AlogP98, PSA, normalised bond flexibility, nitrogen count, oxygen count and chiral centre count were used.

exhibited a large overlap but also differences in the chemical space covered by the drugs and natural products (Figure 2.6).

Tan's analysis of the first two components accounted for 84.2% of the original information, indicating that synthetic drugs and natural products show limited overlap of chemical space.

7 Conclusions

It is clear from these analyses that natural products cover much wider and larger chemical space not covered by combinatorial and synthetic compounds. Natural products contain large numbers of chiral centres, ring fusions and a higher density of functional groups allowing for higher ligand affinity and better specificity to biological targets that is generally not achieved by architecturally flat synthetic drugs. They are often orally bioavailable despite violating many of Lipinski's "rule of five" parameters; examples include FK506 and cyclosporin A.[15]

Significantly, natural products cover biological space not covered by synthetic drugs, for example:

- immunosuppressants such as rapamycin, FK506 and cyclosporin A;
- antibiotics such as vancomycin and thienamycin;
- anticancer drugs such as taxol;
- cholesterol-lowering drugs such as lovastatin and the synthetic version atorvastatin.

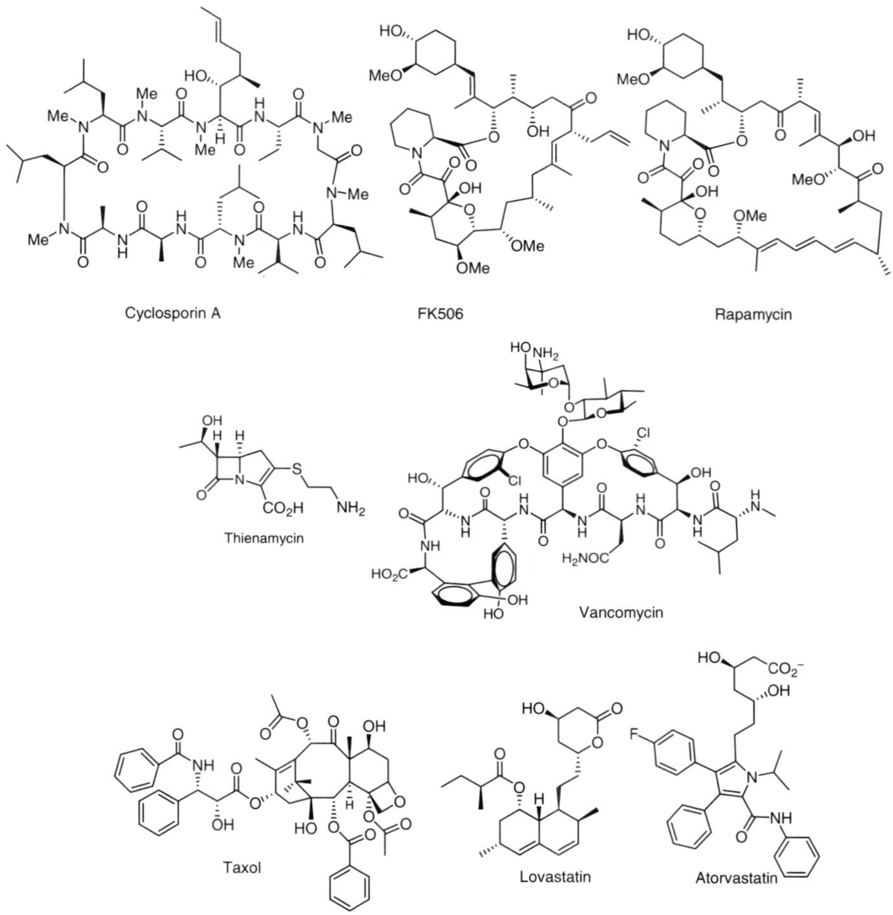

Without the discovery of natural products, these life-saving treatments would not have been available as therapeutic agents. Natural products tend to have properties such as unexpected cell penetration, absorption or solubility that are generally not well understood. This may be due to their co-evolution with the proteins in the cells, which are densely populated with biological materials.[1]

Overall, only a tiny fraction of known compounds are natural products. However, they constitute a higher percentage of marketed drugs, indicating a significantly higher rate of return per molecule.

These data suggest that society would lose out significantly if natural product sources were lost completely and permanently. The loss could potentially be overcome if the following were to happen:

- computing power were to improve where the generation of theoretical structures of all compounds with larger molecular weights were possible in a reasonable timescale;

- the three-dimensional structures of all receptors and enzymes were known;
- computing power and reliability of virtual docking technologies were improved so that reliable docked structures were obtained that predicted reliable binding modes;
- synthetic technologies were improved so that any complex structures could be synthesised in a time and cost effective way.

Tremendous strides have been made in synthetic methodologies where compounds are beginning to be synthesised efficiently without the use of protecting groups.[20] If this type of synthesis were applicable on a wider scale, computers were able to predict structures and modelling was able to provide reliable structures with great binding affinity to protein targets, drug discovery may be expedited. Docking could yield a docked structure that could be chemically synthesised in a laboratory and effectively tested in biological assay systems in wet labs to confirm the activity. At this moment, this is a pipe dream at every level. Today, insufficient computing power exists for the generation of structures larger than 11 atoms, let alone reliable virtual screening and synthetic techniques.[11,12] However, the future is limitless and it is possible that one day this could happen—which would change drug discovery dramatically.

References

1. C. M. Dobson, *Nature*, 2004, **432**, 824.
2. R. S. Bohacek, C. McMartin and W. C. Guida, *Med. Res. Rev.*, 1996, **16**, 3.
3. J. Buckingham, (ed.), *Dictionary of Natural Products*, Chapman and Hall, Oxfordshire, 2008.
4. D. J. Newman, *J. Med. Chem.*, 2008, **51**, 2589.
5. S. B. Singh and F. Pelaez, *Prog. Drug Res.*, 2008, **65**, 141.
6. S. L. Schreiber, *Chem. Eng. News*, 2003, **81**, 51.
7. D. S. Tan, *Nat. Chem. Biol.*, 2005, **1**, 74.
8. R. W. DeSimone, K. S. Currie, S. A. Mitchell, J. W. Darrow and D. A. Pippin, *Comb. Chem. High Throughput Screening*, 2004, **7**, 473.
9. D. A. Erlanson, *Curr. Opin. Biotechnol.*, 2006, **17**, 643.
10. D. A. Erlanson, R. S. McDowell and T. O'Brien, *J. Med. Chem.*, 2004, **47**, 3463.
11. T. Fink, H. Bruggesser and J. L. Reymond, *Angew. Chem. Int. Ed. Engl.*, 2005, **44**, 1504.
12. T. Fink and J. L. Reymond, *J. Chem. Inf. Model.*, 2007, **47**, 342.
13. C. Lipinski, F. Lombardo, B. W. Dominy and P. J. Feeney, *Adv. Drug Del. Rev.*, 1997, **23**, 3.
14. M. Congreve, R. Carr, C. Murray and H. Jhoti, *Drug Discov. Today*, 2003, **8**, 876.
15. C. A. Lipinski, *Drug Discov. Today*, 2003, **8**, 12.

16. T. Henkel, R. M. Brunne, H. Müller and F. Reichel, *Angew. Chem. Int. Ed. Engl.*, 1999, **38**, 643.
17. M. -L. Lee and G. Schneider, *J. Comb. Chem.*, 2001, **3**, 284.
18. M. Feher and J. M. Schmid, *J. Chem. Inf. Comput. Sci.*, 2003, **43**, 218.
19. J. R. Proudfoot, *Bioorg. Med. Chem. Lett.*, 2002, **12**, 1647.
20. P. S. Baran, T. J. Maimone and J. M. Richter, *Nature*, 2007, **446**, 404.

CHAPTER 3
Mechanism of Action Studies

JAMES J. LA CLAIR

Xenobe Research Institute, 3371 Adams Avenue, San Diego, CA 92116, USA

1 Introduction

If asked to write a script to a play or film, one typically begins by generating a list of characters and then defines these characters as they transpose through an ascribed plot. If asked to write on the activity of natural products, I am certain that refined wordsmiths would not only find keen interest in their natural product characters but would also be enthralled to engage their all so devious plots. Natural products are as important to the story of life within both its daily and evolutionary context. It is within the archival access to novel means of regulating the process of life that their unique yet robustly intriguing structural complexity has evolved. While one can spend years attempting to classify natural products according to function, one soon learns that it is the lack of these classifications that illuminate the beauty of the secondary metabolite.

Within the last decade, the development of tools at the synthetic, biochemical, proteomic, genomic and immunological levels has opened access to studies that provide an unprecedented look into the complex roles which natural products play. The following sections provide an overview of studies on a select panel of natural products. The goal is not to glorify these studies or their activities but rather to provide an overview of the wondrous modes of natural product action.

RSC Biomolecular Sciences No. 18
Natural Product Chemistry for Drug Discovery
Edited by Antony D. Buss and Mark S. Butler

Published by the Royal Society of Chemistry, www.rsc.org

2 Some Like It Hot: Esperamicin A_1, Neocarzinostatin and Related Enediyne Antibiotics

In 1985, Konishi and colleagues at Bristol-Myers published the partial structures of a family of broad spectrum antimicrobial and antitumour compounds, called the esperamicins, from cultures of *Actinomadura verrucosospora*, strain H964-62.[1] While unexpected at this time, subsequent structural studies revealed that the esperamicin A_1 contained an unique cyclo[7.3.1] ring system including a trisulfide, a bridgehead olefin contained within an α,β-unsaturated ketone and a distinct 1,5-diyn-3-ene.[2] Soon thereafter, collaborative efforts between Bristol-Myers and the laboratories of James Dabrowiak at Syracuse University discovered a link between esperamicin A_1 and an apparently degraded congener esperamicin Z.[3] It was postulated that reduction of the trisulfide resulted in Michael addition to the α,β-unsaturated ketone, thereby inducing a Bergman cyclisation to form a diradical. In this scenario, the Michael addition of the pendant thiol to the bridgehead olefin reduced the distance between the two alkynes thereby facilitating the Bergman cyclisation.[4] Using a combination of DNA cleavage assays and *in vitro* cytotoxicity assays, the collaborators determined trisulfide reducing agents such as dithiothreitol or glutathione activated esperamycin A_1 inducing both single-stranded and double-stranded DNA cleavage.[5] Subsequent studies indicated that the selection between single and double strand cleavage was regulated by structural motifs within the core and carbohydrate tethers. These observations were confirmed by parallel investigations.[6,7]

Esperamicin A_1 was, however, not the only natural product identified with an enediyne core (Figure 3.1). Studies dating back to 1965[8] identified a chromoprotein containing a neocarzinostatin core that was able to induce comparable

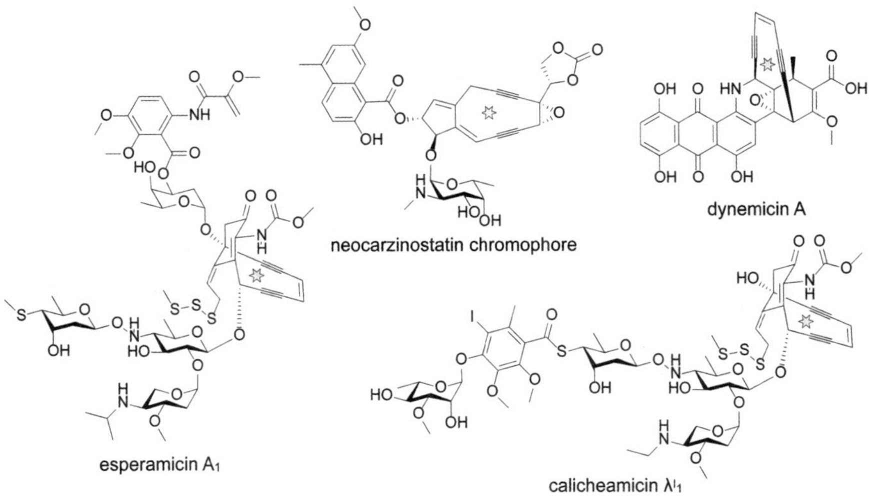

Figure 3.1 Structures of exemplary enediyne natural products. The 'star' notes the location of their reactive enediyne functionality.

DNA cleavage.[9] Subsequent structure elucidation efforts indicated that its chromophore contained a related positioning of alkynes. In 1987, Meyers and Saito determined that this process was also triggered by nucleophilic addition.[10,11]

Parallel studies from 1985–1992 identified comparable enediyne motifs and related mechanisms of DNA cleavage in the calicheamicin, dynemicin and the kedarcidin chromophore.[12–15] Evidence for the selectivity of each of these natural products came through a compendium of structure–activity relationship (SAR) studies enabled by total synthetic efforts. Materials provided by synthetic studies in the laboratories of Myers,[16] Nicolaou,[17] Danishefsky,[18] Schreiber,[19] Kahne,[20] Magus,[21] Wender,[22] Isobe,[23] Takahashi,[24] Hirama[25] and others[26] provided detailed understanding of their mechanistic triggers and the means in which the natural products can regulate DNA damage. Using synthetic methods, a combination of hybrid and synthetic analogues were developed to screen a variety of chemical, thermal, photochemical and redox triggers.[27,28] Combined with semi-synthetic efforts, these studies also provided a comprehensive understanding as to the selectivity of DNA cleavage. Structure–activity relationship studies indicated that components of the carbohydrate motifs were critical to maintain selective single and double strand cleavages. Indications of this selectivity were first apparent during studies on the isolated congeners of esperamycin;[29] however, synthetic and NMR studies were required to provide a complete mechanistic understanding.

NMR studies depicting natural products bound to DNA, conducted by the Patel laboratory, provided a detailed understanding as to how the carbohydrate motifs engaged double-stranded DNA and delivered selectivity.[30] Structures of the DNA complexes of esperamicin A_1 (Figure 3.2a)[31] and calicheamicin λ_1^I (Figure 3.2b)[32] identified multiple interactions between the carbohydrate motifs and the DNA backbone.

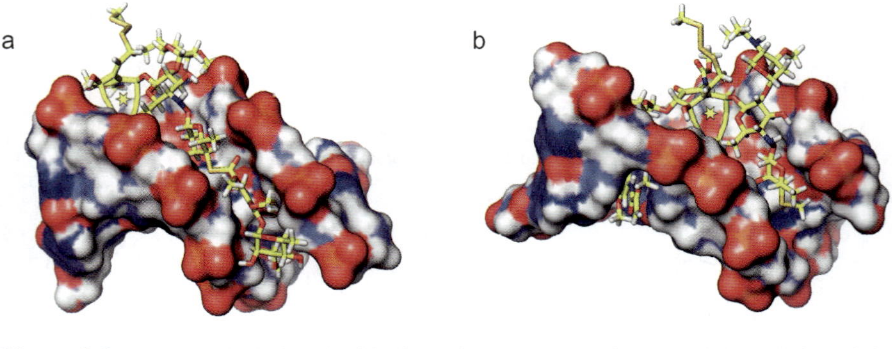

Figure 3.2 Images depicting the binding of (a) esperamycin A_1 and (b) calicheamicin $\lambda^I{}_1$ to double-stranded DNA. The 'star' notes the location of their reactive enediyne functionality. The images were developed from the NMR datasets **1pik** (esperamycin A_1) and **2pik** (calicheamicin $\lambda_I{}^1$), which are readily available from the Protein Data Bank.

Early studies on the neocarzinostatin chromophore and later evaluations of esperamicins and calicheamicins unveiled a novel mechanism of action, wherein the spatial relation between two alkynes is modulated to regulate the formation of a diradical species through either a Bergman,[4] Myers-Saito[10,11] or Schmittel[33] cyclisation. By use of pendant carbohydrate domains, these natural products position their enediyne core in the minor grove such that the formation of a diradical intermediate induces rapid abstraction of an $H^{\bullet}$ from both strands of DNA leading to interception by oxygen and double-stranded cleavage. Modification of the functionality within these carbohydrate motifs alters the positioning of the enediyne core and thereby, leads to incomplete $H^{\bullet}$ abstraction and loss of double-strand cleavage.[34,35] In addition, the carbohydrate motifs provided a selective recognition of DNA sequences.[36] The aryl oligosaccharide unit of calicheamicin λ^{I}_{1} oriented the molecule within the minor grove of DNA, favouring regions containing 5′-TCCT-3′ and 5′-TTTT-3′.[37] The frequency of cutting sites indeed differs with esperamicin cutting at T > C > A > G, calicheamicin at C >> T > A = G and neocarzinostatin T > A > C > G, thereby validating individual selectivities within each structure.[38] The mechanism of action of esperamicin A_1 and the neocarzinostatin chromophore are presented in Figures 3.3 and 3.4 respectively.

Through synthetic studies, this selectivity can now be modulated to provide unique selections within a given DNA target. While complementary, the DNA cutting mode of action of these materials and their unique triggering and activating mechanisms is one of the more beautiful transformations that have been discovered in natural product activity. Their study both at the synthetic and biological level has established the importance of this motif not only as a mechanistic tool but also as forwarded materials that, when used in conjunction with antibody conjugation, provide an effective tool for clinical use.[39]

The actors in this story, as in the film *Some Like It Hot*, while interesting in their own context, enhance the story by their interaction as the plot develops. Here, an intricate balance between the studies on each family of natural product united to provide a story that stands alone. Simply put, there was no single observation that furthered the mode of action of this class, for taken alone, not a single observation offered the story provided by the entire cast. As to their advance in the clinic, the quote, "well, nobody's perfect" adds perspective. While the natural products alone appear to be too toxic, the approval of Mylotarg or Gemtuzumab ozogamicin by the US Food and Drug Administration (FDA) to treat acute myelogenous leukemia provides strong support for the development of the enediynes.[40]

3 To Catch a Mockingbird: Taxol, Epothilone and the Microtubule

The National Cancer Institute (NCI) was started in 1937 by an act of the US Congress to fund and conduct research for the development of projects that

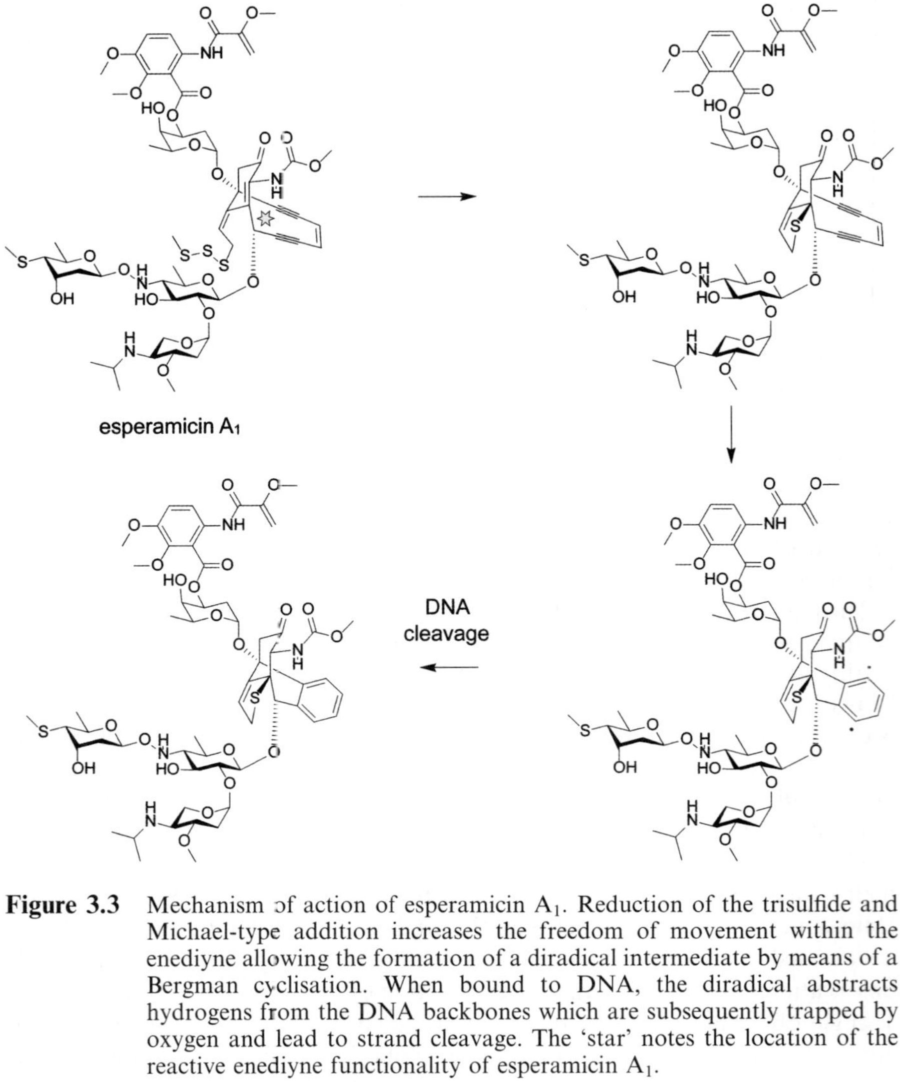

Figure 3.3 Mechanism of action of esperamicin A_1. Reduction of the trisulfide and Michael-type addition increases the freedom of movement within the enediyne allowing the formation of a diradical intermediate by means of a Bergman cyclisation. When bound to DNA, the diradical abstracts hydrogens from the DNA backbones which are subsequently trapped by oxygen and lead to strand cleavage. The 'star' notes the location of the reactive enediyne functionality of esperamicin A_1.

evaluate the causes, prevention, diagnosis and treatment of cancer. In 1955, the NCI launched the Cancer Chemotherapy National Service Center as a vehicle for researchers in both academic and industrial settings to screen materials for anticancer activity.[41] Between 1960 and 1964, studies conducted within this system identified cytotoxic activity within samples from the Pacific yew tree, *Taxus brevifolia*. Isolation efforts in the laboratory of Wani and Wall soon thereafter led to the discovery of taxol (Figure 3.5).[42] Continuing studies at the NCI determined that taxol was active in xenograph models.[43]

In 1979, Susan B. Horwitz initiated a series of studies that unveiled taxol's mode of action. Her studies began by determining that taxol promoted

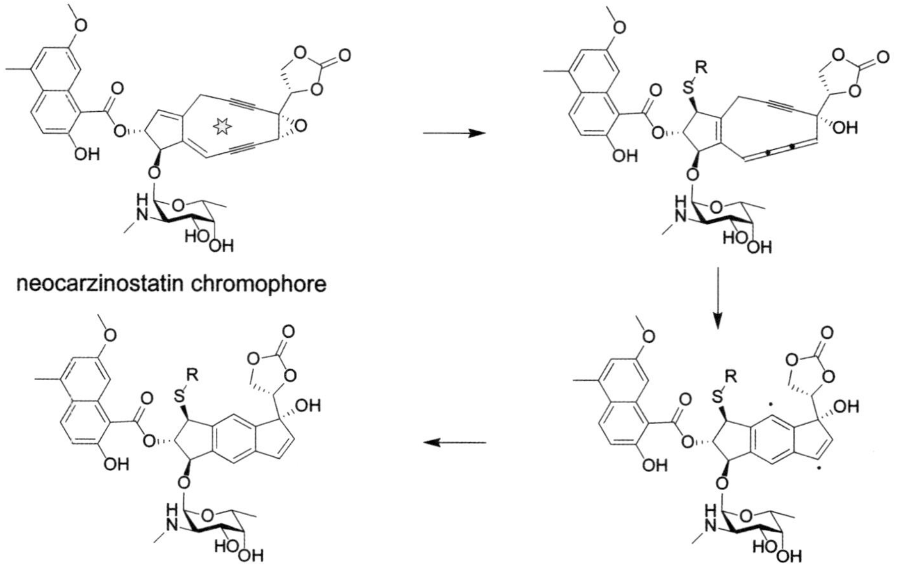

Figure 3.4 Mechanism of action of the neocarzinostatin chromophore. Thiol addition initiates conversion to an allenic intermediate, which undergoes cyclisation to form a diradical intermediate. Formation of this diradical in a DNA-bound state leads to radical abstraction and subsequent strand cleavage. The 'star' notes the location of the reactive enediyne functionality of the neocarzinostatin chromophore.

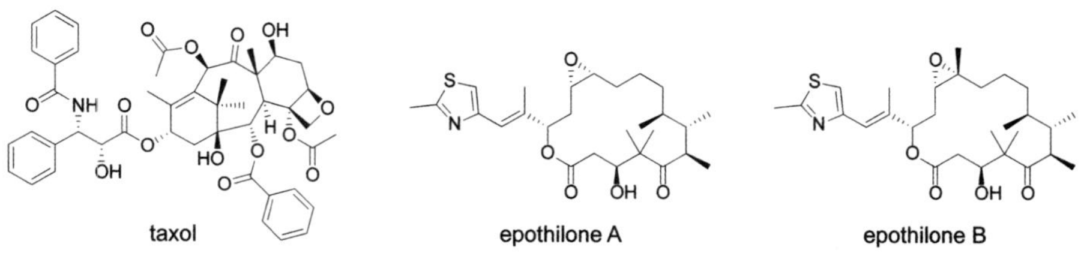

Figure 3.5 Structures of microtubule-binding natural products.

microtubule assembly in an *in vitro* assay.[44] These studies were followed by applications of transmission electron and immunofluorescence microscopy to show that cells treated with taxol underwent formation of characteristic microtubule bundles that failed to depolymerise when incubated at 4 °C.[45] Further studies showed that taxol blocked cell migration and blocked the cell cycle at the G_2 and M phases. Subsequent studies on a variety of mammalian cell lines confirmed that the activity of taxol was not associated with actin, intermediate filaments or DNA, and bound specifically to the tubulin–microtubule assembly, thereby validating its target.[46] Subsequent photoaffinity methods using [^{3}H]taxol,[47] as well as later studies with a fluorescent analogue of taxol, indicated specificity to the β-subunit of tubulin.[48]

These observations were confirmed in 2001 by the Nogales laboratory.[49] By using a combination of NMR spectroscopy, electron crystallography and modelling, taxol was determined to bind to a hydrophobic-binding pocket on zinc-induced tubulin sheets (see Figure 3.6b). This structure also served as the foundation for mechanistic studies which suggested that the binding of taxol to tubulin leads to a reduction in the distance between the distal loops within the tubulin dimer, thereby altering the angle between the α- and β-subunits. Modelling studies combined with analysis of hydrogen/deuterium exchange experiments were then used to predict the means by which taxol modified the interfaces between tubulin dimers in microtubules.[50]

In 1996, Höfle and coworkers at the Gesellschaft für Biotechnologische Forschung (GBF) in Germany reported the isolation of epothilone A and B (Figure 3.5) from culture broths of *Sorangium cellulosum* strain So ce9 obtained from soil collected from the banks of the Zambezi River in South Africa.[51] While these materials were first evaluated as antifungal agents, studies at Merck Research laboratories[52] and the NCI[53] indicated that epothilones delivered comparable induction of tubulin polymerisation *in vivo* and in cells. Further studies indicated these materials blocked the incorporation of [^{3}H]taxol binding and, therefore, likely shared a common binding pocket.[53] This evidence was confirmed in 2004 by applying similar methods to those used to elucidate the binding of taxol to β-tubulin (see Figure 3.6a).[54] These data—combined with comprehensive SAR studies on both taxol[55,56] and epothilone,[57–59] and mutation studies[60] on β-tubulin—provided a definitive model as to how the two materials bind to tubulin.

The attainment of this pocket by quite different structural motifs indicates that the pocket on β-tubulin may be malleable and its moulding by natural product ligands may be the key to unravelling the mechanisms of tubulin assembly.[61] As taxoids and epothilones have proven successful in the clinic,[62,63] advancing the understanding as to the mechanisms involved in regulating

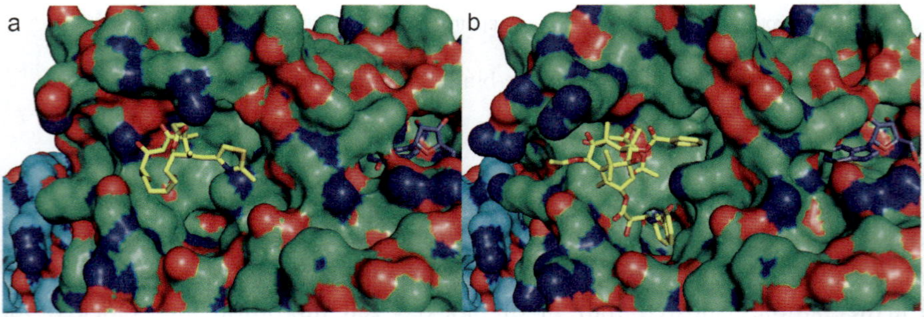

Figure 3.6 Images depicting the binding of (a) epothilone A (yellow) and (b) taxol (yellow) to alpha beta-tubulin (green and cyan). The images were developed from X-ray crystal structure datasets **1tvk** (epothilone A) and **1jff** (taxol), which are readily available from the Protein Data Bank.

tubulin assembly in tumour cells and masses has gained importance. Here several fundamental questions exist. One of the more intriguing queries arises from the mechanistic details of the regulatory process on assembly. How exactly does this lock mechanism work at the atomic level? While modelling studies provide an early prediction, further studies are needed to complete this interpretation. While these data are no longer critical to the current clinical accolades, understanding the refined engineering of microtubule assembly offers a robust foundation for second-generation development of tubulin modifers. In analogy to the Pulitzer Prize winning novel by Harper Lee and the film based thereon, it is not the unique properties of the natural product that delivers the fundamental lesson but rather the lack of structural prejudice within its mode of action that delivers its acclaim.

4 Notorious: Jasplakinolide, Alias Jaspamide and Actin

In February 1986, independent studies led by Crews[64] and a team comprised of Ireland, Faulkner and Clardy[65] elucidated the structure of a hybrid polyketide non-ribosomal peptide obtained from extracts of specimens of sponge *Jaspis* sp.[66,67] (the same metabolite was later found in a variety of sponges including *Auletta* sp.,[68] *Hemiasterella minor*[69] and *Cymbastela* sp.[70]). Using a combination of NMR[64,65] and X-ray crystallography,[65] the structure of this metabolite, known either as jaspamide or jasplakinolide (Figure 3.7a), was shown to contain a propionate unit and two uncommon amino acids—a β-tyrosine and new amino acid, 2-bromoabrine.[71] Twenty months later, its structure was confirmed by total synthesis in the Greco laboratory[72] and more recently by Konopelski,[73] Riccio[74] and Ghosh.[75] Jasplakinolide was first shown to display activity against *Candida albicans* and later broad-spectrum antifungal,[66] insecticidal, anthelminthic and *in vitro* cytotoxicity properties.[76]

Initiated in part by preliminary cytotoxicity assays,[77] collaborative efforts at the NCI demonstrated that jasplakinolide inhibited the binding of phalloidin (Figure 3.7a) to F-actin both *in vitro* and *in vivo*.[78,79] Actin, one of the most conserved and concentrated proteins in mammalian cells, forms one of the three primary structures in the cytoskeleton, providing both mechanical support and a means for motility and locomotion.[80] Its activity is derived through the formation of actin filaments whose organisation and attachments within the cell are modulated through an intricate architecture of structural proteins including actinin, vinculin, cadherin and the catenins.[81] Jasplakinolide modulates the actin assembly process by inducing the formation of polymeric F-actin from monomeric G-actin units, thereby locking the actin within its assembled fibrous state.[82,83]

Other natural products target different aspects of the actin assembly process. Cytochalasin D[84] (Figure 3.7a) binds to the barbed end and inhibits polymerisation and depolymerisation. Latrunculin A (Figure 3.7a),[85] another marine metabolite, binds to monomeric actin and inhibits polymerisation.[86]

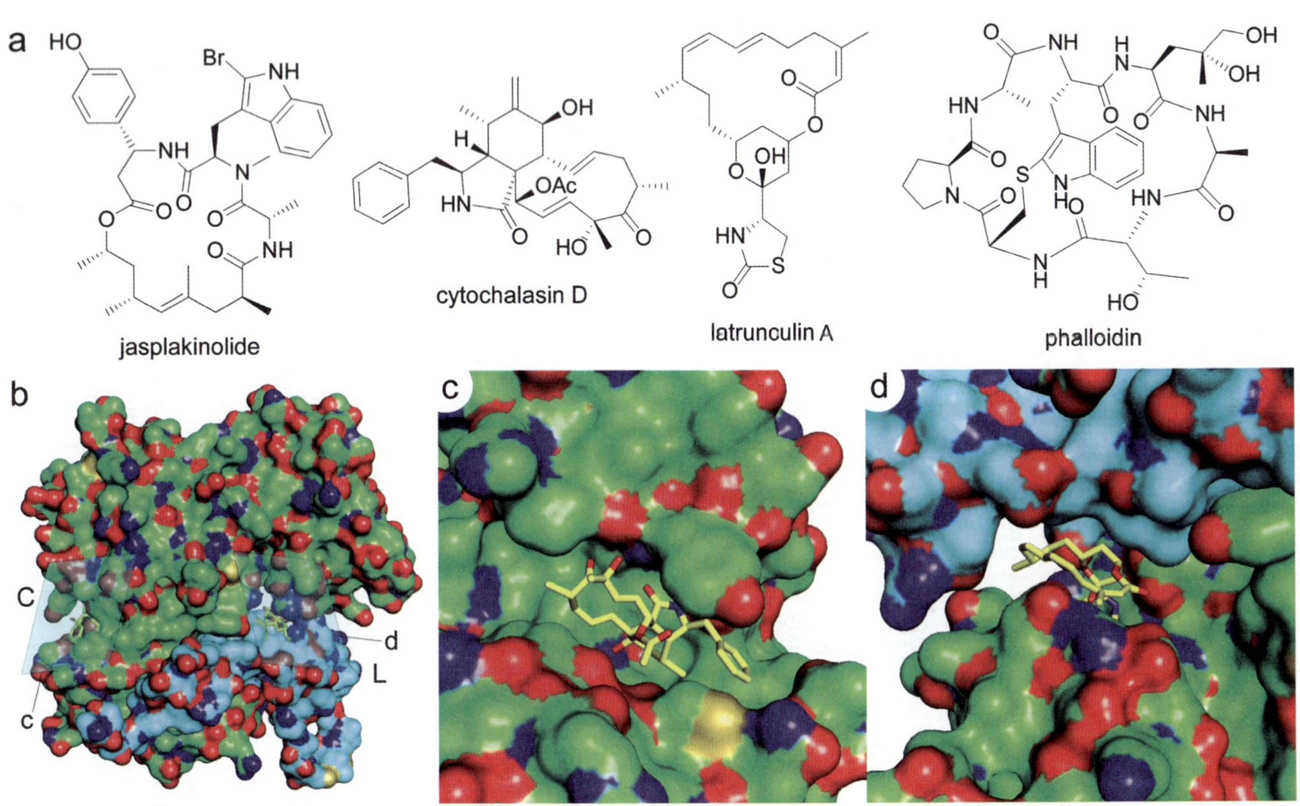

Figure 3.7 Actin-binding natural products. (a) Structures of four actin-binding natural products (b) Cytochalasin D (yellow), noted by letter C and latrunculin A (yellow), noted by letter L, target unique sites on actin (cyan and green), thereby explaining their relative activity on actin assembly. Close-ups of the (c) cytochalasin D (yellow) and (d) latrunculin (green) bound to actin (cyan and green). The images were developed from structure datasets **3eks** (cytochalasin D) and **1ijj** (latrunculin A), which are readily available from the Protein Data Bank.

While these materials share a similar target, their functional activity differs. Understanding this regulation has, in part, become possible due to the development of detailed crystal structure evidence on natural product bound actin complexes. As shown in Figure 3.7b, cytochalasin D and latrunculin A bind to distinct sites on opposite sides of the actin monomer. Cytochalasin D binds to the barbed end of the actin, occupying a hydrophobic cleft between subdomains 1 and 3 (Figure 3.7c).[87] Latrunculin A, on the other hand, binds actin monomers near the nucleotide binding cleft.[88] The story of actin regulation, while well understood, is far from being complete (Figure 3.7d). Most recently, a team in Braunschweig in Germany led by Schubert discovered that, when correctly assigned, the C_2-symmertry in rhizopodin was able to induce dimerisation of actin thereby eliminating two molecules of G-actin from oligomerisation.[89]

Like the film *Notorious* by Alfred Hitchcock, natural products have found diverse and yet complex roles for targeting actin polymerisation. Motivated by their ability to regulate structure and motility at the cellular level, organisms that produce these metabolites gain access to tools that can be used not only to spy on and inhibit the motility of their potential predators and prey but also terminate them by means of regulating their cellular structure. While few molecular pathways have yet to reach the level of understanding as that of actin dynamics, it is clear that the role of the natural product within this story was not only the key to the plot but also delivered an award winning performance.

5 Invasion of the Pathway Snatchers: Artemisinin

The function of some natural products is so unique that, in some ways, one can view them as coming from another world. *Artemisia* has been used as a herbal remedy in Chinese medicine for over 2000 years.[90] In the early 1970s, an antimalarial research programme within the Chinese army led to the discovery of artemisinin in the leaves of *Artemisia annua*. First named Qinghaosu (青蒿素), nearly ten years passed before a publication in a Chinese medical journal revealed its identity.[91] This was immediately followed by international attention,[92] with arguments developing both on validity of the structure of qinghaosu/artemisinin (Figure 3.8) and the viability of a conventionally unstable endoperoxide to appear within a drug motif.[93]

With the Chinese claims validated in 1984 by Klayman and coworkers,[94] studies in the early 1980s rapidly addressed two key issues with the advancement of artemisinin as an antimalarial drug. One, how could it be made commercially and if so, what is its mode of action? Although the former question was addressed,[95,96] the lack in understanding its activity was reported by Gu.[97] Here, evidence from prior studies indicated that, while active analogues could be prepared, including dihydroartemisinin and artemether (Figure 3.8), the endoperoxide function was key to this activity.[98] In their work, Gu and co-workers postulated that the activity arose from modification of protein synthesis within the malaria parasite, *Plasmodium falciparum*.[97] This was followed by a series of studies that indicated that the activity of artemisinin was attenuated by addition

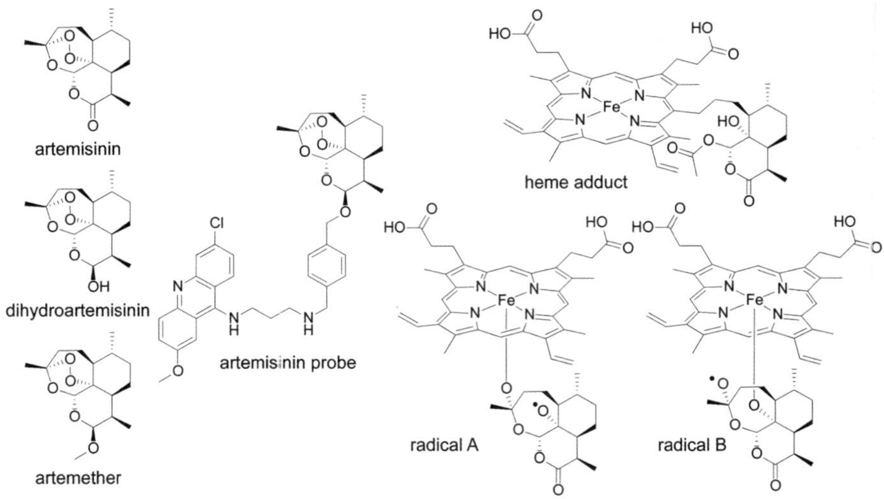

Figure 3.8 Artemisinin and related endoperoxide-containing sesquiterpenes: structures of artemisinin and related congeners (dihydroartemisinin and artemether), the fluorescent artemisinin probe used to identify the targeting of SERCA, and two putative radical intermediates that are formed upon reaction with haem and a haem adduct.

of oxygen[99] and reduced by the presence of heme.[100] Subsequent EPR studies indicated that artemisinin formed radicals only in the presence of heme,[101] thereby suggesting an iron activation mechanism. Since these observations, detailed models suggest the formation of two radical intermediates, radicals A and B (Figure 3.8), both of which can undergo rapid hydrogen abstraction intramolecularly to form carbon radicals or intermolecularly to modify bound proteins.[102,103] Further evidence was obtained by the characterisation of distinct heme insertion products as illustrated by the heme adduct shown in Figure 3.8.[104] The rapid formation of these adducts clearly indicates that, while active against plasmodia in the blood stream, reaction with heme can and does form rapid irreversible degradation of artemisinin. Whether this is fundamental to its activity (or loss in activity) has yet to be determined.[105]

These observations soon led to the discovery that artemisinin, through its ability to generate radicals or more specifically oxygen radicals, was able to target and inhibit digestive processes in vacuoles of the parasite *P. falciparum.*[106] Further studies in yeast models indicated that artemisinin acts on electron transport, generating localised oxygen species and disrupting the mitochondrial membrane.[107]

Subsequent studies indicate that artemisinin and congeners target PfATP6,[108] a SERCA-type enzyme. These studies were further supported by *in vitro* experiments indicating that artemisinin also inhibited sarcoplasmic reticulum Ca^{2+} transport or a SERCA-type enzyme by competing with the sequiterpene lactone inhibitor thapsigargin. The fact that resistance to artemisinin can be induced by a single mutation in PfATP6, both in recombinant systems and in field isolates, provides strong support for PfATP6 being a primary target in the plasmodium.[109,110]

With advances in total synthesis, semi-syntheses and biosynthetic engineering,[111,112] the material access issues will soon be addressed.[113] This along with the ability to prepare novel analogues has led to the evaluation of these materials for cancer treatment.[114,115] As these investigations begin, complementary mode of action studies will be paramount in identifying the key aspects of their specificity or lack thereof.[116] Clearly, the fact that artemisinin derivatives are able to generate radical species suggest that their modification as well as the tailoring of synthetic endoperoxides[117,118] and other endoperoxide-containing natural products[119–121] may provide a general tool for regulating specific pathways within targeted organisms or cells. Akin to the Jack Finney and Don Siegel production, the invasion continues . . .

6 Once Upon a Time in the Immune System: FK-506, Cyclosporin A and Rapamycin

The isolation and structure elucidation of FK-506 (Figure 3.9) was reported in 1987 by a team led by Goto at the Fujisawa Pharmaceutical Company in Japan.[122] FK-506 demonstrated an activity profile similar to that of previously isolated cyclosporin A.[123] The immunosuppressive activity of natural products was first discovered in 1972 by a team at Novartis, which identified cyclosporin A (Figure 3.9)[124,125] by screening small molecule collections using an immunoregulatory assay.[126–127] By 1980, the mechanism of cyclosporin A was beginning to be unveiled indicating that it acted during an early stage of lymphocyte stimulation, targeting human T and B blast cells without effecting cell division.[128,129] By the time FK-506 was identified, evidence was mounting that cyclosporin A regulated interleukin-2 production in targeted lymphocytes.[130] In 1986, studies in the laboratories of Handschumacher[131] determined that cyclosporin A targeted cyclophyllin within the cytosol of targeted lymphocytes. Subsequent studies indicated that this complex inhibits calcineurin,[132,133] thereby inactivating the transcription of interleukin-2.[134] This combined with inhibition

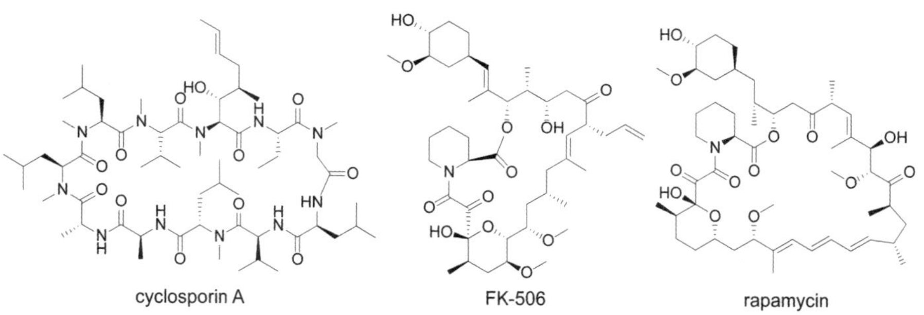

Figure 3.9 Structures of selected immunomodulatory natural products.

of lymphokine production and interleukin release provides a net reduction in the immune response from effector T-cells.[135]

Sparked by the identification of the cyclophyllin–cyclosporin complex, independent studies by Schreiber[136] and Merck & Co[137] led to the discovery of the FK-506 binding protein FKBP. It was soon shown that FK-506 regulates interleukin-2 transcription by reducing the peptidylprolyl isomerase activity by binding to the immunophilin FKBP-12, a FK506 binding protein, creating a new complex with the FK-506–rapamycin binding (FRB) domain (Figure 3.10).[138] The resulting FKBP12-FRB complex then inhibited calcineurin,[139] thus blocking both T-lymphocyte signal transduction and IL-2 transcription.

Subsequent crystallographic studies detailing a related natural product rapamycin (Figure 3.9)[140] provided a molecular depiction of these complexes. In a refined 2.0 Å structure by Clardy,[141] rapamycin provides a two-faced binding wherein the C2–C12 region binds to FKBP12 and the C16–C23 region interacts with FRB.

The function of the natural product in these studies was verified by comparison with the crystal structure of the FKBP12–rapamycin binary complex. Interestingly, while the conformation of the natural product remained the same, the formation of the binding of FRB to FKBP12–rapamycin delivered only a slight shift in the residues of FKBP12 within the rapamycin binding pocket and a more pronounced shift in the residues interacting with FRB with the most pronounced flexibility within two loops at residues 40–47 and 80–89. This reorganisation provides clear evidence that natural products indeed regulate the positioning of residues within protein–protein interactions.

Over four decades, the integral facets of the immune regulation by cyclosporin A, FK-506 and rapamycin were developed through combinations of chemical and biochemical studies.[142] These studies started with the identification of natural products with novel activity and, through a combination of cellular and molecular studies, led to the identification of their mode of action. Mode of action studies now indicate that rapamycin binds the cytosolic protein FK-binding protein 12 (FKBP12) in a manner similar to FK-506. However, unlike the FK-506–FKBP12 complex, which inhibits calcineurin, the rapamycin–FKBP12 complex inhibits the mammalian target of the rapamycin (mTOR) pathway—also called FRAP (FKBP–rapamycin associated protein) or RAFT (rapamycin and FKBP target)—by directly binding the mTOR complex.

Although understanding the mode of action of these natural products was not required to complete preclinical studies, the development of this knowledge not only facilitated their translation into the clinic but also validated a suite of novel targets for ongoing, wider therapeutic development.

7 Back to the Cytoskeleton: the Phorboxazoles

In the mid-1990s, the phorboxazoles (phorboxazole A and B, Figure 3.11) were discovered by Searle and Molinski[143] and soon thereafter by the Capon laboratory[144] from extracts of sponges *Phorbas* sp. collected off the coast of

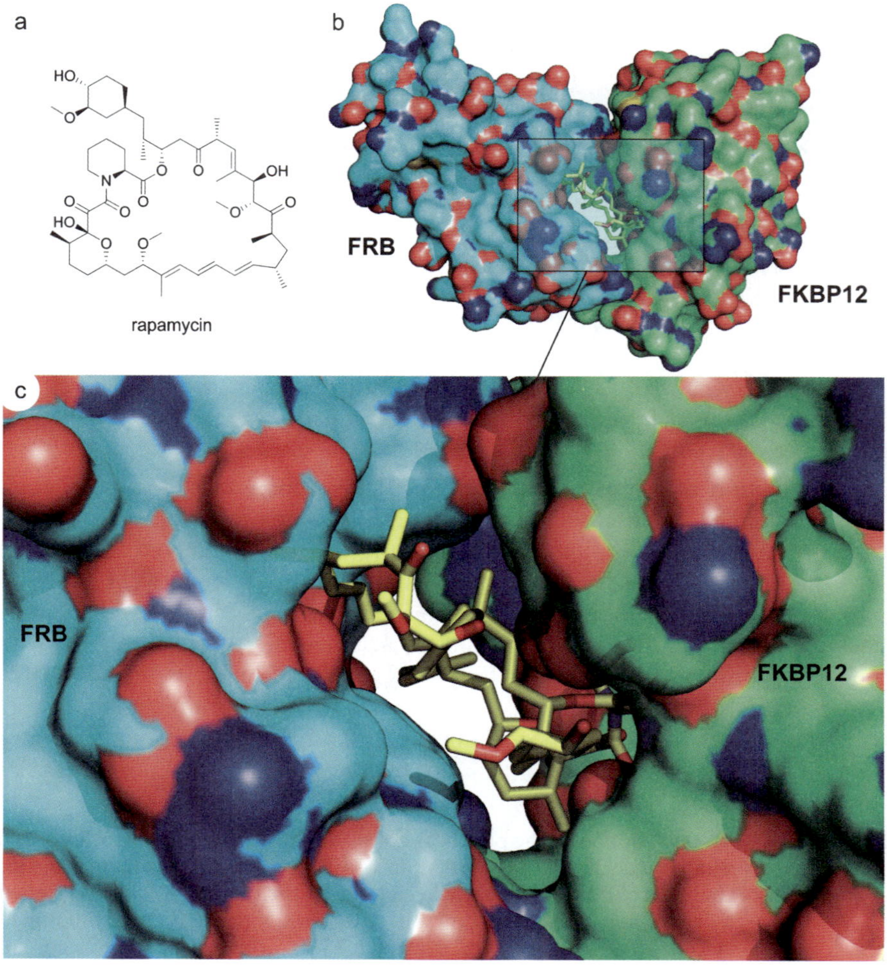

Figure 3.10 The FKBP12 · rapamycin · FRB complex. (a) Structure of rapamycin. (b) A depiction of the natural-product induced interface between FKBP12 and FRB. Each structure contains two proteins, FRB (cyan) and FKBP12 (green) with a bound rapamycin (yellow). (c) A close-up of the natural product binding pocket. The images were developed from structure dataset **4fap**, which is readily available from the Protein Data Bank.

Western Australia. Their unique ability to block cell growth at S phase, as well as display potent sub-nanomolar toxicity over a panel of tumour cell lines, suggested therapeutic potential for cancer.[143,145] Unfortunately, the complexity of their isolation and lack of a manageable producer organism indicated that access to these materials would be a challenge.

Currently, total synthesis is the primary means in which these compounds are produced. A number of impressive campaigns have completed the

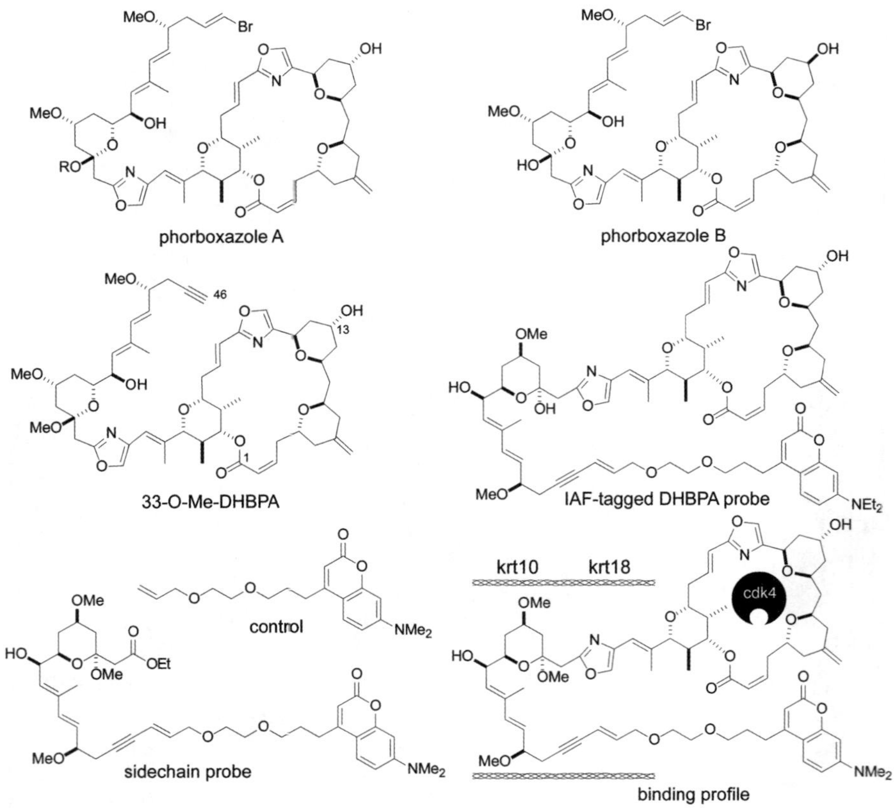

Figure 3.11 Mode of action of the phorboxazoles: structures of phorboxazoles A and B, analogue 33-O-MeDHBPA, an IAF tagged phorboxazole probe, side chains probe and controls. The activity of the phorboxazoles is depicted by a binding profile that shows the side chain of the IAF probe binding to cytokeratins krt10 or krt18, as depicted by a fibre. The macrolide core of the natural product binds cdk4 as shown by a black sphere.

production of the phorboxazoles for preclinical study. These include efforts within the Burke,[146] Forsyth,[147] Lin,[148] Smith[149] and White[150] laboratories. Second generation syntheses have now been completed by both Forsyth[151] and Smith.[152] Most recently, collaborative studies between the Smith and Pettit laboratories provided the potent analogue, (+)-chlorophorboxazole A, displaying sub-picomolar activity and the ability to reduce the growth of solid tumours.[153]

Using access to synthetic materials, an intermediate, 33-OMe-DHBPA was converted into an immunoaffinity fluorescent or IAF probe (Figure 3.11).[154] By developing an antibody against the 7-dimethylaminocoumarin-4-acetamide label, the IAF-tagged DHBPA probe was used both for cellular microscopy and molecular affinity analyses.[155] Using comparative studies against a IAF tag or control (Figure 3.11) and side chain probe (Figure 3.11), the IAF-tagged DHBPA probe was shown to be taken up rapidly in tumour cells and concentrate on filaments within the cytoplasm of the cell. Subsequent co-immunoprecipitation and blot analyses indicated that, while the core of the natural product bound to the cyclin-dependent kinase, cdk4, the side chain or tail of the natural product demonstrated potent affinity to cytokeratins, krt10 and krt18, as illustrated by the binding profile provided in Figure 3.11. The net effect of these binding events resulted in selective recruitment of cdk4 onto the surface of cytokeratin-based intermediate filaments, thereby effectively removing it from the cytosol and preventing nuclear translocation of active cdk4–cyclin D1 complexes required for phosphorylation of the retinoblastoma (Rb) protein and consequent cell cycle progression.[156]

More recent efforts now indicate that this activity is far more complex than the targeting of this single event, as different phenotypic responses readily occur across a panel of cell lines. While at this time it is difficult to determine if this discovery provides a drug target, it has become clear that development of systems that target intermediate filaments and their cytokeratin structures offers a new realm for small molecule discovery.

Like travelling back to the future, mode of action studies on the phorboxazoles and their implication of intermediate filaments suggest a new chapter for further discovery. Given the high success of targeting both the microtubule (Section 3) and actin assemblies (Section 4), one can only imagine as to where the future of the targeting of intermediate filaments exists within the annuals of therapeutic development.

8 It's a Wonderful Target: VTPase and its Targeting by Apicularen A, Salicylihalamide A and Palmerolide A

In 1997, a team from the DTP programme at the NCI reported the isolation of salicylihalamides A and B from the sponge *Haliclona* sp.[157] This new class of natural product contained a 12-membered lactone ring and an enamide side chain (Figure 3.12). Using COMPARE analyses with NCI's 60-cell screening profiles, the team determined that salicylihalamide A displayed a unique biological activity. This report was followed one year later by the isolation of the apicularens A and B from the myxobacteria *Chondromyces robustus* by a team at GBF in Braunschweig, Germany.[158] These structures were soon followed by lobatamides,[159] CJ-12,950,[160] CJ-13,357,[160] and oximidines I and II[161] comprising a family of benzolactone enamide natural products.

In 2000, SAR studies led by De Brabander[162] indicated that the potential protein reactivity of the side chain enamide was critical to maintaining the activity of salicylihalamides, a functional component of the family. Using COMPARE analyses, the DTP team determined that salicylihalamide A, the lobatamides and oximidines correlated with the mean activity of bafilomycin

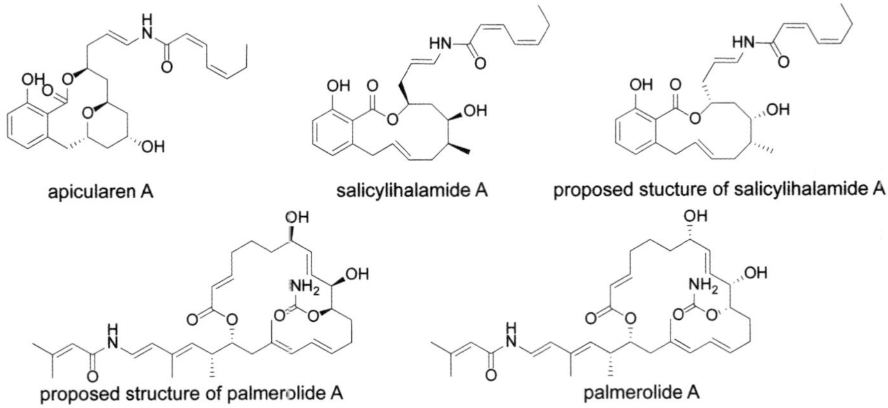

Figure 3.12 Structures of V-ATPase binding benzolactone enamide natural products. Proposed and revised structures are provided for salicylihalamide A and palmerolide A.

A1 and concanamycin A—two established V-ATPase inhibitors.[163] This observation was confirmed by both *in vitro* enzyme assays and *in vivo* activity studies in mutant yeast. These studies indicated that, while the benzolactone enamides were potent V-ATPase inhibitors, a profound selectivity occurred between the mammalian and fungal enzymes. Subsequent studies by De Brabander determined that salicylihalamide A binds irreversibly to the *trans*-membranous proton-translocating domain *via* *N*-acyliminium chemistry.[164]

With this evidence in hand, questions started to arise as to the importance of the entire molecular motif. What if any selectivity was derived from the three-dimensional structure of the natural products? Since the discovery of V-ATPase targeting,[165] a number of natural products were found with comparable activity with palmerolide A[166] (Figure 3.12) bearing one of the more complex structural motifs.[167]

One of the first issues that became clear in these studies is that, while knowing the structure of a natural product is important to its synthetic production, it is not mandatory for determining its mode of action. As their function as V-ATPases inhibitors was developed, the stereochemistry of both salicylihalamide A and palmerolide A was assigned as shown by their proposed structures in Figure 3.12. It was not until after their chemical syntheses that the correct structures of salicylihalamide A[168] and palmerolide A[169] were identified. The fact that one can determine the function of a compound without correctly ascertaining its structure indicates that, while functional descriptions on the mode of action of a natural product do require a structural understanding, perhaps placing an emphasis on structure prior to determining the mode of action of a natural product should not always be taken as precedent.

With analogues of the benzolactone enamides now readily obtained synthetically,[170] materials from these studies are now progressing rapidly toward clinical studies. Here, the structure of the natural products identified during isolation and confirmed or revised during total synthesis was the key to guide material design and implementation.

While now implemented as a "wonderful" target for chemotherapeutic development, V-ATPase modulation has been established in regulating tumour cell proliferation. Its further development now requires a detailed understanding of the recognition elements at the molecular level. Such studies would facilitate detailed understanding of V-ATPase activity, but also would provide a composite to guide further drug development efforts. With recent structures of subunits and sub-unit architecture of V-ATPase from yeast, analogous structural evidence as to the binding domains and regulatory activity of the benzolactone enamides seems a logical next step in the development of this "wonderful target".

9 Double Indemnity: Bistramide A

Between 1989 and 1996, two teams independently reported the isolation of a novel lipid natural product, bistramide A, from specimens of the tunicate *Lissoclium bistratum*.[171–173] Early indications suggested that the compound showed promise as an antitumour agent as it displayed potent antiproliferative activity (GI_{50} values of 20–45 nM) in multiple cell lines[174] and effective tumour reduction in xenograph models.[175] Early investigations in animal models indicated that samples of natural bistramide A inhibited Na^+ conductance and blocked voltage-dependent twitch tension in frog skeletal muscle.[176,177]

In 1996, studies in HL-60 cells indicated that bistramide A activated PKCδ in live cells causing its translation to the nucleus.[178–180] Subsequent studies indicated that bistramide A retained its antiproliferative activity in PKC-depleted cells.[181] Although overlaps in kinase activity can account for this lack of activity, studies led by Kosmin indicated that, while known PKC activators bryostatin and phorbol myristate acetate activated PKC, bistramide A did not have any effect on the enzymes activity *in vitro*.[181] Affinity studies indicated that there is only weak binding between bistramide A and PKCδ, and it only modestly displaces phorbol-12,13-dibutyrate from its PKC binding site. Using an affinity approach involving comparison of biotinylated-bistramide A against a control (Figure 3.13a), the team led by Kosmin identified actin as the primary target of the natural product.[181] Subsequent binding analyses, imaging studies of fluorescent analogues of bistramide A and X-ray crystallography studies[182] confirmed this observation (Figure 3.13b). The latter structural evidence indicates that it shares a comparable pocket with cytochalasin D (see Section 4).

Recently, Kosmin has shown that bistramide A is responsible for severing actin filaments and covalently modifying actin.[183] These studies indicated that, while the spiroketal and amide units of the natural product induced rapid disassembly of F-actin *in vitro*, the enone subunit of bistramide A was able to initiate covalent modification of actin *in vitro* and in live cells. These results indicate that, while PKCδ may not be the primary target of the natural product, it plays a dual role by binding to and severing F-actin and covalently sequestering

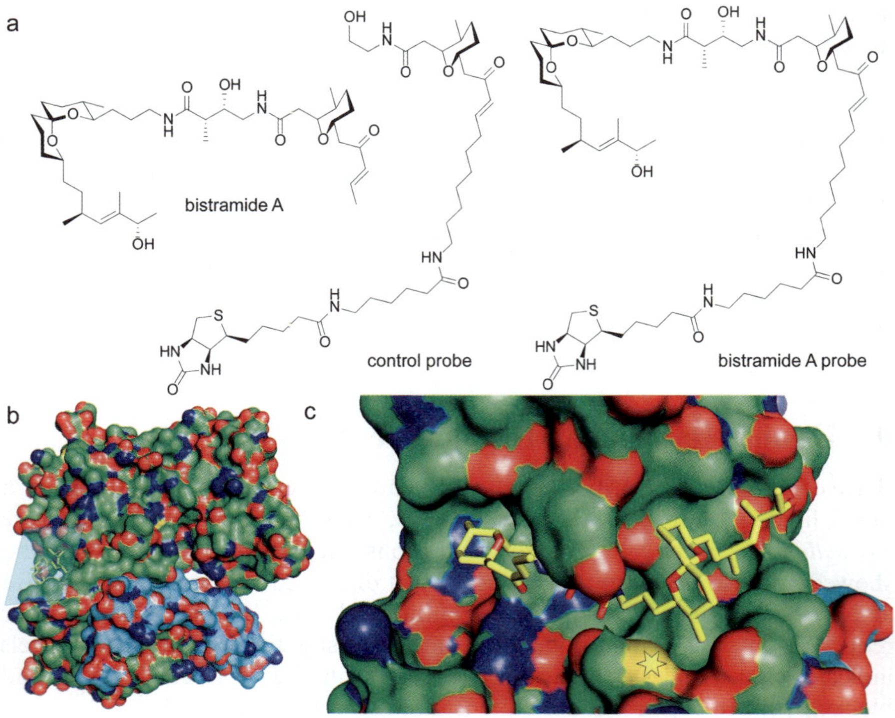

Figure 3.13 Mode of action of bistramide A. (a) Structures of bistramide A, an affinity control and affinity probe. (b) An image depicting the bistramide A (yellow) bound to actin (cyan and green). (c) A close-up of the binding pocket. The thiol that undergoes covalent modification is noted by a 'star'. The Images were developed from structure dataset **2fxu**, which is readily available from the Protein Data Bank.

the monomeric G-actin. It is this unique double "indemnity" that distinguishes this natural product from the other families of actin regulators (see Section 4).

10 The Matrix: the Pladienolides and Splicing Factor SF3b

In 1994, the Taisho Corporation discovered FD-895 by cytotoxicity-guided fractionation of culture extracts from a strain of *Streptomyces hygroscopicus* isolated from a soil sample collected in Okinawa, Japan.[184] Its two-dimensional structure was elucidated by NMR and mass spectroscopy (Figure 3.14). A decade later, seven pladienolides A–G were discovered from cultures of a strain of *S. platensis* collected from a soil sample from Kanagawa, Japan, through a screening programme designed to identify compounds that block the expression of vascular endothelial growth factor (VEGF) genes under hypoxic

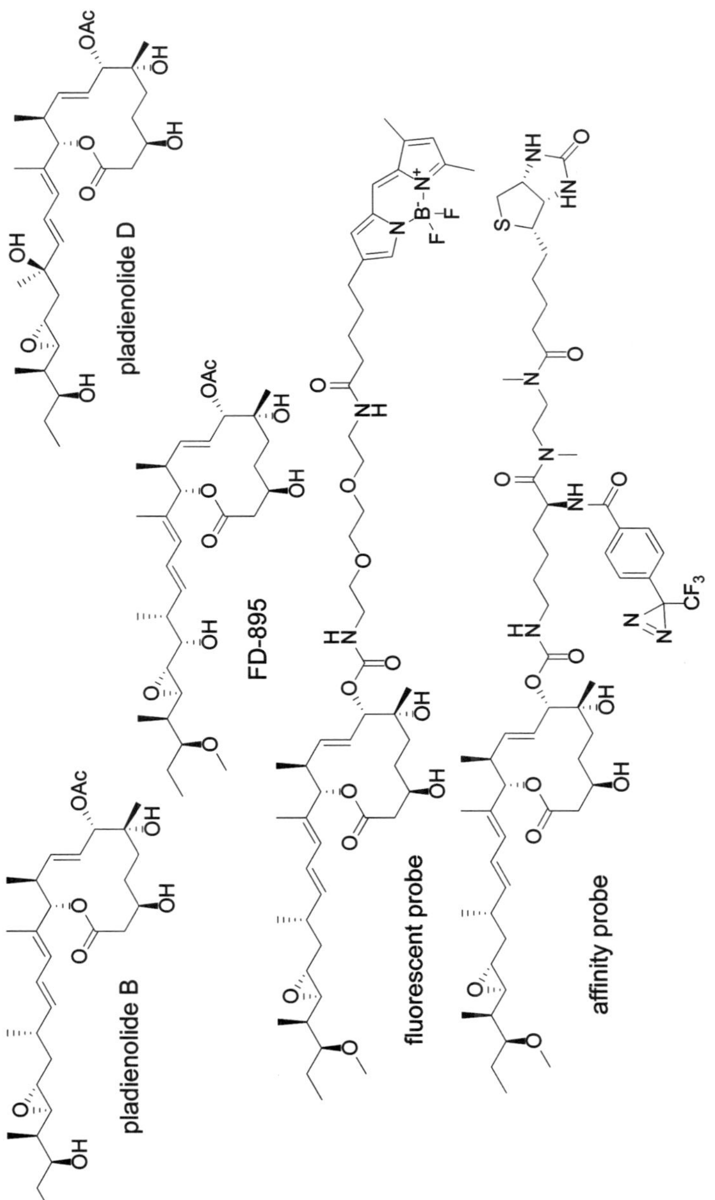

Figure 3.14 Structures of pladienolide B, pladienolide D, FD-895, and fluorescent and affinity pladienolide B probes.

conditions.[185–186] One-dimensional and two-dimensional (2D) NMR techniques as well as mass spectrometry were used to assign their 2D structures, indicating that they were comparable to FD-895.

In mid-2007, the stereochemistry of pladienolides B and D was determined using a combination of NMR methods and degradation experiments.[187] Coupling constants from the ^{1}H-NMR data as well as correlations from 2D experiments provided a first-level approximation of the stereochemistry. Further derivatisation experiments were performed to confirm the relative stereochemistry within pladienolide B. The absolute stereochemistry was determined *via* a modified Mosher method that compared an C3,C21-bis-(R)-MPTA ester to an C3,C21-bis-(S)-MPTA ester. The structural assignment has since been validated by chemical synthesis.[187–189]

Besides having low nanomolar GI50 values against several cancer cell lines *in vitro*, pladienolide B was shown to cause tumour regression in six different mouse xenograft models.[190] These initial results led to extensive research and development efforts guided toward understanding its activity.[191] To find the target protein, Eisai Co. Ltd in Japan developed chemical probes bearing a respective affinity or fluorescent tags (Figure 3.14).[192] Using fluorescence microscopy, the fluorescent probe was found to localise in the nucleus and granular structures around the nucleus. The target of these observations was suggested by conducting immunoprecipitation experiments with tritiated pladienolide B. Antibodies against six nuclear proteins precipitated with the tritiated pladienolide B with various efficiencies.[192]

Despite the fact that the affinity probe was 645-fold less active than pladienolide B, HeLa cells were treated with this probe, fractionated and the nuclear fraction was incubated with the most efficient antibody from the first immunoprecipitation experiment. The precipitate was irradiated with ultraviolet light with the goal of photo-crosslinking the target protein. This enabled an immunoblotting experiment with streptavidin–horseradish peroxidase and resulted in the identification of a single protein band. The same size protein was also obtained using the other five antibodies. Mass spectrometric sequencing of the band identified two candidates, splicing factor SF3b subunits 2 and 3 (both spliceosome associated proteins). Further immunoblotting experiments on green fluorescent protein (GFP) fusion proteins of both subunits resulted in the identification of SF3b subunit 3 as the target protein. There was additional evidence that the probes bind to the protein after it has been assimilated into the splicing factor complex and experiments were also performed to show three specific pre-mRNAs that failed to reach maturity (containing introns) after incubation with pladienolide B.[192]

While the identification of SF3b was an impressive start, there is some concern over the considerable (645-fold) loss of activity in the Eisai probes prepared at C7. While these probes were used to identify the splicing factor SF3b, it is possible the lack in activity of Eisai's probe molecules failed to provide a complete description of the targets of the pladienolides. In addition, the mode of action of pladienolide B has yet to be directly linked to the cellular phenotypes of this family of natural product and related structural motifs.[193] While it is intuitive to assume that blockage of SF3b may lead to loss in VEGF expression, other activities such as the putative V-ATPase activity of FD-895 suggest multiple modes of action. Further studies are now needed to validate the targeting of SF3b and determine its relevance with respect to its activity *in vivo* and in a pharmacological context.

11 The Unusual Suspects: (+)-Avrainvillamide

Avrainvillamide was isolated first from a strain of *Aspergillus* sp. by Fenical and Jensen[194] and later confirmed in the fermentation of *Aspergillus ochraceus*.[195] Subsequent studies at Bristol-Myers Squibb led to the identification of a dimeric structure, called stephacidin B,[196] that displayed enhanced activity compared with avrainvillamide in tumour cell models. While validated by NMR and X-ray crystallographic data, the independent total syntheses of the avrainvillamides and stephacidins by Myers,[197] Baran[198] and Williams[199] were instrumental in validating the interconversion between avrainvillamide and stephacidin B. Here, treatment with base was shown to induce dimerisation to stephacidin B while treatment with SiO_2 or heating reverted the molecule to its monomeric avrainvillamide.[200]

With this indication of dynamic activity, the Myers laboratory prepared a panel of affinity and fluorescent analogues based on natural (+)-avrainvillamide, its enantiomer and simplified congeners (Figure 3.15).[201] Using fluorescence microscopy, a dansylated avrainvillamide conjugate was shown to localise within the nucleoli and the cytosol of mammalian cells. Biotin-tagged affinity analogues were then used by comparison with the appropriate controls to identify nucleophosmin as the target. This was further supported by the lack of this affinity in control experiments.

The Myers team then validated the targeting of nucleophosmin using a clever mutagenesis strategy. Based on preliminary evidence that (+)-avrainvillamide was capable of electrophilic attack by thiols,[201] they prepared nucleophosmin mutants deleted selectively at three cysteine residues and co-expressed with native nucleophosmin in COS-7 cells. These studies indicated that that the mutation of Cys275 to Ala275 effectively reduced affinity isolation of the truncated protein, validating the targeting of avrainvillamide to the Cys275 of nucleophosmin.

These studies, combined with observations that cells treated with avrainvillamide underwent increased p53 expression and that cells silenced with siRNA against nucleophosmin increased sensitivity to avrainvillamide, provided strong evidence that this event was indeed relevant to the cellular response and downstream entry into apoptosis.

While an abundant nucleolar protein, this was the first study to identify a small molecule ligand to nucleophosmin. But although binding has been identified, the role in which dimerisation of avrainvillamide (to stephacidins)

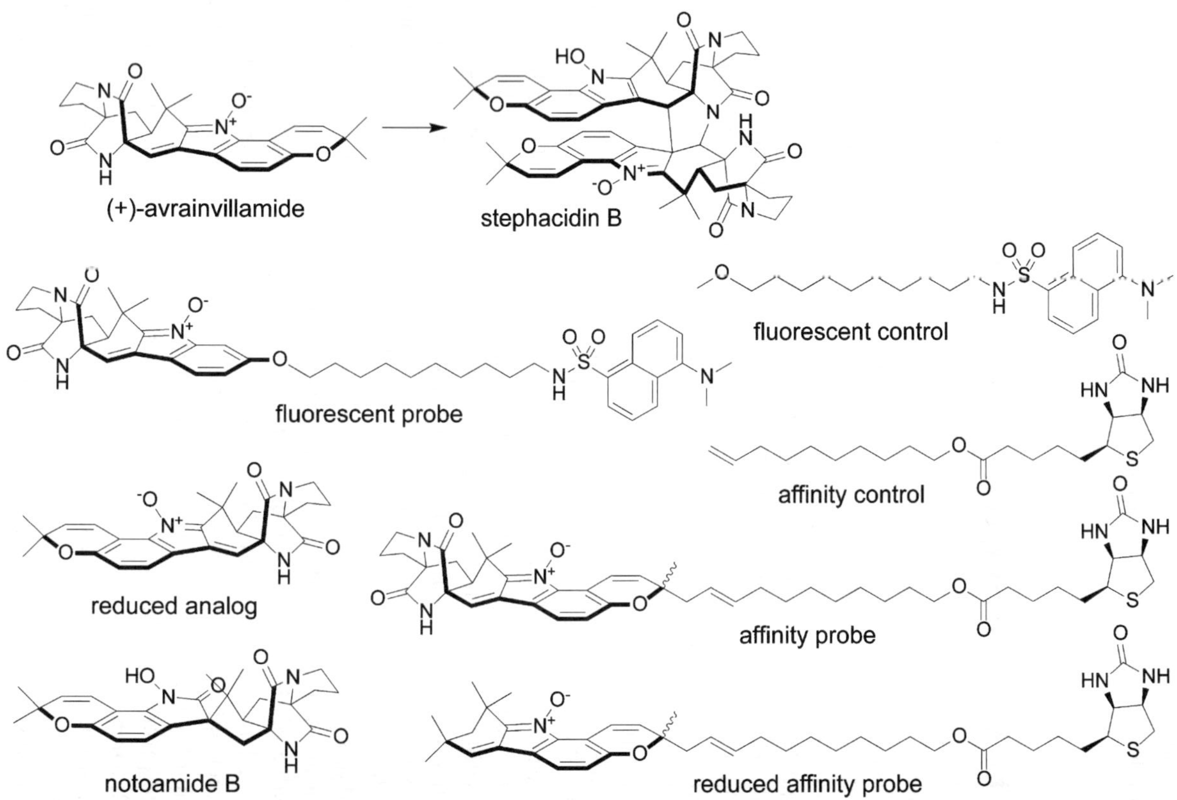

Figure 3.15 Structures of (+)-avrainvillamide, stephacidin B, notoamide B, a reduced analogue of avrainvillamide, and related affinity and fluorescent probes and controls.

modulates this response in both *in vitro* and *in vivo* settings has yet to be determined. Furthermore, the role in which avrainvillamide modifies the numerous cellular processes regulated by nucleophosmin has also yet to be examined. In particular, it is likely that further studies on this avrainvillamide and related analogues may strengthen the understanding as to how nucleophosmin associates with mRNAs and regulates subsequent post-translational events. This in turn will be key to understanding how nucleophosmin regulates ribosome biogenesis[202] and centrosome duplication.[203] Understanding these and related nucleophosmin-mediated events is crucial to developing an understanding of a number of haematological disorders that bear mutations or rearrangements within their nucleophosmin genes.

This discovery is of importance for several fundamental reasons. First, the targeting of nucleophosmin by avrainvillamide suggests a lead for acute myelogenous leukaemia.[204] Second, this suggestion is further supported by the development of refined synthetic efforts that will now allow the production of not only avrainvillamide and stephacidins, but will also allow the preparation of related natural products such as versicolamide and notoamide B,[199] as well as providing discrete protein targets for the preparation of semi-synthetic and synthetic analogues. Finally, these studies further support the vital need to investigate all classes of natural products. Here, the Myers studies show that not only is alkaloid chemistry alive and well but that prenylated indole alkaloids—so-called "unusual suspects"—offer a rich foundation for further chemical biology investigations.

12 Close Encounters of a Third Kind: Ammosamides, Blebbestatin and Myosin

Recent collaborative efforts have demonstrated that mechanism of action studies can be conducted in unison with compound isolation efforts. Ammosamides A and B (Figure 3.16) were isolated during screening efforts to identify novel metabolites in cultures of deep sea actinomycetes.[205] Soon after their isolation and preliminary activity studies, efforts began to label the natural product with an immunoaffinity fluorescent tag. While easily conceived, labelling studies on ammosamide A proved difficult because many of the conditions led either to lack of reactivity or formation of degradation products, one of which was later shown to be ammosamide B. Subsequent synthetic studies in the Fenical laboratory demonstrated the conversion between the thioamide in ammosamide A and amide in ammosamide B. With this knowledge, focused studies provided an effective labelling of ammosamide A delivering an single IAF probe (Figure 3.16).[206]

Without knowledge of its structure, an IAF probe established that the ammosamides were absorbed rapidly in cells, localising throughout the cell and concentrating in the cytosol and lysosomes. Using cell sorting and fluorescent techniques, progression through the cell cycle was examined in media

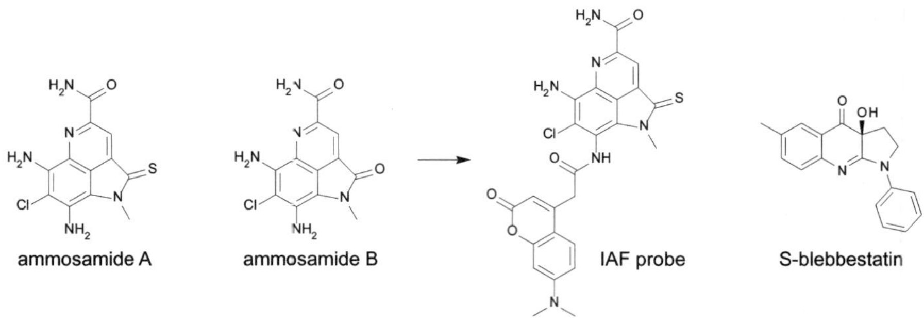

Figure 3.16 Structures of the myosin binding natural products ammosamide A, ammosamide B, an IAF ammosamide A probe, and S-blebbestatin.

containing ammosamide A, ammosamide B and the IAF-tagged ammosamide probe. Unlike many natural products, the response was complex showing multiple stages of inhibition depending on the time and concentration of treatment. Lacking a clear intracellular target or cell cycle response, the team turned to affinity methods to screen for targets.

Using an antibody elicited against the IAF tag, immunoprecipitation studies conducted on cells and lysates treated with the IAF probe identified myosin as a primary target. Interestingly, this protein appeared fluorescent after SDS-PAGE analysis suggesting that a covalent attachment was made between the natural product (or its IAF tag) and the protein. While these effects were not seen in control experiments, myosin was validated as a target of the ammosamides by a combination of *in vitro* analyses and immunoprecipitation studies.

Further studies examining the specification of these probes in live cells and tissues indicated that the ammosamides did not only target myosin II within muscle but also suggested affinity to other classes of myosin. These studies show, in addition to synthetic materials such as blebbistatin, natural products are also capable of targeting myosin.[206]

Additional studies are required to determine the role which these natural products play in regulating myosin activity. Although studies now demonstrate that myosin is a target of natural products, this provides further evidence that other materials may exist that specifically bind to and regulate myosin activity. Given the large diversity within this class of protein, it is quite possible that natural or synthetic materials exist that can specifically regulate movements within the cell by agonising or antagonising specific isoforms or classes of myosin.

These studies, like the film *Close Encounters of the Third Kind*, suggest an additional paradigm within the cytoskeleton that now moves from the well-established targeting of the structural proteins (*i.e.* actin, microtubules or intermediate filaments) but moves further along the process to regulate their motion.

13 The End

Like a good film, natural products tell a story rich in character and plot. There are far more natural products and more wondrous stories on their mode of action that could be covered in this chapter—notable omissions include the recent studies on apratoxin,[207] the fumagillins,[208] fellutamide B,[209] geldanamycin,[210] largazole,[211] sceptrin[212] and others.[213] Furthermore, while not all natural products may be the next blockbuster, watching their stories not only sets the stage for drug discovery efforts, but also furthers the unique manifold in which one can regulate Nature. And although all films have their critiques, so too may the role of the natural product. That aside, few individuals find all feature films to be the same. Following the same argument, few if any natural products share similarity and, therefore, it is time that the perspective of natural products in drug discovery moves from arguing over their importance to simply sitting back and watching them perform.

In our world, it is hard to imagine a day without film or media. In the same venue, what would drug discovery be without the natural product? I predict that, while it may be "quiet on set", drug discovery and its investors are likely to return to natural products as a fundamental resource and the claims that they are "ready to shoot" will again be heard through the discovery sector. Thus, I conclude it is time to stop worrying about the importance or relevance of natural products within the pharmaceutical sector and get out and film.

Lights, camera and action . . .

References

1. M. Konishi, H. Ohkuma, K. Saitoh, H. Kawaguchi, J. Golik, G. Dubay, G. Groenewold, B. Krishnan and T. W. Doyle, *J. Antibiot. (Toyko)*, 1985, **38**, 1605.
2. J. Golik, G. Dubay, G. Groenewold, H. Kawaguchi, M. Konishi, B. Krishnan, H. Ohkuma, K. Saitoh and T. W. Doyle, *J. Am. Chem. Soc.*, 1987, **109**, 3462.
3. B. H. Long, J. Golik, S. Forenza, B. Ward, R. Rehfuss, J. C. Dabrowiak, J. J. Catino, S. T. Musial, K. W. Brookshire and T. W. Doyle, *Proc. Natl. Acad. Sci. USA*, 1989, **86**, 2.
4. R. R. Jones and R. G. Bergman, *J. Am. Chem. Soc.*, 1972, **94**, 660.
5. Y. Sugiura, Y. Uesawa, Y. Takahashi, J. Kuwahara, J. Golik and T. W. Doyle, *Proc. Natl. Acad. Sci. USA*, 1989, **86**, 7672.
6. T. Ebata, H. Hiroaki, S. Uesugi and Y. Sugiura, *Nucleic Acids Symp. Ser.*, 1990, **22**, 19.
7. M. Lu, Q. Guo, B. Krishnan, J. Golik, I. E. Rosenberg, T. W. Doyle and N. R. Kallenbach, *J. Biomol. Struct. Dyn.*, 1991, **9**, 285.
8. N. Ishida, K. Miyazki, K. Kumagai and M. Rikimaru, *J Antibiot. (Tokyo)*, 1965, **18**, 68.
9. Y. Ono, Y. Ito, H. Maeda and N. Ishida, *Biochim. Biophys. Acta*, 1968, **26**, 616.

10. A. G. Myers, E. Y. Kuo and N. S. Finney, *J. Am. Chem. Soc.*, 1989, **111**, 8057.
11. R. Nagata, H. Yainaiiaka, E. Okazaki and I. Saito, *Tetrahedron Lett.*, 1989, **30**, 4995.
12. K. C. Nicolaou, A. L. Smith and E. W. Yue, *Proc. Natl. Acad. Sci. USA*, 1993, **90**, 5881.
13. N. Zein, A. M. Sinha, W. J. McGahren and G. A. Ellestad, *Science*, 1988, **240**, 1198.
14. Y. Sugiura, T. Shiraki, M. Konishi and T. Oki, *Proc. Natl. Acad. Sci. USA*, 1990, **87**, 3831.
15. N. Zein, K. L. Colson, J. E. Leet, D. R. Schroeder, W. Solomon, T. W. Doyle and A. M. Casazza, *Proc. Natl. Acad. Sci. USA*, 1993, **90**, 2822.
16. A. G. Myers, M. E. Kort, S. B. Cohen and N. J. Tom, *Biochemistry*, 1997, **36**, 3903.
17. A. L. Smith and K. C. Nicolaou, *J. Med. Chem.*, 1996, **39**, 2103.
18. S. J. Danishefsky and M. D. Shair, *J. Org. Chem.*, 1996, **61**, 16.
19. R. C. Hawley, L. L. Kiessling and S. L. Schreiber, *Proc. Natl. Acad. Sci. USA*, 1989, **86**, 1105.
20. D. Kahne, *Chem. Biol.*, 1995, **2**, 7.
21. P. Magnus and P. A. Carter, *J. Am. Chem. Soc.*, 1988, **110**, 1626.
22. P. A. Wender, R. C. Kelly, S. Beckham and B. L. Miller, *Proc. Natl. Acad. Sci. USA*, 1991, **88**, 8835.
23. R. Unno, H. Michishita, H. Inagaki, Y. Suzuki, Y. Baba, T. Jomori, M. Moku, T. Nishikawa and M. Isobe, *Bioorg. Med. Chem.*, 1997, **5**, 903.
24. T. Takahashi, H. Tanaka, A. Matsuda, T. Doi and H. Yamada, *Bioorg. Med. Chem. Lett.*, 1998, **8**, 3299.
25. T. Usuki, M. Inoue, M. Hirama and T. Tanaka, *J. Am. Chem. Soc.*, 2004, **126**, 3022.
26. A. Basak, S. K. Roy, B. Roy and A. Basak, *Curr. Top. Med. Chem.*, 2008, **8**, 487.
27. G. B. Jones and F. S. Fouad, *Curr. Pharm. Des.*, 2002, **8**, 2415.
28. J. F. Capitani, S. M. Gaffney, L. Castaldo and A. Mitra, *Curr. Top. Med. Chem.*, 2008, **8**, 470.
29. M. Konishi, H. Ohkuma, K. Saitoh, H. Kawaguchi, J. Golik, G. Dubay, G. Groenewold, B. Krishnan and T. W. Doyle, *J. Antibiot. (Toyko)*, 1985, **38**, 1605.
30. N. Ikemoto, R. A. Kumar, T. T. Ling, G. A. Ellestad, S. J. Danishefsky and D. J. Patel, *Proc. Natl. Acad. Sci. USA*, 1995, **92**, 10506.
31. R. A. Kumar, N. Ikemoto and D. J. Patel, *J. Mol. Biol.*, 1997, **265**, 173.
32. R. A. Kumar, N. Ikemoto and D. J. Patel, *J. Mol. Biol.*, 1997, **265**, 187.
33. M. Schmittel, M. Strittmatter and S. Kiau, *Tetrahedron Lett.*, 1995, **36**, 4975.
34. Y. Sugiura and T. Matsumoto, *Biochemistry*, 1993, **32**, 5548.
35. Y. J. Xu, Y. S. Zhen and I. H. Goldberg, *Biochemistry*, 1994, **33**, 5947.
36. J. Drak, N. Iwasawa, S. Danishefsky and D. M. Crothers, *Proc. Natl. Acad. Sci. USA*, 1991, **88**, 7464.

37. S. Walker, R. Ladovitz, W. D. Ding, G. A. Ellestad and D. Kahne, *Proc. Natl. Acad. Sci. USA*, 1992, **89**, 4608.
38. L. S. Kappen and I. H. Goldberg, *Science*, 1993, **261**, 1319.
39. R. Stasi, M. L. Evangelista, F. Buccisano, A. Venditti and S. Amadori, *Cancer Treat. Rev.*, 2008, **34**, 49.
40. P. F. Bross, J. Beitz, G. Chen, X. H. Chen, E. Duffy, L. Kieffer, S. Roy, R. Sridhara, A. Rahman, G. Williams and R. Pazdur, *Clin. Cancer. Res.*, 2001, **7**, 1490.
41. V. T. DeVita Jr and E. Chu, *Cancer Res.*, 2008, **68**, 8643.
42. M. C. Wani, H. L. Taylor, M. E. Wall, P. Coggon and A. T. McPhail, *J. Am. Chem. Soc.*, 1971, **93**, 2325.
43. S. M. Edgington, *Biotechnology*, 1991, **9**, 933.
44. B. Schiff, J. Fant and S. B. Horwtiz, *Nature*, 1979, **277**, 665.
45. P. B. Schiff and S. B. Horwtiz, *Proc. Natl. Acad. Sci. USA*, 1980, **77**, 1561.
46. J. J. Manfredi, J. Parness and S. B. Horwtiz, *J. Cell. Biol.*, 1982, **94**, 688.
47. S. Rao, S. B. Horwtiz and I. Ringel, *J. Natl. Cancer Inst.*, 1992, **84**, 785.
48. J. J. Manfredi and S. B. Horwtiz, *Pharmacol. Ther.*, 1984, **25**, 83.
49. J. P. Snyder, J. H. Nettles, B. Cornett, K. H. Downing and E. Nogales, *Proc. Natl. Acad. Sci. USA*, 2001, **98**, 5312.
50. H. Xiao, P. Verdier-Pinard, N. Fernandez-Fuentes, B. Burd, R. Angeletti, A. Fiser, S. B. Horwtiz and G. A. Orr, *Proc. Natl. Acad. Sci. USA*, 2006, **103**, 10166.
51. K. Gerth, N. Bedorf, G. Höfle, H. Irschik and H. Reichenbach, *J Antibiot. (Toyko)*, 1996, **49**, 560.
52. D. M. Bollag, P. A. McQueney, J. Zhu, O. Hensens, L. Koupal, J. Liesch, M. Goetz, E. Lazarides and C. M. Woods, *Cancer Res.*, 1995, **55**, 2325.
53. R. J. Kowalski, P. Giannakakou and E. Hamel, *J. Biol. Chem.*, 1997, **272**, 2534.
54. J. H. Nettles, H. Li, B. Cornett, J. M. Krahn, J. P. Snyder and K. H. Downing, *Science*, 2004, **305**, 866.
55. M. L. Miller and I. Ojima, *Chem. Rec.*, 2001, **1**, 195.
56. R. Geney, J. Chen and I. Ojima, *Med. Chem.*, 2005, **1**, 125.
57. K. H. Altmann, *Mini Rev. Med. Chem.*, 2003, **3**, 149.
58. K. C. Nicolaou, A. Ritzén and K. Namoto, *Chem. Commun.*, 2001, 1523.
59. K. H. Altmann and K. Memmert, *Prog. Drug. Res.*, 2008, **66**, 275.
60. B. W. Day, *Trends Pharmacol. Sci.*, 2000, **21**, 321.
61. L. A. Amos and J. Löwe, *Chem. Biol.*, 1999, **6**, 65.
62. I. E. Kuppens, *Curr. Clin. Pharmacol.*, 2006, **1**, 57.
63. S. L. Mooberry, *Methods Mol. Med.*, 2007, **173**, 289.
64. P. Crews, L. V. Manes and M. Boehler, *Tetrahedron Lett.*, 1986, **27**, 2797.
65. T. M. Zabriskie, J. A. Klocke, C. M. Ireland, A. H. Marcus, T. F. Molinski, D. J. Faulkner, C. Xu and J. Clardy, *J. Am. Chem. Soc.*, 1986, **108**, 3123.
66. V. R. Scott, R. Boehme and T. R. Matthews, *Antimicrob. Agents Chemother.*, 1988, **32**, 1154.

67. J. C. Braekman, D. Daloze, B. Moussiaux and R. Riccio, *J. Nat. Prod.*, 1987, **50**, 994.
68. R. N. Sonnenschein, J. J. Farias, K. Tenney, S. L. Mooberry, E. Lobkovsky, J. Clardy and P. Crews, *Org. Lett.*, 2004, **6**, 779.
69. R. Talpir, Y. Benayahu, Y. Kashman, L. Pannell and M. Schleyer, *Tetrahedron Lett.*, 1994, **35**, 4453.
70. C. Chevallier, A. D. Richardson, M. C. Edler, E. Hamel, M. K. Harper and C. M. Ireland, *Org. Lett.*, 2003, **5**, 3737.
71. W. Inman and P. Crews, *J. Am. Chem. Soc.*, 1989, **111**, 2822.
72. P. A. Grieco, Y. S. Hon and A. Perez-Medrano, *J. Am. Chem. Soc.*, 1988, **110**, 1630.
73. K. S. Chu, G. R. Negrete and J. P. Konopelski, *J. Org. Chem.*, 1991, **56**, 5196.
74. S. Terracciano, I. Bruno, G. Bifulco, J. E. Copper, C. D. Smith, L. Gomez-Paloma and R. Riccio, *J. Nat. Prod.*, 1987, **50**, 994.
75. A. K. Gosh and D. K. Moon, *Org. Lett.*, 2007, **9**, 2425.
76. J. Stingl, R. J. Andersen and J. T. Emerman, *Cancer Chemother. Pharmacol.*, 1992, **30**, 401.
77. A. J. Senderowicz, G. Kaur, E. Sainz, C. Laing, W. D. Inman, J. Rodriguez, P. Crews, L. Malspeis, M. R. Grever, E. A. Sausville and K. L. K. Duncan, *J. Nat. Cancer Inst.*, 1985, **87**, 46.
78. M. R. Bubb, A. M. J. Senderowicz, E. A. Sausville, K. L. K. Duncan and E. D. Korn, *J. Biol. Chem.*, 1994, **269**, 14869.
79. M. R. Bubb, I. Spector, B. B. Beyer and K. M. Fosen, *J. Biol. Chem.*, 2000, **275**, 5163.
80. S. Y. Khaitlina, *Int. Rev. Cytol.*, 2001, **202**, 35.
81. R. Li and G. G. Gundersen, *Nat. Rev. Mol. Cell Biol.*, 2008, **9**, 860.
82. J. S. Allingham, V. A. Klenchin and I. Rayment, *Cell Mol. Life Sci.*, 2006, **63**, 2119.
83. G. Fenteany and S. Zhu, *Curr. Top. Med. Chem.*, 2003, **3**, 593.
84. D. C. Aldridge and W. B. Turner, *J. Antibiot. (Toyko)*, 1969, **22**, 170.
85. I. Spector, N. R. Shochet, Y. Kashman and A. Groweiss, *Science*, 1983, **219**, 493.
86. M. Coué, S. L. Brenner, I. Spector and E. D. Korn, *FEBS Lett.*, 1987, **213**, 316.
87. U. B. Nair, P. B. Joel, Q. Wan, S. Lowey, R. A. Rould and K. M. Trybus, *J. Mol. Biol.*, 2008, **384**, 848.
88. M. R. Bubb, L. Govindasamy, E. G. Yarmola, S. M. Vorobiev, S. C. Almo, T. Somasundaram, M. S. Chapman, M. Agbandje-McKenna and R. McKenna, *J. Biol. Chem.*, 2002, **277**, 20999.
89. G. Hagelueken, S. C. Albrecht, H. Steinmetz, R. Jansen, D. W. Heinz, M. Kalesse and W.-D. Schubert, *Angew. Chem. Int. Ed. Eng.*, 2009, **48**, 595.
90. Y. Li and Y. L. Wu, *Med. Trop. (Mars)*, 1998, **58**, 9.
91. Anonymous, *Chin. Med. J. (Engl.)*, 1979, 92, 811.
92. L. J. Bruce-Chwatt, *Br. Med. J. (Clin. Res. Ed)*, 1982, **284**, 767.

93. J. B. Jiang, G. Q. Li, X. B. Guo, Y. C. Kong and K. Arnold, *Lancet*, 1982, **8293**, 285.
94. D. L. Klayman, *Science*, 1985, **228**, 1049.
95. M. S. Nair, N. Acton, D. L. Klayman, K. Kendrick, D. V. Basile and S. Mante, *J. Nat. Prod.*, 1986, **49**, 504.
96. S. A. Elmarakby, F. S. el-Feraly, H. N. Elsohly, E. M. Croom and C. D. Hufford, *J. Nat. Prod.*, 1987, **50**, 903.
97. H. M. Gu, D. C. Warhurst and W. Peters, *Biochem. Pharmacol.*, 1983, **32**, 2463.
98. A. Brossi, B. Venugopalan, G. L. Dominguez, H. J. Yeh, J. L. Flippen-Anderson, P. Buchs, X. D. Luo, W. Milhous and W. Peters, *J. Med. Chem.*, 1998, **32**, 645.
99. S. R. Meshnick, T. W. Tsang, F. B. Lin, H. Z. Pan, C. N. Chang, F. Kuypers, D. Chiu and B. Lubin, *Prog. Clin. Biol. Res.*, 1989, **313**, 95.
100. S. R. Meshnick, A. Thomas, C. M. Ranz, M. Xu and H. Z. Pan, *Mol. Biochem. Parasitol.*, 1991, **49**, 181.
101. S. R. Meshnick, Y.-Z. Yang, V. Lima, F. Kuypers, S. Kamchonwong-piasan and Y. Yuthavong, *Antimicrob. Agent. Chemother.*, 1993, **37**, 1108.
102. A. Robert, Y. Coppel and B. Meunier, *Chem. Comm*, 2002, 414.
103. S. C. Mrian, R. K. Pereira and M. M. C. Ferreira, *J. Chem. Inf. Model.*, 2008, **48**, 85.
104. I. Solomonov, M. Osipova, Y. Feldman, C. Baehtz, K. Kjaer, I. K. Robinson, G. T. Webster, D. McNaughton, B. R. Wood, I. Weissbuch and L. Leiserowitz, *J. Am. Chem. Soc.*, 2007, **129**, 2615.
105. J. Golenser, J. H. Waknine, M. Krugliak, N. H. Hunt and G. E. Grau, *Int. J. Parasitol.*, 2006, **36**, 1427.
106. Y. Maeno, T. Toyoshima, H. Fujioka, Y. Ito, S. R. Meshnick, A. Benakis, W. K. Milhous and M. Aikawa, *Am. J. Trop. Med. Hyg.*, 1993, **49**, 485.
107. W. Li, W. Mo, D. Shen, L. Sun, J. Wang, S. Lu, J. M. Gitschier and B. Zhou, *PLoS Genet.*, 2005, **1**, e36.
108. U. Eckstein-Ludwig, R. J. Webb, I. D. Van Goethem, J. M. East, A. G. Lee, M. Kimura, P. M. O'Neill, P. G. Bray, S. A. Ward and S. Krishna, *Nature*, 2003, **424**, 957.
109. A. C. Uhlemann, A. Cameron, U. Eckstein-Ludwig, J. Fischbarg, P. Iserovich, F. A. Zuniga, M. East, A. Lee, L. Brady, R. K. Haynes and S. Krishna, *Nat. Struct. Mol. Biol.*, 2005, **12**, 628.
110. R. Jambou, E. Legrand, M. Niang, N. Khim, P. Lim, B. Volney, M. T. Ekala, C. Bouchier, P. Esterre, T. Fandeur and O. Mercereau-Puijalon, *Lancet*, 2005, **366**, 1960.
111. P. S. Covello, *Phytochemistry*, 2008, **69**, 2881.
112. P. R. Arsenault, K. K. Wobbe and P. J. Weathers, *Curr. Med. Chem.*, 2008, **15**, 2886.
113. R. K. Haynes, *Curr. Top. Med. Chem.*, 2006, **6**, 509.
114. I. Nakase, H. Lai, N. P. Singh and T. Sasaki T, *Int. J. Pharm.*, 2008, **354**, 28.

115. T. Efferth, *Curr. Drug Targets*, 2006, **7**, 407.
116. F. H. Jansen and S. A. Soomro, *Curr. Med. Chem.*, 2007, **14**, 3243.
117. P. M. O'Neill, *Expert Opin. Investig. Drugs*, 2005, **14**, 1117.
118. J. L. Vennerstrom, S. Arbe-Barnes, R. Brun, S. A. Charman, F. C. Chiu, J. Chollet, Y. Dong, A. Dorn, D. Hunziker, H. Matile, K. McIntosh, M. Padmanilayam, J. Santo Tomas, C. Scheurer, B. Scorneaux, Y. Tang, H. Urwyler, S. Wittlin and W. N. Charman, *Nature*, 2004, **430**, 900.
119. S. Müller, R. Murillo, V. Castro, V. Brecht and I. Merfort, *J. Nat. Prod.*, 2004, **67**, 622.
120. J. J. La Clair, *Angew. Chem. Int. Ed.*, 2006, **45**, 2769.
121. K. Yoshikawa, M. Inoue, Y. Matsumoto, C. Sakakibara, H. Miyataka, H. Matsumoto and S. Arihara, *J. Nat. Prod.*, 2005, **68**, 69.
122. T. Kino, H. Hatanaka, M. Hashimoto, M. Nishiyama, T. Goto, M. Okuhara, M. Kohsaka, H. Aoki and H. Imanaka, *J. Antibiot. (Toyko)*, 1987, **40**, 1249.
123. T. Kino, H. Hatanaka, S. Miyata, N. Inamura, M. Nishiyama, T. Yajima, T. Goto, M. Okuhara, M. Kohsaka, H. Aoki and T. Ochiai, *J. Antibiot. (Toyko)*, 1987, **40**, 1256.
124. T. J. Petcher, H. Weber and A. Rüegger, *Helv. Chim. Acta*, 1976, **59**, 1480.
125. A. Rüegger, M. Kuhn, H. Lichti, H. R. Loosli, R. Huguenin, C. Quiquerez and A. von Wartburg, *Helv. Chim. Acta*, 1976, **59**, 1075.
126. J. F. Borel, C. Feurer, H. U. Gubler and H. Stähelin, *Agents Actions*, 1976, **6**, 468.
127. D. Wiesinger and J. F. Borel, *Immunobiology*, 1980, **156**, 454.
128. D. J. White, R. Y. Calne and A. Plumb, *Transplant Proc.*, 1979, **11**, 855.
129. D. Bunjes, C. Hardt, M. Röllinghoff and H. Wagner, *Eur. J. Immunol.*, 1981, **11**, 657.
130. S. Britton and R. Palacios, *Immunol. Rev.*, 1982, **65**, 5.
131. W. N. Hait, M. W. Harding and R. E. Handschumacher, *Science*, 1986, **233**, 987.
132. D. A. Fruman, C. B. Klee, B. E. Bierer and S. J. Burakoff, *Proc. Natl. Acad. Sci. USA*, 1992, **89**, 3686.
133. N. A. Clipstone and G. R. Crabtree, *Nature*, 1992, **357**, 695.
134. S. J. O'Keefe, J. Tamura, R. L. Kincaid, M. J. Tocci and E. A. O'Neill, *Nature*, 1992, **357**, 692.
135. J. Liu, *Trends Pharmacol. Sci.*, 1993, **14**, 182.
136. M. W. Hardling, A. Galat, D. E. Uehling and S. L. Schreiber, *Nature*, 1989, **341**, 758.
137. J. J. Siekierkam, S. H. Y. Hung, M. Poe, C. S. Lin and N. H. Sigal, *Nature*, 1989, **341**, 755.
138. Y. J. Jin, M. W. Albers, W. S. Lane, B. E. Bierer, S. L. Schreiber and S. J. Burakoff, *Proc. Natl. Acad. Sci. USA*, 1991, **88**, 6677.
139. F. Foor, S. A. Parent, N. Morin, A. M. Dahl, N. Ramadan, G. Chrebet, K. A. Bostian and J. B. Nielsen, *Nature*, 1992, **360**, 682.

140. J. Choi, J. Chen, S. L. Schreiber and J. Clardy, *Science*, 1996, **273**, 239.
141. J. Liang, J. Choi and J. Clardy, *Acta Crystallogr. Section D: Biol. Crystallogr.*, 1999, **55**, 736.
142. M. E. Cardenas, A. Sanfridson, N. S. Cutler and J. Heitman, *Trends Biotechnol.*, 1998, **16**, 427.
143. P. A. Searle and T. F. Molinski, *J. Am. Chem. Soc.*, 1995, **117**, 8126.
144. R. J. Capon, C. Skene, E. H. Liu, E. Lacey, J. H. Gill, K. Heiland and T. Friedel, *Nat. Prod. Res.*, 2004, **18**, 305.
145. P. A. Searle, T. F. Molinski, L. J. Brzezinski and J. W. Leahy, *J. Am. Chem. Soc.*, 1996, **118**, 9422.
146. B. S. Lucas, V. Gopalsamuthiram and S. D. Burke, *Angew. Chem. Int. Ed. Engl.*, 2006, **46**, 769.
147. C. J. Forsyth, F. Ahmed, R. D. Cink and C. S. Lee, *J. Am. Chem. Soc.*, 1997, **120**, 6697.
148. D. R. Li, D. H. Zhang, C. Y. Sun, J. W. Zhang, L. Yang, J. Chen, B. Liu, C. Su, W. S. Zhou and G. Q. Lin, *Chemistry*, 2006, **12**, 1185.
149. A. B. Smith III, K. P. Minbiole, P. R. Verhoest and M. Schelhass, *J. Am. Chem. Soc.*, 2001, **123**, 10942.
150. J. D. White, T. H. Lee and P. Kuntiyong, *Org Lett.*, 2006, **8**, 6043.
151. B. Wang and C. J. Forsyth, *Org. Lett.*, 2006, **8**, 5223.
152. A. B. Smith III, T. M. Razler, J. P. Ciavarri, T. Hirose and T. Iskikawa, *J. Am. Chem. Soc.*, 2001, **123**, 10942.
153. A. B. Smith III, T. M. Razler, R. M. Meis and G. R. Pettit, *J. Org. Chem.*, 2008, **73**, 1201.
154. J. Chen, L. Ying, T. M. Hansen, M. M. Engler, C. S. Lee, J. J. La Clair and C. J. Forsyth, *Bioorg. Med. Chem. Lett.*, 2006, **16**, 901.
155. C. J. Forsyth, L. Ying, J. Chen and J. J. La Clair, *J. Am. Chem. Soc.*, 2006, **128**, 3858.
156. J. M. Paramio, C. Segrelles, S. Ruiz and J. L. Jorcano, *Mol. Cell Biol.*, 2001, **21**, 7449.
157. K. L. Erickson, J. A. Beutler, J. H. Cardellina and M. R. Boyd, *J. Org. Chem.*, 1997, **62**, 8188.
158. B. Kunze, R. Jansen, F. Sasse, G. Höfle and H. Reichenbach, *J Antibiot. (Toyko)*, 1998, **51**, 1075.
159. T. C. McKee, D. L. Galinis, L. K. Pannel, J. H. Cardellina II, J. Lasko, C. M. Ireland, L. Murray, R. J. Capon and M. R. Boyd., *J. Org. Chem.*, 1998, **63**, 7805.
160. K. A. Dekker, R. J. Aiello, H. Hirai, T. Inagaki, T. Sakakibara, Y. Suzuki, J. F. Thompson, Y. Yamaguchi and N. Kojima, *J. Antibiot. (Toyko)*, 1998, **51**, 14.
161. J. W. Kim, K. Shin-Ya, K. Furihata, Y. Hayakawa and H. Seto, *J. Org. Chem.*, 1999, **64**, 153.
162. Y. Wu, O. R. Seguil and J. K. De Brabander, *Org. Lett.*, 2000, **2**, 4241.
163. M. R. Boyd, C. Farinia, P. Belfiore, S. Gagliardi, J. W. Kim, Y. Hayakawa, J. A. Beutler, T. C. McKee, B. J. Bowman and E. J. Bowman, *J. Pharmacol. Exp. Ther.*, 2001, **297**, 114.

164. X. S. Xie, D. Padron, X. Liao, J. Wang, M. G. Roth and J. K. De Brabander, *J. Biol. Chem.*, 2004, **279**, 19755.
165. K. Niikura, *Drug News Perspect.*, 2006, **19**, 139.
166. T. Diyabalange, C. D. Amsler, J. B. McClintock and B. J. Baker, *J. Am. Chem. Soc.*, 2006, **128**, 5630.
167. J. A. Beutler and T. C. McKee, *Curr. Med. Chem.*, 2003, **10**, 787.
168. Y. Wu, X. Laio, R. Wang, X. S. Xie and J. K. De Brabander, *J. Am. Chem. Soc.*, 2002, **124**, 3245.
169. X. Jiang, B. Liu, S. Lebreton and J. K. De Brabander, *J. Am. Chem. Soc.*, 2007, **129**, 6386.
170. S. Lebreton, J. Jaunbergs, M. G. Roth, D. A. Ferguson and J. K. De Brabander, *Bioorg. Med. Chem. Lett.*, 2008, **18**, 5879.
171. B. M. Degan, C. J. Hawkins, M. F. Lavin, E. J. McCaffrey, D. L. Parry and D. J. Watters, *J. Med. Chem.*, 1989, **32**, 1354.
172. J.-F. Biard, C. Roussakis, J.-K. Kornprobst, D. Gouiffes-Babin, J.-F. Verbist, P. Cotelle, M. P. Foster, C. M. Ireland and C. Debitus, *J. Nat. Prod.*, 1994, **57**, 1336.
173. G. Griffiths, B. Garrone, E. Deacon, P. Owen, J. Pongracz, G. Mead, A. Bradwell, D. Watters and J. Lord, *Biochem. Biophys. Res. Comm.*, 1996, **222**, 802.
174. D. Gouiffes, M. Juge, N. Grimaud, L. Welin, M. P. Sauvait, Y. Barbin, D. Laurent, C. Roussakis, J. P. Henichart and J. F Verbist, *Toxicon*, 1988, **25**, 1129.
175. D. Riou, C. Roussakis, J. F. Biard and J. F. Verbist, *J. Anticancer Res.*, 1993, **13**, 2331.
176. M. P. Sauviat and J. F. Gen, *Physiol. Biophys.*, 1993, **12**, 465.
177. M. P. Sauviat, D. Gouiffes-Barbin, E. Ecault and J. F. Verbist, *Biophys. Acta*, 1992, **1103**, 109.
178. C. Stanwell, A. Gescher and D. Watters, *Biochem. Pharmacol.*, 1993, **45**, 1753.
179. G. Griffiths, B. Garrone, E. Deacon, P. Owen, J. Pongracz, G. Mead, A. Bradwell, D. Watters and J. Lord, *Biochem. Biophys. Res. Comm.*, 1996, **222**, 802.
180. D. Watters, B. Garrone, G. Gobert, S. Williams, R. Gardiner and M. Lavin, *Exp. Cell Res.*, 1996, **229**, 327.
181. A. V. Statsuk, R. Bai, J. L. Baryza, V. A. Verma, E. Hamel, P. A. Wender and S. A. Kosmin, *Nat. Chem. Biol.*, 2005, **1**, 383.
182. S. A. Rizvi, V. Tereshko, A. A. Kossiakoff and S. A. Kosmin, *J. Am. Chem. Soc.*, 2006, **128**, 3882.
183. S. A. Rizvi, D. S. Courson, V. A. Keller, R. S. Rock and S. A. Kosmin, *Proc. Natl. Acad. Sci. USA*, 2008, **104**, 4088.
184. M. Seki-Asano, T. Okazaki, M. Yamagishi, N. Sakai, Y. Takayama, K. Hanada, S. Morimoto, A. Takatsuki and K. Mizoue, *J. Antibiot. (Toyko)*, 1994, **47**, 1395.
185. Y. Mizui, T. Sakai, M. Iwata, T. Uenaka, K. Okamoto, H. Shimizu, T. Yamori, K. Yoshimatsu and M. Asada, *J. Antibiot. (Toyko)*, 2004, **57**, 188.

186. N. Asai, Y. Kotake, J. Niijima, Y. Fukuda, T. Uehara and T. Sakai, *J. Antibiot. (Toyko)*, 2007, **60**, 364.
187. R. M. Kanada, D. Itoh, M. Nagai, J. Niijima, N. Asai, Y. Mizui, S. Abe and Y. Kotake, *Angew. Chem. Int. Ed. Engl.*, 2007, **46**, 4350.
188. A. L. Mandel, B. D. Jones, J. J. La Clair and M. D. Burkart, *Bioorg. Med. Chem. Lett.*, 2007, **17**, 5159.
189. P. R. Skaanderup and T. Jensen, *Org. Lett.*, 2008, **10**, 2821.
190. T. Sakai, N. Asai, A. Okuda, N. Kawamura and Y. Mizui, *J. Antibiot. (Toyko)*, 2004, **57**, 180.
191. Y. Mizui, T. Sakai, M. Iwata, T. Uenaka, K. Okamoto, H. Shimizu, T. Yamori, K. Yoshimatsu and M. Asada, *J. Antibiot. (Toyko)*, 2004, **57**, 188.
192. Y. Kotake, K. Sagane, T. Owa, Y. Mimori-Kiyosue, H. Shimizu, M. Uesugi, Y. Ishihama, M. Iwata and Y. Mizui, *Nat. Chem. Biol.*, 2007, **3**, 570.
193. C. Lagisetti, A. Pourpak, Q. Jiang, X. Cui, T. Goronga, S. W. Morris and T. R. Webb, *J. Med. Chem.*, 2008, **51**, 6220.
194. W. Fenical, P. R. Jensen and X. C. Cheng, *US Patent* 6,066,635, 2000.
195. Y. Sugie, H. Hirai, T. Inagaki, M. Ishiguro, Y.-J. Kim, Y. Kojima, T. Sakakibara, S. Sakemi, A. Sugiura, Y. Suzuki, L. Brennan, J. Duignan, L. H. Huang, J. Sutcliffe and N. Kojima, *J. Antibiot. (Toyko)*, 2001, **54**, 911.
196. J. Qian-Cutrone, S. Huang, Y.-Z. Shu, D. Vyas, C. Fairchild, A. Menendez, K. Krampitz, R. Dalterio, S. E. Klohr and Q. Gao, *J. Am. Chem. Soc.*, 2002, **124**, 14556.
197. S. B. Herzon and A. G. Myers, *J. Am. Chem. Soc.*, 2005, **127**, 5342.
198. P. S. Baran, B. D. Hafensteiner, N. B. Ambhaikar, C. A. Guerrero and J. D. Gallagher, *J. Am. Chem. Soc.*, 2006, **128**, 8678.
199. T. J. Greshock, A. W. Grubbs, S. Tsukamoto and R. W. Williams, *Angew. Chem. Int. Ed. Engl.*, 2007, **46**, 2262.
200. J. E. Wulff, S. B. Herzon, R. Siegrist and A. G. Myers, *J. Am. Chem. Soc.*, 2007, **129**, 4898.
201. J. E. Wulff, R. Siegrist and A. G. Myers, *J. Am. Chem. Soc.*, 2007, **129**, 14444.
202. L. B. Maggi Jr., M. Kuchenruether, D. Y. Dadey, R. M. Schwope, S. Grisendi, R. R. Townsend, P. P. Pandolfi and J. D. Weber, *Mol. Cell Biol.*, 2008, **28**, 7050.
203. M. A. Amin, S. Matsunaga, S. Uchiyama and K. Fukui, *Biochem. J.*, 2008, **415**, 345.
204. B. Falini, I. Nicoletti, N. Bolli, M. P. Martelli, A. Liso, P. Gorello, F. Mandelli, C. Mecucci and M. F. Martelli, *Haematologica*, 2007, **92**, 519.
205. C. C. Hughes, J. B. MacMillan, S. P. Gaudêncio, P. R. Jensen and W. Fenical, *Angew. Chem. Int. Ed. Engl.*, 2009, **48**, 725.
206. C. C. Hughes, J. B. MacMillan, S. P. Gaudêncio, W. Fenical and J. J. La Clair, *Angew. Chem. Int. Ed. Engl.*, 2009, **48**, 728.
207. H. Luesch, S. K. Chanda, R. M. Raya, P. D. DeJesus, A. P. Orth, J. R. Walker, J. C. Izpisúa Belmonte and P. G. Schultz, *Nat. Chem. Biol.*, 2006, **2**, 158.

208. Y. Zhang, J. R. Yeh, A. Mara, R. Ju, J. F. Hines, P. Cirone, H. L. Griesbach, I. Schneider, D. C. Slusarski, S. A. Holley and C. M. Crews, *Chem. Biol.*, 2006, **13**, 1001.
209. J. Hines, M. Groll, M. Fahnestock and C. M. Crews, *Chem. Biol.*, 2008, **15**, 501.
210. M. K. Hadden, D. J. Lubbers and B. S. Blagg, *Curr. Top. Med. Chem.*, 2006, **6**, 1173.
211. A. Bowers, N. West, J. Taunton, S. L. Schreiber, J. E. Bradner and R. W. Williams, *J. Am. Chem. Soc.*, 2008, **130**, 11219.
212. A. D. Rodriguez, M. J. Lear and J. J. La Clair, *J. Am. Chem. Soc.*, 2008, **130**, 7256.
213. A. Lopez, A. B. Parsons, C. Nislow, G. Giaever and C. Boone, *Prog. Drug. Res.*, 2008, **66**, 239.

Section 2 Sources of Compounds

CHAPTER 4

The Convention on Biological Diversity and its Impact on Natural Product Research

GEOFFREY A. CORDELL

Natural Products Inc., Evanston, Illinois, USA

1 Introduction

The Earth comprises an incalculable level of genetic resources. Each year new animals (terrestrial and marine), new plants and new organisms in many different phyla are identified for the first time. Efforts to catalogue the enormity of this biological wealth have thus far been only partly successful and a priority for such tasks has yet to reach the level of funding support which would permit authoritative cataloguing of these resources and their abundance on a global basis. At the same time, it must be recognised that each of these organisms has the capacity to produce a vast array of mostly unknown chemical wealth. Consequently, any discussion of biological diversity is also a discussion about chemical diversity and the potential uses of that diversity for the benefit of humankind.

For all of recorded human history, humankind has used these locally available genetic resources for food, for shelter, for furniture, for medicine, for the storing and sharing of information and to meet energy requirements. Over thousands of years, as humans migrated from their original locations on the Earth, they took with them the knowledge of previous generations and accumulated new knowledge through their own personal experiences with the local

RSC Biomolecular Sciences No. 18
Natural Product Chemistry for Drug Discovery
Edited by Antony D. Buss and Mark S. Butler

Published by the Royal Society of Chemistry, www.rsc.org

flora and fauna. Whether that knowledge related to the domestication of animals for food and transportation, the use of particular plant materials as being preferable for construction, or the use of certain plants and animals for the treatment of disease states, it was regarded as being "global" knowledge. Thus, the oldest documented uses of Earth's genetic resources are recorded on Sumerian clay tablets dating to the 3rd millennium BCE[1] and include information relating to the selling of many foods and some plant-based medicinal agents.

Plants have been used as medicinal agents since at least Sumerian times. Over 3000 years ago, Chinese scholars began to document their knowledge of the complexities of the utilisation of biological materials for medicinal purposes in formal texts. Subsequently, several Chinese scholars in the Han, Jin and Tang dynasties compiled and published the use of numerous medicinal plants. The Greeks, the Romans and numerous medieval scholars also recorded their own, different uses of genetic resources in documents that were passed on from generation to generation and whose descriptions moved from "country" to "country" as they changed hands and empires with the passage of time. Similarly, selected genetic materials were moved from their country of origin to serve communities all over the world. The potato, which originated in the lowlands of south-central Chile,[2] is a global commodity; similarly wheat, which originated in the Levant area of the eastern Mediterranean,[3] and rice which came from species native to the lower Yangtze river (Hubei and Anhui Provinces) and to the Niger River delta in Africa.[4]

With the rise of the chemical and biological sciences in the early 19th century and the development of techniques to isolate individual active components of ethnomedical preparations and, subsequently, the ability to modulate the biological effects of the isolated compounds through chemical transformation, new compounds of intense significance were prepared for potential human benefit. One of these was aspirin, a semi-synthetic derivative of a well-established traditional remedy derived from several plants used for treating many forms of pain since at least the time of Hippocrates. This compound was first marketed by Bayer in 1899; today it is available as a generic drug and is the world's most widely consumed medicinal (20 billion tablets per year in the USA alone).

Since that time, the pharmaceutical industry in the developed world has focused primarily on the synthesis of compounds for development as single agent prescription products. These efforts have required very substantial investments and, in the USA alone, investment in pharmaceutical research and development is probably close to $40 billion per year. In order to prevent the "theft" of a drug by an entity that has made no investment in the discovery and development research programme, companies have sought to protect their investments through the patent process.

The notion that an invention having commercial implications could be regarded as intellectual property needing protection probably dates back to at least 500 BCE.[5] The first patent issued in England was granted by King Henry VI in 1449 and related to a particular process for producing the stained glass used in Eton College. Protection of new inventions became more formally developed as a system in the late 15th century when the Republic of Venice in 1474 issued a decree requiring communication of new and inventive devices, once practiced, to be communicated to the Republic to prevent others from using the device without authority.[6]

It was in England, during the reigns of James I (Statute of Monopolies, 1623) and Queen Anne (Statute of Anne, 1709) that products of "new invention" were first required to be documented through written description and where copyrights of works were assigned exclusively to authors rather than printers. The first Patent Act in the USA dates from 1790, with the most significant revisions occurring in 1836 and 1952. These events presaged the development of the concept of patents as we know them today, which have continued to focus on the generation of new knowledge and its protection for the exclusive benefit of the creator. For a pharmaceutical or biotechnology company, patents are their lifeblood and, without their protection for an extended period of time beyond the time when the product is first marketed, investment in research which creates new drugs and new technologies would probably not occur given the significant costs (at least $1 billion) for the development of a new drug.

The ownership of genetic resources was never a part of the discussion with respect to patents since it was made clear that neither a plant nor an animal, nor the knowledge associated with their use (if previously documented) could be patented. Thus, if one discovers a new species of plant from a tropical forest in Papua New Guinea, it cannot be patented. Similarly, it also became clear that (in most cases) global knowledge in the public domain regarding the use of a plant for medicinal or commercial use could not be patented. Thus, the use of *Papaver somniferum* as a treatment for pain could not be protected under patent law. However, the development of a new procedure for the isolation of morphine from *P. somniferum* would be patentable. This distinction made it clear that creativity would be required for there to be economic benefit from a widely available natural source.

Based on the principle of common global heritage, genetic samples (plants, microbes, marine species and animals) for chemical and biological evaluation were, until quite recently, collected in the field *ad libitum*. The prevailing notion was that such resources existed for the general use of humankind and thus, there was an "open access" approach to their acquisition and exportation. No consideration was given to issues relating to compensation either of the country or of an indigenous group in whose territory the genetic material was acquired. Similarly, no thought was given to compensating for the sharing of indigenous knowledge. Sometimes permission to collect was sought from a local farmer or the local indigenous group and frequently a person who knew the location of certain plants being used locally would accompany the collectors. Payments were typically made on a fee for service basis to collectors and samples of collected materials were often deposited in local herbaria. Shipping occurred privately with minimal customs or agricultural controls. Pharmaceutical companies would ask their employees to bring back small samples of soil when they went on vacation, especially overseas, and the samples collected were returned for culturing and antibiotic screening. No ethical considerations were given to the ownership of the biomaterial being provided. In the case of indigenous knowledge, a person who spoke the local language or dialect would accompany an ethnobotanist in the field and be compensated for their time and travel.

In 1983, the United Nations Food and Agriculture Organization (FAO) indicated in its "International Undertaking" that plant genetic resources were a common heritage of humankind and that, as such, they were free of obligation or consideration as to origin and were for the use of all. Other social and ethical movements were stirring in many parts of the world, however. In the early 1990s, sensibilities shifted as developing countries became more aware of the interest in these genetic resources and the possibility that commercial products could result with a resulting significant financial implication for the external corporate entity.

As a result, a number of scientific meetings and international organisations made important, ground-breaking proclamations relating to the conservation of biodiversity and the ethical considerations for compensation for access to genetic materials and traditional knowledge.[7] These included:

- Chiang Mai Declaration from March 1988;[8]
- Declaration of Belem of the International Society of Ethnobiology from July 1988;[9]
- Göteborg Resolution from the International Society of Chemical Ecology in August 1989;[10]
- Hipólito Unanue Agreement from a meeting on "Primary Health Care, Resources and Traditional Medicines in Andean Countries" held in Lima, Peru, in October 1989;[11]
- Kunming Action Plan, also from the International Society of Ethnobiology in October 1990;[12]
- Summary of a Workshop on Drug Development, Biological Diversity and Economic Growth held at the National Institutes of Health, Bethesda, Maryland, in March 1991.[13]

At the Seventh Asian Symposium on Medicinal Plants and Spices and Other Natural Products (ASOMPS VII) held in Manila in February 1992, the scientists present agreed to endorse *The Manila Declaration Concerning the Ethical Utilization of Asian Biological Resources*—the so-called Manila Declaration.[7] This document for the first time brought together the concepts of national requirements for a supply agreement with a local government organisation, specific contractual guidelines with respect to collection of plant materials and provisions for commercial development, mandatory royalty or license agreements and the requirement for the recognition of traditional knowledge as significant intellectual property, among other aspects. At the time and because of the diverse origins of the ASOMPS meeting participants, this Declaration was viewed as being accepted as a model for practice for countries in the East Asian region.

With all these statements of intention and guidelines for practice, the notion of open, uncompensated access to the world's bioresources changed dramatically and a new legal environment evolved. As pointed out by Gollin,[14] who provided a brief overview of some aspects of the effects of those legal changes on natural product research, three sets of rules have now been imposed for the conduct of acquiring genetic resources, *i.e.* international treaties, national laws, and guidelines from professional societies with respect to the conduct of their members.

In the Executive Order 247 of the Republic of the Philippines (see Section 4 and Section 5.3), "bioprospecting" is defined as the *"research, collection and utilization of biological and genetic resources for purposes of applying the knowledge derived therefrom to scientific and/or commercial purposes".*

What is missing from this definition is that the goal of "bioprospecting" is human benefit. Unfortunately, it has a negative connotation and, in the view of many, has become synonymous with "biopiracy". The unethical actions of a minority in the former arena have led to charges in the latter. In certain countries of the world, regulations have been enacted which address "biopiracy" with such vigour that even the majority of ethical researchers are impeded in conducting studies. The many potential benefits of bioprospecting to a society are then ignored. Significantly, there are frequently legal consequences for both real and perceived violators of treaties and laws, the so-called "biopirates", which can bring serious, undesirable public attention to research programmes.[15–17]

Not wishing to be labelled as "biopirates", some corporations have reduced their risk by not being involved in genetic resource collection activities.[18] As discussed in more detail subsequently, expectations for financial returns based on genetic resource acquisition are frequently unrealistic in the source country, leading to misunderstandings about the discovery process (and therefore perceived "biopiracy" claims) and at what point return on investment begins, or where milestone payments may begin, if they are a part of a negotiated agreement. On the other hand, it must also be said that exploitive agreements have existed and that the theft of genetic diversity from many of the biodiversity-rich countries continues today. These unseemly activities also appear in the patent arena, for example, the claim made in 1999 by Japanese scientists with respect to Indian curry[19] and that made with respect to turmeric.[20]

2 Historical Perspective

There exists a "great divide" in the world. For the "North", there is an association with the affluence of millions. The population is regarded as being "global" in outlook. Their use of fossil fuels is considered the cause of climate change. They possess technological knowledge and conduct theory-driven research. They have resource surpluses. For the "South", there is an association with poverty for billions. The population is regarded as being "local" in outlook, as not being responsible for climate change, as possessing and using most of the traditional knowledge, conducting action-driven research and experiencing continuous resource shortages.

In healthcare there is a corresponding great divide and *per capita* investment in countries in the North compared with those in the South is a well-known example. For example, annual *per capita* expenditures on overall healthcare in the USA was $6,094 in 2007, whereas it is $2,560 in the UK, $1,135 in Korea and $21 in Ethiopia.[21] At the same time, very few countries in the South have an industrial infrastructure that can support the discovery and development of medicinal agents—be they synthetic or derived from plant, microbial or marine biomaterials. Consequently, these countries are substantially reliant on imported drugs for primary healthcare—if governments or local residents are able to afford them.

A recent World Health Organization (WHO) report clarifies this situation, from many different perspectives, in considerable detail;[22] only 10 countries of 188 in the world are classified as having a sophisticated pharmaceutical industry with significant research programmes. A further 17 countries have "innovative capability", whereas 42 countries have no pharmaceutical industry structure at all. On the other hand, only 10% of research and development spending goes to diseases that make up 90% of the global disease burden. While the USA spent $34.2 billion on health-related research and development in 2000, the *combined* investment of Thailand, the Philippines and Malaysia was $30 million in 1998. The international trade in medicines grew from $5 billion in 1980 to $120 billion in 1999 and, although the WHO recommends that there be no tariffs for the list of essential medicines,[23] tariffs in many countries added an average of 12% to drug costs. Over the 20-year period from 1980 to 1999, low-income country imports of drugs declined as a percentage of total drug imports globally from 7.2% to 3.4% and from 22.3% to 17.3% in middle-income countries. In parallel, pharmaceutical consumption in the low-income countries declined to 2.9% of the global total from 3.9% in the same period. In terms of annual *per capita* expenditures on pharmaceuticals, people in high-income countries were spending an average of $396 each in 2000 compared with $4.4 per person per year in low-income countries and only $31 per year in middle-income countries. These data would suggest that, in all these low- and middle-income countries, there was a decrease in the importation of pharmaceuticals and an increased reliance on non-pharmaceutically prepared drugs—probably plant-based, traditional medicines. This is a chronic public healthcare situation which needs to be addressed on a global basis from the aspect of the cost of pharmaceuticals and the perspective of the quality control of traditional medicines.[24–29]

The South has control over the access to, and the conservation of, their genetic resources as well as most of the traditional knowledge. It is these resources in which the North has potential commercial interest. On the other hand, the North has control of the technology, the financial base and the patent system to develop the natural resources which the South wishes to see developed. The rapacious urges of various industrial groups in the North to develop these indigenous resources for their own profit and gain, without adequate (or in most instances any) compensation for those resources (genetic or knowledge-based) led to deep resentment which continues today. In addition, emerging nations, becoming aware that their genetic and traditional knowledge-based resources were deemed valuable by academic and industrial groups external to their own country and beginning to acquire the resources to develop them, recognised that there was financial and technology gain to controlling access to those resources. It also became more apparent that, in the past, both microbial and plant resources from various countries in the South had been developed by pharmaceutical companies to provide a number of significant drugs of high commercial value without consideration being given to the provision of any profit-sharing to the resourced country up-front or retroactively.

At the same time, there was a growing concern about the rate of deforestation that was occurring in many areas of rainforest throughout the tropical world and the argument was made that, given the historical origin of many drugs from natural sources, both the diversity and its potential for contributing to the welfare of humankind were being lost at the same time.[30] Consequently, there were incentives and mutual interests to see that preserving the world's biodiversity and developing equitable sharing agreements, relating to both the resources and the knowledge associated with those resources, had substantial merit.

In 1972, the United Nations held its first conference on the environment in Stockholm. At that meeting, the USA led the rest of the world with respect to raising environmental awareness and consciousness. No such leadership was evident at the United Nations Conference on Environment and Development (the Earth Summit) in June 1992 in Rio de Janeiro, where the USA decided not to sign the convention on biodiversity and did not sign the climate convention. It did sign the convention on biodiversity in June 1993, but it remains unratified by the US Senate. At the Rio Summit, 153 countries signed the convention on biodiversity and 154 the one on climate change. At the same time, the USA was significantly outmatched by Japan in donor contributions to environmental initiatives. In addition, a group of industrial leaders, the Business Council for Sustainable Development, proposed that the prices of goods should incorporate the cost of environmental damage and that new economic initiatives relating to pollution taxes and tradable pollution permits be developed. Fossil fuel subsidies were targeted for elimination with a concurrent imposition of carbon taxes.

3 The Convention on Biological Diversity

The Convention on Biological Diversity (CBD) is binding on signatory governments and their agencies, but not on private individuals unless and until there is applicable national legislation. As of April 2009, there are 191 parties to the CBD; Andorra, the Holy See, Iraq, Somalia and the USA are not parties.[31]

The CBD is made up of 42 Articles, of which Articles 20–42 are concerned with financial resources and mechanisms and issues related to arbitration and administrative matters. These do not directly impact the activities of natural product scientists, although several of the Articles do have an influence on the development of access and benefit-sharing agreements. It is Articles 1–19 that have direct relevance to the work of natural product scientists and a detailed discussion is available.[32]

In Article 1, the CBD declared a number of objectives including:

- the conservation of biodiversity;
- the sustainable use of the components of that biodiversity;
- the fair and equitable sharing of the benefits arising from the utilisation of those genetic resources.

Those objectives could be achieved and the sharing of benefits could occur in three main ways:

- appropriate access to genetic resources;
- the appropriate transfer of the relevant technologies, taking into account all rights over those resources;
- appropriate funding.

It was the notion that developing countries were now assured of compensation from bioprospecting that led to their strong support for the CBD.

Article 3 of the CBD clarifies the sovereign right of source countries to investigate their own genetic resources; this is expanded in Article 15. Articles 6–10 are concerned with issues relating to the conservation of biological diversity and, in the event of commercial development, to identify any negative impacts on conservation and sustainable use which might occur, for example from the large scale collection of genetic material. There is an implicit requirement for the deposition of voucher specimens in the source country in Article 9a.

One of the most important articles for natural product researchers is Article 8j, which deals with the protection and respect for local knowledge related to the conservation and sustainable use of biodiversity and, with the sanction and involvement of the knowledge holders, promote its wider application. It also promotes the sharing of benefits that might arise from the utilisation of the indigenous knowledge. Articles 10–14 deal with the sustainable use of the components of biodiversity, incentive measures, research and training, public education and awareness, and impact assessment and minimising adverse impacts, respectively.

The issue of access to genetic resources is presented in Article 15. Article 15.1 describes how each state/government has sovereign rights over the natural resources and that national governments have the authority to determine access. The creation of *"conditions to facilitate access to genetic resources for environmentally sound uses"* is indicated in Article 15.2 with the admonition *"not to impose restrictions that run counter to the objectives of the Convention"*.

Access to materials should follow a negotiation between parties, which would lead to a formal access agreement between the parties involved (Article 15.4). But before any access can be obtained to genetic resources, there is the need for prior informed consent by the designated state regulatory authority (Article 15.5), resulting in the necessity for an agreement to be negotiated. Collaboration between the contracting parties is presented as a desirable goal in Article 15.6, with the results and any commercial benefits being shared through mutually agreed terms according to Article 15.7.

Article 16 is concerned primarily with issues related to the access and transfer of technologies (including biotechnology) relevant to the conservation and sustainable use of genetic resources (Article 16.1) under fair and favourable terms (Article 16.2). This is presumably a two-way process with respect to the technology transfer to the developing country (some of which may be patented or proprietary), receiving in return information regarding indigenous materials and their uses. This transfer includes transfer to government as well as corporate entities (Article 16.4) and respects any issues concerning patent and other intellectual property rights (Article 16.5). All such arrangements are on mutually agreed terms. Given the indication that there is a need for cooperation in this area with respect to national legislation and international law (Article 16.5), it is felt that there is no compulsion to divulge intellectual property or share patents, as had been a concern.[32]

The sharing of information related to the conservation and sustainable use of biological diversity (Article 17.1) and including the results from *"technical, scientific and socio-economic research"* (Article 17.2) is promoted. Article 18 promotes the cooperation between parties for "international technical and scientific cooperation in the field of sustainable use of biological diversity" (Article 18.1) and the development of human resources and institution building (Articles 18.2 and 18.4) and more specifically *"joint research programmes and joint ventures"* (Article 18.5). Article 19 reinforces Article 16 in promoting effective participation in biotechnology research (Article 19.1) and assuring that the developing country party has appropriate access to the results and benefits from the genetic resources provided (Article 19.2).

One can, therefore, consider that there are five main thrusts which the Convention is promoting:[7]

(i) conservation of biodiversity;
(ii) development of socially beneficial (agricultural, pharmaceutical and other industrial products) through the sustainable use of biodiversity;
(iii) promotion of collaborative programmes with respect to the sustainable use of genetic resources in source countries;
(iv) facilitation of access to genetic resources, technology transfer and research and training;
(v) equitable sharing of results and benefits.

In an article in October 1992[33] in response to an invitation from the US National Institutes of Health (NIH), Al Gore posed the question: *"How can we better preserve the biological resources necessary for economic progress?"* His answer was the use of biodiversity to stem agricultural losses rather than using chemical agents as pesticides and fungicides, and to recognise that we do not have the technologies to potentiate biodiversity for benefiting humankind and therefore, that wealth should be left to our children *"with as little damage . . . as possible"*. He indicated that access to native resources and protection of intellectual property are *"complementary concerns"*. He added that *"the most effective way to ensure the sustainable use of genetic resources while enhancing conservation efforts is by establishing international agreements that set standards to which all parties can be held accountable"*.[33]

The CBD was viewed as an important link between the international trade in global genetic resources and biodiversity conservation. It was anticipated that the CBD would promote trade in genetic resources, in part, by increasing the available funding for conservation initiatives. In addition, it was hoped that the biological evaluation of natural products—so-called "screening"—would also raise awareness about the potential economic and beneficial value of the rainforests and create new economic incentives for the conservation of biological diversity. However, that notion requires that the supply of materials for biological evaluation is appropriately compensated and, when subjected to a materials transfer agreement (MTA) rather than true collaboration, the corporate entity typically assesses each sample as having the same worth, irrespective of history of use. For their part, the supplier is perhaps so eager to receive compensation (even at a modest level) for those samples that they effectively agree to ignore the ethnomedical component (with its possible intellectual knowledge complications) and supply the samples. The relationship to supporting conservation through the demonstration of biological potential and relevance to a traditional use is lost. This situation demeans the value of genetic resources (particularly those used in traditional medicine) and diminishes the value of the access that the corporate entity receives from the long-term availability of such unique chemical matrices to their evolving corporate research initiatives. Over time, these acquired sample repositories of natural product extracts (frequently referred to as "libraries") are of immense value and the genetic wealth stored in such a depository can only increase in value.

Seventeen years later, it is probably the case that the anticipated "profits" (royalty stream from an invention) have not yet materialised. What has occurred, as an integral aspect of negotiating access agreements, has been a significant increase in up-front initiatives by both corporations and academic groups to provide compensation in non-financial terms in a timely manner. For example, the use of technology transfer, the building of infrastructure, collaborative research initiatives and training programmes are frequent aspects of agreements to grant access to genetic resources. These initiatives are fully in line with the spirit of the CBD and have provided some significant opportunities for innovative relationships and forms of compensation.[19,34–36] Of special interest in this regard are those initiatives developed with respect to the International Cooperative Biodiversity Group (ICBG) programme (see Section 7.1).

From an academic perspective in the USA, the situation regarding owning and managing inventions changed in 1980 with the passage of the Bayh–Dole Act.[37] This Act allowed universities that were carrying out government-sponsored research to protect their inventions through patent application. The government thereby relinquished its rights to ownership, which might have been assumed as the sponsor of the research. This led to most major universities with significant programmes of federal-sponsored research developing a unit for technology transfer and intellectual property issues. As a result, major universities deal with a full spectrum of reviewing innovations for patent application, copyright and trademark issues, material transfer agreements, licensing inventions to industry, creation of start-up companies, part ownership of new companies in collaboration with academic researchers, *etc*.

A new challenge arose with drug evaluation programmes of natural product samples. Since the late 1970s, the University of Illinois at Chicago (UIC) had recognised the importance of sharing benefits with the source of the genetic materials in the country of origin.[34,37] Prior to the CBD, UIC included the local collector as a co-author in publications, sharing any financial benefits that might accrue from commercialisation and providing the local collector with relevant research results derived from the collected materials. As the programme grew and additional plant collection and drug discovery initiatives were undertaken through government and corporate sponsorship, a well-defined contractual mechanism was employed to acquire samples and to establish reciprocity and benefit-sharing, as well as providing a share (up to 50%) of monetary benefits (such as royalties). It was made clear that, while UIC would be the owner of the technology, the supplier was the owner of the original sample provided under a material transfer agreement (MTA).

Subsequent to the CBD and with additional experience gained locally and through interactions globally, UIC established a well-defined, consistently evolving series of policies, procedures and conditions relating to this type of research.[34,37] However, the Manila Declaration had presented a 50–50 revenue sharing target and collaborators in biologically diverse communities insisted that this was a reasonable collaborative arrangement. The costs of development from the crude plant source to a clinically available product, which could easily reach $1 billion, were not easily factored into the discussion.

When a new International Cooperative Biodiversity Group (ICBG) was funded at UIC in 1998 (see Section 7.1),[38] the need for a more creative approach became apparent.[34] The revised policy embraced source country participants and addressed a number of issues related to ownership, control, possession, use, results sharing, co-authorship, collaboration, reciprocity, training and genetic sourcing issues. As a result, after trust fund and patent expenses are reimbursed from gross technology commercialisation revenues, at least 50% of net revenues are guaranteed to be provided to the source country collaborators. Costs for the patenting process are paid by UIC from royalties. Several scenarios for source country benefit can result from the revised policy depending on other collaborators—such as a corporate partner or a source country collaborator named on an invention—wherein the source country could receive 68% or more of the net royalty revenues.[34]

While the CBD recognised through treaty the rights of sovereign states to control access to their genetic resources, most countries (unless they are single island nations) are relatively recent, politically based, artificial creations (often as a result of war) and thus, their "borders" are biologically porous and have been for millions of years. Consequently, their genetic resources may not be limited to a certain country, or even a particular region of the world. In addition, the indigenous knowledge associated with a given genetic resource may be the same or different from country to country. These factors raise a number of interesting questions. For example, if the medicinal use of a plant in country A is well-known and documented, but no such record exists in country B, what are the intellectual property issues involved in reaching agreements in country B? What is the situation if more than one group within a country or region claims rights over the genetic resources? How is ownership then defined? Do developing countries have the legal, institutional and scientific capacity to appropriately regulate and facilitate access to their genetic resources as required by the CBD? The number of stakeholders in the genetic resources of country A is frequently quite substantial and often conflicting agendas arise. How are these differences resolved between the parties, *i.e.* where does the arbitration take place? Or are negotiations allowed to fail?

The matter of prior informed consent is critical at this point, since the unauthorised collection of samples undermines the position that developing countries are attempting to establish and places the corporate entity in gross violation of both the intent of the CBD and possibly the laws of a particular country. However, such private transactions still occur. At the same time, governments may feel that the threat to genetic resources necessitates that collection be very strictly controlled, thereby possibly having a detrimental effect on the investment in natural product research in their country.

4 Implementation and Regulatory Outcomes of the CBD

The CBD produced a flurry of activity in a number of countries to develop laws that would regulate access to genetic resources. Among the first countries to act were the Republic of the Philippines, Brazil, Colombia and Ecuador.[39,40]

In the Philippines, the President issued Executive Order No. 247 (EO247), which was signed into law on 18 May 1995. Implementing guidelines were approved by the Department of Environment and Natural Resources (DENR) as Department Administrative Order No. 96-20 on 21 June 1996.[41] Implementation of EO247 occurred in July 1997. Of the 33 applications that were submitted for permission to gain access to genetic resources in the seven years following, only two were approved[42]—one to a collaborative group between the Marine Sciences Institute of the University of the Philippines on the

Dilliman Campus and a group at the University of Utah, and the second to a collaborative research programme between the University of the Philippines and the Philippine Government. On 30 July 2001, in the face of substantial criticism of the policy, the Philippines enacted Republic Act No. 9147, known as the Wildlife Resources Conservation and Protection Act. This Act modified EO247 considerably and was designed to facilitate access to genetic resources. Subsequently, approval for the *Guidelines for Bioprospecting Activities in the Philippines* was given in 2005.[43] This document provides the details for making applications to garner access to genetic resources and aims to facilitate the implementation of the Wildlife Act, as well as those sections of EO247 that were not modified.

In Brazil, regulations for access were developed before the CBD in January 1990 and the National Council for Scientific and Technological Development was designated as the agency to provide oversight. Factors involved in regulatory approval were collaboration with a recognised Brazilian institution, information on the source of the financing, and allowing the government access to and rights to limit the export of material and rights to publish research results. Each Brazilian state enacted its own laws relating to access to genetic materials. In Sao Paulo State between 1995 and 1997, any request for access was automatically denied and access requests made since 1997 have been placed on hold. Local scientists collect materials as needed, but are unable to collaborate with international partners because the genetic resources cannot be shipped out of the country.

In July 1999, Glaxo Wellcome (as it was then) and a small Brazilian biotech company, Extracta Moléculas Naturais SA, signed a \$3.2 million contract for Glaxo to evaluate up to 30 000 samples of plant, fungal and bacterial origin from Brazil for their biological potential. The three-year deal included the provision of all research expenses in Brazil, 25% of the royalties from a product going to support community-based conservation, health and education projects and 25% going to the academic unit conducting the research in Brazil. The agreement conformed to the new Brazilian Patent Law and also the principles of the CBD. The therapeutic areas of interest were anti-inflammatory, antibacterial and antifungal agents.[44] The initial phase of the programme yielded seven compounds active on elastase and three compounds active against multidrug-resistant strains of *Staphylococcus aureus*. The programme finished in December 2004. Extracta was the first company in Brazil to receive a licence from the Brazilian Government to access the genetic diversity of the country for commercial purposes and the authorisation was renewed in August 2006 for a further two years.

Although it has only 0.7% of the land mass of the Earth, it is estimated that Colombia has possibly as much as 10% of all plant and animal species. Colombia ratified the CBD in 1994 and implemented Decision 391 in June 1997 (see below). Very few applicants have requested access to Colombia's genetic resources. Thus, between February 1997 and February 2004, 15 bioprospecting protocols were submitted. Two of these involved international collaboration, one was commercial and the rest were for academic and conservation purposes;

as far as is known these applications have not been successful and no projects have been approved.

On 16 September 1992, Ecuador created INEFAN (Instituto Ecuatoriano Forestal y de Areas Naturales y Vida Silvestre; the Ecuadorean Institute for Forests and Natural Areas), which was charged with implementing the Forest Law, with the Ministry of Agriculture being responsible for the authorisation, licensing, collection and export of genetic resources. In September 1996, Ecuador declared that it had sovereign rights over its genetic resources and, in June 1997, implemented Decision 391. UIC attempted in the period January 1997 to August 1998 to reach an agreement to continue to develop a relationship with an academic institution as a part of a National Cooperative Drug Discovery Group (NCDDG) programme, but withdrew its application after the terms of the contract regarding non-compliance were deemed by UIC lawyers to be inappropriate for an academic institution.

Decision 391, the Cartagena Agreement (Régimen Común sobre Acceso a los Recursos Géneticos), also known as the Common Access Regime, was published on 17 July 1996 and implemented in 1997 between the Andean Pact countries (Bolivia, Colombia, Ecuador, Peru and Venezuela) (Peru left the pact in 1997). It was designed to regulate access to the Andean genetic resources and control the use of native species. It establishes minimum standards for the equitable distribution of benefits and guarantees the direct participation of communities in agreements. Confusion and concern are still present because there remains the legal issue with respect to intellectual property rights relating to protection by individuals and whether protection of knowledge or of a developed invention can be held by a community rather than a single individual. In addition, while it is the state that holds the rights over indigenous resources, it is the community that holds the rights over traditional knowledge. Indigenous communities are also not guaranteed to be participants *as signatories* to access and benefit-sharing contracts, which they should be if the materials are being accessed from their communities.[45]

Some countries in the Andean Community have not yet approved access and benefit-sharing contracts because they do not have the necessary infrastructure in terms of national regulations. Even when national regulations do exist, the minimum conditions of the Common Access Regime are not necessarily included; this has led to the continuation of poorly constructed agreements that avoid the legitimate rights of indigenous groups.[45] Among the minimum aspects to be included in an access agreement are:

- description of the material to be collected;
- species involved;
- quantities to be collected;
- how the material will be evaluated, used and maintained as collected material;
- distribution of samples (ex-country and local herbaria, local community, *etc.*);
- how local communities will be informed with respect to benefits;

- what the situation is concerning non-compliance by either side with respect to the agreement;
- how reports of the access and the results from the investigation are reported nationally and to the indigenous group;
- impact of the programme on genetic diversity in the area of collection.[45]

Mention should also be made of the 2001 FAO's International Treaty on Plant Genetic Resources for Food and Agriculture (www.planttreaty.org). One of the treaty's goals relates to harmony with the CBD in terms of conservation, sustainable use and equitable sharing of benefits with respect to genetic resources that are used for food and agriculture. The difference is that, in this treaty, there is a multi-lateral approach for access and benefit-sharing of 65 food and forage crops, whose gene pool may be stored in several key locations in the world, for the benefit of humankind.[39] There is no such international plan for the development of a series of medicinal plant genetic resource depositories in various parts of the world.

5 Assessment of Impact

5.1 An Overview and Some Examples

Following the passage of the Manila Declaration in February 1992, I led a discussion on the subject of sovereignty and intellectual property rights regarding genetic materials at the American Society of Pharmacognosy (ASP) meeting held in July 1992 in San Diego. The aim was to bring awareness to the natural product research attendees at the conference to the new climate for natural product research which was changing rapidly at that point. As Putterman indicates,[46] the topic caused "rancorous debate", with many well-established natural product scientists insisting that all this discussion was groundless since the resources were globally owned; the topic was tabled for further discussion. Subsequently, an *Ad Hoc* Committee on Indigenous Materials was established by the ASP representing a number of regions of the world and the group produced some insight on the new environment that was developing which was published in the society's journal, *Journal of Natural Products*, in September 1995.[7] This publication also includes as appendices the texts of the Manila Declaration, the Code of Ethics for Foreign Collectors from the Botany 2000 Herbarium Curation Workshop held in Perth in 1990 and Articles 1–19 of the CBD. Subsequently, a set of guidelines was published by the ASP[47] which clarified for members an anticipated code of professional standards for developing collaborative relationships with source countries.

Gollin[14] has highlighted some examples of situations where the implications of not following the new rules for natural product research were quite serious. There are several instances where organisations in developing countries have challenged patents based on traditional knowledge in the public domain. One case concerned a 1995 patent on the "use of turmeric in wound healing", which was cancelled in 1998 after it was demonstrated that this was a traditional use in

India. A 1986 patent on a supposed new variety of ayahuasca, *Banisteriopsis caapi*, was challenged on the basis that this was not a new variety of the plant, merely an uncultivated one, and that the described uses were part of a traditional heritage. Dias and da Costa[48] cite the example of maca (*Lepidium peruvianum*), which was studied in the USA by PureWorld Botanicals. The active compounds were isolated and characterised, and patented in July 2001. A consortium group in Peru contested the patent in July 2002 using the argument that, without the prior knowledge of its traditional use for erectile dysfunction, the company would not have chosen it for study. A lack of funds hindered further attempts to contest the patent but it did alert the Peruvian government, as well as governments elsewhere, of the need to track patent applications being filed that might be related to traditional knowledge or the use of genetic resources. In 2004, a National Commission for the Protection of Biodiversity was established in Peru whose aim is to develop a database for tracking genetic resources in the country.

Many different groups of stakeholders have expressed reservations with respect to various aspects of the provisions laid out in the CBD. In 1997, ten Kate and Laird[49] indicated some of the industrial concerns with respect to the conditions and terms of the CBD. In doing so, they recognised that many countries that are rich in biological diversity do not have access to the background and extent of the commercial interest in their genetic resources and traditional knowledge. As a result, drafting effective and appropriate legislation is difficult. Industry does not, or is reluctant to, participate in the drafting of legislation that will facilitate access and share benefits and ease the design of equitable partnership agreements. The study by ten Kate and Laird was conducted with about 60 individuals from 35 companies and industry associations who were mostly engaged in research and development, not in legal affairs. They found a lack of awareness of the CBD and EO247 of the Republic of the Philippines[49] and indicated that, if companies engage in natural product research, they acquire samples mostly through third parties (botanic gardens, academic institutions and brokers). It is those parties that are then responsible for dealing with issues of access and benefit-sharing. Companies do not employ botanists as collectors, although staff from manufacturing and marketing/sales still may send genetic resource samples by mail or courier, or collect them while on holiday.

They noted that companies provide both financial and non-financial compensation to local groups in a variety of ways and indicated that there were few joint ventures where research was being conducted by local personnel in a source country. Most agreements merely involved the acquisition of genetic resources and information. Following their study, ten Kate and Laird[49] identified a number of concerns which interviewees from a variety of industries interested in natural products had expressed.

Firstly, there is the overall lack of clarity of how to regulate access and the undue complexity that results when different countries, even in a single region of the world, adopt quite different regulatory measures. They felt that this level of bureaucracy would mire natural product research in inefficiency and delays, and compromise existing research collaborations. In addition, the application processes involve transaction costs, site fees and renewal fees which are sometimes unreasonable. As a result, ten Kate and Laird felt that the CBD *"could negatively impact natural product research, already under pressure from new technologies"*.[49] Earlier in the article the authors discussed the decline in pharmaceutical interest in natural product research generally, and in acquiring genetic resources specifically, in part because of the strategic opportunities at that time afforded by combinatorial chemistry and computer-assisted drug design.

Although many companies involved in natural product research were not aware of the CBD, those that were provided some telling feedback.[49] Comments from pharmaceutical companies included:

"We used to have 30–40 countries where we collected. . . . We will end up working with a couple of countries at a time."
"Basically, you can't work in Australia, Brazil, the Andean Pact countries . . . you can't get permits."
"The CBD influenced our choice to work in the United States."

ten Kate and Laird[49] felt that:

- the resulting effects on corporate practices would be to reduce collection activities;
- there would be a consolidation of corporate collection resources into fewer countries;
- the use of existing culture collections and compound libraries would supplant collection activities;
- there would be more material transfer agreements (MTAs) rather than collaborative research agreements;
- back-up strategies will be needed to minimise the effect on research programmes if access becomes restricted.

The suggestions from corporate interviewees to regulators of access to mitigate some of these effects and promote activities were:

- not to deter potential partners;
- establish simple procedures with minimum requirements;
- allow the parties to reach their own acceptable terms;
- build the capacity to attract beneficial partnerships;
- not to legislate without a clear strategy for what the expectations of outcomes would be.

For the corporations entering into such agreements, ten Kate and Laird[49] offered some important advice including the need to develop the following:

- a statement of principle and a corporate policy on access and benefit-sharing;

- a commitment to compliance from the company;
- tools for implementation of the commitments;
- verifiable indicators of non-compliance for company employees and/or agents;
- measures for enforcement;
- measures for continuous improvement.

They also recommended that, in the executed MTA, there should be a means to track the origin of all genetic resources acquired and ways to monitor compliance of employees and collaborators.[49]

At least one company has developed a corporate policy statement on biodiversity. In 2003, GlaxoSmithKline provided a clear statement of its position on biodiversity and how the policies and procedures in place meet the requirements of the CBD in spirit and action.[50]

The International Union for the Conservancy of Nature (IUCN) has conducted a survey of the impact of the CBD globally, including an analysis of the regulatory situation.[39] Only 41 countries (22%), out of the 188 parties to the CBD who were analysed in the study had, or were in the process of developing, an access and benefit-sharing policy and all of those countries were still modifying their regulations. Within those countries, only 22 projects had been approved in the period between 1991 and July 2004. The authors of the report acknowledged that the access and benefit-sharing policies had been the target of *"misconceptions, politics and negative publicity"* and added that:

"Biopiracy claims, poorly defined ownership rights over genetic resources, the patenting of life, the protection of traditional knowledge and equity issues have thwarted access issues and have also contributed to the cancellation of bioprospecting projects . . . "[39]

However, the authors do not discuss the stunning impact that having only 22 projects approved within those countries that had developed access and benefit-sharing policies has had on natural product research and both internal and external investment in research in those countries. Researchers in many of those countries have simply become "criminals" by illegally accessing the genetic materials themselves and conducting academic research while waiting for the bureaucracies to act.

In its assessment report, the IUCN discussed the activities in all these 41 countries and described the decision-making processes that have led them to move, in some cases successfully, towards the development of national access and benefit-sharing policies.[51] One of the countries of interest to the IUCN was the Republic of the Philippines because, in 1995, it became the first country to enact an access and benefit-sharing policy.

Executive Order No. 247 was based significantly on the premise and the proposed requirements laid out in the Manila Declaration. Subsequently, the Wildlife Resources Conservation and Protection Act of 2001 (RA 9147) changed the definition of bioprospecting to include only those activities that are for commercial purposes. Academic and scientific collection activities were excluded from coverage under EO247. But the scope of the Act does include all *ex situ* collections of genetic resources obtained from the Philippines and studied for commercial purposes (*e.g.* drug discovery). Commercial bioprospectors must pay $3,000 for each Bioprospecting Undertaking (an access permit) and must pay $1,000 per collection site annually. There is a limit of 3 kg of sample(s) that can be collected at any one site. Access to traditional knowledge in the area of the collection site is not implied. The Implementing Rules and Regulations (IRR) published in 2004[43] and approved on 12 January 2005[42] provide details of how to negotiate and execute the Bioprospecting Undertaking. A chapter in the IUCN volume details the history of EO247 and the Wildlife Act.[52]

One of the conclusions from the IUCN survey was that the breadth of the access and benefit-sharing policies that have been developed have impaired their effective implementation.[39] There is significant controversy over the ownership status of genetic materials acquired before the CBD which may or may not be in the country of origin. Without clarification of their legal status, researchers are reluctant to study those materials. Some national access and benefit-sharing laws (*e.g.* those of the Philippines, Costa Rica, Peru and Samoa) have clearly defined criteria for the minimum benefits they expect to receive. Some regulations place very strict limits on the amount of material that can be accessed.

A number of approaches have been used in various countries to protect the traditional knowledge and the scientific discoveries made following the study of that knowledge. These approaches include traditional and *sui generis* intellectual property rights laws and policies, databases of traditional knowledge, prior informed consent requirements, benefit-sharing agreements and certificates of origin. Many developing countries wish to see the Agreement on Trade-Related Aspects of Intellectual Property Rights (TRIPS) (see Section 6) modified to categorically exclude the patenting of plants, animals and microorganisms. The access and benefit-sharing policies should promote the conservation of biodiversity and impose ecological restrictions on bioprospecting; the Costa Rica experience is that bioprospecting has not been a significant source of funds for conservation.

Monitoring the activities of both approved and unapproved access to genetic resources, whether for commercial or non-commercial use, is both expensive and difficult. No country appears to have an effective system in place. The question of the state being involved in establishing access and benefit-sharing agreements is highly controversial with some countries insisting on maintaining control over the process, while detractors say that this results in (sometimes insurmountable) bureaucratic hurdles and high transaction costs. The complexity of the issues involved in developing and implementing access and benefit-sharing regulations indicates that such policies are likely to be somewhat fluid and require revision over time based on experience.[39]

5.2 An Informal Survey

In summer 2008, I conducted an informal survey in several countries of how the CBD had been implemented in that particular country and what the responders (typically friends and former colleagues) considered was its impact in their country. The questions that were asked were as follows:

1. Do you have regulations in place to cover the access to, and collection of, biological materials in your country?
2. In what year were those regulations approved and introduced?
3. Are plants, microbes and animal species all covered? Is any organism excluded?
4. What is the name of the organisation (ministry, department, *etc.*) which is responsible for overseeing aspects of the implementation of the CBD in your country?
5. Does this ministry/department also have the responsibility of reviewing and approving the access to and acquisition of plant, microbial and animal materials in your country? If not, is there another ministry/department that does?
6. When was the bureaucracy established to implement the regulations?
7. Are there protocols and procedures established for the review and approval of research programmes for the development of local resources in an academic or institutional research environment? If possible, can you send copies (or a link) of those materials?
8. How would you describe the effort needed to comply with these protocols and procedures? (*e.g.* simple and straightforward, onerous, non-functional, *etc.*)
9. Is there a distinction made between academic and industrial requests for research and/or commercial access (*i.e.* large scale access) to biological materials? If so, discuss how this works.
10. Who comprises and appoints the group which reviews the protocols (academics, government officials and industrial representatives)?
11. After application for access is made, are the review procedures working efficiently and effectively such that research protocols are reviewed and approved in a timely manner? If not, what are some of the issues?
12. What are the limits to collection in terms of: (i) time frame (how long does an approval last?); (ii) local involvement (is there a requirement for local personnel to accompany a collector?); and (iii) amount (weight) of sample being collected?
13. Is post-event reporting required? (*i.e.* after the collection is it necessary to report on what was collected?) If so, how detailed is this?
14. How are intellectual property issues of ethnomedical and ethnopharmacological information handled?
15. After 16 years, how would you describe the effect of the CBD: (i) on your own research programme; and (ii) on the investment in research development in your country's biological resources?

16. If you have any other comments that you would like to make regarding the impact of the CBD in your country, please feel free to do so.

Responses were obtained from ten countries; the names of the responders and their countries are given in the Acknowledgements section. A summary of the responses from the various countries are presented below in alphabetical order.

5.3 Survey Results

Indonesia

The national regulations and authorisation for overseas researchers and institutes to have access to and collect plant materials (and other biological materials, including microbes) are administered by the Ministry of Forestry. The Ministry of the Environment is the CBD national focal point. A national act (Undang-undang Perlindungan Sumber Daya Genetik; The National Act for the Protection of Genetic Resources) is still under development. The Ministry of Forestry works with the Ministry of Research and Technology, several other ministries, the Defence Department and the National Police in reviewing and approving access to biological materials. The protocols and procedures for application were described as *"pretty straightforward"*.

There are tougher requirements established for commercial entities applying for access than for academic institutions. A government group, with occasional input from research institutes and academia, reviews the applications for access. The protocols for the research are reviewed at the same time as the application for access is reviewed and this is described as an effective process. The Government has established a maximum time for researchers to stay in the field.

Accessibility to remote areas is becoming more difficult. Post-access event reporting is required and minimally comprises taxonomic name, locality, amount and number of samples collected, sample destination and the preservation method. The level of detail is also determined by the local collaborator. Intellectual property issues are typically handled by the overseas parties in collaboration with the local partners.

It was felt that the CBD had accelerated local efforts to conduct taxonomic inventories. The impact on research breadth and depth and collaborative investment was not clear.

Japan

There are regulations in place to cover the collection of microbes in Japan, but not the collection of higher plants and animals. Since 1968, efforts have been underway to conserve microbial samples and deposit them for patent application in the International Patent Organism Depository (IPOD)[53] within the Patent Office. Protocols and their review are conducted by IPOD and forms are available online. No distinction is made between academic and industrial requests for access to biological resources. Academics are involved in the application for access review process; government representatives are non-official.

There is no institution in Japan that oversees access to ethnomedical information and the associated plants, or is responsible for collecting that information.

Jordan

In 2002, the Government of Jordan, through its Ministry of Environment, introduced Agricultural Law No. 44 which covers access and collection of biological materials. The Ministry of the Agriculture is responsible for implementation and enforcement. The procedures appear to work well in agencies such as the Royal Society for the Conservancy of Nature (RSCN), but are more problematic in some governmental agencies. A distinction is made between academic and industrial requests for access and for protected areas in the country. The review group is comprised of personnel from the Ministry of Agriculture. The scope of the collection programme is specified by the review committee. If the collection is in a reserve, then personnel from RSCN accompany the collectors. RSCN requests feedback on the collection and the work performed on the specimens.

There appear to be no limits or restrictions with respect to the gathering of ethnomedical information on medicinal plants.

Korea

There have been three sets of regulations developed with respect to the access to and conservation of genetic resources in Korea. The Law of the National Parks was passed in 1980, followed by the Law of the Conservation of Endangered Species in 1989 and the Law of the Conservation of the Natural Environment in 1991. The main organisation responsible is the Ministry of the Environment and bureaucracy to review protocols for access was established in 2001. The application documentation is only in English and the procedures were described as being *"simple and straightforward, but sometimes onerous"*. There is no distinction made between academic or corporate requests for access, although mass collection (not defined) is prohibited or strictly controlled. The group reviewing the protocols is comprised mainly of academics with some government officials (details not given). The review procedures appear to be working. There are no specific limits with respect to timeframe, local involvement or amount of sample. There is a requirement, which is not enforced, to report on what was collected. In 2004, in line with the international protection movement, the Korean Intellectual Property Office (KIPO) stimulated the development of a database of Korean traditional medicine, which was compiled between 2005 and 2007 and was operational as a searchable resource in December 2007.

Madagascar

The Government of Madagascar has put in place a Comité Ad Hoc Faune et Flore–Comité Orientation pour la Recherche Environnementale (*Ad hoc* Flora/Fauna Committee/Orientation Committee for Environmental Research; CAFF–CORE), which is also responsible for granting research permits and approving overseas research in Madagascar. In response to the Saint Catherine Convention, from which some elements of the CBD were inspired, a Tripartite Commission was established between the three relevant Ministries—Higher Education, Research, and Water and Forest, Environment and Tourism (MEEFT)—and the regulations were introduced in 1987. Plant and animal species are covered in the regulations, but not microbes. MEEFT (particularly the Directorate General of Environment, Water and Forest; DGEEF) is responsible for overseeing the review of protocols and approving access.

All research proposals must be submitted to CAFF–CORE (in MEEFT), the Association Nationale pour la Gestion des Aires Protégées (ANGAP), the Centre National de Recherche sur l'Environnement (CNRE) and the Departments of Fauna and Flora Biology at Antananarivo University, Parc Botanique et Zoologique de Tsimbazaza (PBZT) for approval. After the research has been carried out, reports must be submitted to MEEFT. The research must be conducted in collaboration with a national public institution and a research proposal has to be submitted to the Ministry no later than one and a half months before the start of the planned field work. A monthly meeting is held to decide on research permit applications. All permits are issued by the DGEEF, except for work at study sites located in a protected area, in which case an approval is required from ANGAP.

The process was described as being relatively simple and functional. Requests for access are submitted by local collaborators and, although they took a significant amount of time initially, renewals and extensions were much easier to obtain. Without in-country assistance, the requirements were described as very onerous for an outside investigator, but not insurmountable.

A distinction is made between academic and industrial requests for access. Commercial requests must be submitted to MEEFT at the CITES (Convention on International Trade in Endangered Species of Wild Fauna and Flora) department and possibly also to the Ministry of Trade. The procedures are quite simple for research purposes (small quantities of biological materials). But for commercial/industrial activities and CITES species, there are specific regulations such as the need to provide proof of the existence of a nursery from which plants have been collected; there is also a quota.

The review process is delegated by MEEFT to ONE (Office National pour l'Environnement). CAFF–CORE includes representatives from ANGAP, DGEEF and the Ministry of Higher Education (represented by the Tsimbazaza Botanical and Zoological Park; PBZT). The process has, thus far, worked reasonably smoothly and effectively.

Once obtained, the collection permit is typically valid for six months. Local personnel must be involved to accompany the collector and the amount of sample collected is limited depending on the type of species and the samples. Five voucher specimens are required for plants and three specimens for animals.

Two kinds of reports are required: a preliminary report immediately after the collecting expedition (including names of collectors, locations, list of plants collected to the species level if possible); and a final report when the research is completed (with the results of the research). Publications based on the research are also requested. Ethnomedical and ethnopharmacological studies exploiting traditional knowledge are not allowed.

This has been a topical issue in Madagascar recently (data exchange, benefit sharing, use of natural resources, information technology and communications, *etc.*), but to date, no regulations have been adopted.

Overall, the experience following the CBD has been positive because biodiversity has been protected through the improvement of scientific knowledge. It also provided the impetus for an ICBG programme to be developed and the research has led to the identification of protected areas.

Pakistan

No act has yet been passed in Pakistan, though one related to access and benefit-sharing is in the process of approval and would cover all organisms. The Directorate of the Ministry of the Environment is responsible for overseeing implementation of the CBD in Pakistan. Some research institutions have a mechanism for the review of access protocols and collection programmes, *e.g.* the Research Review Committee of the National Agricultural Research Council.

The situation with respect to the mandates of the CBD is much better than ten years ago due to extensive efforts to implement the CBD. However, there is a continuing need to build capacity in the academic and research institutions to understand the CBD. The access to ethnomedical and ethnopharmacological information is covered by the Pakistan Intellectual Property Act which recognised the need to enhance education for researchers about the issues in this area.

Peru

There is a law in Peru that protects traditional knowledge and establishes the requirement for prior informed consent and anticipates the realisation of access contracts.

The Peruvian regime on access to genetic resources is regulated by Decision 391 on a *Common Regime for the Access to Genetic Resources* dated 2 July 1996. This standard applies directly and has the normative rank of a law in the countries that belong to the Andean Community—namely Bolivia, Colombia, Ecuador and Peru. The standard has a special relevance for being the first one adopted in this subject worldwide.

It was determined that Decision 391 needed to be developed by a Supreme Decree at the national level to determine the competences and responsibilities of local authorities. To date, this Supreme Decree has not been approved, so no general system on access to biological systems has been put into practice.

The only country of the Andean Community that has developed complementary provisions has been Bolivia. The only provisions that are in place in Peru were approved in relation to access to wild genetic resources and have been included in forestry legislation. All species are included in the regulation. The Ministry of Environment (created in 2008) will be the focal point for the CBD on access to genetic resources. The Instituto Nacional de Recursos Naturales (INRENA), an adjunct to the Ministry of Agriculture and created by decree number 25902, 27 November 1992, is still dealing with the administrative aspects of access to wild diversity. The Vice-Ministry of Fisheries is responsible for access to marine resources. These designations were made in 2001.

Authorisation by INRENA to initiate studies requires the completion of an application form and the foreign party needs a Peruvian contact. *Curriculum vitae* on the researchers are requested, as well as the project describing the research to be undertaken. The authorisation is for one year and, depending on the results attained and presented in a report, may be extended for another year. To export the material, an application form and related documents must be completed.

It was felt that the system was essentially non-functional because there is no infrastructure in the country to control unauthorised access and exportation. The regulations indicate that, if the genetic material is exported, the researchers have no intellectual property rights unless negotiated in an access contract.

The government provides all the personnel for the review process for submitted research access requests; these are typically staff from INRENA. With respect to the review process, it depends on the access requested. If access to traditional knowledge is involved then it becomes a burdensome process. This is because prior negotiations with the communities involved are needed to generate a written permission for access; this document is required by INRENA in order to review the application. If the papers conform perfectly with what is requested, the review process can last one month; if not, it can take up to three months. Obtaining exportation permits has a similar timeframe.

Local involvement in the collection is required, with 50% of the collected material staying in-country. Only if the species are CITES listed are the requirements more stringent.

A final report is requested by INRENA, which is completed by the researcher and the local collaborator.

Philippines

The actions taken in the Republic of the Philippines, as discussed in Section 4, have been the subject of considerable interest as it was a pioneer in developing regulations and protocols for access to indigenous resources and knowledge.

Through the Protected Areas and Wildlife Bureau (PAWB) in the Department of Environment and Natural Resources (DENR), the Inter-Agency Committee on Biological and Genetic Resources (IACBGR) (under EO247, Department Administrative Order 96-20) is responsible for reviewing and approving the access to and acquisition of all plant, insects, microbial and animal materials in the country. There are no excluded organisms.

The Technical Secretariat of IACBGR conducts the initial review and evaluation of the application and documents. The evaluation results, including the draft Research Agreement, are then submitted to IACBGR for final evaluation within 30 days from receipt of all requirements from the principal investigator/collector. IACBGR conducts the final evaluation and submits its recommendation to the agency concerned after receipt of the documents from the Technical Secretariat. The Secretary of the Agency concerned approves the Research Agreement if it is favourably recommended by IACBGR.

In practice, compliance with the published protocols and procedures is not simple and straightforward. Prior informed consent (PIC) is required for both an Academic Research Agreement (ARA) and a Commercial Research Agreement (CRA). However, in the case of the ARA, the application may be processed and the ARA may be executed without the PIC certificate, provided the certificate is acquired by the principal investigator/collector prior to the actual bioprospecting activity.

Securing PIC involves public notification and sector consultation, *e.g.* with the recognised head of the local indigenous peoples (IP) group in cases where the prospecting of biological and genetic resources will be undertaken within their ancestral domains/lands or with a municipal leader or mayor of the local community (LC). Alternatively, the Protected Area Management Board (PAMB) will issue the PIC Certificate upon compliance with all the required documents and activities when the prospecting of biological and genetic resources will be undertaken within a protected area; a private land owner will issue the PIC certificate when the prospecting of biological and genetic resources will be undertaken on private land. Issuance of the PIC certificate takes a minimum of 60 days from the time of submission of the proposal to the IP, LC, PAMB or the private land owner concerned.

Prospecting for biological and genetic resources undertaken by a person, entity or corporation (foreign or domestic) is allowed only if they enter into a Research Agreement with the Philippine government as represented, depending on the nature and character of the prospecting activity, by DENR, the Department of Health (DOH), the Department of Agriculture (DA) and the Department of Science and Technology (DOST). For purposes of EO247, traditional uses of biological resources by indigenous and local communities do not require a Research Agreement.

A CRA is required for research and collection of biological and genetic resources intended, directly or indirectly, for commercial purposes. Private persons and corporations, including all agreements with foreign and international entities, must conform with the CRA's minimum requirements.

On the other hand, if the prospecting of biological and genetic resources is intended primarily for academic purposes, an ARA should be secured. Only recognised Philippine universities and academic institutions, domestic governmental entities and intergovernmental entities can apply for an ARA.

The application should include a research proposal stating the purpose, source of funds, the duration of the programme, a list of biological and genetic materials and the amount to be acquired. A copy of the proposal must be submitted to the recognised head of the local or indigenous cultural community or communities that could be affected in order to secure the prior informed consent (PIC) certificate.

Some of the minimum terms of the CRA and the ARA are as follows (EO247):

a. There must be a limit on samples that the Commercial/Academic Collector may obtain and export. The approved list and amount of the samples taken from the area must be followed strictly.
b. A complete set of all of the specimens collected shall be deposited by the Commercial/Academic Collector with the National Museum or a duly designated governmental entity. The holotypes designated by the author must be maintained at the National Museum.
c. Access to collected specimens and relevant data shall be allowed to all Filipino citizens and the Philippine governmental entities whenever these specimens are deposited in depositories abroad.
d. The Commercial/Academic Collector, or in appropriate cases, its principal investigator, must inform the Philippine Government, as well as the affected local and indigenous cultural communities of all discoveries from the activity conducted in the Philippines, if a commercial product is derived from such activity.
e. The agreement shall include a provision for the payment of royalties to the National Government, the local or indigenous cultural community and the individual person or designated beneficiary in the case that a commercial use is derived from the biological and genetic resources taken. Where appropriate and applicable, other forms of compensation may be negotiated.
f. There shall be a provision allowing the Philippine Government to unilaterally terminate the agreement whenever the Commercial/Academic Collector has violated any of its terms. The Agreement may also be revoked on the basis of public interest and welfare.
g. A status report of the research and the ecological state of the area and/or species concerned shall be submitted to the Inter-Agency Committee regularly as agreed upon.
h. If the Commercial Collector or its Principal is a foreign person or entity, it must be stipulated that scientists who are citizens of the Philippines must be actively involved in the research and the collection process and, where applicable and appropriate as determined by the Inter-Agency

Committee, in the technological development of a product derived from the biological and/or genetic resources taken from any area in the Philippines. This involvement shall be at the cost of the Commercial Collector.

i. The Commercial Collector and/or its Principal shall be encouraged to avail themselves of the services of Philippine universities and academic institutions. Where applicable and appropriate, the Commercial Collector and/or its Principal shall be required to transfer equipment to a Philippine institution or entity.
j. A fixed fee must be paid to DENR in accordance with a schedule of fees formulated by the Inter-Agency Committee.
k. The maximum term for a CRA shall be for three years and renewable upon review by the Inter-Agency Committee.
l. In case of endemic species, there must be a statement that the technology must be made available to a designated Philippine institution and can be used commercially and locally without paying royalty to a Collector or Principal. Provided, however, that where appropriate and applicable, other agreements may be negotiated.

In addition, the following terms are considered in an ARA:

m. The ARA may be comprehensive in scope and cover as many areas as may be projected. It may stipulate that all scientists and researchers affiliated with a duly recognised university, academic institution, governmental and intergovernmental entity need not apply for a different Research Agreement, but may conduct research and collection activities in accordance with an existing ARA. In such cases, the university, academic institution and governmental entity shall ensure that all terms and conditions of the government are complied with by the affiliated scientist or researcher. In all cases, the university institution or governmental entity must ensure that affected communities have given their prior informed consent to the activities to be undertaken.
n. There must be a provision requiring the Academic Collector to apply for a commercial research agreement when it becomes clear that the research and collection being done has commercial prospects.
o. A minimal fee must be paid to the Philippine government in accordance with a schedule of fees by the Inter-Agency Committee.
p. The maximum term for an Academic Research Agreement shall be for five years and renewable upon review by the Inter-Agency Committee.

The inter-agency committee (IACGBR) is composed of:

- an undersecretary of the Department of Environment and Natural Resources designated by the DENR Secretary who shall be the chairperson of the Committee;
- an undersecretary of the Department of Science and Technology (DOST) designated by the DOST Secretary who shall be co-chairperson of the Committee;

- a permanent representative of the Secretary of the Department of Agriculture, who must be knowledgeable about biodiversity or biotechnology;
- two permanent representatives of the Philippine science community from academe who must be experts in any of the following fields: biodiversity, biotechnology, genetics, natural products chemistry or similar disciplines;
- a permanent representative of the Secretary of the Department of Health who must be knowledgeable about pharmaceutical research and development, with emphasis on medicinal plant/herbal pharmaceuticals;
- a permanent representative of the Department of Foreign Affairs who has to facilitate any international linkages relative to bioprospecting;
- a permanent representative of the National Museum who has expertise on natural history and/or biological diversity;
- a representative from a non-governmental organisation (NGO) active in biodiversity protection;
- a representative from a people's organisation (PO) with membership consisting of indigenous cultural communities/indigenous peoples and/or their organisations.

In considering the effectiveness of the system since 2001, the consensus is that the review procedures are not working efficiently and effectively. First, the IACBGR meets only once every quarter. Although a special meeting could be called, generally it is not well attended, which means there is rarely a quorum. It appears that there is a limited fund for the implementation of EO247 and only limited manpower. Also, rapid turnover of the DENR representative at the IACBGR has caused delays in the review process.

The local PIC procedures require at least two visits to the field, a minimum consultation period of 60 days and the provision of animals for rituals, especially when the site of collection is within indigenous territories. The 60-day consultation period to secure the PIC certificate means that obtaining a collection permit takes at least five months, which delays the IACBGR approval process even further.

The Intellectual Property Code of the Philippines (Republic Act No. 8293) does not cover patentable ethnomedical and ethnopharmacological information. It does cover intellectual property protection on derived products or processes and does not preclude the Congress from considering enactment of a law providing *sui generis* protection of plant varieties and animal breeds and a system of community intellectual rights protection. The Philippine Government sees EO247 (and RA9147) as measures to protect the rights of the indigenous cultural communities and other Philippine communities with respect to their traditional knowledge and practices when this information is directly and indirectly put to commercial use.

South Africa

South Africa has a new set of regulations which came into operation on 1 April 2008 to implement the requirements of Chapter 6 of the National

Environmental Management Biodiversity Act No. 10 of 2004. Chapter 6 of the Act deals with the regulation of bioprospecting, access and benefit-sharing, *i.e.* to regulate bioprospecting involving indigenous biological resources; to regulate the export from the Republic of indigenous biological resources for the purpose of bioprospecting or any other kind of research; and to provide for fair and equitable sharing by stakeholders in benefits arising from bioprospecting involving indigenous biological resources.

Apart from this legislation, most of South Africa's nine provinces have provincial ordinances which require researchers to apply for permits to collect indigenous plants in their respective provinces.

Chapter 3 of the Act deals with threatened or protected ecosystems and species and the regulations to implement this chapter came into effect on 1 February 2008. The purpose of these regulations is to:

- further regulate the permit system insofar as the system applies to restricted activities involving specimens of listed threatened or protected species;
- provide for the registration of nurseries, scientific institutions, captive breeding operations, game farms, sanctuaries, rehabilitation facilities and wildlife traders amongst others;
- provide for the regulation of the carrying out of a restricted activity, *e.g.* hunting;
- provide for the prohibition of specific restricted activities involving specific listed threatened or protected species;
- provide for the protection of wild populations of listed threatened species.

The definition of "indigenous biological resources" in the Act is very broad and includes any living or dead animal, plant or other organism of an indigenous species and any derivative of such animal, plant or other organism or any genetic material of such animal, plant or other organism. In Chapter 6 the definition covers:

- any "indigenous biological resource" as defined previously whether gathered from the wild, or accessed from any other source (including animals, plants or other organisms), or an indigenous species cultivated, bred or kept in captivity, or cultivated or altered in any way by means of biotechnology;
- any cultivar, variety, strain, derivative, hybrid or fertile version of any indigenous species or of any animals, plants or other organisms and exotic animals, plants or other organisms whether gathered from the wild or accessed from any other source which through the use of biotechnology have been altered with any genetic material or chemical compound found in any indigenous species or any animals, plants or other organisms.

Excluded is genetic material of human origin, any exotic animals, plants or other organisms provided it has not been biotechnologically altered with genetic material from indigenous species and indigenous biological resources listed in terms of the International Treaty on Plant Genetic Resources for Food and Agriculture.

Management of the environment is a concurrent function in terms of the Constitution of South Africa which means that, although the national government (*i.e.* the Department of Environmental Affairs and Tourism) develops the framework legislation for environmental affairs and represents South Africa at international negotiations on CBD matters, much of the implementation is done at the provincial level by provincial conservation authorities or Provincial Departments of Environmental Affairs which develop their own provincial ordinances to meet the requirements of the national framework legislation.

The national focal point for CBD matters is housed in the Department of Environmental Affairs and Tourism. This department and the provincial conservation authorities, such as the Provincial Departments of Environmental Affairs, deal with all matters that may impact on indigenous biological resources, including overseeing the introduction of alien and potentially invasive species and monitoring the impact of genetically modified biological resources. This cross-sectoral function requires liaison with other government departments such as the National Department of Agriculture and the Department of Water Affairs and Forestry. Importation and exportation of biological resources is also required to comply with phytosanitary requirements controlled by the National Department of Agriculture. Through the National Department of Agriculture, 50 species were declared as "weeds" or "invader plants" in regulations passed in 1984 under the Conservation of Agricultural Resources Act No. 43 of 1983. An amendment to these regulations on alien and invasive species was published in March 2001, but has not yet been approved and has proved controversial. Once approved, it will also take time to put in place the processes to implement the revised regulations.

Under the new biodiversity regulations, existing bioprospecting projects in South Africa were given until 1 September 2008 to submit an application for a bioprospecting permit. Applications for bioprospecting permits and export permits for research other than bioprospecting require that the applicant provide details of:

- the objectives of the bioprospecting project;
- the desired outcomes and benefits that may result;
- the proposed methodology;
- the proposed timeframes (*i.e.* required period of validity of permit);
- any relevant environmental considerations including impacts of the collection of the indigenous biological resources and proposed steps to minimise or remedy those impacts;
- the reporting processes;
- the disposal of the genetic material at the conclusion of the study.

In the case of the export of indigenous biological resources for research other than bioprospecting, the applicant is also required to state the purpose for which the indigenous biological resources are to be exported and to give an indication as to whether the intended research will have some benefit for the conservation of biodiversity in South Africa, the economic development of South Africa and any other matter that is in the public interest. In most cases there is provision for at least annual reporting on progress by the applicants, although the permitting authorities have the power to request more frequent reporting should they so desire.

As a result of some fundamental problems in the legislation itself, the regulations are rather complex and not all that easy to implement. The process of bioprospecting has been divided into two phases—the discovery phase and the commercialisation phase. Although a definition of commercialisation has been provided, there is no clear indication of triggers that would indicate that a research project has become a bioprospecting project and the research being carried out will be regarded as the discovery phase of bioprospecting as opposed to basic research.

The national Department of Environmental Affairs and Tourism has been identified as the competent authority to issue bioprospecting permits, while export permits (*i.e.* for the export of material for research other than bioprospecting) will be issued by the provincial authorities in the province in which the biological resources are collected.

Before a bioprospecting permit will be granted, the applicant is required to obtain the prior consent of any person, including an organ of the state or community, providing access to the indigenous resources to which the application related and to enter into material transfer and benefit-sharing agreements with these stakeholders. The agreements, which have to be in a format laid out by the regulations, must be approved by the Minister of Environmental Affairs before the permit application will be approved. Furthermore, the Act requires that if the applicant has used indigenous knowledge related to the traditional use of the biological resources and that knowledge has initiated or will form part of the bioprospecting or research project to which the permit application refers, then the applicant will be required to enter into benefit-sharing agreements, also approved by the Minister, with the originators of the indigenous knowledge or use.

It has proved very difficult or impossible to identify the potential benefits before the biological resources have been collected and before research has been conducted on the material. There are also high expectations amongst stakeholders, who generally have a limited awareness of the time and cost requirements for research before a product can be put on the market and generate the expected financial rewards.

All the funds generated by bioprospecting are required to be deposited in a National Bioprospecting Trust Fund administered by the Director General of the Department of Environmental Affairs and Tourism. The funds accruing to each stakeholder will be paid out from the National Trust Fund.

A further difficulty is that a great deal of the indigenous knowledge is already in the public domain and it is not always easy or practical to identify the original source of that knowledge and, therefore, the legitimate beneficiaries.

It is felt that monitoring and evaluating the implementation of the processes outlined above could prove very challenging especially in the light of the limited capacity existing within the national and provincial departments for these functions.

The Act and the regulations do not make a clear distinction between academic and corporate requests for access to genetic resources. This is especially because the discovery phase of the bioprospecting project is defined in the regulations as any research on, or development or application of, indigenous biological resources where the nature and extent of any actual or potential commercial or industrial exploitation in relation to the project is not sufficiently clearly known to begin the process of commercialisation. The export permit relates to export for any research other than bioprospecting, while the integrated export/bioprospecting permit will cover export of material for that purpose. However, in a recent proposed amendment of the Act, which has still to be promulgated, the Minister has the authority to declare that this Chapter does not apply to certain categories of research involving indigenous biological resources or commercial exploitation of indigenous biological resources (*e.g.* the cut flower industry).

The implementation plan prepared by the Department of Environmental Affairs and Tourism proposed that an interim expert group will be appointed by the Minister of the National Department of Environmental Affairs and Tourism to assess project proposals and bioprospecting applications and review requests for access. It is too early to make any comment on whether the review process is working because the procedures are not yet fully operational. There is no indication in the regulations as to how long the assessment of the application may take, especially as the applicant may be required to conduct a risk assessment.

Bioprospecting and export permits will be issued only when collaborative initiatives are set up between foreigners and South African collaborators or institutions. The permit application requires information regarding the name of the indigenous resource and the part of the organism to be collected, as well as an indication of the quantity required and full details of the locality where the collection will take place. The timeframe of the permit is not fixed in the regulations, but will be stipulated by each issuing authority. Both the bioprospecting and export permits require the applicant to submit at least annual reports, although in the case of the export permits, the permitting authority can set the dates for the submission of these reports which may be more frequent than on an annual basis.

There is no guidance within the legislation regarding intellectual property rights apart from the definition of commercialisation, which includes:

- the filing of any complete intellectual property application whether in South Africa or elsewhere;
- obtaining or transferring any intellectual property rights or other rights;
- commencing clinical trials and product development, including the conducting of market research and seeking pre-market approval for the sale of resulting products;

- the multiplication of indigenous biological resources through cultivation, propagation, cloning or other means to develop and produce products such as drugs, industrial enzymes, food flavours, fragrances, cosmetics, emulsifiers, oleoresins, colours and extracts.

It is assumed in the regulations that these issues will be dealt with within the benefit-sharing agreements between the applicants and the stakeholders who supply the indigenous biological resources and/or the indigenous knowledge associated with the indigenous biological resources.

The Patents Amendment Act No. 20 of 2005 requires that every applicant lodging an application for a patent needs to lodge with the registrar a statement on whether or not the invention for which the protection is claimed is based on, or derived from, an indigenous biological resource, genetic resource, or traditional knowledge or use. The registrar can call on the applicant to provide proof as to their authority (*i.e.* prior informed consent) to make use of the indigenous biological resource or of the traditional knowledge if the invention for which protection is claimed is based on, or derived from, an indigenous biological resource, genetic resource, or traditional knowledge or use.

With the establishment of the Innovation Fund under the National Research Foundation (by the National Department of Science and Technology), funding in bioprospecting initiatives involving indigenous resources increased. However, that was prior to the implementation of the new legislation and it remains to be seen what the impact of the new legislation on these bioprospecting initiatives will be in the future. International companies are presently reluctant to become involved in new initiatives because of the restrictive and somewhat complicated measures that have been put in place, which enhance their uncertainty.

South Africa has, indeed, seen a number of indigenous biological resources developed and exploited in other countries without receiving any benefit from the use of these resources, which in some cases such as horticulture, have been quite considerable. This took place in a legal vacuum before there was any legislation in place to control access and implement some form of benefit-sharing.

Tanzania

Tanzania's regulations that cover the access to, and collection of, biological resources are under constant review. The general policies concerning biodiversity protection fall under three categories.

There is a policy which deals with the exploration and export of floristic resources of potential medicinal value. Depending on the type of materials required, the policy is regulated by various ministries and departments—in particular the Ministry of Agriculture (Forestry), the Ministry of Natural Resources and Tourism, the Ministry of Trade and Industries, and the

Ministries of Health and Higher Education. There is a policy on fauna or animal substances that deals with the conservation of wildlife and animal substances and sets rules and regulations on the sale, transfer, import and export of wildlife and animal parts. There is also a regulation on marine biological resources that deals with protection, conservation, development, regulation and control of fish, fish products, aquatic flora and products.

The forest and fauna ordinances were introduced in 1958 and the National Agricultural Products Law in 1969. The regulations cover all species except microbes. Depending on the type of organism being investigated, there are at least five Ministries responsible including the Ministry of Education through the Commission for Science and Technology (COSTECH), the Forestry Section at the Ministry of Agriculture, the Ministry of Health for studies of traditional medicine and medicinal plants, and the Ministry of Natural Resources and Tourism for studies in the national parks. The Ministry of Trade and Industry is also involved through its Business Registration and Licensing Agency for issues relating to intellectual property rights.

COSTECH is responsible for the review of most protocols for accessing natural resources. The application forms for access are simple and straightforward. Thus, for a foreign collaborator and a Tanzanian institution, the two parties must complete an Application for Research Clearance before they are allowed to continue and this is attached to the research proposal. There are then fees to be paid and a materials transfer agreement to be established. Academic research requests are processed through COSTECH, but commercial requests, depending on the aim, go through other ministries.

The review committees are typically composed of persons within the institution (COSTECH or a ministry) and can bring in additional local experts as needed. Normally, review procedures work effectively and efficiently, although delays are experienced when there are insufficient members present or there is missing information.

Approval for access lasts one year. It is renewable depending on a satisfactory progress report. Foreign collectors are required to be accompanied by local personnel. Researchers are required to give reports at the end of the project on species collected and location. There is a limit of 500 g for the collection of individual plant samples.

Intellectual property issues have not been dealt with completely and there are no regulations in place. Efforts are being made by the Ministry of Health to register traditional health practitioners and create awareness of the intellectual property issues.

Only if the research proposal indicates that the programme will interview indigenous people does COSTECH request a memorandum of understanding (MOU) between the researchers and the indigenous group.

The experience in Tanzania is that the implementation of the CBD is extremely difficult. There is still rampant biopiracy, particularly involving the indigenous people who are poor and with a low level of education. There have been relatively few opportunities for capacity building with the assistance of foreign collaborators.

5.4 Survey Overview

The brief summaries of the experiences of a number of investigators in various, quite different, countries around the world and in very different economic and social environments, highlight some of the complex issues generated in various countries as they seek to implement the CBD and the impact these issues have on local research communities.

There appear to be no clear instances where the spirit of the CBD (facilitation of access, equity sharing and conservation) has been successfully implemented. Countries are not in any way rushing to enact laws and establish systems that could implement the new laws in a reasonable and effective manner. Those countries which have enacted laws have created systems that are complex, difficult to comprehend and frequently not easy—even for local scientists—to receive approval for their research programmes.

6 The TRIPS Agreement and the CBD

The Agreement on Trade-Related Aspects of Intellectual Property Rights (TRIPS) was negotiated at the Uruguay Round of the General Agreement on Tariffs and Trade (GATT) in 1994 and is administered by the World Trade Organization (WTO). It constitutes the most comprehensive treaty related to intellectual property law and international trade, and calls for nations to meet standards for copyright rights, industrial designs, patents, new plant varieties and trademarks. At the Doha meeting in 2001, clarifying the scope of the TRIPS Agreement, it was concluded that it should be interpreted *"to promote access to medicines for all"*. The notion to link trade policy and intellectual property standards was pushed by the US pharmaceutical industry[54] and subsequently supported by the European Union and Japan. Ratification of TRIPS is a requirement for membership of the WTO and, consequently, most developing countries are required to rewrite or write laws related to intellectual property. In addition, TRIPS provides for enforcement mechanisms.

TRIPS continues to be controversial because of its strict requirements relating to patents and copyrights, and because of its relationship to the Convention on Biological Diversity with which it seems in certain areas to be in opposition—a point discussed below. One aspect of the conflict though has been the controversy over the provision of AIDS drugs to deal with the pandemic in Africa, which resulted in higher healthcare costs. While the Doha Declaration related to how states deal with public health crises, the Pharmaceutical Manufacturer's Association in the USA and several other groups in developed countries, concerned about the impact of significantly lower cost generic drugs in their market, worked to ameliorate the effects of Doha. In 2003, the Bush administration eventually acknowledged that generic drugs could be included in a drug strategy for developing countries.

The relationship between TRIPS and the CBD is a very complex one.[55,56] As pointed out at a CBD/TRIPS workshop held in the Republic of the Philippines in 1999,[56] the underlying premise of TRIPS is that a properly functioning system for the protection of intellectual property rights will provide an appropriate framework that will encourage investment in the development and transfer of technology. Three areas were identified at the workshop as being important in examining the relationship between the CBD and TRIPS:

- the promotion of environmentally sound technologies and the promotion of access to, and transfer of, those technologies;
- the provision of incentives for conservation and sustainable use of genetic resources;
- how environmentally harmful technology is handled.

Only the first two issues are discussed here.

As noted in Section 3, Article 16 of the CBD indicates that access to, and transfer of, technology are essential elements of the objectives of the Convention. In Article 16.5, it is recognised that patents and other intellectual property rights may have an impact on implementation of the CBD and encouraged parties to *"ensure that such rights are supportive of and do not run counter to its [the CBD] objectives"* with the caveat *"subject to national legislation and international law"*.

This implies that cooperative agreements are subject to the TRIPS Agreement, but does not indicate which would take precedence. Thus, if a company had a patent on a technology which would benefit a particular collaborator and meet the objectives of TRIPS, would the patent rights be compromised because of a compulsory licensing requirement? The TRIPS Agreement attempts to balance the objectives of promoting technological innovation and facilitating access to and transfer of technology through the standards of intellectual property protection. It leaves (Articles 1 and 9) governments free to adopt their own standards for protection of intellectual property related to development.

The relationship of the TRIPS Agreement to sustainable use and biodiversity and relevance to the CBD are also of interest. For example, Article 27.3(b) of the TRIPS Agreement provides:

"Members may also exclude from patentability: plants and animals other than microorganisms . . . However, Members shall provide for the protection of plant varieties either by patents or by an effective sui generis *system . . . "*

This implies the development of a unique intellectual property rights system for plant varieties to exclude a non-rights-holder from use and to obtain remuneration from non-rights holders for licensed use. It has been suggested that the plant variety protection, as prescribed by the International Union for the Protection of New Varieties of Plants (UPOV), may provide the best *sui generis* system.[56] UPOV has now been signed off by 68 countries and the Secretariat of the Convention supports the view that UPOV and the CBD *"should be mutually supportive"*.[57]

Although Article 8(j) of the CBD deals with the respect, preservation and maintenance of "*knowledge, innovations and practices of indigenous and local communities*" and encourages "*the equitable sharing of the benefits arising from the utilisation of such knowledge, innovation and practices*", the TRIPS Agreement neither prevents nor promotes any new form of protection for indigenous or local communities. Thus, measures of control over such knowledge which might not be innovative under existing patent law remain to be developed at the discretion of a WTO member government. Intellectual property rights in this area are the subject of various opinions[55,56,58,59] and focus on the narrow approach of assigning such rights as the creation of an original product, the need to identify an individual creator (as opposed to a community group) and the limited duration of such rights. Therefore, can the intellectual property rights system be modified to proved incentives for alternative ways of adding value to a genetic resource and thus promoting conservation?

More formalised approaches to providing access, as required by the CBD, may be a pathway to the enhanced recognition and compensation of the contributions of knowledge from indigenous groups; however, they will not provide longer term protection of rights and an associated remuneration system. The existing intellectual property rights do not protect traditional knowledge against unauthorised commercial use. Some aspects of use may be included in contractual arrangements as confidential proprietary information. However, based on the strong positive relationship between ethnomedical practices and prescription drugs,[60] although traditional knowledge may not be patentable, it can certainly be a substantial asset and investment opportunity for companies to obtain patents based on innovations derived from that knowledge. The TRIPS Agreement does provide for the protection of undisclosed information, but the relationship of traditional knowledge to this avenue for protection is unclear.[56] Countries can develop their own intellectual property rights laws as a WTO member and are then bound to treat their nationals, as well as similar inventions by foreign nationals, in a like manner, which may not be to their advantage.

What are the areas of synergy and conflict between the CBD and the TRIPS Agreement? The two agreements approach intellectual property rights from quite different perspectives, but given the level of commonality of signatories they should result in a mutually supportive outcome for those signatories and there have been efforts made between the two Secretariats. Some synergies[56] include:

- the opportunity in developing an access agreement to deal with intellectual property rights in a TRIPS-compatible manner;
- there could be an effective mechanism for the exchange of intellectual property rights information between the two administering groups, as suggested in the notification requirement in Article 63 of the TRIPS Agreement and Article 18.3 of the CBD;
- requiring patent applications related to genetic resources to indicate the country of origin and the approval information regarding access to the genetic material or the traditional knowledge.

Potential conflicts indicated[56] include:

- national measures to promote access to technology transfer and information on traditional knowledge must assure equal treatment of locals and foreign nationals and must not exceed licensing requirements of the TRIPS Agreement;
- Article 22.1 of the CBD is concerned with situations where the exercising of rights under another international agreement would cause *"a serious damage or threat to biological diversity"*, but it is unclear how that would apply in the case of the TRIPS Agreement;
- no dispute mechanism for matters of conflict between the two agreements exists—TRIPS would seek resolution through the WTO, whereas the CBD would seek resolution with the International Court of Justice.

The Association of Southeast Asian Nations (ASEAN) group of countries met in Indonesia in February 2001 under the sponsorship of the WHO to discuss the relationship between TRIPS, the CBD and traditional medicines. Their report from this meeting is available on the WHO website.[55] They concluded that the options for intellectual property rights protection did not meet the needs for protection of traditional medicine and offered some new forms of intellectual property to suit those who do wish to protect traditional medicine. The contradiction was also raised that ensuring access to medicines at reasonable cost and providing protection to stimulate research and development may be conflicting ideas. There is a complex web of interactions between traditional medicine, biodiversity conservation, protecting indigenous rights, encouraging research and development investment, improving access to medicines and enhancing healthcare. The report urges countries to implement the principles presented in the CBD.

The commercial value of traditional medicines has highlighted the need to protect traditional knowledge from "biopiracy". Some would have traditional medicine protected under new or existing forms of intellectual property rights whereas others object to that concept on ethical, economic or other reasons including, as mentioned above, that protection may limit access at a time when increased levels of healthcare are needed.

The TRIPS Agreement promotes the standardisation of intellectual property rights legislation, which significantly diminishes the possibility that a developing country can design laws appropriate for their level of development and their national priorities. Given that it was the USA that promoted the tightening of patent and copyright laws (as embodied in the TRIPS Agreement) in the first place, this is viewed by some as a form of intellectual property rights hegemony.

Conservation and sustainable use are inseparable within the CBD and the treaty acknowledges the important role of traditional knowledge to communities and within the global healthcare system. However, the provisions are very general; they do not establish standards and they leave it to individual states to develop mechanisms for implementation and for the exercise of rights over their genetic biodiversity and for the protection of traditional knowledge. As the IUCN survey indicates,[39] most states have yet to do this. The TRIPS Agreement on the other hand establishes standards for the intellectual property rights which signatory members *must* incorporate nationally. Signatory members must also develop protocols such that these intellectual property rights can be enforced. The main beneficiary of the patent section of the TRIPS Agreement is the pharmaceutical industry, since the standard period for the validity of a patent is extended to a minimum of 20 years—a significantly higher level of protection for patented drugs.

Prior to TRIPS, many developing countries either did not have patent laws or did not grant patents for drugs. The fear is that TRIPS will further reduce access to drugs in developing nations since the switch from patented drug to generic formulation will be delayed further. The fundamental assumption in the TRIPS Agreement is that the model of intellectual property used in developed countries is also appropriate for developing countries.[55] There is a little flexibility in the laws that can be developed nationally with respect to traditional medicines; in particular, as noted above, TRIPS allows, *but does not require*, plants and animals to be exempted from patentability, while at the same time requiring that plant varieties be protected. The TRIPS Agreement also requires the protection of undisclosed information and trade secrets (traditional knowledge?), but does not require that exclusive rights be given to the holder of the information or the trade secret. The omission here is also that TRIPS does not deal with undisclosed information or trade secrets held collectively (*e.g.* by indigenous communities).

The ASEAN report[55] makes four clear distinctions between the agreements.

(i) While both agreements require countries to enact their own legislation, in one case it is driven by multinational, pharmaceutical or agrochemical companies and in the other by local people with no experience in this area.
(ii) There are significant enforcement mechanisms associated with TRIPS, but none in the case of the CBD.
(iii) The CBD deals mostly with poorly defined public rights, whereas the TRIPS Agreement deals with private (corporate or individual) rights which are well-defined from a legal perspective.
(iv) Whereas the CBD offers general principles and broad guidelines, TRIPS provides a series of precise, minimum standards for implementation by states.

The ASEAN group honed the discussion regarding intellectual property rights and traditional medicine.[55] The most obvious mismatch is that protection of a traditional medicine could limit access, which is not a desirable public health outcome. And even if exclusive rights are sought, the processes are usually not innovative, the actual inventor is probably not known to the holders of the knowledge (since the knowledge has been passed down in the community over generations) and the product is not novel. Copyright is not a solution either since it relates to expression of an idea and, like patents, is typically assigned to an individual and not groups or communities. Trademarks would provide authenticity to a product as originating from a particular community, but indigenous groups typically do not have the resources to develop, obtain, prosecute and promote their trademarked products. Trademarks also do not prevent further development of a product leading to patentable innovations.

The other choice is to disclose traditional knowledge fully in the public domain. Some owners of traditional knowledge object to limiting ownership rights as a matter of principle, seeing it as global knowledge. Others wish to avoid biopiracy and misappropriation of knowledge, since the knowledge is most often held by those who do not have the resources to develop it. They see protection as a way to benefit the developed countries; providing protection is likely to increase costs, thereby reducing access to those who are dependent on traditional medicines for their primary healthcare. Thus, disclosure removes the possibility of patentability. The one glitch is that under US patent law, disclosure through use does not destroy novelty and thus, some patents have been obtained based on traditional knowledge in the USA. The question of publication is sharpened also by discriminating between what is in the public domain through publication in some form and what is not. Again though, disclosure should be a choice and be subject to consideration for compensation from the acquiring group. Publication however, is likely to limit the bargaining power of the indigenous group when it comes to negotiating for benefit-sharing in a mutual agreement.

As discussed elsewhere,[24,27–29,61–63] this raises the matter of where such information is now stored. The answer is in many diverse locations including books, herbaria records, various journal articles, compendia of information held in both oral and written forms by individuals and groups, and the NAPRALERT™ (Natural Products Alert) database. What is needed, as a global initiative funded by a foundation or for example by the World Intellectual Property Organization (WIPO), is a single database of traditional medicinal knowledge that can be made globally available. Such a resource would aid local communities to verify useful plants being used by other groups and also distinguish what is already known in the public domain from what might be previously undisclosed information, should they wish to protect it. In addition, it would allow research groups around the world to prioritise their research efforts on traditional medicinal plants.

Formal protection of traditional knowledge is actually also a form of disclosure, since the patent becomes a globally available public document which others can then improve on and obtain new patents. As a result, a number of suggestions have been made to improve the patent situation for traditional medicines. First, there is the option to raise the standards of innovation, although that may actually make it more difficult for traditional medicines. Secondly, there is the option to explicitly exclude traditional knowledge from patentability, thus assuring continuing local access but possibly limiting investment in further development. Patentability based on use has also been proposed, since some countries do permit the patenting of new uses for a known product or of a substance derived from nature. Under TRIPS, countries are able to decide this at the national level. This may help to curb biopiracy relating to patenting a known use of traditional medicines. However, from the perspective of the developed country, their interest is in seeing that patenting on the basis of new use is not allowable since they are typically promoting prescription drugs.[55]

The ASEAN group[55] developed a number of topics to ponder as tangential thinking with respect to the dilemmas indicated above. Is intellectual property rights (IPR) protection really necessary for the further development of traditional medicines? Do the potential benefits outweigh the disadvantages? In considering this, take into account the importance in local primary care and the anticipated rising costs of imported drugs (in part due to TRIPS). Does intellectual property protection of traditional medicines serve the desired objective or are there other more efficient ways to achieve those objectives? What is proposed to be protected; the economic benefits, the biological aspects and the cultural aspects are all implied and what are the priorities?

The ASEAN group proposed[55] that one way forward would be to try to balance the many diverse objectives and strategies. They selected four areas for comment. The first was to regulate access through establishing the authority of the traditional healers and communities over their knowledge and granting them the right decide how, when and under what conditions to share all or part of that knowledge and to formally regulate access based on conditions as defined and negotiated. The second area was to involve all stakeholders when drafting regulations. The third area was to differentiate between different categories of knowledge (*e.g.* non-contemporary traditional knowledge in the public domain and non-contemporary traditional knowledge that has not been disclosed, *etc.*). The fourth area was regional cooperation where, because traditional medicines and their use may be the same in several neighbouring countries, there could be simplified negotiations for potential collaborators. The recommendations of the ASEAN Workshop arising from the discussion are presented in the group's report.[55]

In November 2000, the European Chemical Industry Council (CEFIC) commented on the legal protection of traditional knowledge in a position paper.[64] CEFIC began by acknowledging that the World Intellectual Property Organization (WIPO) is the appropriate body to deal with this issue and indicated that there are three separate issues to be dealt with:

- traditional knowledge and the protection thereof;
- access to genetic resources;
- the patenting of inventions based on those genetic resources.

In the position paper, CEFIC supported the creation of a system for protecting traditional knowledge following the development of a definition of traditional knowledge, the creation of inventories of traditional knowledge, and clarification of the relationship between protection of traditional knowledge and existing intellectual property rights systems. CEFIC recommended that WIPO should develop the system for the creation of the inventories with a view to establishing access to them from which indigenous communities could generate economic benefit.

In the position paper, CEFIC stated that a *sui generis* system was needed to protect traditional knowledge appropriately if existing intellectual property systems do not apply or are not effective and that, once developed, this system should be incorporated into the TRIPS Agreement.[64] CEFIC also suggested that the rights be registered for a limited period of time and be assignable to a designated person within a community; the rights would also cover traditional knowledge already in the public domain, but only from the point of registration moving forward and would not impact development of traditional knowledge retrospectively.

With respect to access to genetic resources, CEFIC was concerned that "*companies are at a loss as to how access to genetic resources should be obtained in a specific country ... since national laws are sometimes unclear and very often the matter is not regulated at all*". CEFIC recommended that all countries that had signed the CBD should enact defining legislation that would promote its objectives. CEFIC also sought clarification in such legislation with respect to facilitating access, establishing the requirements for terms in a mutual agreement and prior informed consent, establishing at least minimum requirements for benefit sharing and designating a responsible point of contact in the source country.

On the subject of patents related to genetic resources, CEFIC recommended unambiguous legislation that would allow for the patenting of plants and animals "*provided the application of the invention is not technically confined to a single plant or animal variety*". CEFIC supported the inclusion, on a voluntary basis rather than a condition, of an indication of a country of origin in a patent involving genetic material and encouraged its members to indicate the source country. CEFIC also supported the inclusion of proof of prior informed consent in a patent application.[64]

7 Other Aspects and Outcomes

As mentioned previously, there are some significant implications for both academic and corporate institutions as a result of non-compliance with the treaties, national laws and professional guidelines that have been established for the acquisition of genetic resources and traditional knowledge since the introduction of the CBD.

Failure to comply with regulations and laws results in "tainted research", which may make it very difficult or perhaps even impossible to obtain a patent on an invention, or publish the results in certain journals. Even if a patent is obtained, it may be regarded as "weak" and make investment in the invention (such as licensing) difficult to obtain. Clearly demonstrating that acquisition of the biological material indicated as source material in a patent was the result of formal, documented approval for access (prior informed consent), acquisition and export is, therefore, of critical importance. There is also the risk that the source country or its agents could file a law suit against an individual who violates its territorial integrity, as established by the CBD, and removes biological material from its sovereign territory without approval. This is especially true if that country has a full set of regulations and protocols in place for such approval processes.

As materials and the products from them change hands during the conduct of a collaborative research programme, it is now crucial that each party provides documentary evidence that the biological material (or the traditional knowledge) was obtained in an appropriate manner (according to the particular source country involved). Researchers (academic, government, or corporate) working in any laboratory should not accept materials that are not fully certified as being legitimately obtained.

"Biopiracy" can also result in a professional stigma being attached to a research programme, making it more difficult for its members with respect to future collection initiatives, publishing results and obtaining funding. Finally, there is the development of access agreements between the parties themselves; any recognition that one side is not acting appropriately will also result in a stigma being attached to the group seeking access or the bureaucracy in the country responsible for evaluating whether access should be granted.

All these factors place the legal department staff, either at a corporate entity or at an academic institution or research institute, as the key negotiators and arbiters of the process. Such processes are often difficult, time-consuming and highly variable in the predictability of the outcome and level of investment required to conclude a negotiation successfully—in part because of the individuality of the regulations and protocols in each country.

For a variety of reasons in addition to those mentioned, pharmaceutical companies—as well as academic institutions and research institutes—have chosen to be very highly selective about which of the most biodiverse countries to work with in order to achieve their goals of sample access and biological evaluation. This has worked to the detriment of those countries where the most stringent protocols were enacted and where investment, either locally or externally, in the potential of bioresources is now minimal. In addition, there are major pharmaceutical corporations that have decided, at least in the short term, that the complex ethical, financial and legal issues prevalent from sample collection to drug development are too risk intensive and that it is preferable to eliminate newly acquired bioresources from investigation. As a result, they focus only on products that are well-defined and without legal complexities, such as those acquired *via* the catalogue of a chemical supply company.

However, there is one collaborative research initiative that has attracted the interest of some pharmaceutical companies and many other parties around the world.

7.1 The International Cooperative Biodiversity Groups Programme

In spite of the failure of the USA to ratify the CBD, the most successful research programme to bring together its principles is the International Cooperative Biodiversity Groups (ICBG) programme. Initiated in 1992, it derives funding from the US National Institutes of Health (NIH), the National Science Foundation (NSF) and the US Department of Agriculture (USDA) and is administered through the NIH's Fogarty International Center.[65] Within the US systems of grant funding, the programme is a unique effort that seeks to blend natural product drug discovery, biodiversity conservation and sustainable economic growth. These groups are highly collaborative—requiring an industrial partner as well as a local partner—and are typically quite complex to assemble and administer.

In 1999, a special issue of *Pharmaceutical Biology*, with guest editor J. Rosenthal, was devoted to reviews by ICBG groups of their programme initiatives.[66] The papers evolved out of a symposium at the American Society of Pharmacognosy meeting held in Orlando, Florida, in 1998. The ICBG programmes represented research, development and conservation efforts in 12 developing countries. There are presently (since FY2005) seven awards with collaborations in Papua New Guinea, Costa Rica, Panama, Fiji, Madagascar, Jordan, Uzbekistan, Kyrgyzstan, Vietnam and Laos. There have been a number of public discussions since the inception of the programme regarding the various programmes, some of which have had quite limited lifetimes.[67–70]

Dias and da Costa[48] reviewed some of the issues that arose in the development of the ICBG programme between Washington University, Cayetano Heredia University, the Museum of Natural History at San Marcos University and the Aguarana, an indigenous group living in the Amazon region of Peru and represented by the Aguarana-Huambisa Council. Under the contract agreement, which was signed in 1994, plants would be collected and studied in Peru and the USA, where Washington University had a licensing contract with G.D. Searle & Co. In 1995, the Aguarana-Huambisa Council withdrew from the programme and the ICBG developed collaborations with the Central Organization of Aguarana Communities of Alto Maranhão (OCCAAM). Tests were conducted for anti-diabetic and cardiovascular activities. In 1999, Searle/Monsanto cancelled its part of the contract, citing cost-benefit concerns.

The issues and the challenges with establishing and maintaining these programmes have been well delineated by Soejarto and co-workers at the University of Illinois at Chicago (UIC).[70] The specific aims of the UIC-ICBG group are:

"a) to produce a documented inventory of tropical plant diversity of Vietnam and Laos, specifically, the seed plants of the Cuc Phong National Park in Vietnam and the medicinal plants of Laos; b) to discover novel, biologically active molecules

from plants of Vietnam and Laos as possible candidates for drug development for the treatment of malaria, viral (including AIDS), CNS-related diseases, cancer and tuberculosis; (c) to improve the standard of living of members of the communities who participate in the ICBG studies, . . . and to support the development of human resources and the institutional strengthening of research facilities of colleagues . . . in Vietnam and Laos".[70]

The Vietnam–Laos IBCG programme at UIC, which was initiated in the late 1990s, has five main sites of operation:

- UIC;
- Traditional Medicine Research Center, Laos;
- Cuc Phong National Park, Vietnam;
- Institute of Biotechnology, Vietnam;
- Glaxo Wellcome (now GlaxoSmithKline) in the UK and subsequently Bristol-Myers Squibb in the USA (industrial partner).

The complex interactions between the various groups were presented by Soejarto *et al.* in 1999.[71] A five-way MOA was developed which defines the individual centre and group obligations. The members agreed that the plant genetic material belongs to the country where the material originates and that any invention generated from that material should be protected, with the benefits being shared equitably. Benefit-sharing also includes co-authorship and technology transfer. The corporate partner waived its rights to any share of monetary benefits that might result from a royalty stream. Bilateral sub-contractual agreements were established with each of the groups.

The group at UIC has also provided an assessment of the impact of its ICBG programme in the areas of biodiversity inventory and conservation, studies on medicinal plants, drug discovery and development, economic development and intellectual property rights and benefit sharing.[72] Their conclusion was that, in addition to having a research impact, the programme over seven years had a significant positive impact on the institutions involved and the local community. Key issues were considered to be:

- the team's scientific experience;
- a comprehensive agreement that respects the rights of all parties;
- consent to access the genetic resources and traditional knowledge;
- group motivation and trust.

Fluid communications (personal as well as electronic) and regular all-group meetings were also cited as critical components for the operation of a successful programme.

As Rosenthal *et al.* point out,[73] several countries in the ICBG programme have used participation to explore the effectiveness of their own policies and practices for access to genetic materials and benefit-sharing. The programme has demonstrated that bioprospecting is a research process and that, while there are business and legal issues involved, these can be handled with negotiated agreements.

The programmes that have been funded during the course of the scheme reflect a range of approaches and collaborative study designs. These indicate that a single approach to conducting international collaborative programmes and the nature of the research, the breadth of the expertise and the benefit-sharing programmes are all highly variable in successful programmes. These range from developing local herbaria and laboratory resources to assisting in local conservation education programmes.

Thus, access regulations that are elaborate and inflexible will probably harm the interests of both the providers and the investigators of the genetic material and the traditional knowledge. Rosenthal *et al.*[73] state that one of the most important contributions of the programme has been to provide a range of *"important models for governments and other organizations for collaborative (natural product) research that supports multiple objectives, including those of the Convention on Biological Diversity"*.

8 Some Recommendations

It is well-established that the chemical diversity of nature cannot be replicated by humankind. Although only about 150 000 natural products have been characterised, they represent close to 6000 carbon skeletons and most are replete with the functional groups and physical characteristics that are prevalent in the existing drugs of the world.[74]

From both a chemical and biological perspective, it is not logical to eliminate this structural diversity from biological evaluation for current and future healthcare and agrochemical needs. Indeed, it is more reasonable to set scientific effort towards increasing the diversity of natural products in order to enhance the number and breadth of compounds in a given extract which is poised for biological evaluation and future use.[24,25,28,29,61,62,75] Thus, the investigation of what Nature has provided needs to be expanded substantially on behalf of the billions of people in the world who have reduced or no access to contemporary medicinal preparations as delivered in the North, or even validated traditional medicines.

It was St Hildegard of Bingen who indicated back in the 12th century that: *"All nature is at the disposal of humankind. We are to work with it. For without it, we cannot survive"*.[76] Never in our history was that more true than today. Traditional knowledge, particularly as it relates to traditional medicine, is a form of slow throughput clinical screening, rather than the ultra-high throughput screening which is the standard in the pharmaceutical industry today. It has been in clinical practice for thousands of years. It requires scientific evaluation for safety, effectiveness and sustainability.[24–29,75,77,78] For the North, access to natural resources is needed to investigate the provision of new examples of chemical space which can provide templates for new medicinal agents. The core issue is how to balance the considerations of all parties and, thereby, promote improvements in the respective healthcare systems of those parties.

Gollin[32] offered a number of suggestions for those individuals or groups interested in bioprospecting, including:

- develop an access and benefit-sharing agreement for every collection (irrespective of whether the country requires it);
- cultivate long-term relationships with suppliers;
- establish relationships with local regulators to determine local rules and procedures;
- establish local needs (*e.g.* infrastructure development) which could assist in formulating an agreement;
- allow adequate time to obtain permits;
- plan large projects;
- try to minimise restrictions;
- evaluate collection records of all sample providers for necessary approvals;
- disclose all relevant information to local regulators and collaborators, and in any patent applications.

These are certainly sound pieces of advice and practising natural product research in an ethical and fully considerate manner (on both sides) is clearly the key to future successful, long-term programme development.

ten Kate and Laird[49] made a number of recommendations with respect to accessing genetic resources, firstly to the governments regulating access and secondly to the corporations seeking access. In the first instance, they recommended that governments understand the different potential user industries with respect to their demand for access to genetic resources and traditional knowledge, the use made of the resources, and the costs, risks and potential benefits. They recommended a greater understanding of the types of possible partnerships that could be created and the benefits that could be shared. With respect to access they recommended keeping procedures *"simple, speedy and efficient"* and flexible to deal with different genetic resources. Finally, they recommended that governments assume the administrative task of local level prior informed consent (PIC) and benefit-sharing arrangements. For companies, they suggested that they needed to acquire a more accurate understanding of the CBD and to appreciate the priorities of provider countries. The companies should be engaged in the formulation of policy at a national and international level, and should develop company policies and a set of principles and practices regarding access and benefit-sharing which are open and available. Finally, the company should develop tools to see that the policies are implemented and enforced by its staff.

As a result of the ASEAN workshop described in Section 6, a number of important recommendations were made,[55] some of which are indicated here. First, the group recommended the development of a National Traditional Medicine Policy which would strengthen the infrastructure of traditional medicine in the country, integrate with the national healthcare system and increase the funds for research and development. Secondly, they recommended development of national legislation in order to implement both the TRIPS Agreement and the CBD, which could develop creative approaches to the protection of traditional medicine knowledge and prevention of biopiracy. Thirdly, they recommended taking an active role during reviews of the TRIPS Agreement, particularly with respect to seeing that plants and animals were excluded from patentability and developing an ASEAN position on TRIPS. Fourthly, they recommended establishing an information network for traditional medicines among the ASEAN countries. Fifthly, they recommended the reinforcing of technical cooperation between member countries in the area of traditional medicines, including exploring the possibilities for harmonising minimum regulatory standards. Finally, they recommended trying to spearhead, at the international level, the promotion and protection of traditional medicines and traditional medicinal knowledge.

Dias and da Costa[48] in commenting on some examples of international collaborative programme failures in Brazil and Peru had the following observations:

(i) That expectations are out of step with reality in the areas of development of new products and processes, enhancing scientific and technical infrastructure development and sharing of profits.
(ii) That the framework for negotiating agreements in developing countries is unstable.
(iii) That intellectual property rights are at the core of any negotiated agreement.
(iv) That non-governmental organizations, including activist groups, may assume that they are speaking on behalf of indigenous groups when this may not be the case.

The IUCN group, which assessed the impact of the CBD in various countries around the world,[39] also made a series of recommendations. Clarification of ownership rights over genetic resources—particularly in conserved areas, national parks, *etc.*—is a basic requirement for an access and benefit-sharing policy. The ownership of *in situ* and *ex situ* genetic resources needs to be clearly defined. Those seeking to obtain access (both commercial and non-commercial groups) often find the processes long, confusing and frustrating and, therefore, clearly defined offices are needed in each country to handle requests for access. The broad nature of some access and benefit-sharing policies necessitates more careful definition as to what range of activities are covered. Access and benefit-sharing policies that have very well-defined access procedures (*e.g.* the Philippines) should distinguish more carefully between the many industries (both local and international) requesting access, thereby reducing transaction costs for smaller, local industries and stimulating their economic growth.

Prior informed consent procedures should include the national authority and the indigenous group providing the genetic resources or the traditional knowledge. At the same time, those requesting access to genetic resources and traditional knowledge should be sensitive to the cultural, social and economic environment they are entering and explain fully the intellectual property rights issues involved to their hosts. Countries that have defined minimum access and benefit-sharing requirements should examine the application of these across various industries, being careful to define a range of standards.

Protection of traditional knowledge should occur at two levels:

(i) The TRIPS Agreement should require that the origin of samples and/or of traditional knowledge be stated in a patent application.
(ii) These requirements should also be included in national access and benefit-sharing and intellectual property rights policies.

The development of a national biodiversity strategy and action plan before the creation of access and benefit-sharing policies may have the benefit of raising awareness and enhancing local government, local scientist and community discussion and understanding of key issues. Regions of the world that have common ecosystems and significant overlap of genetic resources would benefit from unifying their policies for access and benefit-sharing.[39] This is one of the original intentions of Decision 391 in the Andean Pact countries.

9 A Web of Interconnectedness

There is a web of interconnectedness between numerous major issues faced by the world at the beginning of the 21st century. Burgeoning population, climate change, economic polarisation, security issues, war, energy requirements, healthcare, environment, sustainability, education, women's issues, illegal drugs and globalisation are some of the major concerns that the people of Earth face.

Natural product research is involved in many of those issues directly and indirectly in the remainder. Thus, any regulations (national or international) that impact on natural product research will also have an impact on each of those global issues. Any regulation that facilitates research will assist in addressing the issues at some level; any regulation that impedes research will have a corresponding negative effect. This impact of local regulations is not widely appreciated by those who prepare them. Their thinking is surely that the regulations are concerned with the CBD's aims (access and equity-sharing) and they do not see that beyond these are a plethora of local health, economic, agricultural and sustainability concerns.

There is a complex web of interactions between traditional medicine, biodiversity conservation, protection of indigenous rights, encouraging research and development investment, improving access to medicines and enhancing healthcare. Placing tension at one point in that delicate web can dramatically alter the interactions and the outcomes of natural product research.

Many years ago, I wrote an article entitled, *Pharmacognosy—New Roots for an Old Science*.[79] In it, mention was made of the number of areas of expertise that were required to be involved in various aspects of a pharmacognosy programme whose goal was to start with indigenous knowledge and develop either a fully quality-controlled traditional medicine or a prescription drug. The number of specialties was about 20. Since that time, the number of specialties has probably grown to over 30 as the web of interconnectedness for research on biologically active natural products has grown. Consequently, in order to effect a collaborative programme to address population control (male and female) using medicinal plants derived from traditional knowledge, some of the areas of expertise needed include cultural anthropology, ethnopharmacology, botany, taxonomy, chemotaxonomy, natural product chemistry, analytical chemistry, physical organic chemistry, phytochemistry, terrestrial ecology, biochemistry, enzymology, genetic engineering, molecular pharmacology, clinical pharmacology, formulation pharmaceutics, pharmacokinetics, analytical clinical chemistry, chemical engineering, legal expertise, patent expertise, packaging science, marketing, information technology, agronomy, finance and many more.

Natural product research in this area is a very highly collaborative endeavour; it requires high level communications dealing with numerous ethical, cultural, moral and religious issues. To be successful demands that the programme receives substantial local support and, at the same time, provides substantial local support. The potential is there for advantage to be taken of this web, recognising that its strengths will be greatest when challenged. Looking at those challenges as investments can push these areas of expertise to the highest levels of collaboration and will provide results that will make a vast difference in the future of the Earth and bring the North and the South closer together.

10 A Different World

The world in late 2008 is a very different place to what it was when the CBD was prepared and signed by 153 countries in June 1992. Apart from the political and economic changes in this period, numerous factors have contributed to establishing climate change as a high priority issue for the world to deal with. However, population control (the highest global priority issue in the consideration of this author[27–29]) and the fundamental cause of climate change are still being significantly ignored. In those 16 years, significant loss of rainforest around the world has continued and indeed, in countries like Brazil and Indonesia, those losses appear to be accelerating. Global genetic diversity—an untold and impossible to assess wealth—continues to decline before it can be subjected to study for its potential to benefit humankind. At the same time, the global population has risen from about 5.49 billion in July 1993 to 6.86 billion at the end of 2008. As has been indicated elsewhere,[24,27–29] given that an increasing percentage of the world's population will rely on medicinal agents from natural sources in the future, we are seriously compromising the healthcare of future generations. The need for the sustainable development of medicinal agents has never been clearer. Indeed there is a need to think now of *all* medicinal agents as a global sustainable commodity, *i.e.* as sustainable drugs for healthcare.[28,29,78]

With these aspects in mind, it is apparent that any factor that impedes rather than optimises the evaluation of natural resources for their potential beneficent effects must be mitigated with equity. The CBD has allowed for the development of practices which have protected the sovereign rights of individual states over their genetic resources. It has also fostered the development of protocols and procedures which have proved to be both onerous and stifling on the progress of natural product research in academic, corporate and research institute environments. Serious natural product research programmes around the world have tried for several years to work within the systems that have been established taking, with very few exceptions, the required and ethically appropriate actions to access genetic resources and indigenous knowledge. As a result, natural product research globally has been compromised in scope. For many programmes, the geographic diversity of their collections has been limited because of the difficulty in obtaining approvals in numerous countries at the same time or, in some cases (particularly in the corporate setting of most large pharmaceutical companies), natural product research programmes have been eliminated altogether. These were clearly not the outcomes anticipated by the signatories to the CBD. A more complete review is needed to assess formally at the global level whether the intentions of both the CBD and the TRIPS Agreement are being realised; *i.e.* whether they can be supportive of global goals for enhancing access to quality healthcare and affordable medicines. There is also a need to look at the financial aspects. Has more funding been made available for investment in genetic resources and conservation since 1993? How does the funding compare with the needs for conservation protection globally?

When access is granted to genetic resources, there are insufficient well-established, functional, international collaborative natural product research programmes capable of addressing the healthcare issues for the development of single agent drugs and enhancing the quality control of traditional medicines. Linkages are needed between a number of global agencies including the FAO, International Finance Corporation (IFC)–World Bank, United Nations Environment Programme (UNEP), United Nations Industrial Development Organization (UNIDO), United Nations Development Programme (UNDP), WHO, the European Union, the ASEAN bloc, the South American bloc (Mercosur), the African Union and various international aid agencies and global foundations to fund a source of capital for the development of a series of non-profit centres of excellence, strategically located in various parts of the world. These centres would:

- catalogue traditional knowledge and deal with the intellectual property issues;
- prioritise plants for biological and chemical investigation in areas of health of local significance;

- take materials to a point where patents can be licensed for commercial development.

None of the countries surveyed by the IUCN indicated that, as a part of their regulation and policy-making efforts, the conservation and sustainability of medicinal plants was a priority as part of national strategies which examine how to protect continuing access (locally and regionally) to medicinally important genetic resources for future generations. As indicated earlier, this is a matter of global concern,[80] especially under conditions where very high levels (>90%) of traditional medicines and phytotherapeuticals in many countries are being taken directly from the forest rather than being cultivated in a more sustainable manner. In the USA, for example, the important dietary supplement goldenseal, *Hydrastis canadensis* (Ranunculaceae) is endangered in seven states.[81] One initiative that may alleviate this situation is the wild seed repository being developed at the Millennium Seed Bank by the Royal Botanic Gardens, Kew, at Wakehurst Place in England.[82]

11 Conclusions

The CBD has been a double-edged sword for natural product research. It codified the ethical issues regarding the unapproved acquisition of biological samples and the acquisition of traditional knowledge in sovereign territories and made clear the ownership issues for materials within a country's terrestrial and marine environments. It recommended the facilitation of access to genetic diversity, the equitable sharing of benefits and the conservation of genetic resources.

A number of governments quite quickly introduced regulations based on the CBD, but then established bureaucracies that have made it difficult, or in some cases impossible, for external collaborators to conduct research with natural product groups in those countries and, in some instances, for natural product scientists within their own countries to conduct research. Other countries that over the years since the CBD have developed laws have not provided either the regulatory processes or efficient bureaucracies to handle applications in an expeditious manner.

The international and local costs involved in reaching agreements for access and consideration, the timeliness in which agreements can be reached, the fees involved in filing applications and in maintaining access over meaningful time periods, and the diverse regulations that exist between countries have significantly curtailed both academic and industrial interest in conducting natural product research involving the evaluation of genetic resources for potential commercial significance. It is not clear that, as a result of the CBD, there has been increased funding made available for conservation initiatives. What has become apparent is that, while the intentions of the CBD were noble, the outcome for the development of natural product research both in developing and developed countries has been negative overall; natural product research has been stifled rather than stimulated. The modestly funded ICBG programme is a lone global exception This has occurred at a critically historical time when the study of the surviving global genetic resources for the benefit of humankind should be a high priority, attracting significant investment for the future health of the planet.

Synthetic drugs are not a sustainable commodity for all diseases and many of the chemical reactions used in their production are not "green". What serves as traditional medicine today in most countries of the world has hardly changed in concept in 4000 years. It is now time to change that paradigm. We can, we must do better for the healthcare of future generations. It is time for many of the constraints on the development on national, regional and international collaborations to be resolved. It is time for international agencies and foundations to come together to develop a funding stream (probably of about $1 billion per year) to conduct a 20-year programme based in developing countries on the traditional medicines of the world in order to catalogue and rationalise their present use, their safety, their efficacy and their potential to be developed internationally as a sustainable component of global healthcare. This is not a new vision[26–29,75,83,84] but, as the forests decline, those resources we have taken for granted are diminishing at an unprecedented rate fuelled by the demands of a rapidly growing global population. Twenty years from now, these resources may have diminished by a further 40–50% without dramatically enhanced conservation efforts, closing off more avenues of potential.

We have a window of opportunity,[24,26–29,75] a brief period when the necessary technology and the genetic resources are both available. For the most part, they are literally and figuratively at different "ends" of the Earth. Those "ends"—the North and the South—exhibit substantial political, social, economic and philosophical differences (the "great divide"). Yet, we are a single human species. We are all part of one family of man. We must find creative strategies to span those diverse agendas and to bridge that "great divide" for the sake of humankind, for the sake of the Earth. We must bring together the science and the technology to investigate genetic resources with those resources and that scientific expertise. And this research should be conducted primarily in the developing countries of the world. The people of the world expect effective healthcare for all. It is a human right and our moral duty. The CBD was, in part, trying to open that door, trying to facilitate access and open pathways on both sides to assure that the respective interests of those with the technology and those with the biodiversity could come together. The opportunity is still there to make a difference for present populations of the Earth and for future generations. *Carpe Diem*!

Acknowledgements

In conducting my informal survey of the impact of the CBD in various countries, responses were provided by the following colleagues: Indonesia (Leonardus B. S. Kardono, Indonesian Institute of Sciences), Japan (Shinji

Funayama, Nihon Pharmaceutical University), Jordan (Amal Aboudi, University of Jordan), Korea (Byung-Yoon Sun, Chunbuk National University and Hyeong-Kyu Lee, Korea Research Institute of Bioscience and Biotechnology), Madagascar (David G. I. Kingston, Virginia Tech University, USA and Josette Rahantamalala and Zolalaina Rakotobe, Conservation International, Madagascar), Pakistan (M. Iqbal Choudhary, HEJ Research Institute), Peru (Helena Maruenda, Pontificia Universidad Catolica del Peru), Republic of the Philippines (Maribel Nonato, University of Santo Tomas), South Africa (Maureen Wolfson, South African National Biodiversity Institute) and Tanzania (Charles M. Nshimo, Muhimbili University of Health and Allied Sciences). I am deeply indebted to these colleagues for taking the time to provide their thoughtful and considered input.

References

1. W. von Soden and D. G. Schley, *The Ancient Orient—an Introduction to the Study of the Ancient Near East*, William B. Eerdmans Publishing Company, Grand Rapids, MI, USA, 1994.
2. Anon., Using DNA, Scientists Hunt for the Roots of the Modern Potato, *ScienceDaily*, 4 February 2008, www.sciencedaily.com/releases/2008/01/080129160727.htm, accessed 23 April 2009.
3. L. Gibson and G. Benson, *Origin, History and Uses of Oat (Avena sativa) and Wheat (Tritium aestivum)*, 2002, www.agron.iastate.edu/courses/agron212/Readings/Oat_wheat_history.htm, accessed 23 April 2009.
4. Anon., Rice, in *Wikipedia*, http://en.wikipedia.org/wiki/RICE, accessed 23 April 2009.
5. C. Anthon, *A Classical Dictionary: Containing an Account of the Principal Proper Names Mentioned in Ancient Authors and Intended to Elucidate all the Important Points Connected with the Geography, History, Biography, Mythology and Fine Arts of the Greeks and Romans together with an Account of Coins, Weights and Measures, with Tabular Values of the Same*, Harper & Bros., New York, NY, 1841.
6. H. Schippel, in *Europaische Technik im Mittelalter: 800 bis 1200, Tradition und Innovation*, ed. U. Lindgren, Gebrüder Mann, Berlin, 1996.
7. J. T. Baker, R. P. Borris, B. Carté, G. A. Cordell, D. D. Soejarto, G. M. Cragg, M. P. Gupta, M. M. Iwu, D. R. Madulid and V. E. Tyler, *J. Nat. Prod.*, 1995, **58**, 1325.
8. The Chiang Mai Declaration, in *The Conservation of Medicinal Plants*, ed. O. Akerele, V. Heywood and H. Synge, Cambridge University Press, Cambridge UK, 1991, p. xix.
9. The Declaration of Belem, in *International Traditional Medicine Newsletter*, ed. D. D. Soejarto and C. Gyllenhaal, 1992, **4**, 1.
10. T. Eisner and J. Meinwald, *Chemoecology*, 1990, **1**, 38.
11. M. Balick, E. Elizabetsky and S.A. Laird, *Medicinal Resources of the Tropical Forest*, Columbia University Press, New York NY, 1996 and references therein.

12. The Kunming Action Plan, in *International Traditional Medicine Newsletter*, ed. D.D. Soejarto and C. Gyllenhaal, 1992, **4**, 1.
13. J. Schweitzer, F. G. Handley, J. Edwards, W. F. Harris, M. R. Grever, S. A. Schepartz, G. M. Cragg, K. Snader and A. Bhat, *J. Natl. Cancer Inst.*, 1991, **83**, 1294.
14. M. A. Gollin, *Nature Biotech.*, 1999, **17**, 921.
15. M. Brennan, *Chem. & Eng. News*, March 2, 1998, 10.
16. R. Dalton, *Nature*, 2001, **414**, 685.
17. J. Rosenthal, *Nature*, 2002, **416**, 15.
18. R. J. Nash, *Phytochemistry*, 2001, **56**, 403.
19. Anon., *Bangkok Post* Business Section, 29 November 1999.
20. A. Pollack, *International Herald Tribune*, 27 November 1999.
21. Per Capita Health Expenditures by Country, 2007, www.infoplease.com/ipa/A0934556.html, accessed 23 April 2009.
22. Anon., *The World Medicine Situation*, World Health Organization, Geneva, 2004.
23. Anon., *How to Develop and Implement a National Drug Policy*, World Health Organization, Geneva, 2nd edn, 2001.
24. G. A. Cordell, *Phytochemistry*, 2000, **55**, 463.
25. G. A. Cordell, *Acta Manilana*, 2001, **49**, 1.
26. G. A. Cordell, *Phytochem. Rev.*, 2002, **1**, 261.
27. G. A. Cordell and M. D. Colvard, *J. Ethnopharmacol.*, 2005, **100**, 5.
28. G. A. Cordell and M. D. Colvard, *Arkivoc*, 2007, **7**, 97.
29. G. A. Cordell, *Sci. Cult.*, 2008, **74**, 11.
30. E. O. Wilson, *The Future of Life*, Abacus, Little Brown, London, 2002.
31. List of Parties, Convention on Biological Diversity, www.cbd.int/convention/parties/list/, accessed 23 April 2009.
32. W. Gollin, in *Biodiversity Prospecting*, World Resources Institute, Baltimore, MD, 1993, Annex 3.
33. A. Gore Jr, *J. NIH Res.* 1992, **4**, October, 18.
34. D. D. Soejarto, J. A. T. Sorensen, C. Gyllenhaal, G. A. Cordell, N. R. Farnsworth, H. H. S. Fong, A. D. Kinghorn and J. M. Pezzuto, in *Ethnobiology and Biocultural Diversity, Proceedings of the Seventh International Congress of Ethnobiology*, ed. J. R. Steep, Felice S. Wyndham and R. K. Zarger, The International Society of Ethnobiology, 2002, 21.
35. J. Kloppenburg, Jr., *Cult. Surv. Quart.*, 1991, Summer, 14.
36. For relationships and compensation developed with respect to prostratin see: R. Sanders, *Landmark Agreement Between Samoa and UC Berkeley Could Help Search for AIDS Cure*, http://qb3.org/samoa.htm, accessed 23 April 2009; and The Institute for EthnoMedicine, *The Research: Prostatin*, www.ethnomedicine.org/research/prostratin.asp, accessed 23 April 2009.
37. S. L. Bertha, *J. Ethnopharmacol.*, 1996, **51**, 59.
38. D. D. Soejarto, C. Gyllenhaal, J. C. Regalado, J. M. Pezzuto, H. H. S. Fong, G. T. Tan, N. T. Hiep, L. T. Xuan, D. Q. Binh, T. Q. Bich, N. N. Thin, P. K. Loc, B. M. Vu, B. H. Southavong, K. Sydara, S. Boumanivong,

M. J. O'Neill, J. Lewis, X. M. Xie and G. Dietzman, *Pharm. Biol.*, 1996, **37**(Suppl), 1.
39. S. Carrizosa, S. B. Brush, B. D. Wright and P. E. McGuire, *Accessing Biodiversity and Sharing Benefits: Lessons from Implementing the Convention on Biological Diversity*, International Union for the Conservancy of Nature, Gland, Switzerland and Cambridge, UK, 2004, IUCN Environmental Policy and Law Paper No. 54.
40. Secretariat of the Convention on Biological Diversity, *Handbook of the Convention on Biological Diversity*, Earthscan, London, 2001.
41. A. G. M. La Vina, M. J. A. Caleda and M. L. L. Balon, *Regulating Access to Biological and Genetic Resources in the Philippines. A Manual for Implementation of Executive Order No. 247*, Foundation for Philippine Environment and World Resources Institute, Quezon City, Republic of the Philippines, 1997.
42. C. Dumol, *New Bioprospecting Guidelines in the Philippines*, Philippine Council for Health Research and Development, 28 June 2005, www.pchrd.dost.gov.ph/index.php/news-archive/70, accessed 23 April 2009.
43. Department of Environment and Natural Resources, *Guidelines for Bioprospecting Activities in the Philippines*, Quezon City, Republic of the Philippines, 2004, www.denr.gov.ph/article/view/2332/, accessed 23 April 2009.
44. R. B. Neto and D. Dickson, *Rev. Nature* July 22, 1999. See: R. B. Neto and D. Dickson, *$3m deal launches major hunt for drug leads in Brazil*, www.extracta.com.br/3mdeal_i.htm, accessed 23 April 2009.
45. See: www.sice.oas.org/trade/JUNAC/decisiones/DEC391e.asp
46. D. M. Putterman, *Nature*, 1994, **371**, 553.
47. G. M. Cragg, J. T. Baker, R. P. Borris, B. Carte, G. A. Cordell, D. D. Soejarto, M. P. Gupta, M. M. Iwu, D. A. Madulid and V. E. Tyler, *J. Nat. Prod.*, 1997, **60**, 654.
48. C. C. Dias and M. C. da Costa, *Elect. J. Commun. Inf. Innov. Health*, 2007, **1**, 121.
49. K. ten Kate and S. A. Laird, *Placing Access and Benefits-sharing in the Commercial Context, Summary Report of Phase I*, World Resources Institute and Royal Botanic Gardens, Kew, UK, 1997.
50. GlaxoSmithKilne, *Biodiversity*, www.gsk.com/investors/reps03/EHS03/GSKehs-36.htm, accessed 23 April 2009.
51. S. Carrizosa, in *Accessing Biodiversity and Sharing Benefits: Lessons from Implementing the Convention on Biological Diversity*, ed S. Carrizosa, S. B. Brush, B. D. Wright and P. E. McGuire, International Union for the Conservancy of Nature, Gland, Switzerland and Cambridge, UK, 2004, IUCN Environmental Policy and Law Paper No. 54, p. 9.
52. P. J. Benavidez II, in *Accessing Biodiversity and Sharing Benefits: Lessons from Implementing the Convention on Biological Diversity*, ed. S. Carrizosa, S. B. Brush, B. D. Wright and P. E. McGuire, International Union for the Conservancy of Nature, Gland, Switzerland and Cambridge, UK, 2004, IUCN Environmental Policy and Law Paper No. 54, p, 153.

53. Advanced Industrial Science and Technology (AIST), *International Patent Organism Depositary (IPOD)*, http://unit.aist.go.jp/pod/cie/index.html, accessed 23 April 2009.
54. J. Braithwaite and P. Drahos, *Global Business Regulation*, Cambridge University Press, Cambridge, UK, 2000.
55. K. Timmermans, *TRIPS, CBD and Traditional Medicines: Concepts and Questions*, World Health Organization, Geneva, 2001, NA-DFC, Report of an ASEAN Workshop on the TRIPS Agreement and Traditional Medicine, Jakarta, 13–15 February 2001, www.who.int/medicinedocs/en/d/Jh2996e/, accessed 23 April 2009.
56. Anon., *Strengthening Research and Policy-making Capacity on Trade and Environment in Developing Countries. The CBD/TRIPS Relationship*, Second Workshop held in Los Banos, Republic of the Philippines, 11–13 November, 1999. Discussion paper, draft, www.unctad.org/trade_env/docs/cbd-trip.pdf, accessed 23 April 2009.
57. See letter dated 17 April 2008 from the Secretary-General of UPOV to the Executive Secretary of the Secretariat of the Convention on Biological Diversity, www.upov.int/en/about/pdf/upov_cbd_17_04_2008.pdf, accessed 23 April 2009.
58. J. Caillaux, *J. Environ. Pol. Law Latin Am. Caribbean*, 1994, **1**, 22.
59. D. A. Posey, *J. Ethnopharmacol.*, 2002, **83**, 3.
60. D. S. Fabricant and N. R. Farnsworth, *Environ. Health Perspect.*, 2001, **109** (Suppl. 1), 69.
61. G. A. Cordell, *Phytochemistry*, 1995, **40**, 1585.
62. G. A. Cordell, M. L. Quinn-Beattie and N. R. Farnsworth, *Phytother. Res.*, 2001, **15**, 183.
63. G. A. Cordell, *Rev. Quim.*, 2002, **16**, 5.
64. CEFIC, *The chemical industry comments on the legal protection of traditional knowledge & access to genetic resources-patenting*, CEFIC, Brussels, 2000, Position Paper, http://www.cefic.org/Files/Publications/legal23.11.2000.pdf, accessed 23 April 2009.
65. www.fic.nih.gov/programs/research_grants/icbg/
66. Various papers in *Pharmaceutical Biology*, 1999, **37**, Supplement 1 (see www.informaworld.com/smpp/title ~ content=g714867505 ~ db=all).
67. W. E. Leary, *New York Times*, 8 December 1993.
68. A. M. Rouhi, *Chem. & Eng. News*, 7 April 1997, 14.
69. J. Kelley, *Scientist*, 24 April 2002.
70. D. D. Soejarto, J. M. Pezzuto, H. H. S. Fong, G. T. Tan, H. J. Zhang, P. Tamez, Z. Aydogmus, N. Q. Chen, S. G. Franzblau, C. Gyllenhaal, J. C. Regalado, N. V. Hung, V. D. Hoang, N. T. Hiep, L. T. Xuan, N. V. Hai, N. C. Cuong, T. Q. Bich, P. K. Loc, B. M. Vu, B. H. Southavong, K. Sydara, S. Boumanivong, M. J. O'Neill, J. Lewis and G. Dietzman, *Nat. Prod. Sci.*, 2002, **8**, 1.
71. D. D. Soejarto, C. Gyllenhaal, J. C. Regalado, J. M. Pezzuto, H. H. S. Fong, G. T. Tan, N. T. Hiep, L. T. Xuan, D. Q. Binh, N. V. Hung, T. Q. Bich, N. N. Thin, P. K. Loc, B. M. Vu, B. H. Southavong, K. Sydara, S.

Boumanivong, M. J. O'Neill, J. Lewis, X. Xie and G. Dietzman, *J. Ethnopharmacol.*, 1999, **37**(Supplement), 100.
72. D. D. Soejarto, H. J. Zhang, H. H. S. Fong, G. T. Tan, C. Y. Ma, C. Gyllenhaal, M. C. Riley, M. R. Kadushin, S. G. Franzblau, T. Q. Bich, N. M. Cuong, N. T. Hiep, P. K. Loc, L. T. Xuan, N. V. Hai, N. V. Hung, N. Q. Chen, L. T. Binh, B. M. Vu, H. M. Ly, B. Southavong, K. Sydara, S. Boumanivong, J. M. Pezzuto, W. C. Rose, G. Dietzman, B. E. Miller and T. V. Thuy, *J. Nat. Prod.*, 2006, **69**, 473.
73. J. P. Rosenthal, D. Beck, A. Bhat, J. Biswas, L. Brady, K. Bridbord, S. Collins, G. Cragg, J. Edwards, A. Fairfield, M. Gottlieb, L. A. Gschwind, Y. Hallock, R. Hawks, R. Hegyeli, G. Johnson, G. T. Keusch, E. E. Lyons, R. Miller, J. Rodman, J. Roskoski and D. Siegel-Causey, *Pharmaceut. Biol.*, 1999, **37**(Suppl. 1), 6.
74. C. J. Lipinski, F. Lombardo, B. W. Dominy and P. J. Feeney, *Adv. Drug Disc. Revs.*, 1997, **23**, 3.
75. G. A. Cordell, *Rev. Quim.*, 2004, **19**, 33.
76. Quoted in Matthew Fox, *Original Blessing: A Primer in Creation Spirituality Presented in Four Paths, Twenty-Six Themes and Two Questions*, J.P. Tarcher/Putnam Edition, New York, 2000.
77. G. A. Cordell, in *Proceedings of the Second International Conference on Women's Health and Asian Traditional Medicine*, ed. A. N. Rao, 2006, p. 93.
78. G. A. Cordell and J. Michel, in *Proceedings of the Third International Conference on Women's Health and Asian Traditional Medicine*, ed A.N. Rao, 2007, p. 15.
79. G. A. Cordell, in *Studies in Natural Products Chemistry, Volume 13. Bioactive Natural Products (Part A)*, ed. Atta-ur-Rahman and F.Z. Basha, Elsevier Science Publishers, Amsterdam, 1993, p. 629.
80. O. Akerele, V. Heywood and H. Synge, *The Conservation of Medicinal Plants*, Cambridge University Press, Cambridge, UK, 1991.
81. Natural Resources Conservation Service, PLANTS profile, *Hydrastis candensis* L. goldenseal, in *PLANTS Database*, US Department of Agriculture, http://plants.usda.gov/java/profile?symbol=HYCA, accessed 23 April 2009.
82. *Millennium Seed Bank Project*, Royal Botanic Gardens Kew, London, www.kew.org/msbp/index.htm, accessed 23 April 2009.
83. G. A. Cordell, *ACGC Chem. Res. Commun.*, 2002, **14**, 31.
84. G. A. Cordell, *Sci. Cult.*, 2008, **74**, 11.

CHAPTER 5

Plants: Revamping the Oldest Source of Medicines with Modern Science†

GIOVANNI APPENDINO[a,b] AND FEDERICA POLLASTRO[b]

[a] Indena S.p.A., Viale Ortles 12, 20139 Milano, Italy; [b] Università del Piemonte Orientale, Dipartimento di Scienze Chimiche, Alimentari, Farmaceutiche e Farmacologiche, Via Bovio 6, 28100 Novara, Italy

1 Introduction

The birth of drug discovery is closely connected to the study of plant natural products and was shaped by two seminal events, the isolation of morphine **1** from opium by the pharmacist Sertürner in 1817[1] and the introduction in the clinics of Antipyrin **6** (phenazone) 70 years later, in 1887.[2] The obtaining of a pure compound responsible for the medicinal properties of a crude drug marked the beginning of medicinal chemistry, triggering the transition from botanical extracts to pure molecules and eventually leading to the isolation of the active principle of most "heroic" drugs.

In the wake of the seminal isolation of morphine from opium, emetine, quinine, colchicine, sparteine, caffeine, atropine, codeine and papaverine were

† G. A. would like to dedicate this contribution to the memory of Jasmin Jakupovic (Jaku), the father of thousands of plant natural products, whose extraordinary talent, humanity and friendship will always be remembered.

RSC Biomolecular Sciences No. 18
Natural Product Chemistry for Drug Discovery
Edited by Antony D. Buss and Mark S. Butler

Published by the Royal Society of Chemistry, www.rsc.org

purified between 1820 and 1850.[3] Despite the gap between the isolation and the structure elucidation of these alkaloids, their availability in pure form had immediate clinical translation and was instrumental in fostering the birth of pharmacology. The purification of a series of quintessential active principles of crude drugs also laid the foundation of what next evolved into the concept of "magic bullet" and the current reductionist mantra of modern drug discovery, namely the "one target–one drug–one disease" paradigm. If a complex condition like pain can be managed with a simple molecule like morphine isolated from a complex plant material like opium, then biological processes may be governed by critical and druggable steps and drugs can originate from crude natural materials by the removal of inactive constituents—in a process reminiscent of the alchemists' search for the quintessence of materials and somewhat similar to the genesis of a sculpture from an inform piece of marble.[3]

On the other hand, the discovery of the first synthetic drug, Antipyrin **6**, by the German chemist Ludwig Knorr exemplifies how this process was complex, difficult to plan and critically dependent on serendipity.[2] It also testifies to the replacement of pharmacists by chemists as the leading driving force in drug discovery. Antipyrin was developed as a type of "quinine–morphine" hybrid, whose fuzzy logic of design was the result of a wrong structural assignment of the condensation product of phenylhydrazine **2** and ethyl acetoacetate **3**. Rather than the correct pyrazole structure **4**, this compound was assumed to be the tetrahydroquinoline **7** (Scheme 5.1). A quinoline system was the only structural element of the antipyretic and antimalarial alkaloid quinine **8** known at that time and there was great interest in the bioactivity of quinolines and the treatment of fever, considered at that time as a quintessential evil rather than a physiological way to fight diseases.[3] The alleged tetrahydroquinoline **4** was,

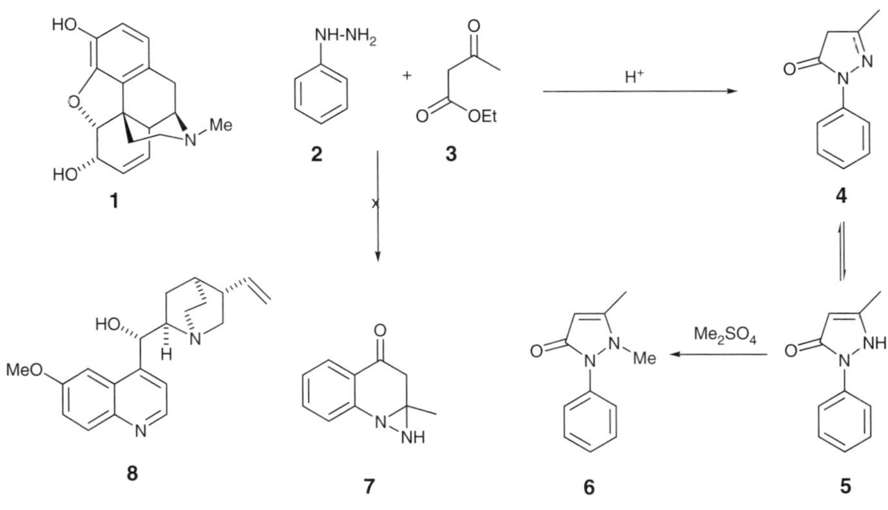

Scheme 5.1 The serendipitous discovery of antipyrine **6**, an alleged morphine **1**/quinine **8** mimic.

therefore, investigated for antipyretic properties, showing only modest activity. Since morphine has an *N*-methyl group, **4** (believed to be **7**), was subjected to *N*-methylation, affording, *via* its tautomer **5**, a compound that became eventually known as Antipyrin **6**. Antipyrin is endowed with potent antipyretic activity and was launched in Germany in the wake of an influenza epidemic that took place in the winter of 1889–1890, with a remarkably short time lag (six years) between synthesis and successful commercialisation.[2]

Over the next century, medicinal chemistry thrived, often serendipitously exploiting a simplistic logic, a flawed rational, or the presence of impurities in drug candidates, but always revolving around natural products. The introduction of fungal sources from the early 1940s came in the wake of the efforts during World War II for the production of penicillin.[3] Indeed, only few major chemotypes unrelated, at least in their genesis, to natural products were discovered prior to the mid-1960s (*e.g.* sulfonamides, benzodiazepines and phenothiazines). Remarkably, a natural products trait was later also discovered in antibacterial sulfonamides and anxiolytic benzodiazepines—the archetypal synthetic drugs. Biochemical and analytical techniques unavailable at the time of the original discoveries identified sulfonamides as biological analogues of *p*-aminobenzoic acid, a key component of folic acid,[4] while anxiolytic benzodiazepines (including diazepam) were discovered to occur naturally.[5] The 1940s to 1970s were undoubtedly the golden age of drug discovery. Despite a limited knowledge of receptor pharmacology and cell regulation, a flurry of original drug chemotypes were discovered, setting the foundation of modern pharmaceutical research and medicinal chemistry.[3]

In the wake of these successes, the next decades saw the birth of rational drug discovery; a process spurred by spectacular progress in the way collections of compounds are assembled (combinatorial chemistry) and screened (robotic high throughput screening; HTS), and druggable targets are identified (genomics). Natural products are few in number, slow to purify and, contrary to vitamins, enzymes and hormones, are generally multi-purpose in their activity. As unfocused and hard-to-obtain exceptions to the "rule of five", natural products have substantially and, probably prematurely, fallen from favour in medicinal chemistry campaigns. This observation is surprising since natural products have continued to afford drugs and drug leads after the advent of HTS.[6] The declining pharmaceutical relevance of plant natural products is especially marked, since only five out of the 23 new natural products or natural product-inspired drugs launched between 2000 and 2005 are of direct plant origin (galanthamine), or are derived from a plant lead (apomorphine, tiotroipium, nitisinone and artheether).[6] If one further considers that galanthamine and apomorphine are old compounds and that the time lag between pharmaceutical development and market launch is at least a decade, this negative trend is presumably bound to exacerbate in the next few years.

The pharmaceutical woes of natural products are echoed by those of the whole "modern" drug discovery process, since there has been a backlash against the unrealistic expectations and promises it prematurely raised, generating (and not only in science-naive investors) a widespread attitude that pharmaceutical research is essentially a high throughput–near zero output process where quantity is more important than quality. This attitude has spurred a general biotech-refocusing of big pharma, where molecular biologists are replacing chemists as the driving force in drug discovery just as chemists replaced pharmacists in the second half of the 19th century.

This chapter critically analyses the reasons why plant natural products, despite accounting for a quarter of all current drugs in terms of origin of inspiration,[7] have fallen out of favour in drug discovery, becoming almost orthogonal to mainstream drug discovery and ending up relegated in the vague area of healthfood and nutraceuticals. Various critical issues for their re-introduction in this process are also evaluated.

2 Plant Secondary Metabolites *vs.* Secondary Metabolites of Other Origin

Arguably, secondary metabolites are easier to obtain from plants compared with other sources, since plants are easily available and store secondary metabolites in major organs (roots, leaves, fruits). Conversely, microorganisms generally secrete their secondary metabolites into the environment, while animals often produce secondary metabolites only on demand, or store them in localised and specific organs. Compared with other sources of natural products, plants have, therefore, been relatively well investigated from a botanical and a biomedical point of view. In fact, it has been calculated that about half the estimated 500 000 plant species in the world have been described, with approximately 10% of them having being investigated chemically, even in a cursory way.[8] For comparison, only 10% of the estimated 1 500 000 fungal species have been described and an even smaller proportion of them has been investigated, while biota like spiders and insects are still largely unclassified and uninvestigated chemically.[8] It is, therefore, relatively simple to outline the major features of plant natural products compared with what has surfaced so far about secondary metabolites from other natural sources.

Several types of secondary metabolites are produced almost exclusively in plants. The most remarkable examples are coumarins, lignans, steroid saponins, several classes of alkaloids (steroid, tropane, pyrrolizidine and *Cinchona*), glucosinolates, *S*-substituted cystein derivatives and the compounds derived from DOPA by aromatic ring cleavage and recyclisation.[9] In addition, flavonoids, stilbenoids and cyanogenic glycosides are found almost exclusively in plants.[9] Limited overlap also exists between most terpenoid and acetogenin plant skeleta and their microbial/animal counterparts. Many structurally unique classes of plant secondary metabolites are derived from biogenetic pathways related to the formation of the plant cell wall and, therefore, are related to differences in primary metabolism. Conversely, several classes of secondary metabolites (*e.g.* polyether toxins) are remarkably absent from plants or are very rare (*e.g.* depsipeptides).[9]

Even though plants and microorganisms produce compounds from the same structural class, in some cases the underlying biogenetic pathways can be different, as often demonstrated by different chemical substitution patterns. Important examples are:[9]

- isoquinoline and piperidine alkaloids (derived from amino acids by plants, but of polyketide origin in microorganisms);
- naphthoquinones (produced from shikimic acid and isoprenoids in plants, but from acetate and amino acids in microorganisms);
- anthraquinones (of polyketide origin in microorganism, but also formed from shikimate and mevanolinc acid in plants).

Remarkably, the same compound can be produced in different ways by different organisms. The best known example is probably the polyketide anthraquinone, chrysophanol **9**, which occurs in both eukaryotes (higher plants, lichens, fungi and insects) and prokaryotes, but is produced through different folding modes of polyketide chains.[10] Similarly, it has also been demonstrated that the biosynthesis of gibberellins involves different metabolic sequences in fungi and plants.[11]

While the overall picture of plant secondary metabolism is apparently clear, a series of recent observations have raised interesting questions about the involvement of microorganisms in the production of plant secondary metabolites, revealing also the occurrence of plant secondary metabolites in non-plant organisms, mammals included. The best known example of a natural product with a puzzling occurrence in Nature is paclitaxel **10**.[12] The distribution of this diterpene alkaloid was long believed to be exclusive of gymnosperms from the genus *Taxus*, but paclitaxel was later isolated as a genuine plant product from the hazelnut tree (*Corylus avellana* L.) as well as from a host of plant microbial symbionts, which *via* horizontal gene transfer, might have acquired the multienzymic machinery required for its biosynthesis from the yew tree and might, in turn, have transferred it to other plant species.[12] These observations are strongly reminiscent of the occurrence of maytansin **11**,[13] a compound structurally similar to microbial polyketides, in plants from the genus *Maytansenus,* or of trichothecenes in plants from the genus *Baccharis.*[14] In the mid-1970s, a heroic effort was undertaken by the National Cancer Institute in the USA to scale up the production of maytansins by isolation from *Maytansenus*—at that time its only known natural source. The clinical failure of maytansine was a tremendous blow to the whole natural product community and, in retrospect, might even have delayed the development of paclitaxel—long perceived as simply "another spindle poison". Maytansins, obtained by microbial fermentation, are currently enjoying a renaissance and are under clinical investigation as peptide conjugates for the treatment of various solid tumours.[15] Similarly, the occurrence of trichothecenes in various Brazilian plants from the genus *Baccharis*, was eventually traced back to a fungal root symbiont.[14] Compounds with typical archeal microbial structural features have also been detected in plants, such as the series of tethered lipids discovered in the Mediterranean umbelliferous plant *Thapsia garganica* L.[16]

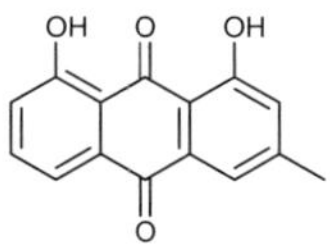

9

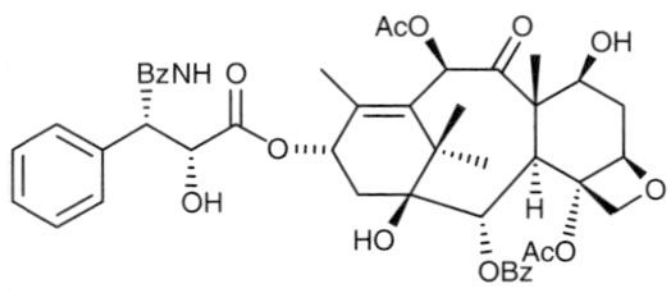

10

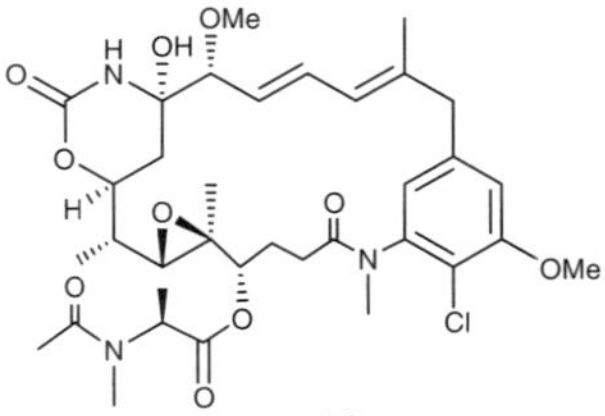

11

Plant products have also been detected in non-plant organisms. Morphine **1**, the archetypal plant alkaloid, has in fact been shown to be a physiological plasma constituent and its production in mammals could be traced to the liver expression of the critical enzymes of its biosynthesis.[17] In addition, the plant hormone abscisic acid **12** has been detected as an endogenous constituent of human brain,[18] while caffeine **13** was isolated from a marine gorgonian (*Paramuricea chamaelon*)[19] and the atisane diterpenoid serofendic acid **14**, an inhibitor of the oxidant-induced mitochondrial death pathway and putative activator of mitoK(ATP) channels, has been characterised from foetal calf serum.[20]

Another observation that might have general relevance in plant physiology is the involvement of microorganisms as elicitors and modulators of terpenoid biosynthesis. Strong evidence for an interaction of this type has been demonstrated in vetiver [*Vetiveria zizzanoides* (L.) Nash].[21] The roots of this graminaceous plant produce a complex mixture of sesquiterpenoids that are used in perfumery. The accumulation of terpenoids is an unusual feature in graminaceous plants and, when grown in a sterile medium, vetiver roots produce only trace amounts of an essential oil devoid in vetivenoids (the most typical terpenoids of vetiver oil). Furthermore, recombinant vetiver sesquiterpene synthases produce sesquiterpenoids different from those contained in vetiver essential oil. These observations strongly suggest a microbial involvement in the production of vetiver sesquiterpenoids and, indeed, several bacteria strains isolated from parenchymatous essential oil-producing vetiver cells have been shown to use sesquiterpenes as a carbon source and to metabolise them to compounds typically found in vetiver oil.[21]

Finally, it should also be mentioned that the active principle of the bark from *Pygeum africanum*, a popular treatment of prostatitis and benign prostate hyperplasia, has recently been identified in the ketide atraric acid **15**, a lichen compound that shows anti-androgenic activity, inhibiting transactivation mediated by ligand-activated androgen receptor.[22] Atraric acid is presumably derived from the lichen compound atranorin, a bitter depside also known to contaminate oak wood and causes considerable problems during wine aging.

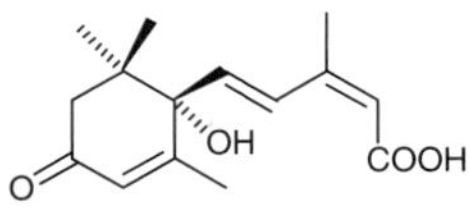

12

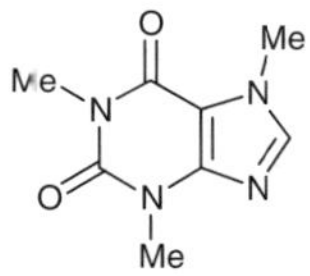

13

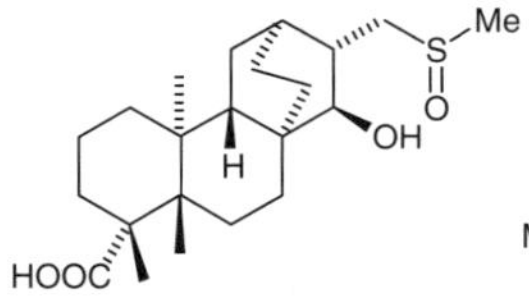

14

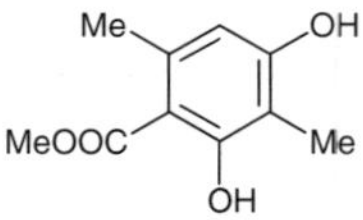

15

3 Unnatural Sources of Plant Secondary Metabolites

In the wake of continuous progress in biotechnology and fermentation, several strategies to "take the Nature out" of plant natural products and produce them in a non-natural way have been developed.[23] These strategies have relevance not only for the mass production of a natural product drug, but also for providing access to natural products-related chemodiversity. Tissue culture of plant cells, tissues or organs and cultures of transgenically engineered microbial cells are the major "unnatural" ways to obtain a plant natural product from biotechnological sources.[23] By securing a continuous supply of a natural product in constant quality and yield, these sources could solve, at least in principle, all the uncertainties (habitat destruction, political turmoil, environmental and climatic effects) involved in obtaining a pharmaceutical product by isolation from a natural biomass. Furthermore, the biotechnological production of secondary metabolites could be "domesticated" by the addition of unnatural substrates and/or the use of modern genomic technologies, modifying a specific biogenetic pathway and expanding the pool of natural products. Last, but not least, the biotransformation of low-cost precursors into more valuable compounds could also be pursued.

In practice, fermentation techniques have been far less successful for plant products than for microbial products. Our limited knowledge of the regulation of plant genes is responsible for the underdevelopment of combinatorial biosynthesis of plant secondary metabolites *vs.* that of microbial natural products.[2] Compared with microbial cells, plant cells are larger, relatively inflexible due to the presence of a rigid cellulose cell wall, slow growing and prone to aggregation. Furthermore, they are also unpredictable in productivity compared with intact plants. The fermentation throughput is therefore low (weeks rather than days for each batch) and sensitivity to shear stress in bioreactors is high. On the other hand, the production of secondary metabolites in cell cultures can be increased by the addition of precursors and elicitors, by careful optimisation of the culture environment,[24] or by mutation-induced strain improvement. This can be affected either chemically (treatment with mutagenic compounds) or physically (ultraviolet radiation).[24]

Plant tissue cultures are generally carried out in a liquid medium, using cell suspensions to provide uniform conditions and support a faster growth in an easily scalable way.[23] The carbon source is a simple carbohydrate (sucrose and glucose), but different conditions are used to support cell proliferation *via* basic primary metabolism and to foster the production of secondary metabolites, since this process is not generally associated with growth.

The first plant product commercially produced by plant cell culture was the prenylated anthraquinone shikonin **16**, from the boraginaceous plant *Lithospermum erythrorhizon* Sieb. et Zucc. (Mitsui Petrochemical Industry Company) in 1983.[25] Shikonin is used as a dye in cosmetics (lipsticks, soaps and lotions) and its production yield from cell cultures was over ten-fold its isolation yield from the intact plant.[25] In practice, eight runs of two weeks each in a 200 L bioreactor could afford the amount of shikonin produced in four years by a 1 ha field of *L. erythrorhizon*![25] Shikonin has an interesting and pleiotropic biological profile, which includes insulin mimicry and interference with protein–protein interactions, but it has not yet found medicinal application.[26]

Ginsenosides have also been produced in commercial scale from *Panax ginseng* L. tissue cultures (Nitto Denko Company),[27] but the most spectacular success of plant cell cultures has been the commercial production of the anticancer diterpenoid paclitaxel **10** by Phyton Biotech and Bristol-Myers Squibb in large-scale fermentators of 75 000 L and under cGMP (current Good Manufacturing Practice).[28] Although being currently phased out, this process exemplifies the potential of the plant tissue culture to produce natural product drugs and was awarded a Presidential Green Chemistry Challenge Award in 2004 by the US Environmental Protection Agency (EPA).[29]

To increase production and facilitate isolation, plant cells have been immobilised on various matrices such as polyurethane foam and calcium alginate gel beads,[24] while elicitation (*i.e.* the induction of a defence response) is generally critical for the production of secondary metabolites. The rationale for the use of elicitors is that plants produce secondary metabolites as part of a defence response to stress, either biotic (pathogen infection) or abiotic (ultraviolet, toxic heavy metals and rare earth ions). Jasmonic acid plays a crucial role in plant stress responses and, along with fungal polysaccharides and heavy metals, is the most widely employed elicitor in plant tissue cultures.[30]

To overcome the genetic instability and the slow growth of plant cell cultures, hairy root cultures have been developed. Hairy roots are plant roots transformed by *Agrobacterium rhizogenes* carrying the Ri T-DNA plasmid and, due to a higher degree of differentiation, are genetically more stable and grow faster than plant cell cultures.[31] The incubation conditions are, in general, much easier and elicitors are generally not essential. The list of plant natural products produced from hairy root cultures is impressive and includes anthraquinones, flavonoids, saponins, alkaloids and all type of terpenoids, including volatile monoterpenoids.[31] Nevertheless, none of these have so far been commercialised.

Natural products can also be obtained from a direct biotechnological route where all the genes involved in the biosynthesis are expressed in a fermentable host. The transgenic production of the antimalarial sesquiterpene lactone artemisinin **17** is currently being investigated as a cheaper alternative to isolation from *Artemisia annua* L. or to total synthesis.[32] A biochemical and chemical precursor of artemisinin **18** (artemisinic acid) has been produced in acceptable yield from the fermentation of an engineered strain of the yeast *Saccharomyces cerevisiae* where the production of farnesyl diphosphate was diverted from the triterpenoid sink to the sesquiterpene pool. The amorphadiene synthase gene and a cytochrome P450 monooxygenase from *A. annua* were then expressed in this engineered yeast, resulting in the conversion of farnesyl diphosphate into artemisinic acid that reached titres of up to 100 mg/L in the fermentation broth.[32]

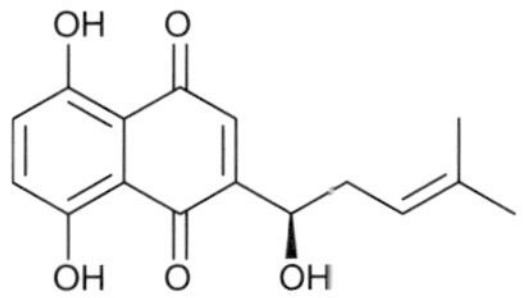

16

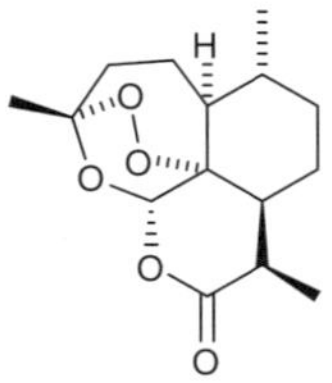

17

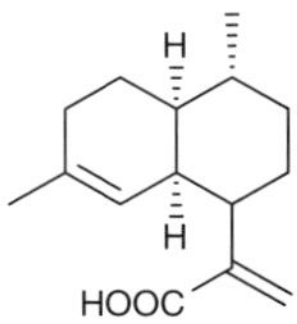

18

Genomic techniques can also be employed to generated patentable transgenic plants. The major drawback of this strategy is the length of time required cultivating plant mutants to assess the effect of a mutation. In this context, the best investigated plant is undoubtedly *Papaver somniferum* L., the only natural source of morphinane alkaloids. The production of these alkaloids turned out to be sensitive to the manipulation of the cytochrome dependent P450 monooxygenase (*S*)-*N*-methylcoclaurine 3′-hydroxylase (CYP80B3), an enzyme on the pathway to the benzylisoquinoline alkaloid branch point intermediate (*S*)-reticuline. Overexpression of cyp80b3 cDNA led to a four-fold increase in the production of alkaloids, while antisense-cyp8030b cDNA expression was detrimental for alkaloid production.[33] Although these changes did not alter the ratio of the individual alkaloids, transgenic lines overexpressing cDNA of the enzyme codeinone reductase (PsCor1.1) showed a significant increase in morphine levels in the capsule alkaloids.[34]

Finally, fermentation of endophytic fungi from higher plants has also been considered for the production of plant natural products. Fungal fermentation is much simpler than plant tissue culture but, at least for paclitaxel, production by fermentation of various *Taxus* endophytic fungi was lower compared with that of plant cells.[35]

Despite the enormous efforts in the biotechnological production of plant secondary metabolites, only three commercial processes have so far been implemented and no genetically modified plant is currently cultivated for the production of secondary metabolites. On the other hand, continuous and rapid advances in plant genomics, transcriptomics and proteomics could make the production of plant natural products by cell culture, transgenic plants or transfected microbial cells much more relevant in the future.

4 Critical Issues in Plant-based Natural Product Drug Discovery

Compared with other sources of secondary metabolites, plants may be considered "easier" to utilise, since their availability is direct (field collection) and they can be propagated and made available in bulk by agricultural procedures (a time-validated technique). Conversely, microbial sources need fermentation expertise, while the direct collection of marine organisms and animal sources is unsustainable. However, several issues complicate the exploitation of plant-based drug discovery campaigns much more than those based on other sources and have undoubtedly contributed to the phasing out of plant secondary metabolites from the drug discovery campaigns of most pharmaceutical companies.

4.1 Intellectual Property (IP) Issues

Political sensitivity regarding access to biodiversity from developing countries is undoubtedly one of the reasons underlying the phasing out of plant natural products from large pharma, since these issues are much more marked with plants than with microbial sources, while marine organisms generally lack ethnopharmacological documentation and the IP issues connected to their collections have an exclusive geographical basis.

The access to plant biodiversity from natural habitats is fraught with legal complication, especially for broad-scale corporate campaigns involving the collection of hundreds, or even thousands, of species. Furthermore, many plants have been employed for ritual or medicinal uses and their "commodification" has sociological implications not addressed by IP considerations. The resolution of these issues, discussed in detail in the Chapter 4, is complicated by the asymmetry between biodiversity-rich developing nations and technology-rich Western countries.

The United Nations Convention on Biological Diversity (CBD) states that countries have sovereign rights over the biological resources within their boundaries and conditions for the preservation and sustainable use of their biodiversity should be established, sharing any commercial benefit resulting from its use.[36] These general claims are very difficult to translate in terms of specific pharmaceutical IP, although their implementation and reinforcement could, in principle, make prospecting an engine for biodiversity conservation.[36] In principle, biodiversity-rich developing societies should interact with technologically advanced developed societies on the basis of a principle of equity. Therefore, countries rich in biological resources should be able to charge companies for bioprospecting for either drugs or genetic information that could lead to new drugs. However, legally binding formulae to control this "trade" are difficult to create and to implement, as exemplified by the so-called Bonn Guidelines on Access and Benefit Sharing. These rules were devised in 2002 by the countries signatory to the CBD to specify how each country should frame licenses to allow companies to access natural resources. This arrangement was fiercely opposed by many environmental and economic organisations and Jeremy Rifkin, who heads the Foundation on Economic Trends, vociferously claimed that "*nobody has the right to enter into exclusive deals over the products of millions of years of evolution*".[37]

As a result of this ambiguity, many countries have placed barriers on the exporting of biological materials, even for non-commercial research. In retrospect, rather than fuelling biodiversity conservation, the CBD has raised unrealistic expectations and has proved inadequate to cope with the complexity of drug discovery.

The anti-HIV diterpenoid prostratin **19** exemplifies this situation. This compound was first isolated in 1977 from a poisonous *Pimelea* species (*P. prostrata*), a New Zealand plant also cultivated in Europe and the USA as an ornamental.[38] Fifteen years later, prostratin was identified as a powerful anti-HIV agent during a campaign launched by the US National Cancer Institute (NCI) to discover new anti-AIDS agents from natural sources.[39] Prostratin is a non-tumour promoting selective activator of protein kinase C (PKC) capable of upregulating latent HIV-1 provirus expression and inhibiting viral infection, potentially leading to the elimination of latent viral reservoirs and to the eradication of HIV-1.[40] Prostratin was isolated from a Samoan medicinal plant [*Homalanthus nutans* (G. Forst) Guill.], but owing to its very low concentration in plant biomass (<3 mg/kg dry weight), its development was hampered by severe shortage problems. The NCI granted a worldwide licence for the use of prostratin in HIV infection to the AIDS Research Alliance (ARA), which included an equitable return of benefits to the Samoan people. Within the options explored to increase the availability of prostratin, in September 2004 the Samoan government signed an agreement with the University of Berkeley related to the isolation from *H. nutans* of the gene sequences for the biosynthesis of prostratin and their transfer into fermentable organisms.[41] The situation was, however, complicated by two unexpected twists, namely the development of a viable semi-synthetic method to produce prostratin from phorbol,[42] a compound more readily available, and the discovery of both semi-synthetic[43] and natural biological analogues of prostratin.[44] It is not clear at this point in time, in case of the clinical development of semi-synthetic prostratin or some of its biological analogues, how an equitable return to the Samoan government could be addressed.[45]

The commodification of traditional knowledge poses problems that also transcend intellectual property considerations since, in indigenous communities, medicinal plants can have cultural, symbolic and ritual values that go beyond a simple medicinal or economic use. Thus, the cultivation of a plant outside its natural habitat and the capture of its medicinal properties into a commercial product can generate mistrust, inequality and betrayal because the loss of the cultural value is not addressed by any monetary compensation. These problems have been exemplified by the development of *Hoodia* gordonii,[46] a sacred "life force" of the South African San, which was turned into a commercial "slimming aid".[47]

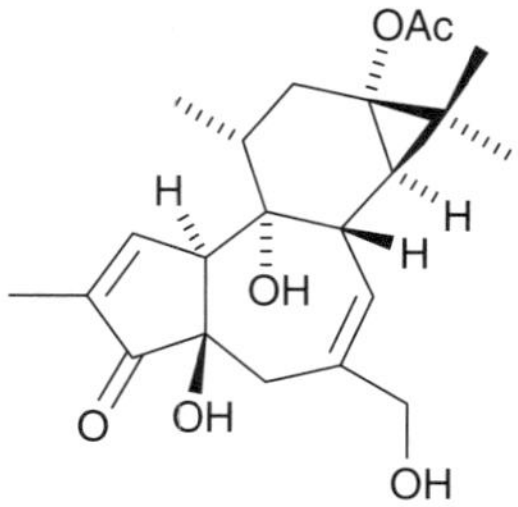

19

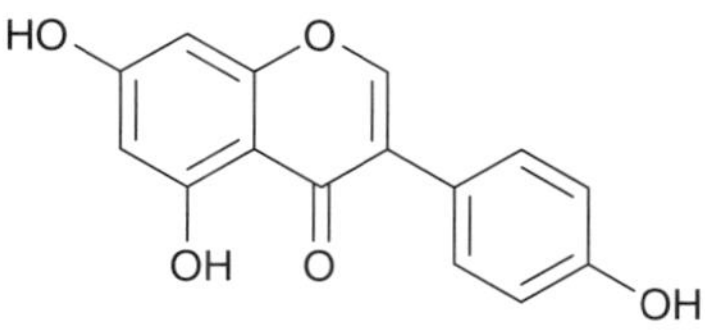

20

4.2 Pleiotropy and Synergy

The active ingredients of certain "heroic" plants bind with high affinity to a single biological target and their potency guarantees significant selectivity. For example, caffeine **13** binds to many macromolecular targets but, at the concentrations normally reached from a dietary or a therapeutic dose, only antagonism at adenosine receptors is relevant.[48]

However, most natural products behave as pharmacologically promiscuous agents and show modest affinity for a host of targets. They are apparently synthesised to address physiological redundancy and pleiotropy—two successful evolutionary mechanisms capable of mitigating the response to a perturbing agent and buffer its cellular effects. Flavonoids are probably the best example of pharmacologically "dirty" plant prototypes and the biological profile of genistein **20**, *per se* one of the more focused agents in the class, gives an idea of the complexity of the pharmacology of these compounds. Thus, genistein binds with modest and comparable affinity to the two estrogen receptors (as well as a multitude of protein kinases), shows anti-oxidant properties, interferes with the cyclo-oxygenase (COX) mediated generation of inflammatory stimuli and has a complicated pharmacokinetics profile that involves the production of active metabolites.[49] The complexity of this molecular profile makes it difficult to predict the clinical activity of genistein, as exemplified by its paradoxical status as both an anticancer agent and a cancer-promoting agent.[50]

Furthermore, plants do not normally contain a single active ingredient but multiple forms of an active ingredient and it is remarkable that escin, a mixture of almost 50 different saponins from horse chestnut, could have found its way into mainstream medicine as a flebotonic agent.[51] The production of a combination of analogues rather than a single major active principle is not exclusive to plants, but occurs in plants with particular frequency and, in a clinical context, can modulate the activity of an active principle (especially its pharmacokinetics) overcoming saturation effects in active transport or extrusion. This is demonstrated by the kavalactones from *Piper methysticum* which are much better adsorbed when given in a mixture than as a single purified agent.[52]

Bioactivity resulting from the synergistic interaction of several compounds to a target is undoubtedly a drawback in HTS, a hit discovery strategy based on the "one target–one drug" assumption. This synergy is often responsible for the disappearance of activity in the course of fractionation of crude extracts into individual chemical components and for the discrepancy between the activity of a series of isolated compounds and that of the extract that contains them. Furthermore, some plant constituents are highly unstable as pure compounds, but stable in an extract mixture; this is especially true of compounds that are oxidised rapidly, since plant extracts generally contain high concentrations of anti-oxidant agents.[53]

Within mixtures of natural products, interactions are the norm rather than the exception and Nature seems to have known very well that there can be strength in numbers and that there are not only "active" compounds but also "supportive" ancillary constituents. Thus, the response of the mammalian immune system to a stimulus depends on the concerted effect of a host of cytokines and not on the activity of just one, while the mandibular pheromone of the queen honey bee is a mixture of nine compounds, all required to elicit its multiple action on worker bees and drones.[54]

In general, synergies are easier to find than to design and, while some drug combinations have a clear mechanistic rationale, such as the association of a β-lactam antibiotic and a lactamase inhibitor, others are mechanistically puzzling and could only be discovered by serendipity or by trial-and-error clinical experience.[53] The association of interferon and ribavirin for the treatment of hepatitis C virus (HCV) infections is an example of "unpredictable" synergy discovered by clinical observation. Thus, ribavirin, a broad-spectrum antiviral, has little intrinsic anti-HCV effect, but was found to significantly improve the clinical responses to interferon and the association of the two agents became a clinical mainstay.[54] Similarly, the discovery that grapefruit juice contains coumarins, which are capable of inhibiting CYP3A4 enzymes and increasing the half-life of a host of drugs, is an example of the serendipitous discovery of synergy in a pre-clinical context, and was found while investigating the ethanol-masking effect of grapefruit juice in oral formulations of 1,4-dihydropyridines calcium blockers.[55]

From a mechanistic standpoint, synergy can take place *via* multivalency or *via* pharmacodynamic and pharmacokinetics effects.[56] Polyvalence takes place when different compounds target distinct elements of a metabolic or signalling pathway and is exemplified by the salicin/acetylsalicylic acid pair. Salicylic acid inhibits the genomic expression of cyclo-oxygenases (COXs), while aspirin inhibits these enzymes directly.[57] Allosteric synergy takes place when two compounds bind distinct sites of an identical target and mutually increase their affinity in a supra-additive way, as observed for sweeteners binding to taste receptors and for the constituents of Synercid for the prokaryotic ribosome.[58] Finally, a pharmacokinetic synergy takes place when one component changes the ability of another to reach its target, either affecting its absorption or metabolism or blocking a resistance mechanism (efflux pump, enzymatic degradation). The apparent disappearance of the antibacterial activity of *Berberis* extract by fractionation is due to an interaction of this type between the antibacterial agent berberine **21**, a lipophilic alkaloid that intercalates DNA and 5′-methoxyhydnocarpin **22**, a non bactericidal flavolignan. Berberine is rapidly and efficiently extruded from bacterial cells by multidrug resistance pumps, but 5′-methoxyhydnocarpin inhibits this efflux, in a classic example of pharmacokinetic synergy at the cell level.[59]

Taken together, these observations on pleiotropism and synergy suggest that a reductionistic approach based on bioassay-directed fractionation against a single target, while providing "target centric" hits, may also miss the activity of many plant extracts.

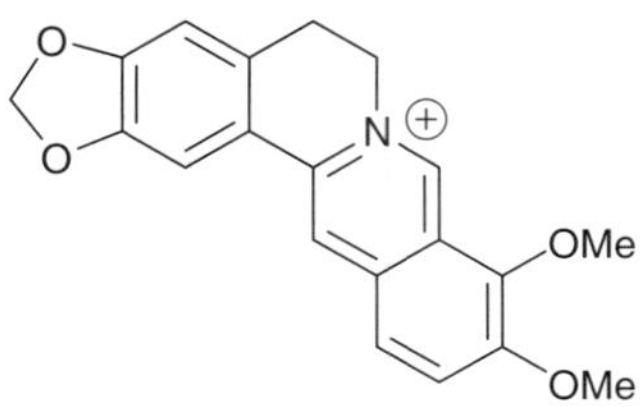

21

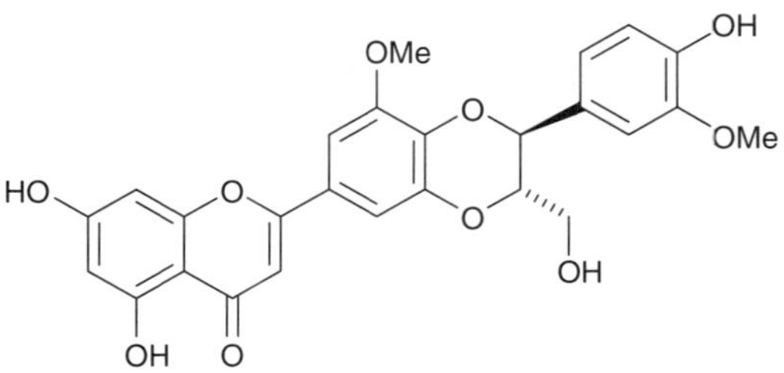

22

4.3 Extract Libraries *vs.* Fraction (Peak) Libraries *vs.* Compound Libraries

Compared with other sources of secondary metabolites, plants are characterised by a higher metabolic profligacy that translates into the production of a host of different types of compounds which span a wide range of polarity. Unsurprisingly, a comparative analysis of plant and microbial natural products sources for the discovery of new secondary metabolites found that plant biomass is by far the best source to scan Nature for new natural products.[60] While the exuberance of plant secondary metabolism makes plant extracts a library of structurally diverse privileged structures, it is nevertheless difficult to directly integrate this diversity into modern drug discovery, since crude extracts are often unsuitable for HTS due to their excessive complexity. False assay read-outs can originate from synergistic interaction or antagonism between the constituents and/or incompatibilities with an assay due to factors such as fluorescence or non-specific interactions with proteins.

Extract libraries contain a large number of compounds and are relatively inexpensive and easy to prepare from plant samples, either by extraction with a single "universal" solvent (*e.g.* methanol or aqueous ethanol) or by sequential extraction with a series of solvents of increasing polarity (*e.g.* hexane, chloroform, methanol). Being characterised by small size and a high chemical diversity, extract libraries require a relatively small screening investment. On the other hand, the possibility of false read-outs is high, minor compounds might not be detected because they are too diluted and multiple dereplication[61] steps are required to translate the activity into a structurally elucidated hit, whose novelty is, at any rate, unpredictable. Hyphenated techniques such as liquid chromatography with ultraviolet detection (LC-UV), liquid chromatography–mass spectrometry (LC-MS), LC-MS/MS or LC-NMR have speeded up dereplication and use only microgram amounts of material, but apart from LC-MS/MS and LC-NMR, they all suffer from an increased possibility of false identifications when the complexity of a mixture increases.[62]

A compromise between the advantages and the disadvantages of testing crude extracts can be achieved by pre-fractionation of extracts by liquid–liquid partition or by solid phase extraction protocols. The so-called Kupchan fractionation is probably the most popular liquid–liquid fractionation scheme for plant extracts, but it is increasingly being replaced by solid phase extraction protocols based on adsorption on a solid matrix and de-adsorption with solvents of growing affinity for the solid phase, *e.g.* petroleum ether/ethyl acetate/acetone for silica gel, mixture of water/methanol (ethanol) with increasing amounts of alcohol for RP-18 silica gel or its cheaper polystyrene resin alternative.[63] Compared with liquid–liquid partition, solid phase extraction is more amenable to automation, but a real comparison between the two methods in terms of efficiency has yet to be reported and is probably dependent on the plant biomass under investigation.

The construction of fractions/peaks libraries has benefited from high throughput purification techniques developed for combinatorial synthesis and a detailed description of a fully automated strategy to generate a library of fractionated extracts has been reported by researchers at Sequoia Sciences Inc.[64] The so called "Sequoia protocol" is based on the combination of automated flash chromatography, solid phase extraction and high-performance liquid chromatography (HPLC) purification that combines isolation, dereplication and identification into a single step. A semi-automated version of a similar process has also been reported and fractionation can also be directly coupled to an assay.[65] Fraction libraries are costly to assemble and screen, but represent a good compromise between extracts and pure natural product libraries. Furthermore, high throughput structure elucidation techniques based on coupling separation and MS or NMR analysis have also been developed, while the advent of relatively cheap, automated flash chromatography instruments has made the construction of small-size fraction libraries amenable to academic groups.[66]

Pure natural product libraries are handled just like any other pure compound library and several of them are commercially available. The largest one (*ca.* 20 000 products, *ca.* 15% of all known natural products) has been built up at AnalytiCon Discovery (Potsdam, Germany) using a strategy based on retrieving from a biological specimen as many novel (non-redundant) compounds as possible.[60,67] The construction of an in-house library of pure natural products requires high investment in terms of both financial and human resources and access to biomass from several hotbeds of plant biodiversity. Alternatively, these libraries can be built by acquisition from natural product scientists in academia, as undertaken by professional providers such as BioSPECS and InterBioScreen. The construction of natural product libraries has become popular in small- or middle-sized biotechnology companies and when big pharma companies shed their natural product drug discovery campaigns, their collections were acquired by smaller biotechnology companies (InterMed Discovery for Bayer, MerLion Pharmaceuticals for GlaxoSmithKline, Albany Molecular Research for Lilly).[66,67]

A library of pure natural products will always be less chemically diverse compared to a library of extracts and will not generally benefit from the bioactivity clues that the ethnopharmacological selection of the starting plant material can afford.

4.4 Removal of Interfering Compounds

Given their complexity, plant extracts may contain compounds that interfere with certain molecular assays. For instance, tannins, a class of typical plant constituents, at least in the gallic and catechic forms, are characterised by a host of hydrogen-bond interactions that lead to the possibility of interaction with several protein targets, especially proline-rich proteins. For this reason, they are often removed from extracts by various techniques—polyamide filtration, polyvinylpyrrolidone (PVP) complexation and caffeine precipitation.[68] While there is hardly any doubt about the poor druggability of large tannins, their simpler forms such as galloylate sugars or dimeric catechins have an excellent bioactivity profile. Thus, pentagalloyl-D-glucopyranose behaves as an orally bioavailable insulin analogue and has been used as a lead structure for the discovery of non-protein mimics of insulin,[69] while procyanidin B2 **23**, a powerful activator of PPARγ2, is absorbed following topical administration and is a powerful lipolytic agent.[70,71]

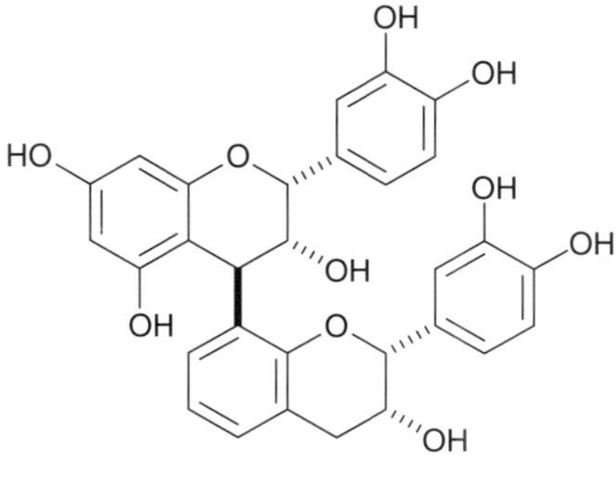

23

Polysaccharides are another class of macromolecules that can interfere with cellular and enzyme assays, as demonstrated during the NCI campaign to discover anti-HIV agents from natural sources. Anionic polysaccharides show anti-HIV activity and, to avoid a too high hit rate, had to be removed by 50% aqueous ethanol precipitation.[72]

For these considerations, the selective removal of certain constituents from plant extracts can be a double-edged sword. On the one hand, it reduces the detection of false positives during HTS campaigns, but it may also reduce or even destroy bioactivity and great care should be exerted in the simplification of plant extracts. For instance, linoleic acid (a ubiquitous compound found in plants) inhibits COXs and can give false positive results in a host of cellular and biochemical anti-inflammatory assays.[73] Although linoleic acid can be easily removed by filtration over neutral alumina, phenolics and other carboxylic acids will also be removed.

5 Selection Strategies for Plant-Based Natural Product Drug Discovery

Bioprospecting is multi-faceted and complex, and requires international cooperation, multidisciplinary attention and a careful planning in terms of geographical focus and methodology. Although unfocused natural product-based drug discovery projects based on the simple assembly of a "peak library" do not require any pre-selection of the starting biomass and are, therefore, relatively easier to implement, selection criteria other than random screening are necessary to limit the variables and uncertainties of more focused programmes of biomedical prospecting. Clues to the presence of bioactivity could in fact be obtained from various sources.

5.1 Ethnopharmacology

The discovery of several important drugs can be traced back to the medicinal or ritual use of specific plants in a non-Western culture. Alkaloids like quinine, emetine, physostigmine and tubocurarine were all discovered because their plant producer had been identified for a particular activity by a native culture.[74] The World Health Organization (WHO) has estimated that 80% of the world's population rely mainly on plant-based traditional medicine for their primary health.[75] There is, therefore, an enormous past and current literature on the human use of plants.

Enthopharmacology is undoubtedly an invaluable source of information for drug discovery, being a sort of validated clinical testing and a shortcut to bioactivity, but its integration into mainstream drug discovery is difficult. Indeed, ethnobotanical information is too often just like the Etruscan language: we can read it, but we do not understand its meaning. Thus, only a limited number of activity categories (vulnerary, antipruritic, antitussive, antiparasitic, contraceptive, haemostatic and laxative) can be directly translated from ethnopharmacology into clear clinical indications.[74] Conversely, modern drug discovery is currently firmly focused on chronic degenerative diseases like cancer, Alzheimer's and atherosclerosis—all conditions that do not translate well into symptomatic observations and which are poorly defined in terms of traditional medicine and folklore, both of which focus on the symptoms rather than cause of diseases.

But despite these limitations, the supplementation of phytochemical data with ethnopharmacological information has been demonstrated to lead to higher hit rates in both the virtual and the actual screening of natural product libraries.[76]

5.2 Zoopharmacy and Animal Toxicology

There is considerable controversy in the literature regarding the self-medicative behaviour of animals and its medicinal value.[77] The discovery of the medicinal properties of certain plants has been traditionally ascribed to animals, with coffee being the best-known example. Furthermore, herbivores depend entirely on plants for their nutrition and well-being, and must have evolved mechanisms to avoid certain foods and prefer others, especially when sick. Even though most observations regarding the ability of animals to self-medicate are anecdotal and equivocal,[77] there is, nevertheless, a general consensus that sick animals seek substances not normally included in their diet and that might indeed contain active ingredients capable of improving their health.[78]

The best documented example of zoopharmacy involves the consumption of *Vernonia amygdalina*, a bitter non-dietary plant, by wild chimpanzees suffering from parasite-related diseases. *V. amygdalina* has a rich phytochemistry that includes sesquiterpene lactones and steroid glycosides endowed with anti-parasitic activity[79] sufficient, at the amounts ingested, to exert anti-nematode activity.[79] It has also been suggested that the high soil consumption by African elephants is related to the need to "detoxify" a diet very rich in secondary metabolites and there is evidence of self-medication in a wide range of animals from bears to geese, leopards, dogs and rhinos.[78] In this context, natural pastures are both nutrition centres and pharmacies for herbivores, and the simplification of the current agricultural systems, which view animals as "machines" to convert agricultural cheap commodities like corn and soy into valuable meat, is highly regrettable.

On the other hand, the observation of the poisoning effects of plants in animals can lead to the discovery of interesting drug leads. The hemorrhagic coumarins, dicoumarol **24** from fermented sweet clover and ferulenol **25** from giant fennel, are classic examples of drug prototypes discovered because of observations from veterinarian toxicology.[80] Important recent examples are:

- cyclopamine **26**,[81] a teratogenic triterpene alkaloid that blocks the activation of hedgehog (Hh) responses by directly binding to and inhibiting one of its critical components (Smo);[82]
- swainsonine **27**,[82] an indolizidine alkaloid that inhibits α-mannosidases;
- castanospermine **28**, a polyhydroxylated alkaloid that inhibits acid α-glucosidase;[82]
- calystegines, a class of polyhydroxylated tropanes that inhibit β-glucosidases as well as α- and β-galactosidases.[82]

Overall, veterinarian observations are, undoubtedly, a significant source of inspiration for natural product-based drug discovery campaigns.

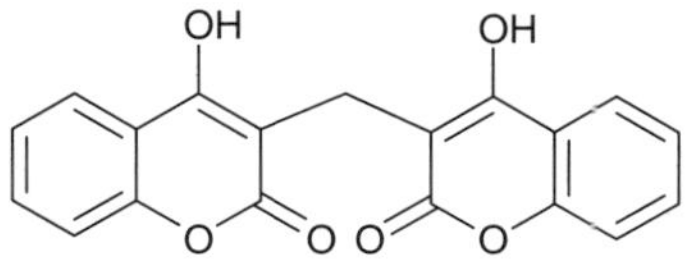

24

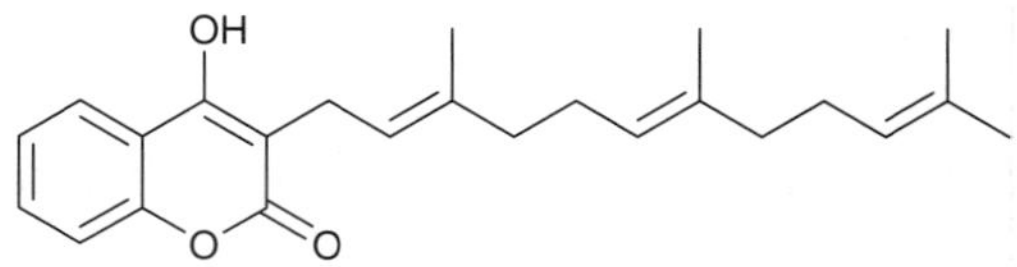

25

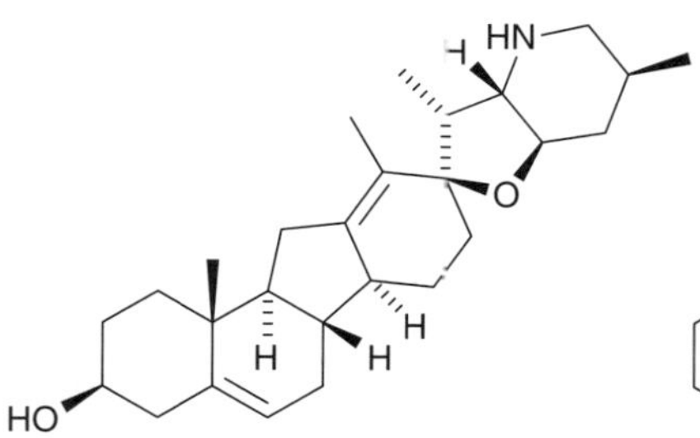

26

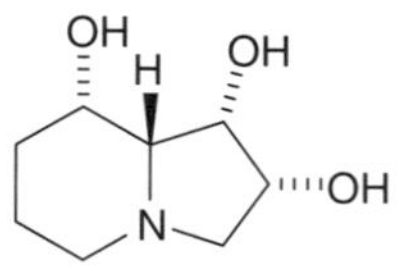

27

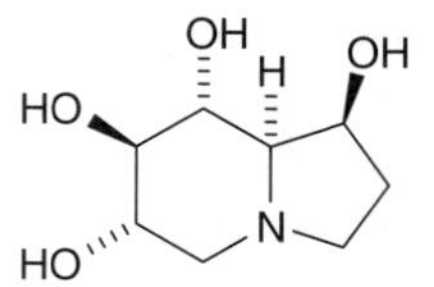

28

5.3 Traditional Medicine

Medicinal plants were identified in grave sites dating back at least to the Middle Palaeolithic period (*ca.* 60 000 years ago) [83] and their use is documented in the medicinal literature of all major civilisations. The number of plants whose medicinal use has been codified is unknown, but over 1000 medicinal plants are used in the Chinese medicinal system.[84] Compared with ethnopharmacological observations, medicinal documentations are generally easier to translate into definite clinical pathologies and our debt to the great plant pharmacognosists of the past (Theophrastus, Dioscorides, Mattioli and Gerard) can hardly be overestimated.

Plants from the Greek–Roman–Arabic tradition have been the cornerstone of our medicinal system, while Ayurvedic medicine, traditional Chinese medicine (TCM) and Kampo medicine are conceptually very different from and more difficult to integrate into mainstream Western medicine.[85] Over 100 000 multi-drug formulas and over 12 000 crude drugs have been recorded in TCM and their study under the reductionistic study of Western medicine is, therefore, unconvincing.[86] Not surprisingly, only a few plant natural products from the Indian and the Chinese medicinal system have been developed as mainstream drugs, with ephedrine and reserpine being the best examples.

Medicinal plants have had a sort of continuous and critically controlled clinical trial and, therefore, represent a primary source for the discovery of new drugs. It is, therefore, amazing that many medicinal plants from the

Mediterranean (Greek–Latin–Arabic) tradition were overlooked for so long by modern medicine. Compounds like resiniferatoxin **29** (RTX) and thapsigargin **30** have become mainstay pharmacological probes, but their plant sources (*Euphorbia resinifera* Berg. and *Thapsia garganica* L.) were investigated only recently despite their relevance in the ancient medicinal literature.[87] Scanning ancient texts for clues to identify new bioactive compounds is very attractive, but it requires a truly a multidisciplinary expertise which is difficult to assemble as it involves both classicists and scientists.[88]

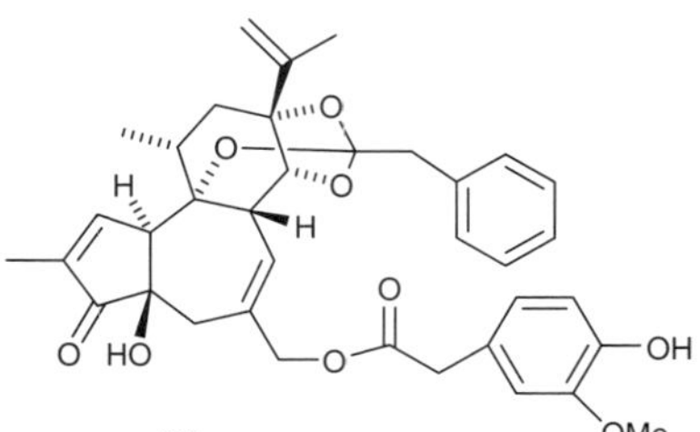

29

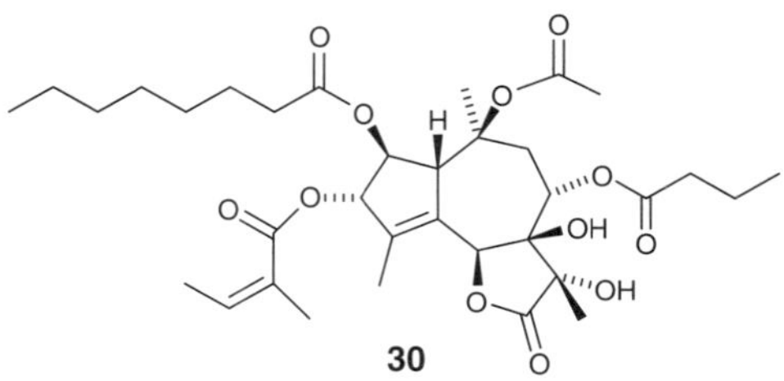

30

5.4 Dietary Plants and Spices

There are close connections between drug discovery and nutrition, as testified by the observation that the first controlled clinical trial was carried out using food plants and not medicinal plants. Thus, in search of a cure for scurvy, the British naval surgeon James Lind in 1747 took 12 sick sailors and divided them into six groups putting them on a fixed diet to which different supplements were added (cider, seawater, vinegar, diluted acids and citrus fruits). After six days, a couple of sailors who had eaten lemons and oranges could be sent back on duty, thus identifying citrus fruits as a source of an anti-scurvy principle.[89]

We are exposed daily to a multitude of secondary metabolites from edible plants and spices that have accompanied us during evolution, playing a role in the shaping of our genome and making us not what we eat, but rather what our ancestors have eaten.[90] Many dietary secondary metabolites appear to play a role, still undefined in molecular terms, for the maintenance of health and there is, therefore, great interest in their identification and in the characterisation of their biological profiles.

Dietary compounds are represented in a number of highly successful drugs such as lovastatin **31** and salicylic acid, the archetypal statin and non-steroid anti-inflammatory drugs, respectively. Lovastatin occurs in the red yeast of rice (*Monascus ruber*),[91] an ingredient of Eastern cuisine used to give a red colour to the Pekinese duck, while salicylic acid is ubiquitous in plants.[92] Other important dietary drug candidates are curcumin **32** from turmeric and capsaicin from hot pepper **33**,[93] while traces of pharmaceutical benzodiazepines (including diazepam) occur in common edible plants such as potatoes and cherries.[5]

Furthermore, dietary observations underlie the discovery of some important drugs. The anti-asthmatic properties of theophylline **34**, a caffeine metabolite and a minor constituent of tea, were discovered because of the improvement on the breathing problems of asthmatic patients who consumed strong black coffee[94] and resveratrol **35** came under the limelight because of the alleged protective effect of red wine in the fat-rich French diet (French paradox).[95] Resveratrol was recently granted orphan drug status for the treatment of encephalomyopathy, presumably because of its sirtuin-binding activity.[96]

In addition, negative dietary correlations can afford clues to drug discovery. Thus, the potent immunosuppressant dammarane triterpenoid **36** was discovered because of epidemiological correlations between the incidence of cancer and the consumption of Palmyrah flour (*Borassus flabellifer*), a staple for Sri Lankan Tamils.[97]

One important issue to be taken into consideration is that one cannot assume a good profile of safety simply because a compound comes from an edible plant. Until the advent of refrigeration and the global market, most fruits and vegetable were only consumed during a limited time of the year and, therefore, chronic toxicity could have easily escaped observation. This issue came dramatically in evidence during the development of annonaceous acetogenins from the paw-paw tree.[98] In the USA, the fruits of this tree are only consumed during a few weeks in the year and no detrimental health effects have seemingly ever been recorded. Conversely, in tropical countries, a severe Parkinson-like neuropathy (tropical Parkinsonism) was correlated to the consumption of annonaceous fruits and the exposure to their acetogenins.[99]

The major limitation of the many dietary clues is that they are often difficult to interpret, since the observed beneficial or detrimental effects of food resist a reductionistic analysis, being the results of a combination of compounds and their bacterial and hepatic metabolites.

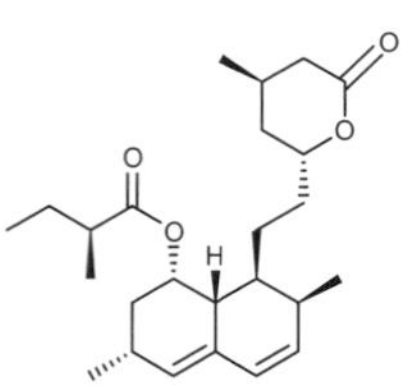

31

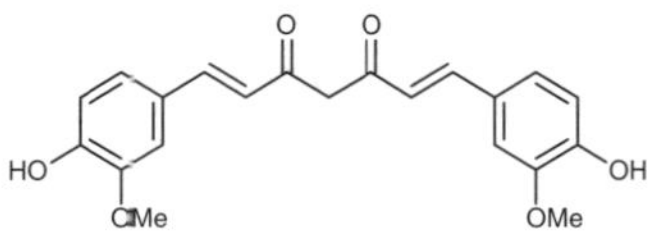

32

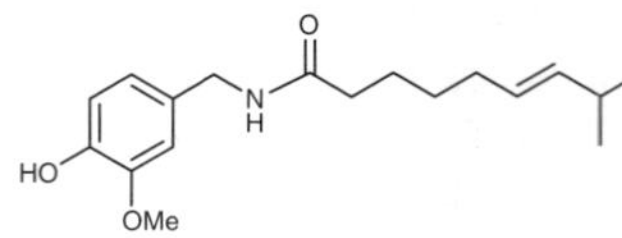

33

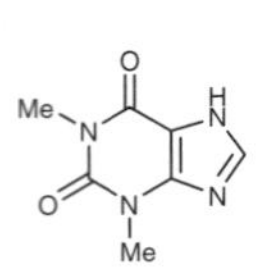

34

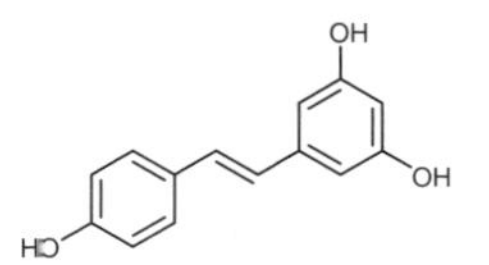

35

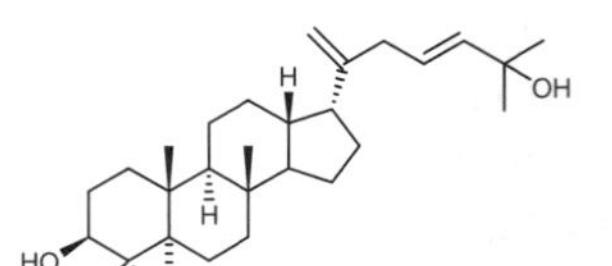

36

6 The Pharmaceutical Relevance of Plants

"Targeted drug discovery"—the "one drug–one target–one disease" paradigm—has been the central dogma in the pharmaceutical industry for over two decades and is the easiest way for a medicinal chemist to approach the issue of discovering novel medicines. Natural products have, probably prematurely, been dropped from drug discovery campaigns because they do not fit easily into HTS programmes. This indispensable element of modern drug discovery is highly automated and labour efficient. Conversely, isolation of pure compounds and structure elucidation are labour-intensive and time-consuming processes that require skilled operators and *ad hoc* solutions.[100] These considerations are at the basis of the constant decline of natural products, and especially those from plants, in programmes of drug discovery. However, the past few years have seen a resurrection of interest for plant extracts, with the establishment of special regulatory status for them both in the USA (botanical drugs) and Europe (traditional drugs).

Although there is no shortage of reviews on the relevance of plants as a source of drugs,[100] drug leads and pharmaceutical intermediates, the area of plant extracts as mainstream drugs is much less well covered. Therefore, the discussion below focuses mainly on this issue.

6.1 Plants as a Source of Lead Structures and Drugs

The contribution of plants to modern medicine can hardly be overstated. Many plant natural products are used in medicine in their native, undomesticated form and the anti-Alzheimer's alkaloid galanthamine and the anticancer diterpenoid paclitaxel are additions to the inventory of plant-derived drugs obtained by direct isolation.[101]

The major systematic project of screening of plant extracts for drug discovery was carried out by the National Cancer Institute between 1960 and 1986. During this titanic effort, 108 330 extracts were screened (not in HTS fashion) for their antitumour potential using murine tumours as end-points. The programme was terminated because it was apparently unsuccessful, although two important classes of drugs later emerged from this activity (taxanes and camptothecins) and others, like maytansins, are currently under development.[102] Given the success rate of the current highly focused medicinal chemistry programmes (one lead every 120 000 compounds screened)[103] the hit rate of the "unfocused" NCI programme was outstanding and its premature phase-out exemplifies the long time required by classic natural product drug discovery projects to bear fruit.

However, the majority of natural products were not created to meet human needs and "domestication" in the form of chemical manipulation is required to increase potency, improve selectivity and reach a clinically acceptable pharmacokinetic and safety profile. Thus, while antibiotics are generated by microorganisms to fight other microorganisms and can be employed clinically to do so, natural cytotoxic agents are not produced to kill cancer cells and they do so only because cancer cells share targets with normal (primary) cells; hence, the need for "domestication" by chemical manipulation to improve the clinical profile and bioactivity of these compounds.

Aspirin was the first "domesticated" natural product introduced into clinical use and the plant chalchone phlorizin **37** is a recent and more sophisticated example of a natural product that underwent extensive "domestication" campaigns.[104] Phlorizin is an inhibitor of glucose transport, both at the intestinal and the renal level, and it is not unreasonable to assume that it is produced to deter predation, since its net effects are detrimental from a nutritional standpoint. On the other hand, the discovery of differences between the intestinal (GLUT4a) and the renal (GLUT4b) glucose transporters has spurred the search for analogues of phlorizin capable of distinguishing between these isophorms and selectively inhibiting the intestinal glucose transporter. The availability of such agents will be very useful for the management of diabetes and these efforts eventually led to the discovery of the *C*-glucoside dapagliflozin **38**, a compound currently undergoing Phase III clinical trials as an antidiabetic agent.[104]

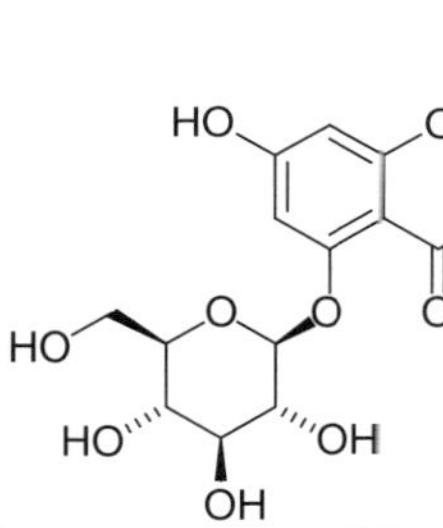

37

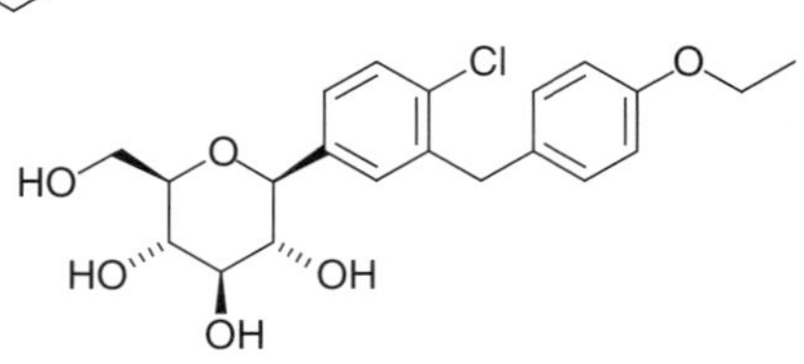

38

Owing to the ready availability of plant biomasses and the high concentration that certain secondary compounds can attain in their natural producer, plant products can also contribute to drug discovery as molecular platforms for the tailored expansion of chemical diversity by combinatorial-style modification.

Natural products contain a wealth of functional groups that can be manipulated by intramolecular reactions or by reaction with simple chemicals. Alternatively, natural products can also be combined in a modular way, generating new natural-product like biodiversity. Thus, 1,3-dienes like the alkaloids thebaine **39** and colchicine **40** could be reacted with 2-nitrosopyridine in a regioselective Diels–Alder fashion, generating a series of conformationally restricted natural product analogues,[105] while transannular cyclisation—a strategy exploited by Nature to create the wealth of sesquiterpenoids—could be extended to diterpenoid precursors, going beyond the duplication of Nature and generating a host of unnatural cyclised skeleta.[106]

Several attempts have also been reported to integrate natural products and combinatorial synthesis, either by using natural products as scaffolds for combinatorial modification or by building libraries of natural product-like compounds by diversity-oriented synthesis. The first approach, as reported for instance for the diterpenoid paclitaxel[107] and for 14-deoxyandrographolide,[108] involves linking a natural product to a solid support[108] or to a radiofrequency tag[107] and then exploiting the presence of reactive groups to carry out extensive derivatisation (generally acylative or alkylative).[107,108] The second approach involves the solid phase generation of a library of analogues of a natural compound, often using a biomimetic reaction to assemble the "privileged" skeleton of the lead structure. The most remarkable successes of this strategy were the identification of a perturbator of mammalian secretory pathways (secramine **41**) from a 2527 member library of analogues of the plant alkaloid galanthamine, a compound *per se* devoid of activity on cellular protein trafficking,[109] and of a highly potent non-steroidal farnesoid X receptor (FXR) agonist from a library of benzopyrans inspired by plant chromones and coumarins.[110]

The pursuit of a dual strategy, where libraries of fractionated extracts and chromatographic peaks are assembled and bulk availability of multifunctional compounds is exploited to expand the natural chemodiversity, is probably the best way to translate biodiversity into biomedically useful chemodiversity. In this way, the uncertainties related to bioprospecting are "buffered" and the biological relativism that characterises combinatorial libraries, where all molecular possibilities are equally possible to succeed, is tempered by the intelligent design associated with natural products. Nature is indeed preferential rather than probabilistic and the failure of the early unbiased combinatorial libraries to deliver leads of biological relevance can be easily understood.

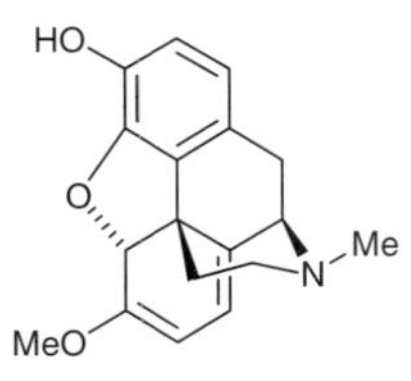

39

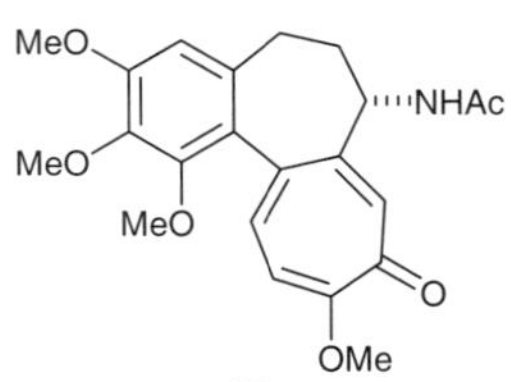

40

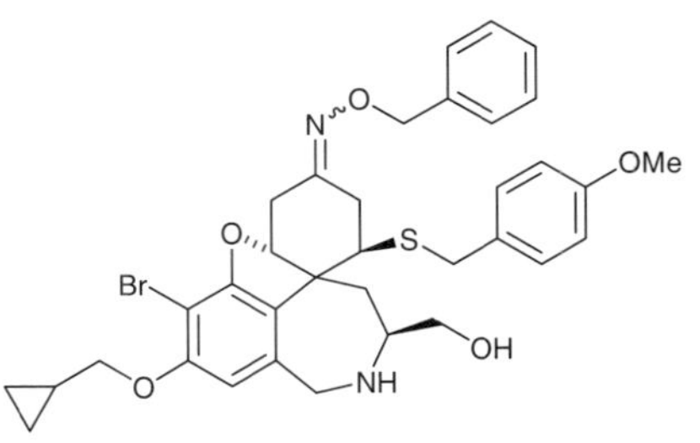

41

6.2 Plants as a Source of Standardised Extracts

The transition from plant extracts to pure compounds was spurred by the desire to overcome three major drawbacks of crude extracts:

- the geographical, seasonal and ecological variation of the contents of the active constituents of a plant;
- the co-occurrence of undesirable compounds capable of modulating bioactivity in an unpredictable way;
- the changes of bioactivity during storage and extraction.

Compared with extracts, pure compounds undoubtedly guarantee a more precise and reproducible dosage and are the only acceptable form of administration of active principles characterised by an extremely narrow therapeutic window, such as the plant alkaloid colchicine **40** from *Colchicum autumnale* L. Lethal poisoning from colchicine has been reported upon ingestion of 3–4 times the therapeutic dose of this compound[111] and it would, therefore, be impractical to use colchicine-rich extracts rather than their active principle.

For these reasons, the transition from extracts to pure compounds has been advocated since the birth of modern medicine. Thus, the eminent pharmacist Charles Louis Cadet de Gassicourt (le *premièr pharmacien de Napoléon*) already pledged this in the inaugural issue of the *Bulletin de Pharmacie* in 1809[112] and, thanks to the efforts of generations of phytochemists, the transition could be considered complete, at least for heroic drugs a century later.

At first sight it seems, therefore, anachronistic that, one century later, plant extracts are again being considered for mainstream pharmaceutical development. Nevertheless, sound reasons underlie this resurrection of interest. The major ones are undoubtedly the clinical success of therapeutic regimes that comprise more than one active principle and the realisation that most chronic and degenerative diseases are characterised by a network complexity and a pathway redundancy that is better addressed by the simultaneous perturbation of several targets rather than a single one. In this context, the dramatic clinical failures of "magic bullets" such as COX-2 inhibitors and CB1 antagonists stand as a sober reminder of the complexity of biological systems and the intrinsic danger of approaching the field with a reductionist strategy.

It is becoming increasingly clear that cell regulation is a tangle of networks of cross-talking nuclear, cytosolic and membrane receptors that share partners, ligands and DNA response elements and doubts have been raised on the suitability of treating diseases caused by cell disregulation with a puritan ideal of target specificity and exclusivity.[113] Inflammation, Alzheimer's disease, obesity and cancer are fundamentally system problems and system solutions are probably required to address them and the use of plant extracts is considered a possible option. Remarkably, a system approach is also advocated for single-gene diseases such as cystic fibroses or sickle cell anaemia, in principle the ideal arena for magic-bullet strategies since, despite their simple genetic cause, system properties confound these pathologies.

Plant extracts offer the possibility to integrate multi-target actions at a patient level. Different compounds can impact distinct targets in the same or different regulatory pathways, resulting in a combined effect that transcends that of each single agent and might provide a clinical benefit. Valerian (*Valeriana officinalis* L.)[114] and St John's wort (*Hypericum perforatum* L.)[115] are two classic examples. The anxiolytic and sleep-inducing properties of valerian are the result of the activity of the sesquiterpenoid valerenic acid **42** on GABA and melatonin receptors, the benzodiazepine-like activity of a series of flavonoid constituents and the spasmolytic action of the iridoid valeropotriates. Extracts from St John's wort are used to treat mild-to-moderate depression and show efficacy comparable with that of the synthetic selective serotonic reuptake inhibitors—a class of compound exemplified by fluoxetine (Prozac®). However, the activity of St John's wort extracts transcends that of every single constituent. The phloroglucinol hyperforin **43**, a TRPC2 inhibitor that shows good brain penetration,[116] interferes with the reuptake of a host of biogenic amines (serotonin, norepinephrine and dopamine), while the naphthodianthrone hypericin **44** inhibits monoamine oxidase (MAO), another antidepressant target, though its oral absorption requires the presence of procyanidins. It seems, therefore, that the clinical efficacy of St John's wort is the result of the combined action of at least three constituents of the plant.[115]

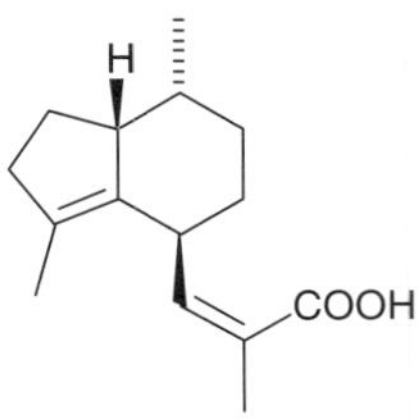

42

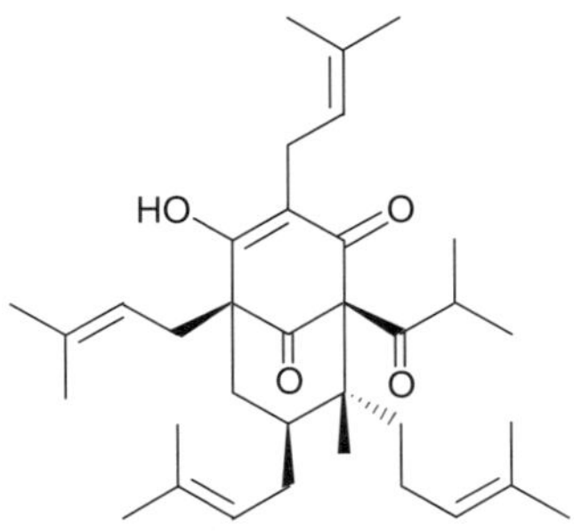

43

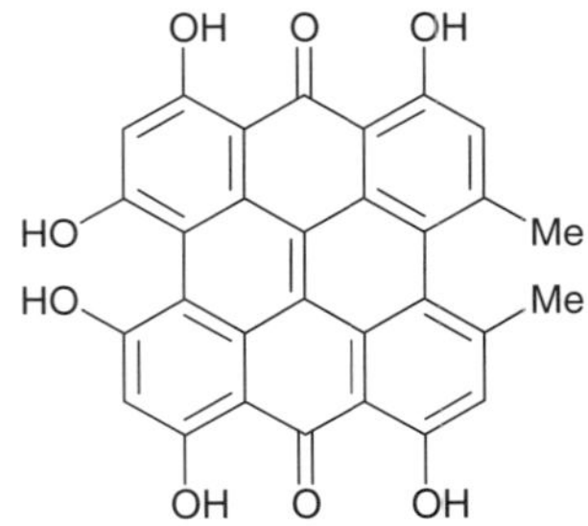

44

The transition from a "magic bullet" to a "magic shotgun" (*i.e.* from a single active ingredient to a combination of active ingredients) can be addressed within a mainstream pharmaceutical context, but the clinical development of extracts (*i.e.* mixtures of structurally and biologically unrelated agents) cannot be accommodated easily under current pharmaceutical thinking.[117] Most natural products are already intrinsically multi-target agents and their use as mixtures, as in extracts, dramatically increases the number of metabolic systems that can be perturbed, attaining a degree of biological complexity that modern medicine is fundamentally unable to address mechanistically on a reductionistic "target centric" approach. Simply put, extracts contain too many keys for too many different locks. Realising the unique nature of plant extracts, the US Food and Drug Administration (FDA) published a special regulatory policy in 2004 to accommodate them into mainstream medicine.[117] In short, if a plant extract or a combination of plant extracts has been legally marketed in the USA or elsewhere without safety problems, Investigational New Drug (IND) applications for a "botanical drug" can be submitted with reduced documentation of preclinical safety and chemistry, manufacture and control (CMC) to support initial clinical studies of activity. When Phase III is reached, the distinction between a drug and a botanical drug becomes much less marked and detailed CMC information is required. Most importantly, neither a full characterisation of all constituents of an extract is required nor the mechanistic elucidation of their interaction, while a certain variation in the final composition of the extract can also be tolerated. Because of their previous usage, botanical drugs cannot generally be protected by patents of composition, at least when a single extract is involved, but protection by methods of use or preparation is possible.

Just like biological drugs, botanical drugs are also relatively safe from generic competition, since a bioequivalence proof will be required to develop a non-proprietary copy. An analysis of the R&D pipeline has shown that over 100 botanical drugs are under development by 80 different companies,[118] but so far only one botanical drug, the antiviral topical cream Veregen®, has been approved by the FDA under the new regulation. Veregen®, approved in autumn 2006 as a prescription drug, is based on Polyphenon E, a catechin-rich extract from green tea developed by the German company MediGen, and is used for the topical treatment of venereal warts. Another product enjoying botanical drug status is Senokot, an over-the-counter laxative containing a fixed dose (8.6 or 17.2 mg) of sennosides from Senna leaves.

A second possibility for the development of plant extracts as mainstream pharmaceuticals is as medical foods. Medical foods are prescribed by physicians to address special nutritional needs or to manage disease conditions that require a special diet and are very different from nutritional supplements. Just like nutritional supplements, medical foods must be given orally, but they are not meant to be used by the general public, may not be available in stores or supermarkets and must be produced under pharmaceutical conditions (cGMP). Furthermore, while no health claims are possible for nutritional supplements, medical foods can have such claims although these must be proven by clinical trials. The ingredients of medical foods are, in fact, food ingredients and, therefore, must have a Generally Recognised as Safe (GRAS) status. Some flavonoid-rich extracts have been developed as medicinal foods. Limbrel (flavocoxid)[119] is based on baicalin and catechin—two phenolics widespread in edible plants (soy, peanuts, cauliflower, kale, apples and green tea)—and is used to treat osteoarthritis. Its dual mechanism of activity involves inhibition of oxidative metabolisation of arachidonic acid by COXs and lipoxygenases (LOXs) and antioxidant activity. Another medical food is Fosteum,[120] an association of genistein **20** from soy, vitamin D and chelated zinc, used for the management of osteoporosis.

Since extracts contain multiple active constituents that may act in combination, the chemical and biological profiling of an extract is much more difficult than that of a single drug and the development of innovative methods for validation, characterisation and standardisation are necessary to develop extracts as mainstream drugs. "Omic" sciences and systems biology might provide the means to decipher the complex molecular interactions responsible for the biological profile of an extract, but the task is enormous and we might never claim a level of knowledge comparable to that of a unimolecular drug. A plant extract might well be akin to the Zen garden in Kyoto that contains 15 large stones that cannot be all seen from any point of the garden. Wherever one sits, at least one stone is blocked by another and the whole can never be perceived in its complexity

The daunting difficulties involved in the pharmacological development of an extract become combinatorial when more than one extract is used. It is, therefore, hardly surprising that most botanical drugs under development are single extracts and not combinations of extracts. On the other hand, advantages could derive when combinations of extracts are used. Ayahuasca, a sacred visionary beverage from the Amazons, is probably the best investigated example. Ayahuasca is a mixture of two plants, *Psychotria viridis* and *Banisteriopsis caapi*. Neither plant is active alone, but their combination is. *P. viridis* contains the psychoactive amine dimethyltryptamine **45** (DMT), a compound not orally active because of its rapid metabolic oxidation by MAO. On the other hand, *B. caapi* contains the powerful MOA-inhibitor, harmine **46**, which makes DMT orally bioavailable.[121]

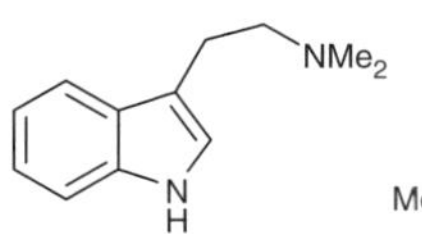

45

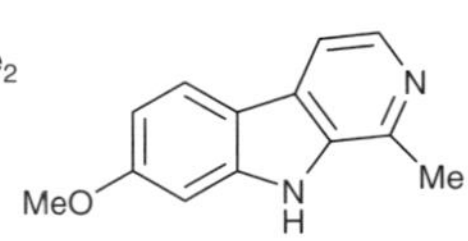

46

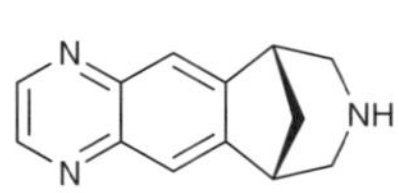

47

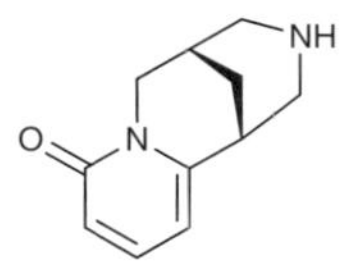

48

7 Conclusions

Over the past century, plants have substantially lost their role as a source of ready-to-use medicines (crude extracts and active principles) and, with a few notable exceptions, their pharmaceutical relevance has nowadays shifted to the realm of drug discovery and process development where they act as a source of drug leads and/or starting material for drug synthesis. The development of varenicline **47**, a drug to treat nicotine abuse, from the alkaloid cytisine **48**,[122] is a recent example of this trend. On the other hand, advances in analytical techniques and "omic" sciences have now provided the basis for the development of purified plant extracts as mainstream drugs. While natural products are rarely obtained with a pharmacodynamic and pharmacokinetic profile sufficient to meet the stringent current standards of clinical development, plant extracts might provide a better opportunity of intervention, especially for multifactorial chronic diseases hardly amenable to management with a single active principle.

Plant natural products are our pharmaceutical cultural heritage as their exquisite biological specificity has laid the foundations of modern medicine. Nevertheless, research on plant natural products has lost the high profile status it long enjoyed in the realm of drug discovery, which is currently dominated by flamboyant new strategies constantly in the news but rarely in the clinic and where plant natural products are perceived as "old" rather than validated. On the other hand, as Longfellow wrote: *"Age is opportunity no less than youth itself, though in another dress, and as the evening twilight fades away, the sky is filled with stars, invisible by day."*

Rational drug discovery in its stellar enthusiasm seems to have adopted the Olympic motto (*citius, altius, fortius*), aiming at obtaining, in the shortest time, the highest number of compounds and with the strongest potency. In frantically doing so, modern drug discovery might be missing the many stars that plants could bring into the limelight of biomedical research and acting like a farmer who eats its seeds rather than planting them. Compared with synthetic compounds, natural products are undoubtedly expensive in terms of manpower, generation and technology and the major challenge facing natural product chemists in the next few years will be to dispel the anti-Olympic status of secondary metabolites, often perceived as too slow to obtain (*lentius*), too low in number (*profundius*) and too mild (*suavius*) in activity for drug discovery. In 1994, it was estimated that three months of work (as well as $50 000) were required for the deconvolution of a single plant extract.[123] Just five years later, Analyticon, in collaboration with Aventis, generated a library of 4000 pure and non-redundant natural products in only 18 months[60] and libraries of over 10 000 compounds can nowadays be screened routinely against more than one target in one week.

Surely, we have never been in a better position to leverage on plant biodiversity to discover new drugs. What is missing is mainly the appreciation that capitalising on natural products takes time and that, unlike so many medical novelties *du jour*, natural products are more suitable for the clinics than for the news.

References

1. Sertürner's claims to have already obtained morphine in 1806 seem largely unsubstantiated, since he probably isolated what became later known as meconic acid and he unequivocally reported the isolation of a narcotic opium basic compound only in 1817. For a discussion, see: F. A. Chast, in *The Practice of Medicinal Chemistry*, ed. C. G. Wermuth, Elsevier, Amsterdam, 3rd edn, 2008, pp. 3–62.
2. L. Knorr, *Chem. Ber.*, 1884, **17**, 2032.
3. W. Sneader, *Drug Prototypes and Their Exploitation*, Wiley, Chichester, 1996, p. 81.
4. W. Sneader, *Drug Prototypes and Their Exploitation*, Wiley, Chichester, 1996, p. 589.
5. U. Klotz, *Life Sci.*, 1991, **48**, 209.
6. D. J. Newman, G. M. Cragg and K. M. Snader, *J. Nat. Prod.*, 2003, **66**, 1022.
7. I. Raskin and C. Ripoll, *Curr. Pharm. Des.*, 2004, **10**, 3419.
8. D. I. Hawksworth, *Stud. Mycol.*, 2004, **50**, 9.
9. P. M. Dewick, *Medicinal Natural Products: a Biosynthetic Approach*, Wiley, New York, 2nd edn, 2002.
10. G. Bringman and A. Irmer, *Phytochem. Rev.*, 2008, **7**, 499.
11. P. Hedden, A. L. Phillips, M. C. Rojas, E. Carrera and B. Tudzynski, *J. Plant Growth Regul.*, 2002, **20**, 319.

12. L. Ottaggio, F. Bestoso, A. Armirotti, A. Balbi, G. Damonte, M. Mazzei, M. Sancandi and M. Miele, *J. Nat. Prod.*, 2008, **71**, 58.
13. J. M. Cassady, K. K. Chan, H. G. Floss and E. Leistner, *Chem. Pharm. Bull.*, 2004, **52**, 1.
14. B. B. Jarvis, J. O. Midiwo, G. A. Bean, M. B. Aboul-Nasr and C. S. Barros, *J. Nat. Prod.*, 1988, **51**, 736.
15. W. C. Widdison, S. D. Wilhelm, E. E. Cavanagh, K. R. Whiteman, B. A. Leece, Y. Kovtun, V. S. Goldmacher, H. XIe, R. M. Steeves, R. J. Lutz, R. Zhao, L. Wang, W. A. Blaetter and R. V. Chari, *J. Med. Chem.*, 2006, **49**, 4392.
16. H. Liu, C. E. Olsen and S. B. Christensen, *J. Nat. Prod.*, 2004, **67**, 1439.
17. G. B. Stefano, P. Cadet, R. M. Kream and W. Zhu, *Neurochem. Res.*, 2008, **33**, 1933.
18. S. Bruzzone, I. Moreschi, C. Usai, L. Guida, G. Damonte, A. Salis, S. Scarfi, E. Millo, A. De Flora and E. Zocchi, *Proc. Natl. Acad. Sci. USA*, 2007, **104**, 5759.
19. S. Imre, A. Oztuinc, T. Celik and H. Wagner, *J. Nat. Prod.*, 1987, **50**, 1187.
20. M. Akao, T. Takeda, T. Kita, T. Kume and A. Akaike, *Cardiovasc. Drug Rev.*, 2007, **25**, 333.
21. L. Del Giudice, D. R. Massardo, P. Pontieri, C. M. Bertea, D. Mombello, E. Carata, S. M. Tredici, A. Talà, M. Mucciarelli, V. I. Groudeva, M. Di Stefano, V. Vigliotta, M. E. Maffei and P. Alifano, *Environ. Microbiol.*, 2008, **10**, 2824.
22. M. Papaioannou, S. Schleich, I. Prade, S. Degen, D. Roell, U. Schubert, T. Tanner, F. Claessens , R. Matusch and A. Baniahmad, *J. Cell Mol. Med.*, 2008, (in press).
23. L. G. Zhou and J. Y. Wu, *Nat. Prod. Rep.*, 2006, **23**, 789.
24. S. C. Roberts, *Nat. Chem. Biol.*, 2007, **3**, 387.
25. M. Tabata and Y. Fujita, in *Biotechnology in Plant Sciences*, ed. M. Tabata and Y. Fujita, Academic Press, New York, 1985, pp. 207–218.
26. X. Chen, L. Yang, J. J. Oppenheim and M. Z. Howard, *Phytother. Res.*, 2002, **16**, 199.
27. T. Furuya and K. Ushiyama, in *Biotechnological Applications of Plant Cultures*, ed. P. D. Shargool and T. T. Ngo, CRC Press, Boca Raton, 1994, pp. 1–35.
28. K. Venkat, *Pure Appl. Chem.*, 1998, **70**, 2127.
29. H. Tabata, *Curr. Drug Targets*, 2006, **7**, 453.
30. F. Di Cosmo and M. Misawa, *Biotechnol. Adv.*, 1995, **13**, 425.
31. S. Guillon, J. Tremouillaux-Guiller, P. K. Pati, M. Rideau and P. Gantet, *Curr. Opin. Plant Biol.*, 2006, **9**, 341.
32. D. K. Ro, E. M. Paradise, M. Quellet, K. J. Fisher, K. L. Newman, J. M. Ndungu, K. A. Ho, R. A. Eachus, T. S. Ham, J. Kirby, M. C. Chang, S. T. Withers, Y. Shiba, R. Sarpong and J. D. Keasling, *Nature*, 2006, **440**, 940. The technology to produce artemisinic acid in engineered yeasts was licensed to Sanofi-Aventis in March 2008. The introduction of artemisinic

acid into the supply chain of artemisinin is scheduled for 2010 (www.amyris.com).

33. S. Frick, R. Kramell and T. M. Kutchan, *Metab. Eng.*, 2007, **9**, 169.
34. P. J. Larkin, J. A. Miller, R. S. Allen, J. A. Chitty, W. L. Gerlach, S. Frick, T. M. Kutchan and A. J. Fist, *Plant Biotechnol. J.*, 2007, **5**, 26.
35. G. A. Strobel, *Crit. Rev. Biotechnol.*, 2002, **22**, 315.
36. (a) G. A. Cordell, *Phytochemistry*, 1995, **40**, 1585; (b) D. Mays and K. D. Mazan, *J. Ethnopharmacol.*, 1996, **51**, 93.
37. (a) See: https://www.cbd.int/doc/articles/2003/A-00457.pdf; (b) F. Pearce, *New Scientist*, 18 April 2002, 14.
38. S. Zayed, A. Hafez, W. Adolf and E. Hecker, *Experientia*, 1977, **33**, 1554.
39. K. R. Gustafson, J. H. Cardellina 2nd, J. B. McMahon, R. J: Gulakowsi, J. Ishitoya, Z. Szallasi, N. F. Lewin, P. M. Blumberg, O. S. Weislow and J. A. Beutler, *J. Med. Chem.*, 1992, **35**, 1978.
40. M. Hezareh, *Drug News Perspect.*, 2005, **18**, 496.
41. H. E. Johnson, S. A. Banack and P. A. Cox, *J. Nat. Prod.*, 2008, **71**, 2041.
42. P. A. Wender, J. M. Kee and J. M. Warrington, *Science*, 2008, **320**, 649.
43. N. Marquez, M. A. Calzado, G. Sánchz-Duffhues, M. Pérez, A. Minassi, A. Pagani, G. Appendino, L. Diaz, M. A. Muñoz-Fernández and E. Muñoz, *Biochem. Pharmacol.*, 2008, **75**, 1370.
44. L. M. Bedoya, N. Márquez, N. Martinez, S. Gutiérrez-Eisman, A. Alvarez, M. A. Calzado, J. M. Rojas, G. Appendino, E. Muñoz and J. Alcamí, *Biochem. Pharmacol.*, 2009, **77**, 965.
45. P. A. Cox, H. E. Johnson and G. Tavana, *Science*, 2008, **320**, 1589.
46. F. R. van Heerden, *J. Ethnopharmacol.*, 2008, **119**, 434.
47. S. Vermeylen, *App. Geogr.*, 2008, **28**, 224.
48. S. H. Snyder and P. Sklar, *J. Psychiatr. Res.*, 1984, **18**, 91.
49. For recent reviews, see: (a) K. Szkydelska and L. Nogowski, *J. Steroid Biochem. Mol. Biol.*, 2007, **105**, 37; and (b) R. A. Dixon and D. Ferreira, *Phytochemistry*, 2002, **60**, 205.
50. S. Banerjee, Y. Li, Z. Wang and F. H. Sarkar, *Cancer Lett.*, 2008, **269**, 226.
51. R. W. Frick, *Angiology*, 2000, **51**, 197.
52. J. Keledjian, P. H. Duffield and D. D. Jamieson, *J. Pharm. Sci.*, 1988, **77**, 1003.
53. P. J. Houghton, M. J. Howes, C. C. Lee and G. Steventon, *J. Ethnoparmacol.*, 2007, **110**, 391.
54. K. T. Curtis, A. A. Borisy and B. R. Stockwell, *Nat. Rev. Drug Discov.*, 2005, **4**, 71.
55. T. Uno and N. Yasui-Furukori, *Curr. Clin. Pharmacol.*, 2006, **1**, 157.
56. For recent reviews, see: (a) X. J. Li and H. Y. Zhang, *Trends Pharmacol. Sci.*, 2008, **29**, 331; (b) H. Wagner, *Phytomedicine,* 2006, **13** (Suppl. 5), 122; and (c) K. Spelman, *J. Herb. Pharmacother.*, 2005, **5**, 31.
57. K. K. Wu, *Thromb. Haemost.*, 2006, **96**, 417.
58. D. Beyer and K. Pepper, *Expert Opin. Investig. Drugs*, 1998, **7**, 591.

59. F. R. Stermitz, P. Lorenz, J. N. Tawara, L. A. Zenewicz and K. Lewis, *Proc. Natl. Acad. Sci. USA*, 2000, **97**, 1433.
60. K. U. Bindseil, J. Jakupovic, D. Wolf, J. Lavayre, J. Leboul and D. van der Pyl, *Drug Discov. Today*, 2004, **6**, 840.
61. In general terms, dereplication refers to the differentiation of a compound of interest from nuisance compounds, *e.g.* a novel structure from known natural products.
62. G. Lang, N. A. Mayhuding, M. I. Mitova, L. Sun, S. van der Sar, J. W. Blung, A. L. J. Cole, G. Ellis, H. Laatsch and M. H. G. Munro, *J. Nat. Prod.*, 2008, **71**, 1595.
63. E. L. Ghisalberti, in *Bioactive Natural Products: Detection, Isolation and Structure Determination*, ed. S. M. Colegate and R. J. Molyneux, CRC Press, Boca Raton, 2008, pp. 11–76.
64. G. R. Eldridge, H. C. Vervoort, C. M. Lee, P. A. Cremin, C. T. Williams, S. M. Hart, M. G. Goering, M. O'Neil-Johnson and L. Zeng, *Anal. Chem.*, 2002, **74**, 3963.
65. U. Abel, C. Koch, M. Speitling and F. G. Hansske, *Curr. Opin. Chem. Biol.*, 2002, **6**, 453.
66. Y. Wang, *Phytochem. Rev.*, 2008, **7**, 395.
67. F. E. Koehn, *Prog. Drug Res.*, 2008, **65**, 177.
68. (a) W. D. Loomis and J. Battaile, *Phytochemistry*, 1966, **5**, 423; (b) M. E. Wall, H. Taylor, L. Amrosio and K. J. Davis, *Pharm. Sci.*, 1969, **58**, 839; (c) G. T. Tan, J. M. Pezzuto, A. D. Kinghorn and S. H. Hughes, *J. Nat. Prod.*, 1991, **54**, 143.
69. Y. Ren, K. Himmeldirk and X. Chen, *J. Med. Chem.*, 2006, **49**, 2829.
70. D. M. Chen, X. Cai, C. L. Kwik-Uribe, R. Zeng and X. Z. Zhu, *J. Cardiovasc. Pharmacol.*, 2006, **48**, 54.
71. J. C. Rocha, F. Pedrochi, F. L. Hernandes, J. C. de Mello and M. L. Baesso, *Anal. Chim. Acta*, 2007, **587**, 132.
72. H. J. H. Cardellina, M. H. G. Munro, R. W. Fuller, K. P. Manfredi, T. C. McKee, M. Tischler, H. R. Nokesh, K. R. Gustafson, J. A. Beutler and M. R. Boy, *J. Nat. Prod.*, 1993, **56**, 1123.
73. T. Ringbom, U. Huss, A. Stenholm, S. Flock, L. Skattebøl, P. Perera and L. Bohlin, *J. Nat. Prod.*, 2001, **64**, 745.
74. (a) P. A. Cox, in *Ethnobotany and the Search for New Drugs*, ed. G. Prance and J. Marsh, Wiley, Chichester, 1994, p. 25; (b) M. J. Balick and P. A. Cox, *Plants, People and Culture. The Science of Ethnobotany*, Scientific American Library, New York, 1996.
75. World Health Organization, *The World Health Report. Bridging the Gaps*. WHO, Geneva, 1995, p. 1.
76. (a) J. M. Rollinger, S. Haput, H. Stuppner and T. Langer, *J. Chem. Inf. Comput.*, 2004, **44**, 480; (b) J. M. Rollinger, A. Hornick, T. Langer, H. Stuppner and H. Prast, *J. Med. Chem.*, 2004, **47**, 6248.
77. M. A. Huffman, *Proc. Nutr. Soc.*, 2003, **62**, 371.
78. J. J. Villalba and F. D. Provenza, *Animal*, 2007, **9**, 1360.

79. K. Koshimizu, H. Ohigashi and M. A. Huffman, *Physiol. Behav.*, 1994, **56**, 1209.
80. For a discussion, see: G. Appendino, E. Mercalli, N. Fuzzati, L. Arnoldi, Stavri, S. Gibbons, M. Ballero and A. Maxia, *J. Nat. Prod.*, 2004, **67**, 2180.
81. R. F. Keeler, *Phytochemistry*, 1968, **7**, 303.
82. R. J. Molyneux, S. T. Lee, D. R. Gardner, K. E. Panter and L. F. James, *Phytochemistry*, 2007, **68**, 2973.
83. J. Lietava, *J. Ethnopharmacol.*, 1992, **35**, 263.
84. J. A. Duke and E. S. Ayensu, *Medicinal Plants of China*, Research Publications, Algonac, MI, 1985.
85. D.-X. Kong, X.-J. Li, G.-Y. Tang and H.-Y. Zhang, *ChemMedChem*, 2008, **3**, 233.
86. Y. D. Yi and L. M. Chang, *Evid. Based Complement Alternat. Med.*, 2004, **1**, 125.
87. G. Appendino and A. Szallasi, *Life Sci.*, 1997, **60**, 681.
88. B. K. Holland, *Nature*, 1994, **369**, 702.
89. U. Tröhler, *J. Royal Soc. Med.*, 2005, **98**, 51.
90. G. Appendino and O. Taglialatela-Scafati, in *Dietary Supplements of Plant Origin.*, ed. M. Maffei, Taylor & Francis, London, 2003, pp. 43–74.
91. A. Endo, *J. Antibiot. (Tokyo)*, 1979, **32**, 852.
92. I. Raskin, *Plant Physiol.*, 1992, **99**, 799.
93. G. Appendino, in *Modern Alkaloids: Structure, Isolation, Synthesis and Biology*, ed. E. Fattorusso and O. Taglialatela-Scafati, Wiley, Weinheim, 2007, pp. 75–112.
94. H. Salter, *Asthma. Its Pathology and Treatment*, Churchill, London, 1860, p. 181.
95. J. C. Espin, M. T. Garcia-Conesa and F. A. Tomás-Barberán, *Phytochemistry*, 2007, **68**, 2986.
96. Anon., *Scrip*, **3362**, 16 May 2008, 19.
97. L. Révész, P. Hiestand, L. La Vecchia, R. Naef, H. U. Naegeli, L. Overer and H. J. Roth, *Bioorg. Med. Chem. Lett.*, 1999, **9**, 1521.
98. J. L. McLaughlin, *J. Nat. Prod.*, 2008, **71**, 1311.
99. D. Caparros-Lefebvre, J. Steele, Y. Kotake and S. Ohta, *Mov. Disord.*, 2006, **21**, 1769.
100. J. D. Phillipson, *Phytochemistry*, 2007, **68**, 2960.
101. In fact, galanthamine is now better obtained by total synthesis rather than by direct isolation and paclitaxel by semi-synthesis from 10-deacetylbaccatin III.
102. G. M. Cragg, D. J. Newman and S. S. Yang, *J. Nat. Prod.*, 2006, **69**, 488.
103. I. Muegge, *Chemistry*, 2002, **8**, 1976.
104. W. Meng, B. A. Ellsworth, A. A. Nirschl, P. J. McCann, M. Patel, R. N. Girotra, G. Wu, P. M. Sher, E. P. Morrison, S. A. Biller, R. Zahler, P. P. Deshpande, A. Pullockaran, D. L. Hagan, N. Morgan, J. R. Taylor, M. T. Obermeier, W. G. Humphreys, A. Khanna, L. Discenza, J. G.

Robertson, A. Wang, S. Han, J. R. Wetterau, E. B. Janovitz, O. P. Flint, J. M. Whaley and W. N. Washburn, *J. Med. Chem.*, 2008, **51**, 1145.
105. F. Li, B. Yang, M. J. Miller, J. Zjicek, B. C. Noll, U. Moellmann, H.-M. Dahse and P. A. Miller, *Org. Lett.*, 2007, **9**, 2923.
106. G. Appendino, G. C. Tron, T. Jarevaeng and O. Sterner, *Org. Lett.*, 2001, **3**, 1609.
107. X.-J. Xiao, Z. Parandoosh and M. P. Nova, *J. Org. Chem.*, 1997, **62**, 6029.
108. M. A. F. Biabani, R. K. Grover, S. K. Singh, S. Kumar, K. Raj, R. Roy and B. Kundu, *Tetrahedron Lett.*, 2001, **42**, 7119.
109. H. E. Pelish, N. J. Westwood, Y. Feng, T. Kirchhausen and M. D. Shair, *J. Am. Chem. Soc.*, 2001, **123**, 6740.
110. K. C. Nicolaou, R. M. Evans, A. J. Roecker, R. Hughes, M. Downes and J. A. Pfefferkorn, *Org. Biomol. Chem.*, 2003, **1**, 908.
111. For a recent review, see: V. Jayaprakash, G. Ansell and D. Galler, *N. Z. Med. J.*, 2007, **120**, U2402.
112. W. Sneader, *Drug Prototypes and Their Exploitation*, Wiley, Chichester, 1996, p. 6.
113. R. Verpoorte Y. H. Choi and H. K. Kim, *J. Ethnopharmacol.*, 2005, **100**, 53.
114. P. J. Houghton, *J. Pharm. Pharmacol.*, 1999, **51**, 505.
115. M. Spinella, *Altern. Med. Rev.*, 2002, **7**, 130.
116. K. Leuner, V. Kazanski, M. Müller, K. Essin, B. Henke, M. Gollasch, C. Harteneck and W. E. Müller, *FASEB J.*, 2007, **21**, 4101.
117. B. M. Schmidt, D. M. Ribnicky, O. E. Lipsky and I. Raskin, *Nat. Chem. Biol.*, 2007, **3**, 360.
118. According to the FDA, over 200 investigational new drug (IND) applications are pending (quoted in ref. 117).
119. www.limbrel.com.
120. www.fosteum.com.
121. R. S. Gable, *Addiction*, 2007, **102**, 24.
122. J. T. Hays and J. O. Ebbert, *N. Engl. J. Med.*, 2008, **359**, 2018.
123. D. G. Corley and R. C. Durley, *J. Nat. Prod.*, 1994, **57**, 1484.

CHAPTER 6

Macromarines: A Selective Account of the Potential of Marine Sponges, Molluscs, Soft Corals and Tunicates as a Source of Therapeutically Important Molecular Structures

JENNIFER CARROLL[a] AND PHILLIP CREWS[b]

[a] Department of Chemistry and Biochemistry, California Polytechnic State University, San Luis Obispo, California 93407, USA; [b] Department of Chemistry and Biochemistry and Institute for Marine Sciences, University of California Santa Cruz, Santa Cruz, California 95064, USA

1 Introduction

This chapter reviews a selection of marine invertebrates and the compounds isolated from them which have therapeutic potential. The major phyla of macromarines discussed include the marine Porifera, Mollusca, the soft corals of Cnidaria and the tunicates which belong to the sub-phylum of Urochordates. The isolation, synthesis, preclinical data and structure–activity relationships (SARs) are discussed for a selection of pharmacologically important chemical structures for each of the four groups.

RSC Biomolecular Sciences No. 18
Natural Product Chemistry for Drug Discovery
Edited by Antony D. Buss and Mark S. Butler

Published by the Royal Society of Chemistry, www.rsc.org

1.1 Macroorganisms: Outstanding Success in Producing Viable Drug Leads

The number of biologically active compounds isolated from marine sponges, molluscs, tunicates and soft corals continues to grow with a wide variety of molecular scaffolds and biological activities appearing in the literature each year.[1,2] The quest to develop such marine natural products as clinically useful agents is a well-developed field of research.[3,4] A number of these inspirational compounds are currently undergoing development in preclinical trials and will serve as a pipeline in the development of novel modes of chemotherapy and molecular probes. The following review is a selective account of these molecular structures and congeners with a focus on their development in *in vitro* and *in vivo* studies. As the knowledge increases regarding the interaction of these agents with their molecular targets, many of the leads developed in this research will become the pharmacophore scaffolds for future drugs.

1.2 Setting that Ara A and Ara C Story Straight

Before PharmaMar's recent success in Europe with ecteinascidin 743, (trabectedin), no other approved anticancer agents had been based on an actual marine natural product. However, one clinical drug, cytosine arabinoside (Ara C), has been available as a synthetic derivative of a sponge natural product. In the early 1950s, a series of publications entitled *Contributions to the Study of Marine Products* described the isolation of a number of unique structures including nucleosides.[5,6] Three of these compounds—spongouridine (Figure 6.1), spongothymidine[7,8] and spongosine[9]—were discovered by Bergmann from the Caribbean marine sponge *Cryptotethia crypta*. These compounds were exceptional in that they contained an arabinose pentose rather than the traditional ribose isomer. At this point in time, traditional nucleosides composed of deoxyribose subunits were being examined as treatments for

Spongouridine: R = 1
Ara A: R = 2
Ara C (Cytarabine): R = 3

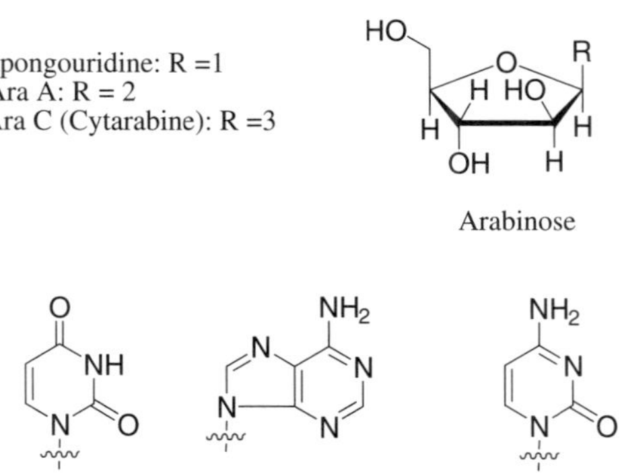

Figure 6.1 Spongouridine, Ara A and Ara C structures.

cancer by way of DNA synthesis inhibition. The discovery of spongouridine piqued the interest of medicinal chemists as an alternative pentoside.[10]

In 1959, syntheses were reported of Ara C and, in 1960, of adenine arabinoside (Ara A, vidarabine)[11] (Figure 6.1). In 1961, Ara C was reported to have antitumour activity[12] and later in the 1980s, both Ara A and Ara U were isolated from the gorgonian *Eunicella cavolini*.[13] Today, Ara C is used to treat Hodgkin's lymphoma and acute myelocytic leukaemia,[14] while Ara A is marketed as an antiviral drug.

1.3 The Potential Role of Invertebrate Associated Microorganisms and Secondary Metabolite Production

Macromarine invertebrates provide safe environments with plentiful nutrient supplies for microorganism growth and reproduction. In fact, it has been well-established that marine sponges are prolific in terms of their associated microorganisms. Some of these associations are species specific. For example, the marine sponge *Theonella swinhoei*, which is the source of the dimeric marcrolide and actin inhibitor swinholide A **20** (Scheme 6.1 Part 3), has an obvious association with the alga *Symploca* cf. sp. and *Geitlerinema* sp. which also produce this compound.[15]

An analysis of the known microorganisms associated with sponges[16] and soft corals[17] is shown in Figure 6.2. The major contributing microorganisms are the Gram-negative proteobacteria, which include among others, nitrogen-fixing bacteria, bioluminescent bacteria and gliding myxobacteria. Interestingly the microbial communities of other marine invertebrate populations remain an area of research that is largely neglected.[18,19] These marine microbe communities and their hosts are undoubtedly a developing source for molecular diversity.[20,21]

1.4 Macromarine Evolution

The Cambrian explosion which occurred between 580 and 490 million years ago (MYA) provided the major marine invertebrate phyla including the sponges, molluscs and soft corals.[22] The hypothetical ancestor of the modern invertebrate phyla, the Urmetazoa, is shown in Figure 6.3 and is attributed with the earliest production of secondary metabolites responsible for apoptosis mediation, control of morphogenesis, and immune and cell adhesion molecules.

The first phyla to arise at the beginning of Cambrian explosion, the Archaeocyatha, resembled simple sponges with holdfasts and pore structures and were among the earliest reef builders. Mysteriously, these early "ancient cups" became extinct at about 520 MYA. The Porifera (sponge) groups, Hexactinellida and Demospongidae, appeared following the evolution of silicic acid skeletons and were followed by the Calcarea, which possess a skeleton of calcium carbonate. The Cnidaria and Ctenophora arrived following the development of further complexity including the appearance of oral and aboral

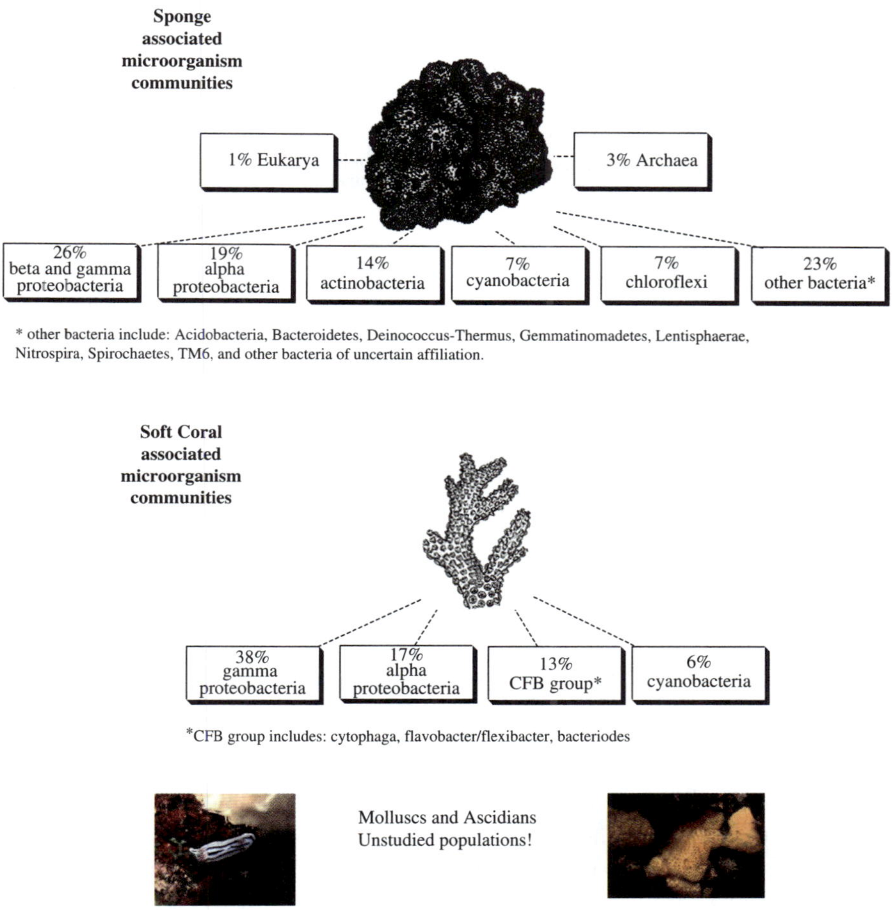

Figure 6.2 Macromarine associated microorganisms demystified.

axis and radial symmetry. Another hypothetical ancestor, the Uribilateria, appeared near the end of the Cambrian explosion and preceded the bilateral symmetry of the modern Phyla of Chordata and Mollusca.

Over 580 million years of evolution has followed the first suggestion of chemical differentiation resulting in an amazing number of biochemical tools for the production of secondary metabolites in marine invertebrates.

2 Sponges

2.1 Natural History of Sponges—a Primitive Phylum with Remarkable Biosynthetic Capabilities

Animals of the phylum Porifera, meaning pore bearing, have an amazingly simple yet perfectly functional body plan which consists of collection of cells

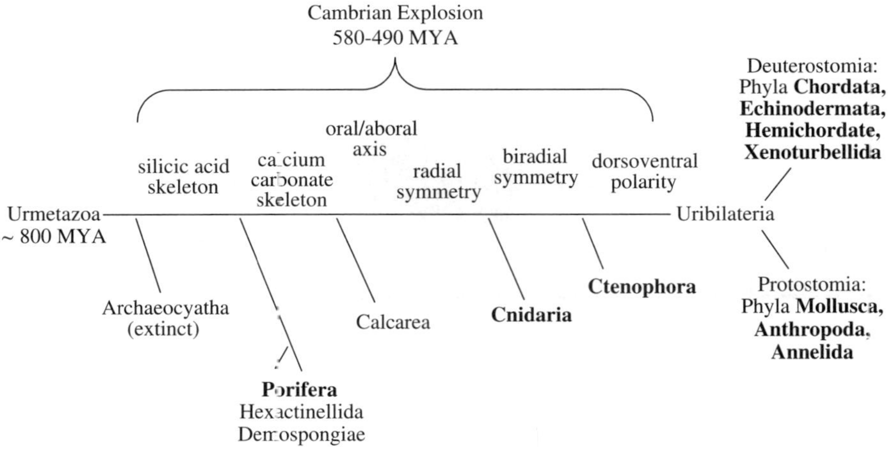

Figure 6.3 Evolution of macromarines (adapted from Muller 2003). Modern Phyla are in bold.

arranged around a series of canals and chambers.[23] These continuously filter the water column (as much as one litre per cubic centimetre per hour) for food by the use of flagellated cells called choanocytes. Sponges are also unique animals in that they have no conventional nerves or muscle tissue and their mobility is only attained by individual cellular movement.

While the higher systematics of sponges remains a widely debated field, the three main classes of Porifera are generally well known. Spicules, the main skeletal and defensive structures within sponges, are the basis for this classification. Within this are the demosponges which contain collagenous spongin fibres, the calcareous which contain bony or calcium carbonate spicules and hexactinellid which contain spicules of silica. The three basic morphologies include the stalk-like asconoid, the more complex body walls of the syconoid and the leuconoid sponges with the flaggellated chonanocytes known as collar cells.

The earliest fossil records of sponges are from the Late Precambrian period (580 MYA).[24] Marine sponges are found in all of the world's oceans and at all depths. Currently, approximately 7000 species of sponges are recognised as taxonomically valid and new species are still frequently discovered.[25] However, molecular biodiversity studies indicate that the majority are under sampled and unidentified and that the actual number of species could possibly be up to twice that number.[26] This amazing natural history, combined with the considerable number of biologically active pure compounds isolated from marine sponges, makes them worthy of investigation.

Current examination of the MarinLit database indicates that 6076 articles have been published with 6980 unique chemical structures from marine sponges.[27] This accounts for approximately 22% of the estimated number of species available.[27] The number of total structures in this well-known database has now reached over 20 000 compounds. Of the marine phyla tested

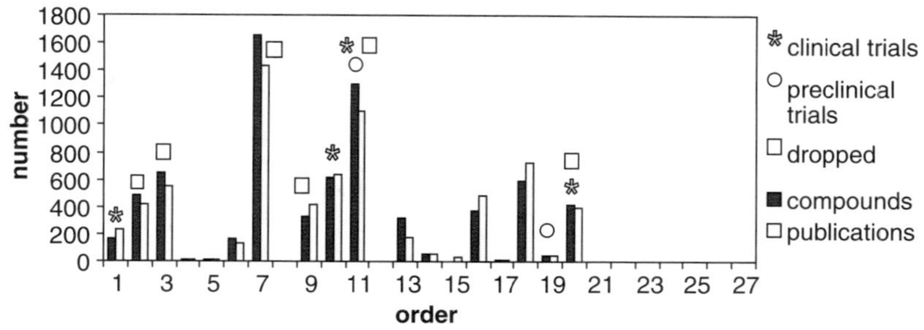

Figure 6.4 Number of publications and pure compounds from Porifera (sponges) (MarinLit 1951–2008). Orders: 1. Agelasida, 2. Astrophorida, 3. Axinellida, 4. Choristida, 5. Clathrinida, 6. Dendroceratida, 7. Dictyoceratida, 8. Epipolasida, 9. Hadromerida, 10. Halichondrida, 11. Haplosclerida, 12. Hexasterophora, 13. Homosclerophorida, 14. Leucettida, 15. Leucosolenida, 16. Lithistida, 17. Petrosida, 18. Poecilosclerida, 19. Spirophorida, 20. Verongida

thus far, sponges appear to be the most prolific in terms of the percent bioactivity in US National Cancer Institute (NCI) anticancer screens (16%).[2,28] Figure 6.4 presents the 20 different orders within the phylum Porifera and their relative productivity based upon the number of publications and pure compounds found in MarinLit. The most prolific in terms of number of pure compounds and publications is the order of Dictyoceratida. The runner up, Haplosclerida, currently has compounds in both preclinical and clinical trials. Also apparent from Figure 6.4 is the number of "forgotten" orders where, due to difficulty in collections or rarity of species, there are far fewer publications.

No other single phylum has been more productive than that of the marine Porifera in producing leads for drug discovery and molecular tools. The remarkable activity of the cytotoxic compounds **1–19** (Scheme 6.1 Parts 1 and 2) have led to the development of marine natural products and their congeners as leads in numerous anticancer clinical trials. The actin inhibitors **20–26** (Scheme 6.1 Part 3) have led to the use of sponge-derived agents as probes in the study of cellular processes. In addition, a number of unique structures such as **27–31** (Scheme 6.1 Part 4) have potential as leads in the treatment of other persistent maladies such as asthma, pain and malaria.

Dictyostatin

Dictyostatin 1 **4**, a sponge macrolide originally isolated from an Indian Ocean *Spongia* sp. in 1994,[162,163] was also isolated from a Caribbean *Corallistidae* sp. in 2003.[41] The relative stereochemistry of dictyostatin 1 was established through *J*-based NMR analysis and molecular modelling studies; it shares common stereochemical configurations with discodermolide **3**.[164] Initial reports indicated that dictyostatin 1 inhibited P388 murine lymphocytic leukaemia cells, inhibited the growth of selected cancer cell lines in the NCI human cancer cell line panel

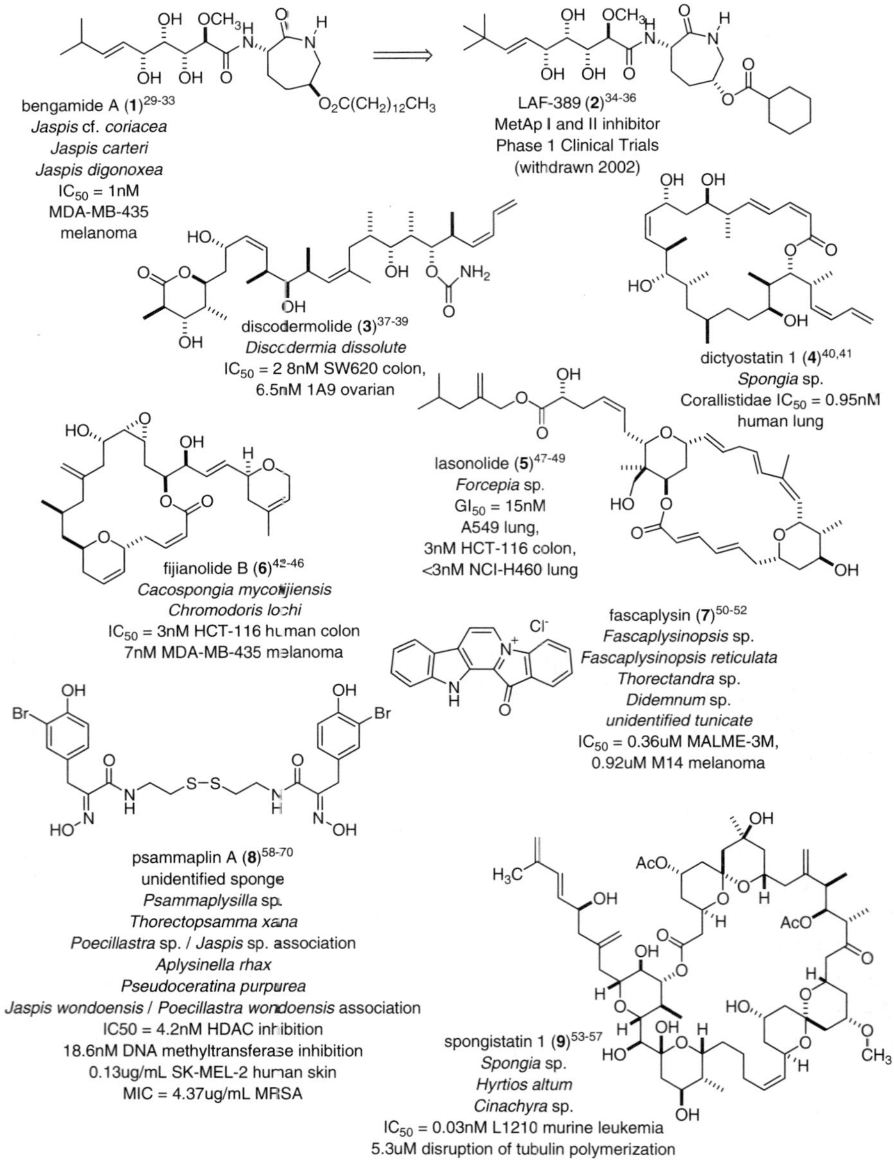

Scheme 6.1 (Part 1) Selected Porifera (sponge) compounds: sources and activity.

and induced tubulin polymerisation in cells expressing the P-glycoprotein efflux pump. Dictyostatin 1 inhibits the binding of discodermolide to microtubules[165] and has stronger affinity than either epothilone B or paclitaxel.[166]

Synthetic efforts have produced a number of reports on the partial synthesis,[167–174] total synthesis,[175–178] and a synthesis and SAR review.[179] Two independent synthetic groups, one led by Paterson at the University Chemical

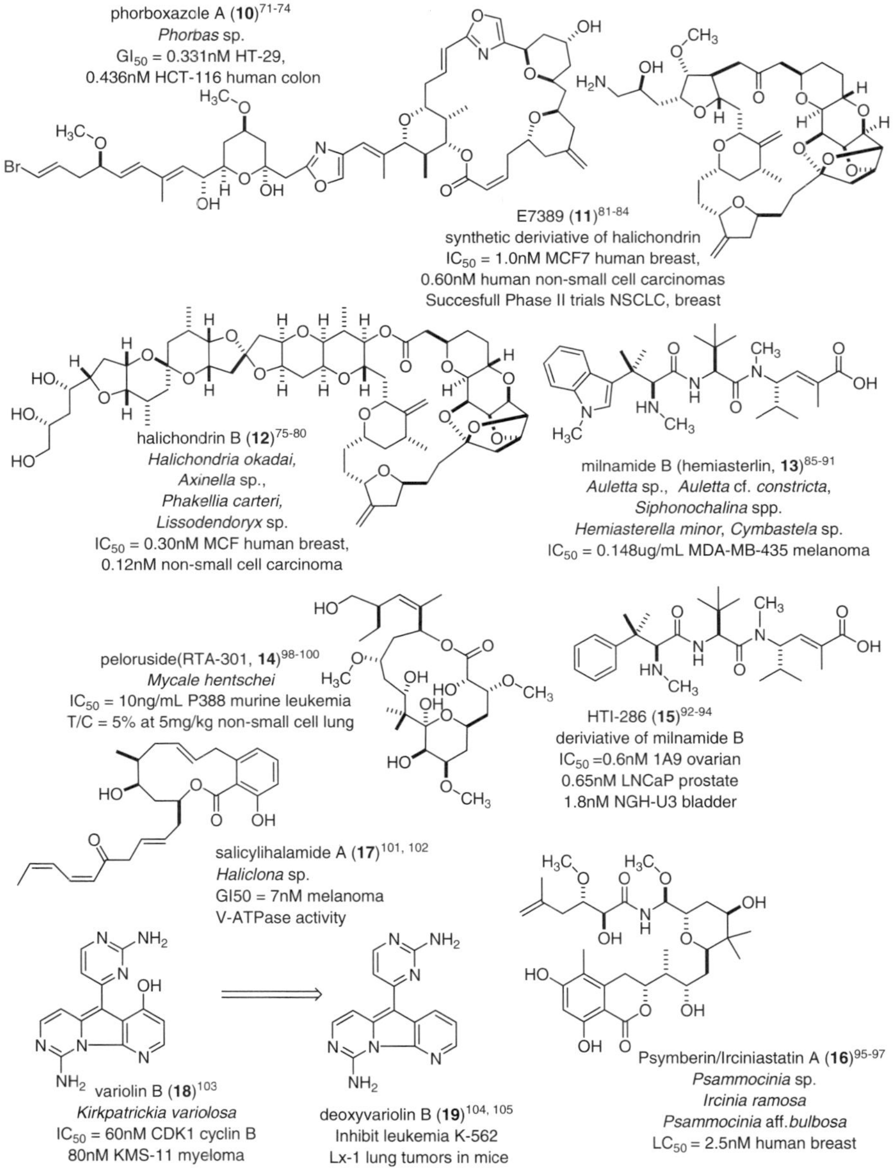

Scheme 6.1 (Continued; Part 2).

Laboratory in Cambridge and the second directed by Day and Curran at the University of Pittsburgh, have taken a lead in the development of dictyostatin 1 analogues. A number of these analogues have retained nanomolar potency,[180–184] while others are less active than the natural product.[185–190] A number of patents exist for the development of dictyostatin 1 both as analogues[191,192] and as the natural product.[163,193]

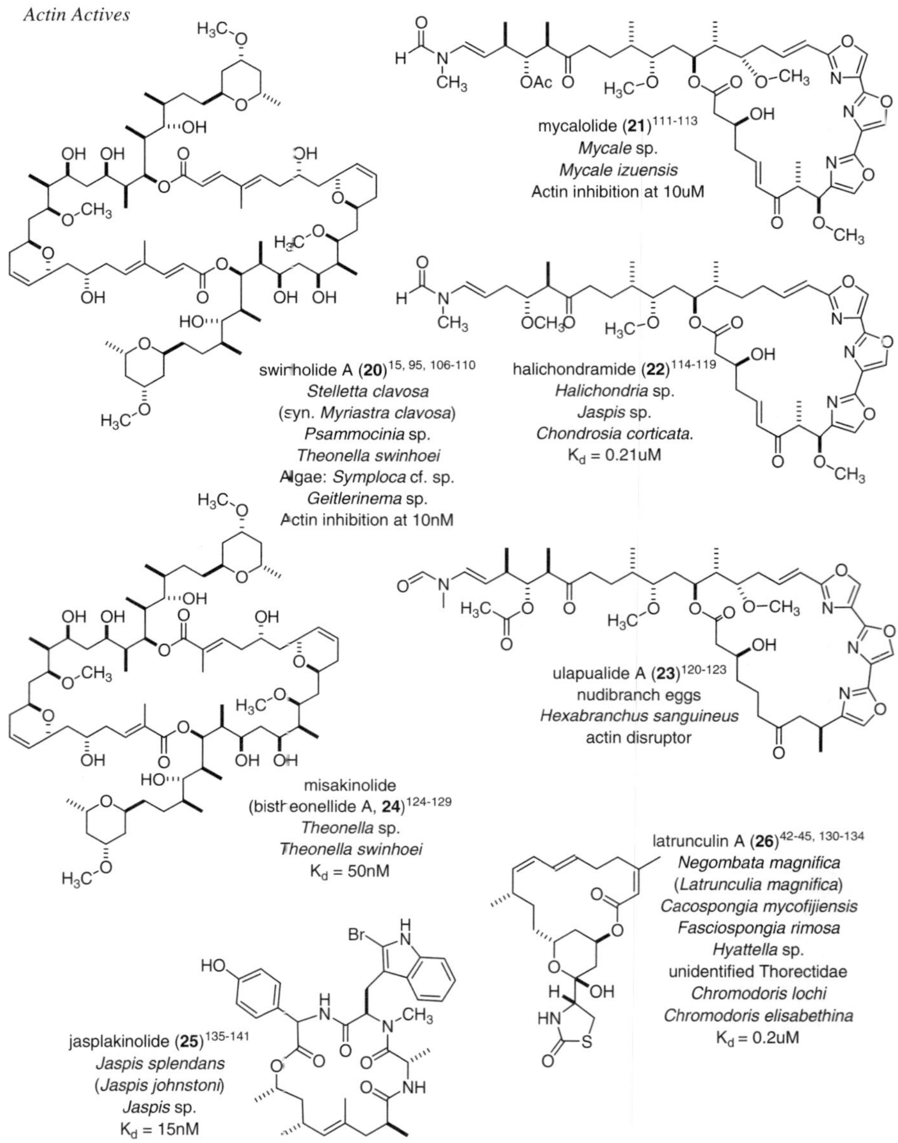

Scheme 6.1 (Continued; Part 3).

Fijianolide B (laulimalide)

Fijianolide B **6** (Scheme 6.1 Part 1) was first reported concurrently by two laboratories.[42,43] Our group isolated fijianolides A and B from a sample of *Cacospongia mycofijiensis*, whereas the Scheuer group obtained identical compounds from an Indonesian sponge and its nudibranch predator,

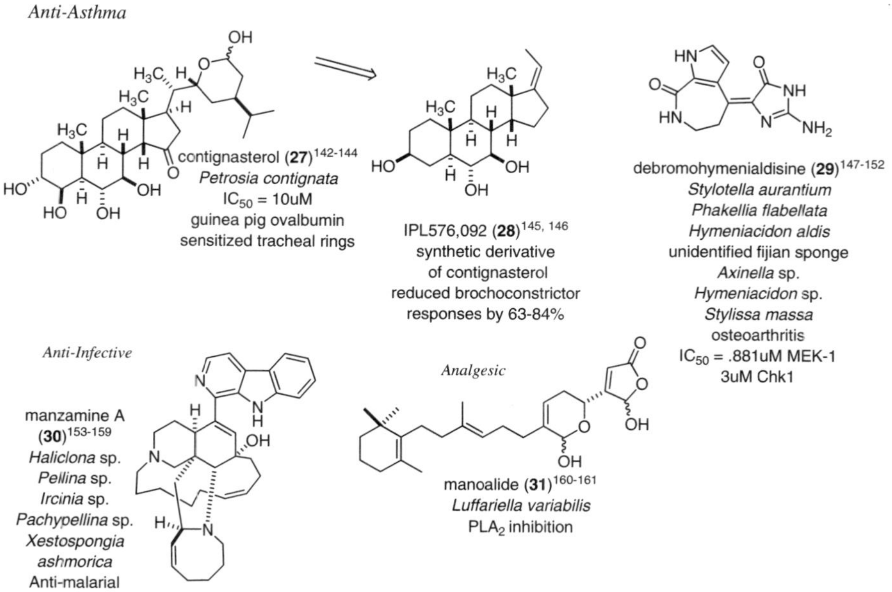

Scheme 6.1 (Continued; Part 4).

Chromodoris lochi. Initial activity of fijianolide B indicated that it was a highly potent KB cell line inhibitor with an IC_{50} of 50 ng/mL. The absolute configuration of fijianolide B was then established in 1996 by Higa and Bernardinelli using X-ray analysis.[45] Two partial syntheses of fijianolide B were published in the following year.[194,195]

Interest in fijianolide B increased markedly following the report that this compound has the ability to stabilise microtubules causing abnormal tubulin bundles to form.[196] With this exciting discovery, synthetic efforts dramatically increased with four partial syntheses published in 2000.[197–200] These were followed by the first total synthesis in 2000 by Ghosh[201] and ten other synthetic routes were developed by eight different groups including Paterson,[202] Crimmins,[203] Ghosh,[204] Mulzer,[205–207] Nelson,[208] Uenishi,[209] Williams[210] and Wender.[211] Additional work followed generating a number of synthetic publications[212–219] and three reviews.[220–222]

It was soon established that fijianolide B does not bind to the taxoid site of tubulin.[223] This prompted an investigation into the similarities and synergistic effects of fijianolide B and other taxoid site agents.[224,225] Unlike epothilone A, fijianolide B was found to act synergistically with taxol at lower temperatures.

Synthetic fijianolide B analogues began to appear in 2002, allowing for the cytotoxicity structure analysis of 35 different compounds;[228–238] recently, six novel naturally occurring fijianolides have been isolated by our group.[239] Although none of these have exhibited greater cytotoxicity than that of fijianolide B, testing of fijianolide B *in vivo* in an HCT-116 human colon tumour model using severe combined immuno-deficiency (SCID) mice demonstrated significant inhibition of tumour growth when dosed at 25 mg/kg over five days.[239] In another study, the *in vivo* efficacy of fijianolide B was tested on the MDA-MB-435 cell line (which has been newly characterised as a Melanoma cell line[240]) and HT-1080 Fibrosarcoma solid tumours in mice, showing minimal growth inhibition and high toxicity.[241] Further preclinical studies are warranted for fijianolide B and their derivatives to resolve these two conflicting scenarios.

Peloruside A

Peloruside A **14** (Scheme 6.1 Part 2) was isolated from a New Zealand *Mycale hentschei* marine sponge and initially showed activity against P388 murine leukaemia cells at 10 ng/mL.[98] Peloruside's cytotoxicity profile and structural similarity to bryostatin led to the examination of protein kinase C (PKC) as a possible mode of action.[242] This was determined to be incorrect and it was soon established that the remarkable activity of peloruside was through the stabilisation of microtubules at a site distinct from the taxoid site.[243]

However, determination of the location of the exact site of peloruside A's binding on tubulin was initially problematic. NMR spectroscopy studies of peloruside A using TR-NOSEY experiments run by the Jiménez-Barbero group indicated a binding position for peloruside A on the α-monomer of tubulin.[244] Another study using computational docking and quantitative structure–activity relationship (QSAR) analysis suggested a favourable interaction of peloruside A with both the taxoid site and another site on α-tubulin.[245] Recently, the Schriemer group reported its binding on the β-monomer using hydrogen–deuterium exchange comparisons measured by mass spectroscopy.[246]

In any case, peloruside A continues to be an exciting preclinical lead due to its ability to preferentially cause apoptosis in oncogene-transformed cells over non-transformed cells[247] and cause tumour regression (T/C = 5% and 16% at 10 and 5 mg/kg respectively)[99] in non-small cell lung carcinoma tumours. Peloruside A works synergistically with the taxoid site drugs paclitaxel and epothilone A,[225,248] and small doses of paclitaxel or vinorelbine convert peloruside A resistant tumours into peloruside A sensitive tumours in the reduction of non-small cell lung carcinoma.[100] Also of note is the possible anti-inflammatory activity of peloruside A in that it decreases the production of proinflammatory mediators by lipopolysaccharide stimulated murine macrophages.[249]

To obtain a supply of this natural product for biological and preclinical development and in an attempt to develop derivatives for SAR, various groups soon began work toward the synthesis of peloruside A. This resulted in a number of notable publications on the partial synthesis[250–259] and three total synthesis by De Bradander,[260] Taylor[261] and Ghosh.[262] In addition, one review of the synthetic methods, microtubule stabilisation and computational analyses of peloruside A has been completed recently.[263]

Psymberin/Irciniastatin A

Psymberin **16** (Scheme 6.1 Part 2), a highly potent pedrin-like molecule, was first isolated from a *Psammocinia* sp. marine sponge collected in Papua New Guinea.[95] An identical structure, irciniastatin A was also isolated from *Ircinia ramosa* collected in Borneo.[96] The researchers reported potent biological activity, including activity against human breast cancer cell lines. The isolation of psymberin in these two taxonomically distinct species prompted an evaluation of the morphotypes and geographic locations for psymberin production in *Psammocina*.[97] These results determined a reliable source of the natural product from the *Psammocinia* aff. *bulbosa* sponge collected in Amphlett, Papua New Guinea.

The determination of the psymberin amide side chain stereochemistry was accomplished through synthesis, X-ray crystallographic analysis and a comparison of the ^{1}H and ^{13}C NMR data.[264] Soon afterwards, partial synthesis of the C_1–C_6 and N_7–C_{25} fragments was completed.[265,266] The total synthesis of psymberin was reported by De Brabander in 2005[267] and was followed by two others in 2007.[268,269] A number of derivates of psymberin have been formulated including psympederin, a psymberin–pedrin hybrid, with dramatically reduced activity ($\sim$1000 fold) and two epimeric psymberins; 8-*epi*-psymberin and 4-*epi*-psymberin with KM12 colon tumour IC_{50} values of 37 and 126 nM respectively.[270] Also of note is a synthetically produced *seco*-psymberin which is reported to be inactive.[271] Preclinical efficacy data of psymberin against HCT-116 tumour bearing SCID mice was reported to be minimal but encouraging.[272]

Salicylihalamide A

The isolation of the benzolactone enamide salicylihalamide A **17**[101] (Scheme 6.1 Part 2) was guided by the NCI's Drug Discovery Research and Development, Developmental Therapeutics Program screen for differential cytotoxicity, which showed that the Australian sample of *Haliclona* possessed a unique 60 cell line profile. Nanomolar potency was evident in the melanoma cell lines (GI_{50} = 7 nM) and a COMPARE analysis of the 60 cell line data indicated vascular ATPase (V-ATPase) activity potential. Further study of this structure class determined that salicylihalamide A is equipotent to the V-ATPase standards bafilomycin A and concanamycin A,[273] but unlike these standards, selectively inhibits mammalian V-ATPases.

In an attempt to explore the structural–activity relationships and to map the binding site for salicylihalamide A, a number of synthetic derivatives have been produced.[274–278] Thus far, SAR studies have established that: the hexadienoyl fragment is not vital for activity, sterically large amide nitrogen groups can be accommodated and that there is a size restriction with the *N*-acyl fragment which indicates that it is involved in more than just lipophilic membrane anchoring.[279]

In addition, lactone structural simplifications or ring size changes do not increase the activity of salicylihalamide A. Later biochemical work established that the binding site for salicylihalamide A resides in the V_0 domain, the proton channel of V-ATPase rather than the catalytic V_1 domain.[102]

The potential for salicylihalamide A to become a tool for the study of V-ATPases has led to a number of synthetic investigations[275,280–287] including total syntheses by five groups,[277,279,288–291] formal total syntheses by Georg,[292–294] Rizzacasa,[295] Maier[296] and Yadav[297] and has been the subject of a review.[298] Also of interest is the temperature and temporal variability of salicylihalamide A in the sponge *Haliclona* sp., which indicates environmental and physiological variations in its production.[299]

Variolins

The difficult to collect Antarctic sponge, *Kirkpatrickia variolosa*, is the source of the cytotoxic variolins.[300,301] The most active of which, variolin B **18** (Scheme 6.1 Part 2), has shown activity against herpes simplex type I and P388 leukaemia cells. Variolin B induced apoptosis in multidrug resistant leukaemia and epithelial cancer cell lines.[302–305] Due to the difficulty of Antarctic collection, synthetic efforts have been pursued resulting in the partial synthesis of the tricyclic core of variolin B,[306–310] three total syntheses of variolin B[311–313] and the development of variolin derivatives.[314–317]

Both variolin B and its 5-deoxy derivative **19** (Scheme 6.1 Part 2) have shown promise in pharmacokinetic and *in vivo* studies. These compounds have been shown to have long terminal half-lives and low normal cell toxicity, however the 5-deoxy derivative demonstrated better C_{max}, plasma clearance and terminal plasma half-life.[318,319] Both are effective against human lung carcinoma cell lines in nude mice.[104] The deoxy-variolin B showed growth inhibitory activity against human leukemic cell lines.[105] A standardised method for liquid chromatography-mass spectrometry (LC-MS)/MS analysis of plasma has been developed to monitor the results of the *in vivo* studies.[320]

Two very recent patents have been obtained by PharmaMar S.A. for the synthesis and development of the derivatives of variolin B as anticancer agents.[321,322] The meriolins, a hybrid of variolin B and the meridianins isolated from the Ascidian *Aplidium meridianum*, have shown enhanced selectivity for cyclin-dependent kinases and promising *in vivo* results against LS174T colorectal carcinoma and A4573 Ewing's sarcoma.[323]

3 Molluscs

3.1 Natural History of Molluscs—the Source of Numerous Preclinical Drug Leads

The characteristics of the phylum Mollusca include the presence of unsegmented coelomate protostomes and many examples are bilaterally symmetrical.[23] The body plan includes a head with eyes or tentacles, a muscular foot often with a flattened creeping sole, and the body cavity. This body cavity is covered by a thick epidermal–cuticular sheet of skin called the mantle, which forms a blood-filled space that houses many organs. Mantles may also include shell glands which secrete calcareous epidermal spicules, shell plates, or shells. Molluscs also have a complete digestive tract with a mouth and anus, and posses a rasplike feeding structure called a radula. These chitinous radula are unique to molluscs and vary in form and function from the brush-like fine teeth of abalone, to the hard beak of squid, to the toxin injecting barbs of the cone snail.

There are seven living classes of molluscs,[23] including the worm-like Aplacophora, the chitons of the Polyplacophora, the limpet-like creatures of the Monoplacophora and the Gastropods which includes the abalone, marine snails, slugs, nudibranchs and conch. The next three are comprised of the Cephalopods (cuttlefish, squid and octopus), the Bivalves (clams, oysters) and the Scaphopoda (Tusk shells). Figure 6.5 shows that the most prolific order within Mollusca is the Aplysiomorpha (Anaspidea) with 507 articles and 384 structures published since 1951. There have been a total of 1684 publications and 1225 chemical structures reported from the order Mollusca. One of the most well-known structure classes from molluscs are the dolastatins, which are from the Anaspidea order.

The structural diversity of molluscan chemistry can be seen in compounds **32–40** in Scheme 6.2 and includes a number that have been included in clinical studies such as the kahalalides and the dolastatins **34** and **37**. An example of material exchange from prey to predator can be shown in kahalide F **32**, which can be found in both the sea hare *Elysia* sp. and its diet of *Bryopsis* sp. algae. In addition, the structural similarity of ulapualide A **23** and that of the compounds mycalolide **21** and halichondramide **22** (Scheme 6.1 Part 3), found in a diverse array of sponges, indicates a possible alternate source for this structure class.

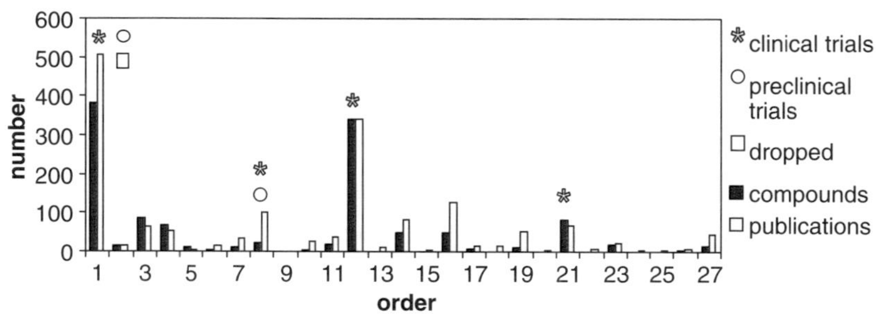

Figure 6.5 Number of publications and pure compounds from Mollusca (MarinLit 1951–2008). Orders: 1. Anaspidea, 2. Archaeogastropoda, 3. Bassomatophora, 4. Cephalaspidea, 5. Eulamelibranchiata, 6. Mesogastropoda, 7. Mytiloida, 8. Neogastropoda, 9. Neoloricata, 10. Neotaenioglossa, 11. Notaspidea, 12. Nudibranchia, 13. Octopoda, 14. Ostreoida, 15. Patellogastropoda, 16. Prinodesmacea, 17. Prosobranchia, 18, Prosota, 19. Pterioida, 20, Pulmonata, 21. Saccoglossa, 22. Sepiida, 23. Systellommatophora, 24. Teleodesmacea, 25. Teuthida, 26. Unionoida, 27. Veneroida

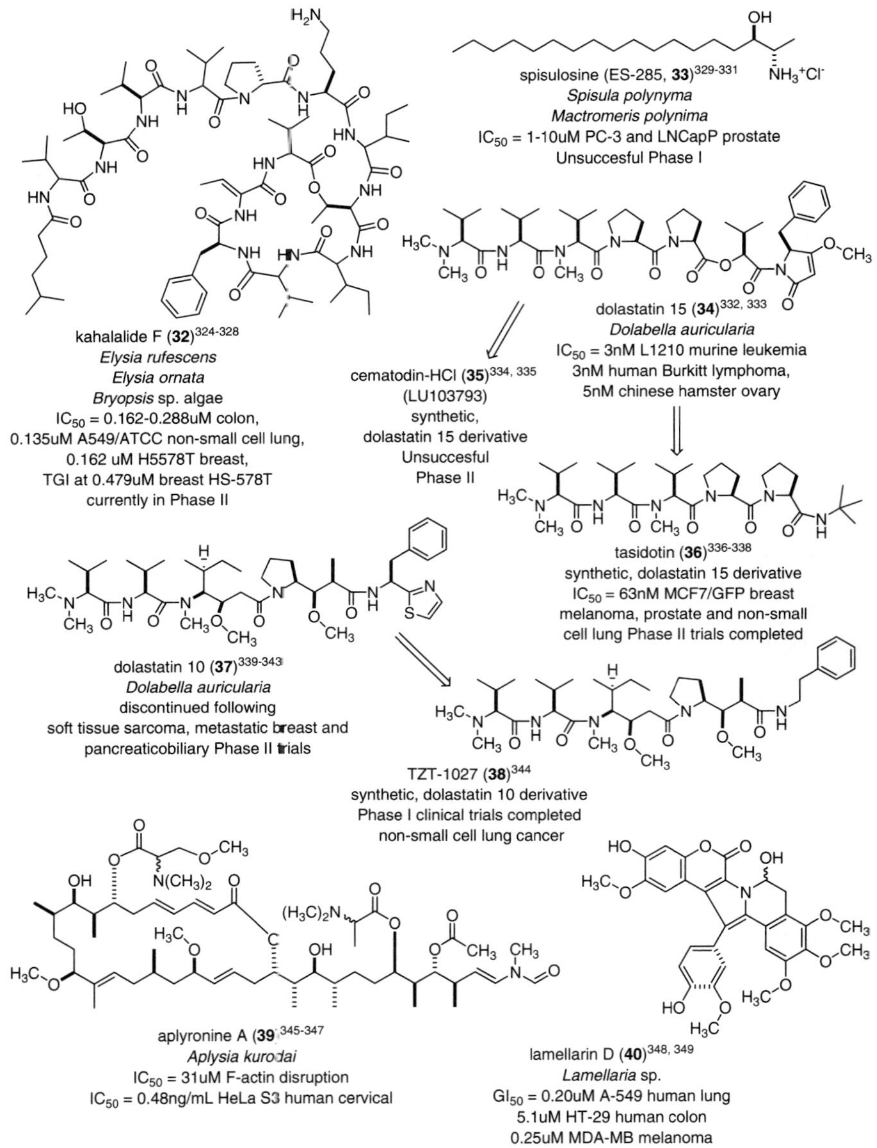

Scheme 6.2 Selected Mollusc compounds: sources and activity.

Ulapualide A

In the late 1980s, ulapualide A **23** was isolated from the egg masses of the nudibranch *Hexabranchus sanguineus*[120] and was shown to inhibit the growth of *Candida albicans* and L1210 leukaemia cell proliferation. A molecular mechanics study initially suggested that the relative stereochemistry was related to that of the halichondramides and mycalamides,[350] however, the correct absolute stereochemistry was later established by X-ray crystallography.[121] The synthetic groups of Panek,[351–354] Inoue[355] and Pattenden[356–362] worked on the partial synthesis of both the natural product and its stereoisomer. The total synthesis of the diastereomer of ulapualide A was completed in 1998,[363] followed by the synthesis of the correct structure in 2007[364] and 2008.[365]

Activity studies have shown that ulapualide A and other related trisoxazole macrolides act as small molecule biomimetics of the actin-binding protein gelsoin.[122,366] The X-ray crystal structures of ulapualide A and related actin-binding macrolides indicates a region of common binding for these compounds that comprises a flexible macrocyclic segment and a relatively restricted tail portion.[123] Synthetic analogues, including the diasteromer of ulapualide A[367] and tail region mimetics of these macrolides,[368] were less active than the natural product.

4 Soft Corals

4.1 Natural History of Cnidarians—the "Stinging Nettle" of the Sea

Cnidarians are united by the presence of specialised cells called nematocytes, which carry stinging nematocyst organelles.[23] They encompass a wide array of forms including the corals, hydroids and box jellies. The basic body plan of Cnidaria is that of a gastrovascular sac with a single opening that functions as both a mouth and anus. This is surrounded by tentacles which contain the stinging cells used to capture prey. Other groups such as the corals utilise zooxanthellae, symbiotic dinoflaggelates for the production of carbohydrates. Cnidarians posses a simple nervous system composed of bare and largely nonpolar neurons. Sexual reproduction results in the formation of planulae, which are motile and ciliated larvae.

The four main classes of Cnidaria are the Anthozoa (corals and sea anemones), Cubozoa (sea wasps or box jellyfish), the Hydrozoa (hydroids, fire coral and the Portuguese man o'war) and the Scyphozoa (jellyfish).

The literature of Cnidara includes 2444 articles comprised of 3744 chemical structures.[27] Included within this impressive list is the commercially available diterpenoid glycosides, the pseudopterosins,[369,370] with remarkable anti-inflammatory and analgesic properties, eleutherobin and the sarcodyctins. Figure 6.6 shows that study within Cnidaria has focused on two orders, that of the soft corals, the Alcyonacea and the Gorgonacea.

Scheme 6.3 illustrates a few of the main structural classes of soft coral compounds. Included in this list are the tubulin interactive agents, eleutherobin **41** and sarcodictyin A **42**, which show nanomolar potency against cancer cell lines. Also of interest are the pseudopterosin A **44** terpenoids which reduce inflammation. Two other terpenoids, sarcophytol A **43** and sinulodurin A **45**, have shown micromolar activity against BALB/3T3 and mammary epithelial cells respectively.

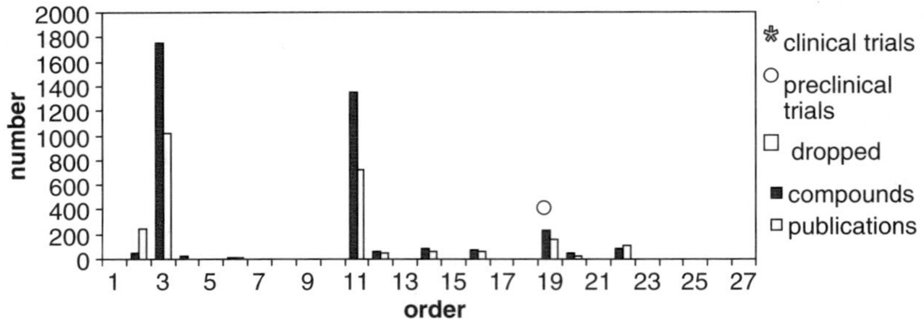

Figure 6.6 Number of publications and pure compounds from Cnidaria (MarinLit 1951–2008). Orders: 1. Actinaria, 2. Actiniaria, 3. Alcyonacea, 4. Antipatharia, 5. Capitata, 6. Coenothecalia, 7. Conica, 8. Corallimorpharia, 9. Cubomedusae, 10. Discomedusae, 11. Gorgonacea, 12. Hydroida, 13. Nynantheae, 14. Pennatulacea, 15. Rhizostomeae,16. Scleractina, 17. Semaeostomeae, 18. Siphonophora, 19. Stolonifera, 20.Telestacea, 21. Thecatae, 22. Zoanthiniaria

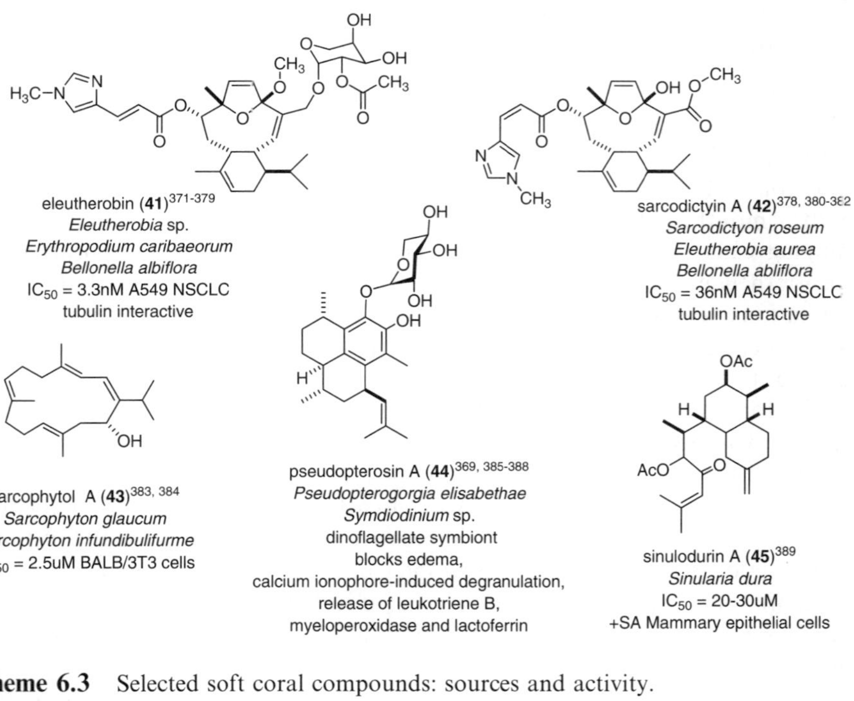

Scheme 6.3 Selected soft coral compounds: sources and activity.

Eleutherobin

The diterpene glycoside eleutherobin **41**[373,374] was isolated from the soft coral *Eleutherobia* sp. collected from Bennett's shoal in Western Australia. Eleutherobin, along with the erythrolides A and B, has also been found in the cultured and common gorgonian *Erythropodium caribaeorum*.[390] Exciting initial biological activity was reported for eleutherobin, including similar tumour type selectivity to paclitaxel in the NCI's COMPARE protocol and an IC_{50} in the 10–15 nM range against a variety of cancer cell lines. The tubulin polymerisation and cytotoxicity of eleutherobin was found to be similar to that of paclitaxel,[379,391] although there is debate over the report that the two share a common pharmacophore.[375,392] A number of SAR studies have been completed on the eleutherobin/sarcodictyin structural core including alternation of the carbohydrate chain, changes to the urocanic acid ester and simplification of the eleutherobin skeleton—none of which enhance the cytotoxicity of eleutherobin.[393–398] Alteration at the C4 ketal to larger groups had no effect on the antimitotic activity, indicating that the small group at this position is not required for activity.[396] A number of groups have worked on the synthesis of the eleutherobin core[399–409] and three independent groups led by Nicolaou,[410] Danishefsky[411] and Gennari[174,412,413] have completed total syntheses.

Sarcodictyins

The isolation of sarcodictyin A **42** (Scheme 6.3) began with a stolonifer coral *Sarcodictyon roseum* collected in 1986 off Cap Bear on the French Mediterranean coast.[380] Soon afterwards, a series of sarcodictyins C, D, E and F[414] were also found from the same species. Two other soft coral sources of sarcodictyins were later discovered—one from a South African *Eleutherobia aurea*[381] and another from *Bellonella albiflora*[378] collected in Southern Japan, which also yielded the (*Z*) diastereomer of sarcodictyin A.

The sarcodictyins show paclitaxel-like microtubule stabilisation,[415] but are ten-fold less cytotoxic that eleutherobin **41**[395] and show moderate cytotoxicity against 1A9 human ovarian carcinoma cells.[382] The structural similarities that exist between sarcodictyin A and eleutherobin are evident, with minor differences including the C3 glycoside and a methyl group at C15 in eleutherobin which are replaced by a C3 methyl ester and hydrogen, respectively, in sarcodictyin A.

In an attempt to elucidate the differences in microtubule stabilisation between sarcodictyin A and eleutherobin, a number of derivatives of sarcodictyin A have been produced. These examinations have led to the conclusions that[395,416,417] large C4 ketal substitutions are well tolerated, the glycoside at C15 is a requirement for the disruption of microtubule assembly, but has no effect on the cytotoxicity, reduction of the ester to the alcohol is not tolerated, the replacement of the urocanic side chain disrupts activity and that both nitrogens in the imidiazole heterocycle are required.

Another analogue, with an opened C4–C7 bridge, also shows similar activity to sarcodictyin A, yet again lower than that of eleutherobin.[418]

Two total syntheses have been reported of sarcodictyin A[419,420] and a number of articles on the partial synthesis[399,421–423] have been published. Also of interest are the use of sarcodictyin A and eleutherobin in the development of cell-based screens for antimitotic activity[424] and their use in improving the resolution of microtubule structures.[425]

5 Tunicates

5.1 Natural History of Tunicates—Our Closest Marine Invertebrate Relations

Tunicates belong to the phylum Chordata (subphylum Urochordata) and are also saclike filter feeders.[23] Their tunic is composed of a polysaccharide matrix of tunicin, a complex similar to cellulose which is attached to the substrate by a holdfast. Incurrent and outcurrent siphons allow for the flow of nutrient-rich water to pass through the pharnx where particles are filtered out. The movement of water is controlled by both cilia and muscular contraction. It is this fast contraction of the tunic which causes a jet of water to exit the outcurrent siphon that gives tunicates their nickname of "sea squirt". They can be either solitary or colonial and the colonial variety will often share a siphon. During reproduction, sperm are released for distribution whereas the eggs are kept within the body cavity until larvae are matured. These tadpole-like larvae most closely resemble the chordates and have a primitive spinal cord called a notochord. Once settled, the notochord is reabsorbed into the body within minutes.

Among the Urochordata are four main classes: the Ascidiacea (sea squirts), the Thaliacea (salps), the Appendicularia (pelagic urochordates) and the Sorberacea (benthic urochordates). There have been 1458 articles published on the phylum Chordata and 1014 chemical structures reported.[27] Among the Ascidians there are three orders—Aplousobranchia, Enterogona and Pleurogona. As seen in Figure 6.7, the Enterogona is the most abundant in terms of both biologically active leads and number of publications.

Diazonamide A **46** (Scheme 6.4) is toxic to human colon HCT-116 cancer cells at 15 ng/mL. Other structures include cyclic peptides such as didemnin B **47** which has been discontinued in Phase II clinical trials, aplidin **48** which is currently in Phase II clinical studies for multiple myeloma and vitilevuamide **50** which effects tubulin polymerisation at micromolar concentrations. One of the

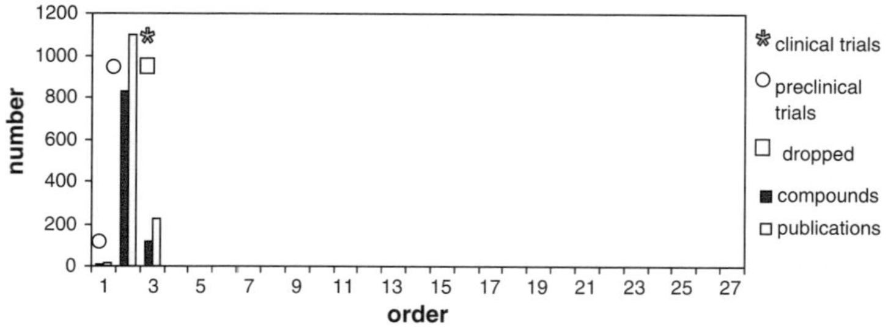

Figure 6.7 Number of publications and pure compounds from tunicates (MarinLit 1951–2008). Orders: 1. Aplousobranchia, 2. Enterogona, 3. Pleurogona

most well-known marine compounds, ecteinascidin 743 **49** (trabectedin),[426] is undergoing Phase III clinical trials in the USA for the treatment of various cancers and has been approved for commercial use in Europe.

Diazonamide A

A protonated nitrogen, rather than an oxygen on an X-ray diagram, led to the publication of the incorrect initial chemical structure of diazonamide

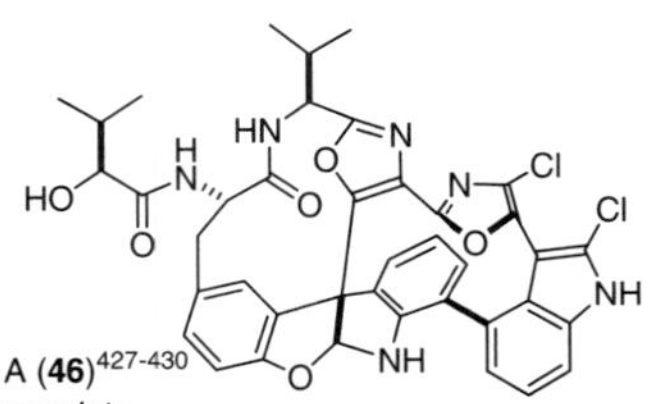

diazonamide A (**46**)[427-430]
Diazona angulata
IC_{50} = < 15ng/mL
HCT-116 human colon
B-16 murine leukemia

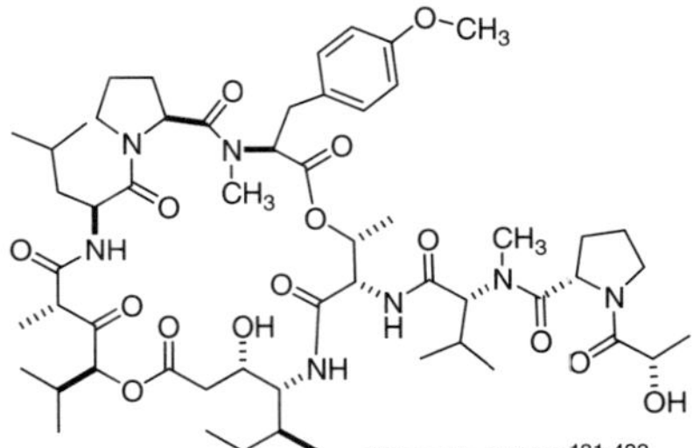

didemnin B (**47**)[431-433]
Trididemnum sp.
ID_{50} = 11ng/mL L1210 leukemia
Phase II discontinued

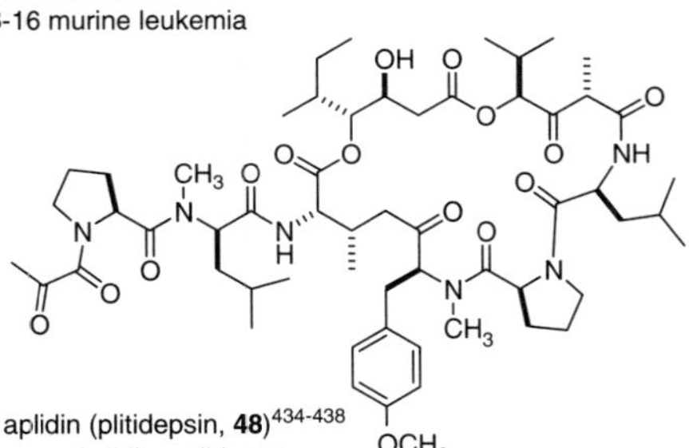

aplidin (plitidepsin, **48**)[434-438]
Aplidium albicans
IC_{50} = 0.2-0.5ng/mL P388 leukemia, HT-29 colon, MEL-28 melanoma.
Phase II multiple myeloma.

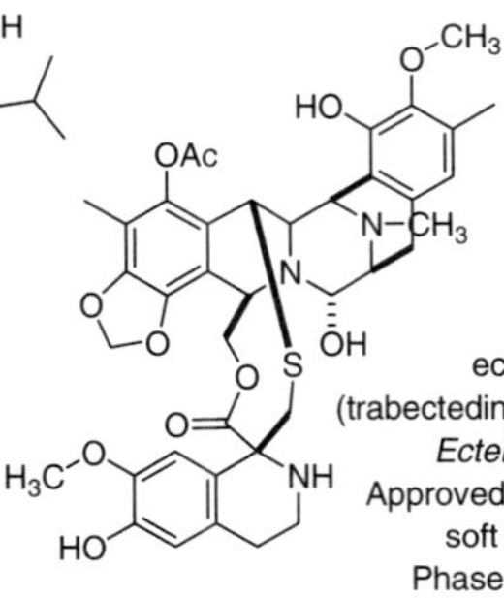

ecteinascidin 743
(trabectedin, yondelis, **49**)[426, 439, 440]
Ecteinascidia turbinata
Approved (Europe)for advanced soft tissue carcinoma
Phase III breast, prostate, paediatric sarcoma

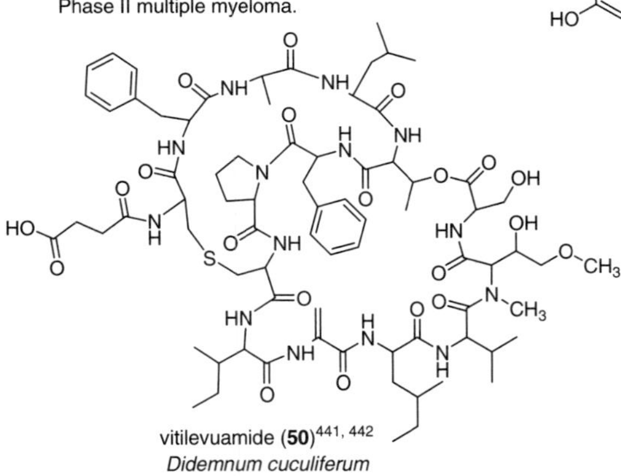

vitilevuamide (**50**)[441, 442]
Didemnum cuculiferum
Polysyncranton lithostrontum
IC_{50} = 2.5uM C6 rat glioma
2uM tubulin polymerization

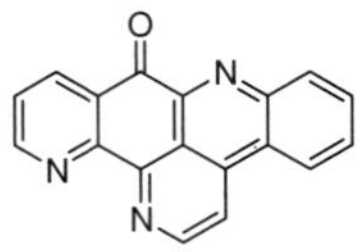

ascididemin (**51**)[443-446]
Didemnum sp.
Eudistoma sp.
IC_{50} = 0.39ug/mL
L1210 murine leukemia

Scheme 6.4 Selected tunicate compounds: sources and activity.

A **46** (Scheme 6.4).[447] This led to a number of synthetic efforts resulting in the total synthesis of the incorrect structure.[448] Soon after, the total synthesis of the correct structure was published by Harran,[449,450] Nicolaou[451–454] and Magnus.[455] Interestingly, both diazonamide A and this incorrect diazonamide A are potent inhibitors of tubulin assembly.[456] These appear to either have a binding site on tubulin that is different from the vinca or dolastatin 10 binding sites, or bind weakly to unpolymerised tubulin but strongly to microtubule ends. The majority of synthetically produced derivatives were not as potent as the natural product;[454,457,458] however, the derivative AB-5 showed promise in HCT116 colon, PC3 prostate and MDA-MB-435 melanoma *in vivo* studies.[459]

6 Conclusions

A number of historically significant marine natural products have established macromarines as a valuable source of physiologically active molecular scaffolds. In addition, many of the compounds illustrated in this work are currently in clinical development. Unfortunately, the journey from initial isolation to clinical development has ended for a number of notable compounds. Among the list found in Table 6.1 are leads obtained from the macromarines such as sponges, tunicates and molluscs. Although synthetic derivatives often attempt

Table 6.1 Discontinued compounds.

Clinical trial	*Compound name*	*Source*	*Target*	*Discovering lab*
Phase II (<2004)	Dolastatin 10	Sea hare	Tubulin	Pettit
Phase II (<1999)	Didemnin B	Tunicate	Antineoplastic	Rinehart
Phase II (<2004)	Cemadotin (dola-15 insp.)	Synthetic	Tubulin	BASF (Pettit)
Phase II (<2002)	Cryptophycin 52 (≈ arenastatin)	Synthetic	Tubulin	Lilly (Valeriote, Moore)
Phase I (2006)	Taltobulin (aka. HTI286, hemiasterlin insp.)	Synthetic	Tubulin	Wyeth (Andersen)
Phase I (2004)	Discodermolide	Sponge	Tubulin	Novartis (HBOI)
Phase I (2002)	LAF 389 (bengamide insp.)	Synthetic	MetAP	Novartis (Crews)
Phase I (<2006)	LAQ 824 (psammaplin insp.)	Synthetic	HDAC	Novartis (Crews)
Phase I (<2000)	Girroline (aka. girodazole)	Sponge	Protein synthesis	Potier

to reduce toxicity of natural products, taltobulin, LAF 389 and LAQ 824 have been unable to go beyond Phase I clinical trials.

One of the biggest issues hindering the utilisation of macromarine compounds as drugs is that of supply. Total synthesis, partial synthesis and the semi-synthesis of natural products may reduce the demand pressure during commercial development however, the complexity of natural products extends the timeline for the realistic production of material needed for clinical testing. Sensible mariculture techniques will undoubtedly help to preserve the natural sources for these compounds while also allowing for testing to continue.

The supply problem may be alleviated if the true producers of these biologically active compounds could be identified. Once this can be established, the next step is to isolate the gene clusters responsible for these novel structural scaffolds. Combined with biosynthetic engineering, a solution to the supply problem could be readily at hand allowing a greater number of macromarine agents through the pharmaceutical development pipeline.

References

1. J. W. Blunt and M. H. G. Munro, *MarinLit Marine Literature Database*, University of Canterbury, Christchurch, New Zealand, 2007.
2. J. W. Blunt, B. R. Copp, W. P. Hu, M. H. G. Munro, P. T. Northcote and M. R. Prinsep, *Nat. Prod. Rep.*, 2008, **25**, 35.
3. D. J. Newman and G. M. Cragg, *J. Nat. Prod.*, 2007, **70**, 461.
4. R. S. M. Singh, P. Joshi and D. S. Rawat, *Anti-Cancer Agents Med. Chem.*, 2008, **8**, 603.
5. W. Bergmann and R. J. Feeney, *J. Am. Chem. Soc.*, 1950, **72**, 2809.
6. W. Bergmann and R. J. Feeney, *J. Org. Chem.*, 1951, **16**, 981.
7. W. Bergmann and D. C. Burke, *J. Org. Chem.*, 1955, **20**, 1501.
8. W. Bergmann and D. F. Burke, *Angew. Chem.*, 1955, **67**, 127.
9. W. Bergmann and D. C. Burke, *J. Org. Chem.*, 1956, **21**, 226.
10. S. S. Cohen, *Perspect. Biol. Med.*, 1963, **6**, 215.
11. W. W. Lee, A. Benitoez, L. Goodman and B. R. Baker, *J. Am. Chem. Soc.*, 1960, **82**, 2648.
12. J. S. Evans, H. J. Hunter, K. R. Forsblad, E. A. Musser and G. D. Mengel, *Proc. Soc. Exp. Biol. Med.*, 1961, **106**, 350.
13. G. Cimino, S. Derosa and S. Destefano, *Experientia*, 1984, **40**, 339.
14. T. W. North and S. S. Cohen, *Pharmacol. Ther.*, 1979, **4**, 81.
15. E. H. Andrianasolo, H. Gross, D. Goeger, M. Musafija-Girt, K. P. Mcphail, R. M. Leal, S. L. Mooberry and W. H. Gerwick, *Org. Lett.*, 2005, **7**, 1375.
16. M. W. Taylor, R. Radax, D. Steger and M. Wagner, *Micro. Mol. Biol. Rev.*, 2007, **71**, 295.
17. F. Rohwer, F. Azam and N. Knowlton, *Abstr. Gen. Meet. Am. Soc. Microbiol.*, 2002, **102**, 334.
18. P. R. Jensen and W. Fenical, *Annu. Rev. Microbiol.*, 1994, **48**, 559.

19. S. Engel, P. R. Jensen and W. Fenical, *J. Chem. Ecol.*, 2002, **28**, 1971.
20. C. E. Salomon, N. A. Magarvey and D. H. Sherman, *Nat. Prod. Rep.*, 2004, **21**, 105.
21. J. Piel, *Curr. Med. Chem.*, 2006, **13**, 39.
22. W. E. G. Muller, F. Brummer, R. Batel, I. M. Muller and H. C. Schroder, *Naturwissenschaften*, 2003, **90**, 103.
23. R. C. Brusca and G. J. Brusca, *Invertebrates*, Sinauer Associates, Sunderland, MA, 2003.
24. W. E. G. Muller, J. H. Li, H. C. Schroder, L. Qiao and X. H. Wang, *Biogeosciences*, 2007, **4**, 219.
25. G. Worheide, A. M. Sole-Cava and J. N. A. Hooper, *Integr. Comp. Biol.*, 2005, **45**, 377.
26. P. R. Berquist, R. W. M. Van Soest, T. M. G. Van Kempen and J. C. Braekman, *Sponges in Time and Space: Biology, Chemistry, Paleontology*, Balkema, Rotterdam, 1994, Proceedings of the Fourth International Porifera Congress, Amsterdam, Netherlands, 19–23 April 1993, p. xiii.
27. J. W. Blunt, B. R. Copp, W. P. Hu, M. H. G. Munro, P. T. Northcote and M. R. Prinsep, *Nat. Prod. Rep.*, 2007, **24**, 31.
28. M. J. Garson, in *Sponges in Time and Space: Biology, Chemistry, Paleontology*, ed. P. R. Berquist, R. W. M. Van Soest, T. M. G. Van Kempen and J. C. Braekman, Balkema, Rotterdam, 1994, Proceedings of the Fourth International Porifera Congress, Amsterdam, Netherlands, 19–23 April 1993, p. 427.
29. E. Quinoa, M. Adamczeski, P. Crews and G. J. Bakus, *J. Org. Chem.*, 1986, **51**, 4494.
30. A. Rudi, Y. Kashman, Y. Benayahu and M. Schleyer, *J. Nat. Prod.*, 1994, **57**, 829.
31. M. V. D'Auria, C. Giannini, L. Minale, A. Zampella, C. Debitus and M. Frostin, *J. Nat. Prod.*, 1997, **60**, 814.
32. R. Fernandez, M. Dherbomez, Y. Letourneux, M. Nabil, J. F. Verbist and J. F. Biard, *J. Nat. Prod.*, 1999, **62**, 678.
33. Z. Thale, F. R. Kinder, K. W. Bair, J. Bontempo, A. M. Czuchta, R. W. Versace, P. E. Phillips, M. L. Sanders, S. Wattanasin and P. Crews, *J. Org. Chem.*, 2001, **66**, 1733.
34. F. R. Kinder, R. W. Versace, K. W. Bair, J. M. Bontempo, D. Cesarz, S. Chen, P. Crews, A. M. Czuchta, C. T. Jagoe, Y. Mou, R. Nemzek, P. E. Phillips, L. D. Tran, R. M. Wang and S. Weltchek, *J. Med. Chem.*, 2001, **44**, 3692.
35. P. Crews, W. H. Gerwick, F. Schmitz, D. France, K. Bair, A. Wright and Y. Hallock, *Pharm. Biol.*, 2003, **41**, 39.
36. H. Dumez, H. Gall, R. Capdeville, C. Dutreix, A. T. Van Oosterom and G. Giaccone, *Anti-Cancer Drugs*, 2007, **18**, 219.
37. S. P. Gunasekera, M. Gunasekera, R. E. Longley and G. K. Schulte, *J. Org. Chem.*, 1990, **55**, 4912.

38. S. P. Gunasekera, M. Gunasekera, R. E. Longley and G. K. Schulte, *J. Org. Chem.*, 1991, **56**, 1346.
39. R. J. Kowalski, P. Giannakakou, S. P. Gunasekera, R. E. Longley, B. W. Day and E. Hamel, *Mol. Pharmacol.*, 1997, **52**, 613.
40. G. R. Pettit, Z. A. Cichacz, F. Gao, M. R. Boyd and J. M. Schmidt, *J. Chem. Soc., Chem. Commun.*, 1994, 1111.
41. R. A. Isbrucker, J. Cummins, S. A. Pomponi, R. E. Longley and A. E. Wright, *Biochem. Pharmacol.*, 2003, **66**, 75.
42. E. Quinoa, Y. Kakou and P. Crews, *J. Org. Chem.*, 1988, **53**, 3642.
43. D. G. Corley, R. Herb, R. E. Moore, P. J. Scheuer and V. J. Paul, *J. Org. Chem.*, 1988, **53**, 3644.
44. N. K. Gulavita, S. P. Gunasekera and S. A. Pomponi, *J. Nat. Prod.*, 1992, **55**, 506.
45. C. W. Jefford, G. Bernardinelli, J. Tanaka and T. Higa, *Tetrahedron Lett.*, 1996, **37**, 159.
46. T. A. Johnson, K. Tenney, R. H. Cichewicz, B. I. Morinaka, K. N. White, T. Amagata, B. Subramanian, J. Media, S. L. Mooberry, F. A. Valeriote and P. Crews, *J. Med. Chem.*, 2007, **50**, 3795.
47. P. A. Horton, F. E. Koehn, R. E. Longley and O. J. McConnell, *J. Am. Chem. Soc.*, 1994, **116**, 6015.
48. H. Y. Song, J. M. Joo, J. W. Kang, D. S. Kim, C. K. Jung, H. S. Kwak, J. H. Park, E. Lee, C. Y. Hong, S. Jeong and K. Jeon, *J. Org. Chem.*, 2003, **68**, 8080.
49. J. M. Joo, H. S. Kwak, J. H. Park, H. Y. Song and E. Lee, *Bioorg. Med. Chem. Lett.*, 2004, **14**, 1905.
50. D. M. Roll, C. M. Ireland, H. S. M. Lu and J. Clardy, *J. Org. Chem.*, 1988, **53**, 3276.
51. N. L. Segraves, S. Lopez, T. A. Johnson, S. A. Said, X. Fu, F. J. Schmitz, H. Pietraszkiewicz, F. A. Valeriote and P. Crews, *Tetrahedron Lett.*, 2003, **44**, 3471.
52. N. L. Segraves, S. J. Robinson, D. Garcia, S. A. Said, X. Fu, F. J. Schmitz, H. Pietraszkiewicz, F. A. Valeriote and P. Crews, *J. Nat. Prod.*, 2004, **67**, 783.
53. G. Pettit, Z. Cichacz, F. Gao, C. Herald, M. Boyd, J. Schmidt and J. Hooper, *J. Org. Chem.*, 1993, **58**, 1302.
54. M. Kobayashi, S. Aoki, H. Sakai, K. Kawazoe, N. Kihara, T. Sasaki and I. Kitagawa, *Tetrahedron Lett.*, 1993, **34**, 2795.
55. M. Kobayashi, S. Aoki and I. Kitagawa, *Tetrahedron Lett.*, 1994, **35**, 1243.
56. R. Bai, G. Taylor, Z. Cichacz, C. Herald, J. Kepler, G. Pettit and E. Hamel, *Biochemistry*, 1995, **34**, 9714.
57. N. S. K. Fusetani and S. Matsunaga, *J. Am. Chem. Soc.*, 1993, **115**, 3977.
58. E. Quinoa and P. Crews, *Tetrahedron Lett.*, 1987, **28**, 3229.
59. L. Arabshahi and F. Schmitz, *J. Org. Chem.*, 1987, **52**, 3584.
60. A. Rodriguez, R. Akee and P. Scheuer, *Tetrahedron Lett.*, 1987, **28**, 4989.

61. C. Jimenez and P. Crews, *Tetrahedron*, 1991, **47**, 2097.
62. J. Jung, C. Sim and C. Lee, *J. Nat. Prod.*, 1995, **58**, 1722.
63. J. Shin, H. Lee, Y. Seo, J. Rho, K. Cho and V. Paul, *Tetrahedron*, 2000, **56**, 9071.
64. N. Pham, M. Butler and R. Quinn, *J. Nat. Prod.*, 2000, **63**, 393.
65. J. Tabudravu, V. Eijsink, G. Gooday, M. Jaspars, D. Komander, M. Legg, B. Synstad and D. Van Aalten, *Bioorg. Med. Chem.*, 2002, **10**, 1123.
66. I. Pina, J. Gautschi, G. Wang, M. Sanders, F. Schmitz, D. France, S. Cornell-Kennon, L. Sambucetti, S. Remiszewski, L. Perez, K. Bair and P. Crews, *J. Org. Chem.*, 2003, **68**, 3866.
67. Y. Park, Y. Liu, J. Hong, C. Lee, H. Cho, D. Kim, K. Im and J. Jung, *J. Nat. Prod.*, 2003, **66**, 1495.
68. D. Kim, J. Shin and H. Kwon, *Exp. Mol. Med.*, 2007, **39**, 47.
69. K. Nicolaou, R. Hughes, J. Pfefferkorn, S. Barluenga and A. Roecker, *Chem.-Eur. J.*, 2001, **7**, 4280.
70. K. Nicolaou, R. Hughes, J. Pfefferkorn and S. Barluenga, *Chem.-Eur. J.*, 2001, **7**, 4296.
71. P. A. Searle and T. F. Molinski, *J. Am. Chem. Soc.*, 1995, **117**, 8126.
72. P. A. Searle and T. F. Molinski, *J. Am. Chem. Soc.*, 1996, **118**, 9422.
73. T. F. Molinski, *Tetrahedron Lett.*, 1996, **37**, 7879.
74. C. J. Forsyth, L. Ying, J. Chen and J. J. La Clair, *J. Am. Chem. Soc.*, 2006, **128**, 3858.
75. Y. Hirata and D. Uemura, *Pure Appl. Chem.*, 1986, **58**, 701.
76. G. Pettit, C. Herald, M. Boyd, J. Leet, C. Dufresne, D. Doubek, J. Schmidt, R. Cerny, J. Hooper and K. Rutzler, *J. Med. Chem.*, 1991, **34**, 3339.
77. G. Pettit, R. Tan, F. Gao, M. Williams, D. Doubek, M. Boyd, J. Schmidt, J. Chapuis, E. Hamel, R. Bai, J. Hooper and L. Tackett, *J. Org. Chem.*, 1993, **58**, 2538.
78. G. Pettit, F. Gao, R. Cerny, D. Doubek, L. Tackett, J. Schmidt and J. Chapuis, *J. Med. Chem.*, 1994, **37**, 1165.
79. M. Litaudon, J. Hart, J. Blunt, R. Lake and M. Munro, *Tetrahedron Lett.*, 1994, **35**, 9435.
80. M. Litaudon, S. Hickford, R. Lill, R. Lake, J. Blunt and M. Munro, *J. Org. Chem.*, 1997, **62**, 1868.
81. M. J. Towle, K. A. Salvato, J. Budrow, B. F. Wels, G. Kuznetsov, K. K. Aalfs, S. Welsh, W. J. Zheng, B. M. Seletsky, M. H. Palme, G. J. Habgood, L. A. Singer, L. V. Dipietro, Y. Wang, J. J. Chen, D. A. Quincy, A. Davis, K. Yoshimatsu, Y. Kishi, M. J. Yu and B. A. Littlefield, *Cancer Res.*, 2001, **61**, 1013.
82. W. J. Zheng, B. M. Seletsky, M. H. Palme, P. J. Lydon, L. A. Singer, C. E. Chase, C. A. Lemelin, Y. C. Shen, H. Davis, L. Tremblay, M. J. Towle, K. A. Salvato, B. F. Wels, K. K. Aalfs, Y. Kishi, B. A. Littlefield and M. J. Yu, *Bioorg. Med. Chem. Lett.*, 2004, **14**, 5551.
83. A. Das, A. Spira, N. Iannotti, M. Savin, E. Zang, K. Datta, G. Shapiro, M. Patturajan and S. Silberman, *J. Clin. Oncol.*, 2006, **24**, 390S.

84. J. Blum, L. Forero, M. Heiskala, N. Meneses, K. Chandrawansa, F. Fang, G. Shapiro, S. Fields, S. Silberman and L. Vahdat, *J. Clin. Oncol.*, 2006, **24**, 41S.
85. P. Crews, J. J. Farias, R. Emrich and P. A. Keifer, *J. Org. Chem.*, 1994, **59**, 2932.
86. R. Talpir, Y. Benayahu, Y. Kashman, L. Pannell and M. Schleyer, *Tetrahedron Lett.*, 1994, **35**, 4453.
87. J. E. Coleman, E. D. Desilva, F. M. Kong, R. J. Andersen and T. M. Allen, *Tetrahedron*, 1995, **51**, 10653.
88. J. Coleman, B. Patrick, R. Andersen and S. Rettig, *Acta Crystallogr., Sect. C: Cryst. Struct. Commun.*, 1996, **52**, 1525.
89. W. R. Gamble, N. A. Durso, R. W. Fuller, C. K. Westergaard, T. R. Johnson, D. L. Sackett, E. Hamel, J. H. Cardellina and M. R. Boyd, *Bioorg. Med. Chem.*, 1999, **7**, 1611.
90. C. Chevallier, A. Richardson, M. Edler, E. Hamel, M. Harper and C. Ireland, *Org. Lett.*, 2003, **5**, 3737.
91. R. N. Sonnenschein, J. J. Farias, K. Tenney, S. L. Mooberry, E. Lobkovsky, J. Clardy and P. Crews, *Org. Lett.*, 2004, **6**, 779.
92. J. Nieman, J. Coleman, D. Wallace, E. Piers, L. Lim, M. Roberge and R. Andersen, *J. Nat. Prod.*, 2003, **66**, 183.
93. B. Hadaschik, S. Ettinger, R. Sowery, A. Zoubeidi, R. Andersen, M. Roberge and M. Gleave, *Int. J. Cancer*, 2008, **122**, 2368.
94. B. Hadaschik, H. Adomat, L. Fazli, Y. Fradet, R. Andersen, M. Gleave and A. So, *Clin. Cancer Res.*, 2008, **14**, 1510.
95. R. H. Cichewicz, F. A. Valeriote and P. Crews, *Org. Lett.*, 2004, **6**, 1951.
96. G. R. Pettit, J. P. Xu, J. C. Chapuis, R. K. Pettit, L. P. Tackett, D. L. Doubek, J. N. A. Hooper and J. M. Schmidt, *J. Med. Chem.*, 2004, **47**, 1149.
97. B. K. Rubio, S. J. Robinson, C. E. Avalos, F. A. Valeriote, N. J. De Voogd and P. Crews, *J. Nat. Prod.*, 2008, **71**, 1475.
98. L. M. West, P. T. Northcote and C. N. Battershill, *J. Org. Chem.*, 2000, **65**, 445.
99. C. Meyer, D. Ferguson, M. Krauth, M. Wick and P. Northcote, *EJC Supplements*, 2006, **4**, 192.
100. M. J. Peyton, S. Weber, R. Greer, L. Girard, A. F. Gazdar, J. De Brabander and J. D. Minna, *Proc. Am. Ass. Cancer Res.*, 2007, **48**, 336.
101. K. L. Erickson, J. A. Beutler, J. H. Cardellina II and M. R. Boyd, *J. Org. Chem.*, 1997, **62**, 5.
102. X. S. Xie, D. Padron, X. B. Liao, J. Wang, M. G. Roth and J. K. De Brabander, *J. Biol. Chem.*, 2004, **279**, 19755.
103. N. B. Perry, L. Ettouati, M. Litaudon, J. W. Blunt, M. H. G. Munro, S. Parkin and H. Hope, *Tetrahedron*, 1994, **50**, 3987.
104. M. V. Cespedes, M. J. Guillen, M. Parreno, I. Casanova, M. A. Pavon, R. Mangues, G. Faircloth, C. Cuevas and P. Aviles, *Proc. Am. Ass. Cancer Res.*, 2005, **46**, 152.

105. M. V. Cespedes, M. J. Guillen, M. Parreno, I. Casanova, M. A. Pavon, R. Mangues, G. Faircloth, C. Cuevas and P. Aviles, *Proc. Am. Ass. Cancer Res.*, 2005, **46**, 151.
106. S. Carmely and Y. Kashman, *Tetrahedron Lett.*, 1985, **26**, 511.
107. M. Kobayashi, J. Tanaka, T. Katori, M. Matsuura and I. Kitagawa, *Tetrahedron Lett.*, 1989, **30**, 2963.
108. I. Kitagawa, M. Kobayashi, T. Katori, M. Yamashita, J. Tanaka, M. Doi and T. Ishida, *J. Am. Chem. Soc.*, 1990, **112**, 3710.
109. C. J. Wegerski, D. France, S. Cornell-Kennon and P. Crews, *Bioorg. Med. Chem.*, 2004, **12**, 5631.
110. M. R. Bubb, I. Spector, A. D. Bershadsky and E. D. Korn, *J. Biol. Chem.*, 1995, **270**, 3463.
111. N. Fusetani, K. Yasumuro, S. Matsunaga and K. Hashimoto, *Tetrahedron Lett.*, 1989, **30**, 2809.
112. P. Phuwapraisirisan, S. Matsunaga, R. W. M. Van Soest and N. Fusetani, *J. Nat. Prod.*, 2002, **65**, 942.
113. S. Y. Saito, S. Watabe, M. Hori, H. Ozaki, N. Fusetani and H. Karaki, *Jpn. J. Pharmacol.*, 1994, **64**, 347P.
114. M. R. Kernan and D. J. Faulkner, *Tetrahedron Lett.*, 1987, **28**, 2809.
115. M. R. Kernan, T. F. Molinski and D. J. Faulkner, *J. Org. Chem.*, 1988, **53**, 5014.
116. J. Kobayashi, O. Murata, H. Shigemori and T. Sasaki, *J. Nat. Prod.*, 1993, **56**, 787.
117. J. Kobayashi, M. Tsuda, H. Fuse, T. Sasaki and Y. Mikami, *J. Nat. Prod.*, 1997, **60**, 150.
118. J. Shin, H. S. Lee, J. Y. Kim, H. J. Shin, J. W. Ahn and V. J. Paul, *J. Nat. Prod.*, 2004, **67**, 1889.
119. I. Spector, F. Braet, N. Shochet and M. Bubb, *Microscopy Res. Tech.*, 1999, **47**, 18.
120. J. A. Roesener and P. J. Scheuer, *J. Am. Chem. Soc.*, 1986, **108**, 846.
121. J. S. Allingham, J. Tanaka, G. Marriott and I. Rayment, *Org. Lett.*, 2004, **6**, 597.
122. V. A. Klenchin, J. S. Allingham, R. King, J. Tanaka, G. Marriott and I. Rayment, *Nat. Struct. Biol.*, 2003, **10**, 1058.
123. J. S. Allingham, Z. Zampella, M. V. D'Auria and I. Rayment, *Proc. Natl. Acad. Sci. USA*, 2005, **102**, 14527.
124. R. Sakai, T. Higa and Y. Kashman, *Chem. Lett.*, 1986, 1499.
125. Y. Kato, N. Fusetani, S. Matsunaga, K. Hashimoto, R. Sakai, T. Higa and Y. Kashman, *Tetrahedron Lett.*, 1987, **28**, 6225.
126. J. Tanaka, T. Higa, M. Kobayashi and I. Kitagawa, *Chem. Pharm. Bull.*, 1990, **38**, 2967.
127. A. Qureshi and D. Faulkner, *J. Nat. Prod.*, 2000, **63**, 841.
128. J. Kobayashi, S. Tsukamoto, A. Tanabe, T. Sasaki and M. Ishibashi, *J. Chem. Soc., Perkin Trans. 1*, 1991, 2379.
129. D. R. Terry, I. Spector, T. Higa and M. R. Bubb, *J. Biol. Chem.*, 1997, **272**, 7841.

130. Y. Kashman, A. Groweiss and J. Shmueli, *Tetrahedron Lett.*, 1980, **21**, 3629.
131. A. Groweiss, U. Shmueli and Y. Kashman, *J. Org. Chem.*, 1983, **48**, 3512.
132. R. Okuda and P. Scheuer, *Experientia*, 1985, **41**, 1355.
133. Y. Kakou, P. Crews and G. Bakus, *J. Nat. Prod.*, 1987, **50**, 482.
134. M. Coue, S. Brenner, I. Spector and E. Korn, *FEBS Lett.*, 1987, **213**, 316.
135. P. Crews, L. Manes and M. Boehler, *Tetrahedron Lett.*, 1986, **27**, 2797.
136. T. Zabriskie, J. Klocke, C. Ireland, A. Marcus, T. Molinski, D. Faulkner, C. Xu and J. Clardy, *J. Am. Chem. Soc.*, 1986, **108**, 3123.
137. J. Braekman, D. Daloze, B. Moussiaux and R. Riccio, *J. Nat. Prod.*, 1987, **50**, 994.
138. M. Bubb, A. Senderowicz, E. Sausville, K. Duncan and E. Korn, *J. Biol. Chem.*, 1994, **269**, 14869.
139. F. Lazaro-Dieguez, C. Aguado, E. Mato, Y. Sanchez-Ruiz, I. Esteban, J. Alberch, E. Knecht and G. Egea, *J. Cell Sci.*, 2008, **121**, 1415.
140. M. Sanders, M. C. Diaz and P. Crews, *Mem. Queensl. Mus.*, 1999, **44**, 525.
141. L. Murray, A. Johnson, M. Diaz and P. Crews, *J. Org. Chem.*, 1997, **62**, 5638.
142. D. Burgoyne, R. Andersen and T. Allen, *J. Org. Chem.*, 1992, **57**, 525.
143. L. Yang and R. Andersen, *J. Nat. Prod.*, 2002, **65**, 1924.
144. A. Bramley, J. Langlands, A. Jones, D. Burgoyne, Y. Li, R. Andersen and H. Salari, *Br. J. Pharmacol.*, 1995, **115**, 1433.
145. Y. Shen and D. Burgoyne, *J. Org. Chem.*, 2002, **67**, 3908.
146. C. Kasserra, P. Harris, G. Stenton, W. Abraham and J. Langlands, *Pulm. Pharmacol. Ther.*, 2004, **17**, 309.
147. G. Sharma, J. Buyer and M. Pomerantz, *J. Chem. Soc. Chem. Commun.*, 1980, 435.
148. I. Kitagawa, M. Kobayashi, K. Kitanaka, M. Kido and Y. Kyogoku, *Chem. Pharm. Bull.*, 1983, **31**, 2321.
149. F. Schmitz, S. Gunasekera, V. Lakshmi and L. Tillekeratne, *J. Nat. Prod.*, 1985, **48**, 47.
150. G. Pettit, C. Herald, J. Leet, R. Gupta, D. Schaufelberger, R. Bates, P. Clewlow, D. Doubek, K. Manfredi, K. Rutzler, J. Schmidt, L. Tackett, F. Ward, M. Bruck and F. Camou, *Can. J. Chem.*, 1990, **68**, 1621.
151. D. Curman, B. Cinel, D. Williams, N. Rundle, W. Block, A. Goodarzi, J. Hutchins, P. Clarke, B. Zhou, S. Lees-Miller, R. Andersen and M. Roberge, *J. Biol. Chem.*, 2001, **276**, 17914.
152. D. Tasdemir, R. Mallon, M. Greenstein, L. Feldberg, S. Kim, K. Collins, D. Wojciechowicz, G. Mangalindan, G. Concepcion, M. Harper and C. Ireland, *J. Med. Chem.*, 2002, **45**, 529.
153. R. Sakai and T. Higa, *J. Am. Chem. Soc.*, 1986, **108**, 6404.
154. H. Nakamura, S. Z. Deng, J. Kobayashi, Y. Ohizumi, Y. Tomotake, T. Matsuzaki and Y. Hirata, *Tetrahedron Lett.*, 1987, **28**, 621.
155. K. Kondo, H. Shigemori, Y. Kikuchi, M. Ishibashi, T. Sasaki and J. Kobayashi, *J. Org. Chem.*, 1992, **57**, 2480.

156. T. Ichiba, J. M. Corgiat, P. J. Scheuer and M. Kellyborges, *J. Nat. Prod.*, 1994, **57**, 168.
157. J. Kobayashi, M. Tsuda, N. Kawasaki, T. Sasaki and Y. Mikami, *J. Nat. Prod.*, 1994, **57**, 1737.
158. R. A. Edrada, P. Proksch, V. Wray, L. Witte, W. E. G. Muller and R. W. M. Van Soest, *J. Nat. Prod.*, 1996, **59**, 1056.
159. K. K. H. Ang, M. J. Holmes, T. Higa, M. T. Hamann and U. A. K. Kara, *Antimicrob. Agents Chemother.*, 2000, **44**, 1645.
160. E. D. De Silva and P. J. Scheuer, *Tetrahedron Lett.*, 1980, **21**, 1611.
161. D. Lombardo and E. A. Dennis, *J. Biol. Chem.*, 1985, **260**, 7234.
162. G. R. Pettit, Z. A. Cichacz, F. Gao, M. R. Boyd and J. M. Schmidt, *J. Chem. Soc., Chem. Commun.*, 1994, 2.
163. G. R. Pettit and Z. A. Cichacz, *US Patent* 5,430,053, 1995.
164. I. Paterson, R. Britton, O. Delgado and A. E. Wright, *Chem. Commun.*, 2004, 632.
165. C. Madiraju, M. C. Edler, E. Hamel, B. S. Raccor, R. Balachandran, G. Y. Zhu, K. A. Giuliano, A. Vogt, Y. S. Shin, J. H. Fournier, Y. H. Fukui, A. M. Bruckner, D. P. Curran and B. W. Day, *Biochemistry*, 2005, **44**, 15053.
166. C. Madiraju, B. S. Raccor, Y. Shin, D. P. Curran and B. W. Day, *Clin. Cancer Res.*, 2005, **11**, 9092S.
167. G. W. O'Neil and A. J. Phillips, *Tetrahedron Lett.*, 2004, **45**, 4253.
168. C. O. Kangani, A. M. Bruckner and D. P. Curran, *Org. Lett.*, 2005, **7**, 379.
169. E. Prusov, H. Rohm and M. E. Maier, *Org. Lett.*, 2006, **8**, 1025.
170. J. Jagel and M. E. Maier, *Synlett*, 2006, 693.
171. O. Sharon, C. Monti and C. Gennari, *Tetrahedron*, 2007, **63**, 5873.
172. A. K. Dilger, V. Gopalsamuthiram and S. D. Burke, *J. Am. Chem. Soc.*, 2007, **129**, 16273.
173. V. Saibaba, A. Sampath, K. Mukkanti, J. Iqbal and P. Das, *Synthesis (Stuttgart)*, 2007, **18**, 2797.
174. C. Gennari, D. Castoldi and O. Sharon, *Pure Appl. Chem.*, 2007, **79**, 173.
175. I. Paterson, R. Britton, O. Delgado, A. Meyer and K. G. Poullennec, *Angew. Chem., Int. Ed. Engl.*, 2004, **43**, 4629.
176. Y. Shin, J. H. Fournier, Y. Fukui, A. M. Bruckner and D. P. Curran, *Angew. Chem., Int. Ed. Engl.*, 2004, **43**, 4634.
177. G. W. O'Neil and A. J. Phillips, *J. Am. Chem. Soc.*, 2006, **128**, 5340.
178. P. V. Ramachandran, A. Srivastava and D. Hazra, *Org. Lett.*, 2007, **9**, 157.
179. G. J. Florence, N. M. Gardner and I. Paterson, *Nat. Prod. Rep.*, 2008, **25**, 342.
180. B. S. Raccor, A. Vogt, R. P. Sikorski, C. Madiraju, R. Balachandran, K. Montgomery, Y. Shin, Y. Fukui, W. H. Jung, D. P. Curran and B. W. Day, *Mol. Pharmacol.*, 2008, **73**, 718.
181. W. H. Jung, C. Harrison, Y. Shin, J. H. Fournier, R. Balachandran, B. S. Raccor, R. P. Sikorski, A. Vogt, D. P. Curran and B. W. Day, *J. Med. Chem.*, 2007, **50**, 2951.

182. Y. Fukui, A. M. Bruckner, Y. Shin, R. Balachandran, B. W. Day and D. P. Curran, *Org. Lett.*, 2006, **8**, 301.
183. Y. Shin, J. H. Fournier, R. Balachandran, C. Madiraju, B. S. Raccor, G. Zhu, M. C. Edler, E. Hamel, B. W. Day and D. P. Curran, *Org. Lett.*, 2005, **7**, 2873.
184. C. Madiraju, M. C. Edler, B. S. Raccor, R. Balachandran, E. Hamel, K. A. Giuliano, A. Vogt, Y. Shin, J. H. Fournier, Y. Fukui, A. M. Bruckner, D. P. Curran and B. W. Day, *Proc. Am. Ass. Cancer Res.*, 2005, **46**, 932.
185. Y. Shin, N. Choy, R. Balachandran, C. Madiraju, B. W. Day and D. P. Curran, *Org. Lett.*, 2002, **4**, 4443.
186. Y. Shin, N. Choy, T. R. Turner, R. Balachandran, C. Madiraju, B. W. Day and D. P. Curran, *Org. Lett.*, 2003, **5**, 603.
187. Y. Shin, J. H. Fournier, A. Bruckner, C. Madiraju, R. Balachandran, B. S. Raccor, M. C. Edler, E. Hamel, R. P. Sikorski, A. Vogt, B. W. Day and D. P. Curran, *Tetrahedron*, 2007, **63**, 8537.
188. I. Paterson, N. M. Gardner, K. G. Poullennec and A. E. Wright, *Bioorg. Med. Chem. Lett.*, 2007, **17**, 2443.
189. I. Paterson and N. M. Gardner, *Chem. Commun.*, 2007, 49.
190. I. Paterson, N. M. Gardner, K. G. Poullennec and A. E. Wright, *J. Nat. Prod.*, **71**, 364.
191. D. P. Curran, Y. Shin, N. Choy, B. W. Day, R. Balachandran, C. Madiraju and T. Turner, *US Patent* 7,122,686, 2006.
192. D. P. Curran, Y. Shin, J. H. Fournier, J. Mancuso, B. W. Day, A. Bruckner and Y. Fukui, *US Patent* 7,321,046, 2008.
193. A. E. Wright, J. L. Cummins, S. A. Pomponi, R. E. Longley and R. A. Isbrucker, *US Patent* 6,677,370, 2004.
194. A. K. Ghosh, M. Packiarajan and J. Cappiello, *Tetrahedron Lett.*, 1997, **38**, 2427.
195. A. Shimizu and S. Nishiyama, *Tetrahedron Lett.*, 1997, **38**, 6011.
196. S. L. Mooberry, G. Tien, A. H. Hernandez, A. Plubrukarn and B. S. Davidson, *Cancer Res.*, 1999, **59**, 653.
197. E. K. Dorling, E. Ohler and J. Mulzer, *Tetrahedron Lett.*, 2000, **41**, 6323.
198. A. K. Ghosh and Y. Wang, *Tetrahedron Lett.*, 2000, **41**, 2319.
199. A. K. Ghosh and Y. Wang, *Tetrahedron Lett.*, 2000, **41**, 4705.
200. J. Mulzer and M. Hanbauer, *Tetrahedron Lett.*, 2000, **41**, 33.
201. A. K. Ghosh and Y. Wang, *J. Am. Chem. Soc.*, 2000, **122**, 2.
202. I. Paterson, C. De Savi and M. Tudge, *Org. Lett.*, 2001, **3**, 3149.
203. M. T. Crimmins, M. G. Stanton and S. P. Allwein, *J. Am. Chem. Soc.*, 2002, **124**, 5958.
204. A. K. Ghosh, Y. Wang and J. T. Kim, *J. Org. Chem.*, 2001, **66**, 8973.
205. V. S. Enev, H. Kaehlig and J. Mulzer, *J. Am. Chem. Soc.*, 2001, **123**, 10764.
206. M. R. Pitts and J. Mulzer, *Tetrahedron Lett.*, 2002, **43**, 8471.
207. J. Mulzer and M. Hanbauer, *Tetrahedron Lett.*, 2002, **43**, 3381.
208. S. G. Nelson, W. S. Cheung, A. J. Kassick and M. A. Hilfiker, *J. Am. Chem. Soc.*, 2002, **124**, 13654.

209. J. Uenishi and M. Ohmi, *Angew. Chem., Int. Ed. Eng.*, 2005, **44**, 2756.
210. D. R. Williams, L. Mi, R. J. Mullins and R. E. Stites, *Tetrahedron Lett.*, 2002, **43**, 4841.
211. P. A. Wender, S. G. Hegde, R. D. Hubbard and L. Zhang, *J. Am. Chem. Soc.*, 2002, **124**, 4956.
212. I. Paterson, C. De Savi and M. Tudge, *Org. Lett.*, 2001, **3**, 213.
213. G. T. Nadolski and B. S. Davidson, *Tetrahedron Lett.*, 2001, **42**, 797.
214. B. T. Messenger and B. S. Davidson, *Tetrahedron Lett.*, 2001, **42**, 801.
215. A. K. Ghosh and Y. Wang, *Tetrahedron Lett.*, 2001, **42**, 3399.
216. A. Sivaramakrishnan, G. T. Nadolski, I. A. Mc Alexander and B. S. Davidson, *Tetrahedron Lett.*, 2002, **43**, 213.
217. S. R. Punna, A. Prabhakar, G. V. M. Sharma and A. C. Kunwar, *Lett. Org. Chem.*, 2006, **3**, 447.
218. J. Uenishi, M. Ohmi, K. Matsui and M. Iwano, *Tetrahedron*, 2005, **61**, 1971.
219. H. W. Lee and J. Y. Hong, *Bull. Korean Chem. Soc.*, 2003, **24**, 1569.
220. M. T. Crimmins, *Curr. Opin. Drug Discov. Devel.*, 2002, **5**, 944.
221. J. Mulzer, E. Ohler, V. S. Enev and M. Hanbauer, *Adv. Synth. Catal.*, 2002, **344**, 573.
222. J. Mulzer and E. Ohler, *Chem. Rev.*, 2003, **103**, 3753.
223. D. E. Pryor, A. O'brate, G. Bilcer, J. F. Diaz, Y. F. Wang, Y. Wang, M. Kabaki, M. K. Jung, J. M. Andreu, A. K. Ghosh, P. Giannakakou and E. Hamel, *Biochemistry*, 2002, **41**, 9109.
224. E. J. Gapud, R. L. Bai, A. K. Ghosh and E. Hamel, *Mol. Pharmacol.*, 2004, **66**, 113.
225. E. Hamel, B. W. Day, J. H. Miller, M. K. Jung, P. T. Northcote, A. K. Ghosh, D. P. Curran, M. Cushman, K. C. Nicolaou, I. Paterson and E. J. Sorensen, *Mol. Pharmacol.*, 2006, **70**, 1555.
226. H. Lu and E. L. Schwartz, *Proc. Am. Ass. Cancer Res.*, 2004, **45**, 597.
227. H. Y. Lu, J. Murtagh and E. L. Schwartz, *Mol. Pharmacol.*, 2006, **69**, 1207.
228. A. Ahmed, E. K. Hoegenauer, V. A. S. Enev, M. Hanbauer, H. Kaehlig, E. Ohler and J. Mulzer, *J. Org. Chem.*, 2003, **68**, 3026.
229. P. A. Wender, S. G. Hegde, R. D. Hubbard, L. Zhang and S. L. Mooberry, *Org. Lett.*, 2003, **5**, 3507.
230. I. Paterson, H. Bergmann, D. Menche and A. Berkessel, *Org. Lett.*, 2004, **6**, 1293.
231. S. L. Mooberry, D. A. Randall-Hlubek, R. M. Leal, S. G. Hegde, R. D. Hubbard, L. Zhang and P. A. Wender, *Proc. Natl. Acad. Sci. USA*, 2004, **101**, 8803.
232. B. M. Gallagher, F. G. Fang, C. W. Johannes, M. Pesant, M. R. Tremblay, H. J. Zhao, K. Akasaka, X. Y. Li, J. K. Liu and B. A. Littlefield, *Bioorg. Med. Chem. Lett.*, 2004, **14**, 575.
233. I. Paterson, D. Menche, R. Britton, A. E. Hakansson and M. A. Silva-Martinez, *Tetrahedron Lett.*, 2005, **46**, 3677.

234. I. Paterson, D. Menche, A. E. Hakansson, A. Longstaff, D. Wong, I. Barasoain, R. M. Buey and J. F. Diaz, *Bioorg. Med. Chem. Lett.*, 2005, **15**, 2243.
235. B. M. Gallagher, H. J. Zhao, M. Pesant and F. G. Fang, *Tetrahedron Lett.*, 2005, **46**, 923.
236. P. A. Wender, M. K. Hilinski, N. Soldermann and S. L. Mooberry, *Org. Lett.*, 2006, **8**, 1507.
237. P. A. Wender, M. K. Hilinski, P. R. Skaanderup, N. G. Soldermann and S. L. Mooberry, *Organic Lett.*, 2006, **8**, 4105.
238. C. Faveau, M. Mondon, J. P. Gesson, T. Mahnke, S. Gebhardt and U. Koert, *Tetrahedron Lett.*, 2006, **47**, 8305.
239. T. A. Johnson, K. Tenney, R. H. Cichewicz, B. I. Morinaka, K. N. White, T. Amagata, B. Subramanian, J. Media, S. L. Mooberry, F. A. Valeriote and P. Crews, *J. Med. Chem.*, 2007, **50**, 3795.
240. J. M. Rae, C. J. Creighton, J. M. Meck, H. R. Bassem and M. D. Johnson, *Breast Cancer Res. Treat.*, 2007, **104**, 13.
241. X. Jiang, B. Liu, S. Lebreton and J. De Brabander, *J. Am. Chem. Soc.*, 2007, **129**, 6386.
242. K. A. Hood, B. T. Backstrom, L. M. West, P. T. Northcote, M. V. Berridge and J. H. Miller, *Anti-Cancer Drug Des.*, 2001, **16**, 155.
243. K. A. Hood, L. M. West, B. Rouwe, P. T. Northcote, M. V. Berridge, S. J. Wakefield and J. H. Miller, *Cancer Res.*, 2002, **62**, 3356.
244. J. Jimenez-Barbero, A. Canales, P. T. Northcote, R. M. Buey, J. M. Andreu and J. F. Diaz, *J. Am. Chem. Soc.*, 2006, **128**, 8757.
245. O. Pineda, J. Farras, L. Maccari, F. Manetti, M. Botta and J. Vilarrasa, *Bioorg. Med. Chem. Lett.*, 2004, **14**, 4825.
246. J. T. Huzil, J. K. Chik, G. W. Slysz, H. Freedman, J. Tuszynski, R. E. Taylor, D. L. Sackett and D. C. Schriemer, *J. Mol. Biol.*, 2008, **378**, 1016.
247. J. H. Miller, B. Rouwe, T. N. Gaitanos, K. A. Hood, K. P. Crume, B. T. Backstrom, A. C. La Flamme, M. V. Berridge and P. T. Northcote, *Apoptosis*, 2004, **9**, 785.
248. A. Wilmes, K. Bargh, C. Kelly, P. T. Northcote and J. H. Miller, *Mol. Pharm.*, 2007, **4**, 269.
249. K. P. Crume, J. H. Miller and A. C. La Flamme, *Exp. Biol. Med.*, 2007, **232**, 607.
250. I. Paterson, M. E. Di Francesco and T. Kuhn, *Org. Lett.*, 2003, **5**, 599.
251. A. K. Ghosh and J. H. Kim, *Tetrahedron Lett.*, 2003, **44**, 3967.
252. A. K. Ghosh and J. H. Kim, *Tetrahedron Lett.*, 2003, **44**, 7659.
253. R. E. Taylor and M. Z. Jin, *Org. Lett.*, 2003, **5**, 4959.
254. M. K. Gurjar, Y. Pedduri, C. V. Ramana, V. G. Puranik and R. G. Gonnade, *Tetrahedron Lett.*, 2004, **45**, 387.
255. B. Liu and W. S. Zhou, *Organic Lett.*, 2004, **6**, 71.
256. D. W. Engers, M. J. Bassindale and B. L. Pagenkopf, *Org. Lett.*, 2004, **6**, 663.
257. E. Roulland and M. S. Ermolenko, *Org. Lett.*, 2005, **7**, 2225.

258. R. M. Owen and W. R. Roush, *Org. Lett.*, 2005, **7**, 3941.
259. F. Vaghefi, L. B. Chen and M. L. Sherman, *US Patent Application* US32717 2006033913, 2006.
260. X. B. Liao, Y. S. Wu and J. K. De Brabander, *Angew. Chem., Int. Ed. Engl.*, 2003, **42**, 1648.
261. M. Z. Jin and R. E. Taylor, *Org. Lett.*, 2005, **7**, 1303.
262. A. K. Ghosh, X. Xu, J. H. Kim and C. X. Xu, *Org. Lett.*, 2008, **10**, 1001.
263. D. R. Williams, P. P. Nag and N. Zorn, *Curr. Opin. Drug Discov. Devel.*, 2008, **11**, 251.
264. S. Kiren and L. J. Williams, *Org. Lett.*, 2005, **7**, 2905.
265. M. E. Green, J. C. Rech and P. E. Floreancig, *Org. Lett.*, 2005, **7**, 4117.
266. J. C. Rech and P. E. Floreancig, *Org. Lett.*, 2005, **7**, 5175.
267. X. Jiang, J. Garcia-Fortanet and J. K. De Brabander, *J. Am. Chem. Soc.*, 2005, **127**, 11254.
268. N. Shangguan, S. Kiren and L. J. Williams, *Org. Lett.*, 2007, **9**, 1093.
269. X. H. Huang, N. Shao, A. Palani, R. Aslanian and A. Buevich, *Org. Lett.*, 2007, **9**, 2597.
270. X. Jiang, N. Williams and J. K. De Brabander, *Org. Lett.*, 2007, **9**, 227.
271. X. H. Huang, N. Shao, A. Palani, R. Aslanian, A. Buevich, C. Seidel-Dugan and R. Huryk, *Tetrahedron Lett.*, 2008, **49**, 3592.
272. S. J. Robinson, K. Tenney, D. F. Yee, L. Martinez, J. E. Media, F. A. Valeriote, R. W. M. Van Soest and P. Crews, *J. Nat. Prod.*, 2007, **70**, 1002.
273. M. R. Boyd, C. Farina, P. Belfiore, S. Gagliardi, J. W. Kim, Y. Hayakawa, J. A. Beutler, T. C. Mckee, B. J. Bowman and E. J. Bowman, *J. Pharmacol. Exp. Ther.*, 2001, **297**, 114.
274. S. Charron, D. Labrecque, S. Lamothe, G. Attardo, F. Ouellet, B. Grouix and L. Leblond, *Abstr. Pap. Am. Chem. Soc.*, 2001, **222**, MEDI124.
275. A. Fuerstner, G. Seidel and N. Kindler, *Tetrahedron*, 1999, **55**, 8215.
276. Y. S. Wu, O. R. Seguil and J. K. De Brabander, *Org. Lett.*, 2000, **2**, 4241.
277. A. B. Smith and J. Y. Zheng, *Tetrahedron*, 2002, **58**, 6455.
278. S. S. Tang and K. L. Erickson, *J. Nat. Prod.*, 2008, **71**, 898.
279. Y. S. Wu, X. B. Liao, R. F. Wang, X. S. Xie and J. K. De Brabander, *J. Am. Chem. Soc.*, 2002, **124**, 3245.
280. A. Furstner, O. R. Thiel and G. Blanda, *Org. Lett.*, 2000, **2**, 3731.
281. B. B. Snider and F. B. Song, *Org. Lett.*, 2000, **2**, 407.
282. M. Bauer and M. E. Maier, *Org. Lett.*, 2002, **4**, 2205.
283. J. T. Feutrill, G. A. Holloway, F. Hilli, H. M. Hugel and M. A. Rizzacasa, *Tetrahedron Lett.*, 2000, **41**, 8569.
284. S. Lebreton, X. S. Xie, D. Ferguson and J. K. De Brabander, *Tetrahedron*, 2004, **60**, 9635.
285. R. S. Coleman and P. H. Liu, *Org. Lett.*, 2004, **6**, 577.
286. C. Herb, F. Dettner and M. E. Maier, *Eur. J. Org. Chem.*, 2005, 728.
287. J. S. Yadav and P. S. R. Reddy, *Synthesis (Stuttgart)*, 2007, **7**, 1070.
288. D. Labrecque, S. Charron, R. Rej, C. Blais and S. Lamothe, *Tetrahedron Lett.*, 2001, **42**, 2645.

289. B. B. Snider and F. B. Song, *Org. Lett.*, 2001, **3**, 1817.
290. A. Furstner, T. Dierkes, O. R. Thiel and G. Blanda, *Chem. Eur. J.*, 2001, **7**, 5286.
291. C. Herb, A. Bayer and M. E. Maier, *Chem. Eur. J.*, 2004, **10**, 5649.
292. K. L. Yang, Y. M. Ahn, B. Blackman, W. E. Diederich, P. T. Flahertry, C. J. Mossman, S. Roy and G. I. Georg, *Abstr. Pap. Am. Chem. Soc.*, 2002, **224**, U245.
293. K. L. Yang, B. Blackman, W. Diederich, P. T. Flaherty, C. J. Mossman, S. Roy, Y. M. Ahn and G. I. Georg, *J. Org. Chem.*, 2003, **68**, 10030.
294. T. Haack, K. Haack, W. E. Diederich, B. Blackman, S. Roy, S. Pusuluri and G. I. Georg, *J. Org. Chem.*, 2005, **70**, 7592.
295. G. A. Holloway, H. M. Hugel and M. A. Rizzacasa, *J. Org. Chem.*, 2003, **68**, 2200.
296. C. Herb and M. E. Maier, *J. Org. Chem.*, 2003, **68**, 8129.
297. J. S. Yadav and P. Srihari, *Tetrahedron Asymmetry*, 2004, **15**, 81.
298. L. Yet, *Chem. Rev.*, 2003, **103**, 4283.
299. D. A. Abdo, C. A. Motti, C. N. Battershill and E. S. Harvey, *J. Chem. Ecol.*, 2007, **33**, 1635.
300. N. B. Perry, L. Ettouati, M. Litaudon, J. W. Blunt, M. H. G. Munro, S. Parkin and H. Hope, *Tetrahedron*, 1994, **50**, 3987.
301. G. Trimurtulu, D. J. Faulkner, N. B. Perry, L. Ettouati, M. Litaudon, J. W. Blunt, M. G. H. Munro and G. B. Jameson, *Tetrahedron*, 1994, **50**, 3993.
302. E. Erba, G. Balconi, M. Faretta, D. Bergamaschi, G. Guidi, J. Jimeno, G. Faircloth and M. D'incalci, *Proc. Am. Ass. Cancer Res.*, 1996, **37**, 28.
303. E. Erba, G. Tognon, J. Jimeno, G. T. Faircloth and M. D'incalci, *Eur. J. Cancer*, 2002, **38**, S33.
304. E. Erba, M. Simone, J. Jimeno, G. T. Faircloth and M. D'incalci, *Proc. Am. Ass. Cancer Res.*, 2003, **44**, 127.
305. E. Erba, M. Simone, G. Damia, F. Vikhanskaya, A. Di Francesco, J. Jimeno, C. Cuevas, G. Faircloth and M. D'incalci, *Clin. Cancer Res.*, 2003, **9**, 6222S.
306. J. Mendiola, A. Baeza, J. Alvarez-Builla and J. J. Vaquero, *J. Org. Chem.*, 2004, **69**, 4974.
307. J. Mendiola, J. M. Minguez, J. Alvarez-Builla and J. J. Vaquero, *Org. Lett.*, 2000, **2**, 3253.
308. A. Baeza, C. Burgos, J. Alvarez-Builla and J. J. Vaquero, *Tetrahedron Lett.*, 2007, **48**, 2597.
309. R. J. Anderson and J. C. Morris, *Tetrahedron Lett.*, 2000, **42**, 311.
310. P. M. Fresneda, P. Molina, S. Delgado and J. A. Bleda, *Tetrahedron Lett.*, 2000, **41**, 4777.
311. R. J. Anderson and J. C. Morris, *Tetrahedron Lett.*, 2001, **42**, 8697.
312. P. Molina, P. M. Fresneda, S. Delgado and J. A. Bleda, *Tetrahedron Lett.*, 2002, **43**, 1005.
313. A. Ahaidar, D. Fernandez, O. Perez, G. Danelon, C. Cuevas, I. Manzanares, F. Albericio, J. A. Joule and M. Alvarez, *Tetrahedron Lett.*, 2003, **44**, 6191.

314. P. Molina, P. M. Fresneda and S. Delgado, *J. Org. Chem.*, 2003, **68**, 489.
315. A. Ahaidar, D. Fernandez, G. Danelon, C. Cuevas, I. Manzanares, F. Albericio, J. A. Joule and M. Alvarez, *J. Org. Chem.*, 2003, **68**, 10020.
316. R. J. Anderson, J. B. Hill and J. C. Morris, *J. Org. Chem.*, 2005, **70**, 6204.
317. P. M. Fresneda, S. Delgado, A. Francesch, I. Manzanares, C. Cuevas and P. Molina, *J. Med. Chem.*, 2006, **49**, 1217.
318. J. Yin, P. Aviles, M. J. Guillen, C. Ly, W. Lee, S. Munt, C. C. Merchante and G. T. Faircloth, *Proc. Am. Ass. Cancer Res.*, 2004, **45**, 1245.
319. J. Yin, M. J. Guillen, C. Ly, W. Lee, S. Munt, C. Cuevas, P. Aviles and G. Faircloth, *Proc. Am. Ass. Cancer Res.*, 2005, **46**, 985.
320. J. M. Yin, P. Aviles, C. Ly, W. Lee, M. J. Guillen, S. Munt, C. Cuevas and G. Faircloth, *J. Chromatogr., B: Analyt. Technol. Biomed. Life Sci.*, 2006, **832**, 268.
321. M. Alvarez, J. L. F. Puentes and D. F. Bleda, *US Patent* 7,329,666, 2008.
322. J. C. Morris, R. J. Anderson, M. Remuinan and I. Manzanares, *US Patent* 7,320,981, 2008.
323. K. Bettayeb, O. M. Tirado, S. Marionneau-Lambot, Y. Ferandin, O. Lozach, J. C. Morris, S. Mateo-Lozano, P. Drueckes, C. Schaechtele, M. H. G. Kubbutat, F. Liger, B. Marquet, B. Joseph, A. Echalier, J. A. Endicott, V. Notario and L. Meijer, *Cancer Res.*, 2007, **67**, 8325.
324. M. T. Hamann and P. J. Scheuer, *J. Am. Chem. Soc.*, 1993, **115**, 5825.
325. M. Ashour, R. Edrada, R. Ebel, V. Wray, W. Watjen, K. Padmakumar, W. E. G. Muller, W. H. Lin and P. Proksch, *J. Nat. Prod.*, 2006, **69**, 1547.
326. Y. Suarez, L. Gonzalez, A. Cuadrado, M. Berciano, M. Lafarga and A. Munoz, *Mol. Cancer Ther.*, 2003, **2**, 863.
327. J. M. Rademaker-Lakhai, S. Horenblas, W. Meinhardt, E. Stokvis, T. M. De Reijke, J. M. Jimeno, L. Lopez-Lazaro, J. A. L. Martin, J. H. Beijnen and J. H. M. Schellens, *Clin. Cancer Res.*, 2005, **11**, 1854.
328. A. E. D. Scheuer, M. T. Hamann and D. G. Gravalos, *US Patent* RE039496, 2007.
329. K. L. Rinehart, N. L. Fregeau and R. A. Warwick, *US Patent* 6,107,520, 2000.
330. J. M. Padron and G. J. Peters, *Invest. New Drugs*, 2006, **24**, 195.
331. R. Baird, A. Planting, A. Reid, J. Kitzen, S. Reade, P. Clarke, L. Welsh, L. L. Lazaro, B. D. L. Heras, I. Judson, C. Pico, P. Workman, F. Eskens, S. Kaye, J. De Bono and J. Verweij, *Mol. Cancer Ther.*, 2007, **6**, 3381S.
332. G. R. Pettit, Y. Kamano, C. Dufresne, R. L. Cerny, C. L. Herald and J. M. Schmidt, *J. Org. Chem.*, 1989, **54**, 6005.
333. R. Bai, S. J. Friedman, G. R. Pettit and E. Hamel, *Biochem. Pharmacol.*, 1992, **43**, 2637.
334. P. Kerbrat, V. Dieras, N. Pavlidis, A. Ravaud, J. Wanders, P. Fumoleau and E. E. C. S. G. New, *Eur. J. Cancer.*, 2003, **39**, 317.
335. R. S. Marks, D. L. Graham, J. A. Sloan, S. Hillman, S. Fishkoff, J. E. Krook, S. H. Okuno, J. A. Mailliard, T. R. Fitch and F. Addo, *Am. J. Clin. Oncol.-Cancer Clin. Trials*, 2003, **26**, 336.

336. A. Ray, T. Okouneva, T. Manna, H. P. Miller, S. Schmid, L. Arthaud, R. Luduena, M. A. Jordan and L. Wilson, *Cancer Res.*, 2007, **67**, 3767.
337. C. Cunningham, L. J. Appleman, M. Kirvan-Visovatti, D. P. Ryan, E. Regan, S. Vukelja, P. L. Bonate, F. Ruvuna, R. J. Fram, A. Jekunen, S. Weitman, L. A. Hammond and J. P. Eder, *Clin. Cancer Res.*, 2005, **11**, 7825.
338. A. C. Mita, L. A. Hammond, P. L. Bonate, G. Weiss, H. McCreery, S. Syed, M. Garrison, Q. S. C. Chu, J. S. Debono, C. B. Jones, S. Weitman and E. Rowinsky, *Clin. Cancer Res.*, 2006, **12**, 5207.
339. G. R. Pettit, Y. Kamano, C. L. Herald, A. A. Tuinman, F. E. Boettner, H. Kizu, J. M. Schmidt, L. Baczynskyj, K. B. Tomer and R. J. Bontems, *J. Am. Chem. Soc.*, 1987, **109**, 6883.
340. G. R. Pettit, S. B. Singh, F. Hogan, P. Lloydwilliams, D. L. Herald, D. D. Burkett and P. J. Clewlow, *J. Am. Chem. Soc.*, 1989, **111**, 5463.
341. M. Von Mehren, S. P. Balcerzak, A. S. Kraft, J. H. Edmonson, S. H. Okuno, M. Davey, S. Mclaughlin, M. T. Beard and A. Rogatko, *Sarcoma*, 2004, **8**, 107.
342. H. L. Kindler, P. K. Tothy, R. Wolff, R. A. McCormack, J. L. Abbruzzese, S. Mani, K. T. Wade-Oliver and E. E. Vokes, *Invest. New Drugs*, 2005, **23**, 489.
343. E. A. Perez, D. W. Hillman, P. A. Fishkin, J. E. Krook, W. W. Tan, P. A. Kuriakose, S. R. Alberts and S. R. Dakhil, *Invest. New Drugs*, 2005, **23**, 257.
344. J. Horti, E. Juhasz, Z. Monostori, K. Maeda, S. Eckhardt and I. Bodrogi, *Cancer Chemother. Pharmacol.*, 2008, **62**, 173.
345. K. Yamada, M. Ojika, T. Ishigaki and Y. Yoshida, *J. Am. Chem. Soc.*, 1993, **115**, 11020.
346. M. Ojika, H. Kigoshi, T. Ishigaki, I. Tsukada, T. Tsuboi, T. Ogawa and K. Yamada, *J. Am. Chem. Soc.*, 1994, **116**, 7441.
347. H. Kigoshi, K. Suenaga, M. Takagi, A. Akao, K. Kanematsu, N. Kamei, Y. Okugawa and K. Yamada, *Tetrahedron*, 2002, **58**, 1075.
348. R. J. Andersen, D. J. Faulkner, H. Cun-Heng, G. D. Van Duyne and J. Clardy, *J. Am. Chem. Soc.*, 1985, **107**, 5492.
349. D. Pla, F. Albericio and M. Alvarez, *Anti-Cancer Agents Med. Chem.*, 2008, **8**, 746.
350. J. Maddock, G. Pattenden and P. G. Wight, *J. Comput. Aided Mol. Des.*, 1993, **7**, 573.
351. J. S. Panek, R. T. Beresis and C. A. Celatka, *J. Org. Chem.*, 1996, **61**, 6494.
352. J. S. Panek and R. T. Beresis, *J. Org. Chem.*, 1996, **61**, 6496.
353. C. A. Celatka, P. Liu and J. S. Panek, *Tetrahedron Lett.*, 1997, **38**, 5449.
354. C. A. Celatka and J. S. Panek, *Tetrahedron Lett.*, 2002, **43**, 7043.
355. H. Araki, T. Katoh and M. Inoue, *Tetrahedron Lett.*, 2007, **48**, 3713.
356. D. W. Knight, G. Pattenden and D. E. Rippon, *Synlett*, 1990, **1**, 36.
357. S. K. Chattopadhyay and G. Pattenden, *Tetrahedron Lett.*, 1995, **36**, 5271.
358. S. K. Chattopadhyay and G. Pattenden, *J. Chem. Soc., Perkin Trans. 1*, 2000, 2429.

359. S. K. Chattopadhyay and G. Pattenden, *Synlett*, 1997, 1342.
360. S. K. Chattopadhyay and G. Pattenden, *Synlett*, 1997, 1345.
361. J. Kempson and G. Pattenden, *Synlett*, 1999, 533.
362. S. K. Chattopadhyay, J. Kempson, A. Mcneil, G. Pattenden, M. Reader, D. E. Rippon and D. Waite, *J. Chem. Soc., Perkin Trans. 1*, 2000, 2415.
363. S. K. Chattopadhyay and G. Pattenden, *Tetrahedron Lett.*, 1998, **39**, 6095.
364. G. Pattenden, N. J. Ashweek, C. A. G. Baker-Glenn, G. M. Walker and J. G. K. Yee, *Angew. Chem., Int. Ed. Engl.*, 2007, **46**, 4359.
365. G. Pattenden, N. J. Ashweek, C. A. G. Baker-Glenn, J. Kempson, G. M. Walker and J. G. K. Yee, *Org. Biomol. Chem.*, 2008, **6**, 1478.
366. J. Tanaka, Y. Yan, J. Choi, J. Bai, V. A. Klenchin, I. Rayment and G. Marriott, *Proc. Natl. Acad. Sci. USA*, 2003, **100**, 13851.
367. E. Vincent, J. Saxton, C. Baker-Glenn, I. Moal, J. D. Hirst, G. Pattenden and P. E. Shaw, *Cell. Mol. Life Sci.*, 2007, **64**, 487.
368. R. D. Perrins, G. C. Cecere, I. Paterson and G. Marriott, *Chem. Biol.*, 2008, **15**, 287.
369. S. A. Look, W. Fenical, G. K. Matsumoto and J. Clardy, *J. Org. Chem.*, 1986, **51**, 5140.
370. R. S. Jacobs and W. H. Fenical, *US Patent* 4,745,104. 1988.
371. L. D. Mydlarz, R. S. Jacobs, J. Boehnlein and R. G. Kerr, *Chem. Biol.*, 2003, **10**, 1051.
372. V. Roussis, Z. D. Wu, W. Fenical, S. A. Strobel, G. D. Vanduyne and J. Clardy, *J. Org. Chem.*, 1990, **55**, 4916.
373. W. H. Fenical, P. R. Jensen and T. Lindel, *US Patent* 5,473,057, 1995.
374. T. Lindel, P. R. Jensen, W. Fenical, B. H. Long, A. M. Casazza, J. Carboni and C. R. Fairchild, *J. Am. Chem. Soc.*, 1997, **119**, 8744.
375. B. Cinel, B. O. Patrick, M. Roberge and R. J. Andersen, *Tetrahedron Lett.*, 2000, **41**, 2811.
376. R. Britton, M. Roberge, H. Berisch and R. J. Andersen, *Tetrahedron Lett.*, 2001, **42**, 2953.
377. O. Taglialatela-Scafati, U. Deo-Jangra, M. Campbell, M. Roberge and R. J. Andersen, *Org. Lett.*, 2002, **4**, 4085.
378. Y. Nakao, S. Yoshida, S. Matsunaga and N. Fusetani, *J. Nat. Prod.*, 2003, **66**, 524.
379. B. H. Long, J. M. Carboni, A. J. Wasserman, L. A. Cornell, A. M. Casazza, P. R. Jensen, T. Lindel, W. Fenical and C. R. Fairchild, *Cancer Res.*, 1998, **58**, 1111.
380. M. D'ambrosio, A. Guerriero and F. Pietra, *Helv. Chim. Acta*, 1987, **70**, 2019.
381. S. Ketzinel, A. Rudi, M. Schleyer, Y. Benayahu and Y. Kashman, *J. Nat. Prod.*, 1996, **59**, 873.
382. R. M. Buey, I. Barasoain, E. Jackson, A. Meyer, P. Giannakakou, I. Paterson, S. Mooberry, J. M. Andreu and J. F. Diaz, *Chem. Biol.*, 2005, **12**, 1269.

383. M. Kobayashi, T. Nakagawa and H. Mitsuhashi, *Chem. Pharm. Bull.*, 1979, **27**, 2382.
384. M. Suganuma, S. Okabe, E. Sueoka, N. Iida, A. Komori, S. Kim and H. Fujiki, *Cancer Res.*, 1996, **56**, 3711.
385. S. A. Look, W. Fenical, R. S. Jacobs and J. Clardy, *Proc. Nat. Acad. Sci. USA*, 1986, **83**, 6238.
386. S. A. Look and W. Fenical, *Tetrahedron*, 1987, **43**, 3363.
387. C. E. Moya and R. S. Jacobs, *FASEB J.*, 2007, **21**, A427.
388. A. Ata, R. G. Kerr, C. E. Moya and R. S. Jacobs, *Tetrahedron*, 2003, **59**, 4215.
389. M. M. Radwan, S. P. Manly, K. A. El Sayed, V. B. Wali, P. W. Sylvester, B. Awate, G. Shah and S. A. Ross, *J. Nat. Prod.*, 2008, **71**, 1468.
390. O. Taglialatela-Scafati, U. Deo-Jangra, M. Campbell, M. Roberge and R. J. Andersen, *Org. Lett.*, 2002, **4**, 4085.
391. B. H. Long, A. J. Wasserman, J. M. Carboni, S. Oh, L. A. Cornell, R. W. Peterson, K. A. Johnston, W. Fenical and C. A. Fairchild, *Proc. Annu. Meet. Am. Assoc. Cancer Res.*, 1998, **39**, 165.
392. I. Ojima, S. Chakravarty, T. Inoue, S. Lin, L. He, S. B. Horwitz, S. D. Kuduk and S. J. Danishefsky, *Proc. Natl. Acad. Sci. USA*, 1999, **96**, 4256.
393. H. M. McDaid, S. K. Bhattacharya, X. T. Chen, L. He, H. J. Shen, C. E. Gutteridge, S. J. Danishefsky and S. B. Horwitz, *Proc. Am. Ass. Cancer Res.*, 1999, **40**, 623.
394. H. M. McDaid, S. K. Bhattacharya, X. T. Chen, L. F. He, H. J. Shen, C. E. Gutteridge, S. B. Horwitz and S. J. Danishefsky, *Cancer Chem. Pharm.*, 1999, **44**, 131.
395. E. Hamel, D. L. Sackett, D. Vourloumis and K. C. Nicolaou, *Biochemistry*, 1999, **38**, 5490.
396. R. Britton, E. D. De Silva, C. M. Bigg, L. M. McHardy, M. Roberge and R. J. Andersen, *J. Am. Chem. Soc.*, 2001, **123**, 8632.
397. S. Chandrasekhar, V. Jagadeshwar, C. Narsihmulu, M. Sarangapani, D. R. Krishna, J. Vidyasagar, D. Vijay and G. N. Sastry, *Bioorg. Med. Chem. Lett.*, 2004, **14**, 3687.
398. G. C. H. Chiang, A. D. Bond, A. Ayscough, G. Pain, S. Ducki and A. B. Holmes, *Chem. Commun.*, 2005, 1860.
399. S. Ceccarelli, U. Piarulli and C. Gennari, *Tetrahedron Lett.*, 1999, **40**, 153.
400. A. Baron, V. Caprio and J. Mann, *Tetrahedron Lett.*, 1999, **40**, 9321.
401. R. Carter, K. Hodgetts, J. McKenna, P. Magnus and S. Wren, *Tetrahedron*, 2000, **56**, 4367.
402. K. P. Kaliappan and N. Kumar, *Tetrahedron Lett.*, 2003, **44**, 379.
403. G. Scalabrino, X. W. Sun, J. Mann and A. Baron, *Org. Biomol. Chem.*, 2003, **1**, 318.
404. J. D. Winkler, K. J. Quinn, C. H. MacKinnon, S. D. Hiscock and E. C. McLaughlin, *Org. Lett.*, 2003, **5**, 1805.
405. K. P. Kaliappan and N. Kumar, *Tetrahedron*, 2005, **61**, 7461.

406. S. Chandrasekhar, C. Narsihmulu, V. Jagadeshwar and S. S. Sultana, *Arkivoc*, 2005, 92.
407. H. Bruyere, S. Samaritani, S. Ballereau, A. Tomas and J. Royer, *Synlett*, 2005, 1421.
408. H. Bruyere, C. Dos Reis, S. Samaritani, S. Ballereau and J. Royer, *Synthesis-Stuttgart*, 2006, 1673.
409. S. Dhulut, A. Bourin, M. I. Lannou, E. Fleury, N. Lensen, E. Chelain, A. Pancrazi, J. Ardisson and J. Fahy, *Eur. J. Org. Chem.*, 2007, 5235.
410. K. C. Nicolaou, F. Van Delft, T. Ohshima, D. Vourloumis, J. Xu, S. Hosokawa, J. Pfefferkorn, S. Kim and T. Li, *Angew. Chem., Int. Ed. Engl.*, 1997, **36**, 2520.
411. X. T. Chen, S. K. Bhattacharya, B. Zhou, C. E. Gutteridge, T. R. R. Pettus and S. J. Danishefsky, *J. Am. Chem. Soc.*, 1999, **121**, 6563.
412. D. Castoldi, L. Caggiano, L. Panigada, O. Sharon, A. M. Costa and C. Gennari, *Angew. Chem., Int. Ed. Engl.*, 2005, **44**, 588.
413. D. Castoldi, L. Caggiano, L. Panigada, S. Sharon, A. M. Costa and C. Gennari, *Chem. Eur. J.*, 2005, **12**, 51.
414. M. Dambrosio, A. Guerriero and F. Pietra, *Helv. Chim. Acta*, 1988, **71**, 964.
415. M. Ciomei, C. Albanese, W. Pastori, M. Grandi, F. Pietra, M. D'ambrosio, A. Guerriero and C. Battistini, *Proc. Am. Ass. Cancer Res.*, 1997, **38**, 5.
416. K. C. Nicolaou, N. Winssinger, D. Vourloumis, T. Ohshima, S. Kim, J. Pfefferkorn, J. Y. Xu and T. Li, *J. Am. Chem. Soc.*, 1998, **120**, 10814.
417. K. C. Nicolaou, J. Pfefferkorn, J. Y. Xu, N. Winssinger, T. Ohshima, S. H. Kim, S. Hosokawa, D. Vouruloumis, F. Van Delft and T. Li, *Chem. Pharm. Bull.*, 1999, **47**, 1199.
418. J. Telser, R. Beumer, A. A. Bell, S. M. Ceccarelli, D. Monti and C. Gennari, *Tetrahedron Lett.*, 2001, **42**, 9187.
419. K. C. Nicolaou, J. Y. Xu, S. Kim, T. Ohshima, S. Hosokawa and J. Pfefferkorn, *J. Am. Chem. Soc.*, 1997, **119**, 11353.
420. N. Ritter and P. Metz, *Synlett*, 2003, **1**, 2422.
421. K. By, P. A. Kelly, M. J. Kurth, M. M. Olmstead and M. H. Nantz, *Tetrahedron*, 2001, **57**, 1183.
422. C. Sandoval, E. Redero, M. A. Mateos-Timoneda and F. A. Bermejo, *Tetrahedron Lett.*, 2002, **43**, 6521.
423. L. Caggiano, D. Castoldi, R. Beumer, P. Bayon, J. Telser and C. Gennari, *Tetrahedron Lett.*, 2003, **44**, 7913.
424. M. Roberge, B. Cinel, H. J. Anderson, L. Lim, X. Jiang, L. Xu, C. M. Bigg, M. T. Kelly and R. J. Andersen, *Cancer Res.*, 2000, **60**, 5052.
425. P. Meurer-Grob, J. Kasparian and R. H. Wade, *Biochemistry*, 2001, **40**, 8000.
426. K. L. Rinehart, T. G. Holt, N. L. Fregeau, J. G. Stroh, P. A. Keifer, F. Sun, L. H. Li and D. G. Martin, *J. Org. Chem.*, 1990, **55**, 4512.
427. N. Lindquist, W. Fenical, G. D. Vanduyne and J. Clardy, *J. Am. Chem. Soc.*, 1991, **113**, 2303.

428. K. C. Nicolaou and S. A. Snyder, *Angew. Chem., Int. Ed. Engl.*, 2005, **44**, 1012.
429. G. M. Cragg and D. J. Newman, *J. Nat. Prod.*, 2004, **67**, 232.
430. G. L. Wang, L. B. Shang, A. W. G. Burgett, P. G. Harran and X. D. Wang, *Proc. Nat. Acad. Sci. USA*, 2007, **104**, 2068.
431. K. L. Rinehart, J. B. Gloer, R. G. Hughes, H. E. Renis, J. P. McGovren, E. B. Swynenberg, D. A. Stringfellow, S. L. Kuentzel and L. H. Li, *Science*, 1981, **212**, 933.
432. K. L. Rinehart, J. B. Gloer, J. C. Cook, S. A. Mizsak and T. A. Scahill, *J. Am. Chem. Soc.*, 1981, **103**, 1857.
433. M. D. Vera and M. M. Joullie, *Med. Res. Rev.*, 2002, **22**, 102.
434. J. L. Urdiales, P. Morata, I. N. Decastro and F. Sanchez-Jimenez, *Cancer Lett.*, 1996, **102**, 31.
435. R. Sakai, K. Rinehart, V. Kishore, B. Kundu, G. Faircloth, J. Gloer, J. Carney, M. Namikoshi, F. Sun, R. Hughes, D. Gravalos, T. Dequesada, G. Wilson and R. Heid, *J. Med. Chem.*, 1996, **39**, 2819.
436. M. A. Izquierdo, A. Bowman, M. Garcia, D. Jodrell, M. Martinez, B. Pardo, J. Gomez, J. A. Lopez-Martin, J. Jimeno, J. R. Germa and J. F. Smyth, *Clin. Cancer Res.*, 2008, **14**, 3105.
437. C. Peschel, J. T. Hartmann, A. Schmittet, C. Bokemeyer, F. Schneller, U. Keilholz, D. Buchheidt, S. Millan, M. A. Izquierdo and R. D. Hofheinz, *Lung Cancer*, 2008, **60**, 374.
438. E. M. Ocio, C. Mitsiades, M. V. Mateos, P. Maiso, F. Mollinedo, M. Garayoa, C. Gajate, J. Blade, F. Prosper, J. J. Lahuerta, N. Mitsiades, C. J. McMullan, N. C. Munshi, T. Hideshima, D. Chauhan, C. Cuevas, P. Aviles, G. Faircloth, P. G. Richardson, A. Pandiella, K. C. Anderson and J. F. S. Miguel, *Blood*, 2007, **110**, 357A.
439. K. L. Rinehart, *Med. Res. Rev.*, 2000, **20**, 1.
440. PharmaMar Pipeline, www.pharmamar.com/pipeline.aspx, accessed 28 April 2009.
441. A. M. Fernandez, H. Y. He, L. A. McDonald, P. Lassota, C. Discafani, E. F. Sorensen, M. C. Edler, L. R. Barrows, J. C. Clardy and C. M. Ireland, *Pure Appl. Chem.*, 1998, **70**, 2130.
442. M. C. Edler, A. M. Fernandez, P. Lassota, C. M. Ireland and L. R. Barrows, *Biochem. Pharmacol.*, 2002, **63**, 707.
443. S. J. Bloor and F. J. Schmitz, *J. Am. Chem. Soc.*, 1987, **109**, 6134.
444. J. Kobayashi, J. F. Cheng, H. Nakamura, Y. Ohizumi, Y. Hirata, T. Sasaki, T. Ohta and S. Nozoe, *Tetrahedron Lett.*, 1988, **29**, 1177.
445. F. S. De Guzman and F. J. Schmitz, *Tetrahedron Lett.*, 1989, **30**, 1069.
446. H. Y. He and D. J. Faulkner, *J. Org. Chem.*, 1991, **56**, 5369.
447. N. Lindquist, W. Fenical, G. D. Van Duyne and J. Clardy, *J. Am. Chem. Soc.*, 1991, **113**, 2303.
448. J. Li, S. Jeong and P. G. Harran, *Angew. Chem., Int. Ed. Engl.*, 2001, **40**, 4765.
449. J. Li, A. W. G. Burgett, L. Esser, C. Amezcua and P. G. Harran, *Angew. Chem., Int. Ed. Engl.*, 2001, **40**, 4770.

450. A. W. G. Burgett, Q. Li, Q. Wei and P. G. Harran, *Angew. Chem., Int. Ed. Engl.*, 2003, **42**, 4961.
451. K. C. Nicolaou, M. Bella, D. Y. K. Chen, X. Huang, T. Ling and S. A. Snyder, *Angew. Chem., Int. Ed. Engl.*, 2002, **41**, 3495.
452. K. C. Nicolaou, R. P. Bheema, J. Hao, M. V. Reddy, G. Rassias, X. Huang, D. Y. K. Chen and S. A. Snyder, *Angew. Chem., Int. Ed. Engl.*, 2003, **42**, 1753.
453. K. C. Nicolaou, D. Y. K. Chen, X. Huang, T. Ling, M. Bella and S. A. Snyder, *J. Am. Chem. Soc.*, 2004, **126**, 12888.
454. K. C. Nicolaou, J. Hao, M. V. Reddy, R. P. Bheema, G. Rassias, S. A. Snyder, X. Huang, D. Y. K. Chen, W. E. Brenzovich, N. Giuseppone, A. O'Brate and P. Giannakakou, *J. Am. Chem. Soc.*, 2004, **126**, 15316.
455. C. M. Cheung, F. W. Goldberg, P. Magnus, C. J. Russell, R. Turnbull and V. Lynch, *J. Am. Chem. Soc.*, 2007, **129**, 12320.
456. Z. Cruz-Monserrate, H. C. Vervoort, R. Bai, D. J. Newman, S. B. Howell, G. Los, J. T. Mullaney, M. D. Williams, G. R. Pettit, W. Fenical and E. Hamel, *Mol. Pharmacol.*, 2003, **63**, 1273.
457. K. C. Nicolaou, J. Hao, M. V. Reddy, P. B. Rao, G. Rassias, S. A. Snyder, X. Huang, D. Y. K. Chen, W. E. Brenzovich, N. Giuseppone, A. O'Brate and P. Giannakakou, *J. Am. Chem. Soc.*, 2004, **126**, 12897.
458. J. Li, X. Chen, A. W. G. Burgett and P. G. Harran, *Angew. Chem., Int. Ed. Engl.*, 2001, **40**, 2682.
459. N. S. Williams, A. W. G. Burgett, A. S. Atkins, X. D. Wang, P. G. Harran and S. L. McKnight, *Proc. Natl. Acad. Sci. USA*, 2007, **104**, 2074.

CHAPTER 7

Microorganisms: Their Role in the Discovery and Development of Medicines

CEDRIC PEARCE,*[a] PETER ECKARD,[b] IRIS GRUEN-WOLLNY[c] AND FRIEDRICH G. HANSSKE[b]

[a] Mycosynthetix Inc, 505 Meadowlands Drive, Suite 103, Hillsborough, North Carolina 27278, USA; [b] BioFocus DPI AG, Gewerbestr. 16, CH-4123 Allschwil, Switzerland; [c] Labor Gruen-Wollny, Versailler Str.1, D-35394 Giessen, Germany

1 Introduction

Prokaryotic and eukaryotic microorganisms have provided some of the most potent and effective medicines and agrochemicals in the modern era (Table 7.1) starting over a half century ago with the introduction of the antibiotics penicillin, a metabolite from a fungus *Penicillium notatum*,[1] and streptomycin from the bacterium *Streptomyces griseus*.[2] These were staggering advances in the treatment of infectious diseases and both discoveries were honoured with the Nobel Prize in Medicine or Physiology: Sir Alexander Fleming was awarded the Prize in 1945 (together with Ernst Boris Chain and Sir Howard Walter Florey) and Selman Waksman received the Prize in 1952.

Reviews have been published[3,4] covering the importance of microbial products as a source of medicines which include the statins, arguably one of the most successful class of drugs in the history of medicine. A list of commercialised

RSC Biomolecular Sciences No. 18
Natural Product Chemistry for Drug Discovery
Edited by Antony D. Buss and Mark S. Butler

Published by the Royal Society of Chemistry, www.rsc.org

Table 7.1 Some commercialised pharmaceutical and agricultural microbial products and their sources.

Natural product	*Producing microorganism*	*Application*
Penicillin G	*Penicillium notatum*	Antibacterial
Cephalosporin	*Acremonium* spp.	Antibacterial
Streptomycin	*Streptomyces griseus*	Antibacterial
Neomycin	*Streptomyces fradiae*	Antibacterial
Gentamicin	*Micromonospora purpurea*	Antibacterial
Kanamycin	*Streptomyces kanamycetus*	Antibacterial
Vancomycin	*Amycolatopsis orientalis*	Antibacterial
Daptomycin	*Streptomyces roseosporus*	Antibacterial
Tetracycline	*Streptomyces* spp.	Antibacterial
Erythromycin	*Saccharopolyspora erythraea*	Antibacterial
Nisin	*Lactococcus lactis*	Antibacterial
Chloramphenicol	*Streptomyces venezelae*	Antibacterial
Nystatin	*Streptomyces noursei*	Antibacterial
Lincomycin	*Streptomyces linconensis*	Antibacterial
Fusidic acid	*Fusidium coccineum*	Antibacterial
Pleuromutilin	*Pleurotus mutilus*	Antibacterial
Rifamycin	*Amycolatopsis mediterranei*	Antibacterial
Clavulanic acid	*Streptomyces clavuligerus*	β-Lactamase inhibitor
Amphotericin	*Streptomyces nodosus*	Antifungal
Pneumocandin	*Glarea lozoyensis*	Antifungal
Strobilurins	*Strobilurus tenacellus*	Agriculture fungicides
Bleomycin	*Streptomyces verticillus*	Antitumour
Doxorubicin	*Streptomyces peucetius*	Antitumour
Calicheamycin	*Micromonospora echinospora*	Antitumour
Compactin and related statins	*Aspergillus terreus, Monascus ruber, Penicillium* spp.	Cholesterol biosynthesis inhibitor
Cyclosporin	*Tolypocladium* spp.	Immunosuppressant
Rapamycin	*Streptomyces hygroscopicus*	Immunosuppressant
Mycophenolic acid	*Penicillium* spp.	Immunosuppressant
Ergotamine	*Claviceps* spp.	Migraine relief
Lipstatin	*Streptomyces toxytricin*	Lipase inhibition
Spinosyn	*Saccharopolyspora spinosa*	Insecticidal
Avermectin	*Streptomyces avermitilis*	Anthelmintic
Milbemycin	*Streptomyces hygroscopicus*	Anthelmintic
Tylosin	*Streptomyces fradiae*	Feed additive
Avilamycin	*Streptomyces viridochromogenes*	Feed additive
Monensin	*Streptomyces cinnamonensis*	Feed additive

microbial products from bacteria and fungi is presented in Table 7.1. We estimate that the total worldwide annual market value for these compounds and their derivatives was approximately US$ 60 billion in 2006, with antibiotics and statins comprising three-quarters of this.

Microorganisms have been studied for more than 60 years as sources for natural products and continue to be very reliable sources of novel lead compounds. Besides various classes of fungi and myxobacteria, actinomycetes remain a major source of novel, therapeutically useful natural products.

A number of exciting new compounds that have progressed into clinical trials are reviewed within this book, including the salinosporamides, daptomycin and micafungin, representing metabolites from marine and terrestrial bacteria and fungi.

The use of microbes as a source for new compounds depended upon the development of the relatively modern science of microbiology, followed by the systematic collection and testing of material collected around the world. This is in contrast to the use of vascular plants as a source for bioactive materials which is based, in part, on traditional medicine. Unlike plants which depend on seasonal changes or are protected under the Convention on Biological Diversity, which limits substantial recollection (see Chapter 4), one of the greatest advantages of using microorganisms for finding new medicines and lead compounds is that they are renewable, *i.e.* they can be cultured in the laboratory and at the large scale necessary for manufacture, making additional field collection unnecessary.

To continue the successful discovery of biologically active metabolites from microbial sources, new approaches must be taken in order to reduce the probability of rediscovering known compounds. In this regard rare actinomycetes and so-called uncultivable (or unculturable) actinomycetes represent a promising source of novel pharmacologically active compounds.

Furthermore, the success of a screening campaign is strongly dependent on both screening methods and the diversity of chemical libraries. The character of the chemical collection has an essential role for the developmental potential of the generated hits and leads. Most current synthetic libraries are not optimal to generate hits and leads by either phenotypical/high content screens or new target/high throughput screening. Corporate compound collections, as well as commercial combinatorial libraries, still lack the three-dimensional structure and the polyfunctionality of bioactive natural products that are synthesised by their producing organisms to address biological targets (see Chapter 2). Natural product libraries derived from a genetically diverse strain collection provide the best coverage of scarcely populated areas of bioactive chemical space and, therefore, are a perfect complementation to "conventional" libraries—together covering a large part of the chemical and bioactivity diversity space.

In this chapter we address a number of basic questions regarding the use of microorganisms as a source for structurally unique organic compounds with pharmacological activity. Our initial focus is on prokaryotic and eukaryotic organisms, which produce what are generally considered as unique bioactive products: we address the issue of how an understanding of the nature of these organisms can help in the further development of the area, *i.e.* which groups of microbes are the best source for compounds. This leads to the question of whether, after 60 years of investigation into microbial metabolites, there are still good prospects for finding new microorganisms and new compounds. Molecular biology has been used to demonstrate that there are many microbes that are difficult to culture (*i.e.* so-called uncultivable organisms) as well as showing that, in a number of cases, the organisms that have been cultured have many unexpressed sequences within their genome which code for compounds that belong to groups of bioactive agents. Finally we discuss methods that can be used to expand the metabolite profile of known microbes.

2 Bacteria

The local and global scale of microbial diversity and its relationship to plants, animals and geology is poorly understood. The spatial scaling of microbial biodiversity is still in its infancy.[5] In the 1930s, Dutch microbiologist Lourens G. M. Baas Becking wrote that *"everything is everywhere, but, the environment selects"*.[6,7] The importance of biodiversity for innovations in biotechnology has been well documented.[8]

Soil contains a wide array of microhabitats which create an enormous variation in the appearance and survival strategies of soil-borne microbes. The various groups of microbes must compete with one another for the variety of nourishing organic matter. Based on the highly complex and diverse soil composition, an almost infinite microbial diversity has evolved.

To study and quantify microbial diversity one has to define what a species is. This is not a trivial objective since bacteria reproduce asexually and interbreeding cannot be applied for traditional species definition. While individual species of higher organisms may differ by as little as 1% in their respective DNA, bacterial strains differ by up to 30%. Beside the so called "species problem" in microbiology, it is a widely accepted concept that bacteria with 16S rRNA gene sequences with similarity more than 98.7% (checked by DNA–DNA hybridisation) belong to the same species.

Bacterial/microbial diversity remains a scarcely understood phenomenon. The environmental factors governing and controlling the distribution of soil-borne bacteria are very different from those impacting plants and animals. Furthermore, prokaryotes play essential functions in biochemical processes, *e.g.* decomposition of organic matter, nitrogen fixation, *etc.* There is a growing interest in understanding the pattern of distribution not only of the taxa, but of the traits those taxa possess. The development of environmental genomics may place trait-based biogeography into a situation where micro- and macro-organisms will be studied in concert with the results of targeted isolation of highly successful natural product sources.

Although soil microbes have been studied for decades, fundamental biological questions still remain unanswered. As stated by Noah Fierer of the University of Colorado: *"We probably know more about the organisms in the deepest ocean trenches than we know about the organisms living in our soil in our backyard"*.[9]

A recent publication presented evidence that soil pH correlates most with microbial diversity: soils with similar pH exhibit similar bacterial populations regardless of geography and distance.[9] Another interesting finding is the inverse correlation between microbial and plant diversity: semi-arid ecosystems express the highest diversity of bacteria.[9] However, it is still not known to what extent plant diversity affects the diversity of bacterial communities. Furthermore, one has to distinguish four major microbial habitats which determine functional diversity and microbial phenotypes: anoxic bulk soil, oxic surface soil, partially oxic rhizosphere and decomposing plant debris.[10]

Bacteria inhabit every environment on the planet earth. They can be found in temperate as well as tropical areas, in deserts, ice caps, alkali lakes, oceans and hot springs. In one gram of soil, approximately 10^8 bacteria are present[11] and these are estimated to represent over 10 000 species. It is estimated that 50% of the biomass on this planet is microbial. Microorganisms represent by far the richest repertoire of functionally and chemically diverse producers of secondary metabolites in nature.

Early attempts to quantify the species richness and complexity in environmental samples were restricted to bacteria that could be isolated using conventional growth media. Genetic analyses gave hints that the abundance of bacteria was higher in the vicinity of roots than in the bulk soil, but only a small percentage of bacteria can be assessed by these classical methods due to the fact that the overwhelming majority still resist cultivation. Estimations are that between 0.1 and 15% of the total bacterial community in defined soil samples have been cultivated.[12–14] The method of choice for these examinations relies on 16S rRNA analysis while computer-based tools complement the analyses of the species composition of a community.[15,16] In respect to interesting sources of natural products, however, these tools do not provide relevant information on the potential to generate useful secondary metabolites.

The current list of known bacteria contains approximately 7000 species. This does not reflect the true number, which is estimated to be between 50 000 and 3 000 000.[17]

A genetic classification for prokaryotic species proposed by Woese in 1987 is based on the nucleotide sequence of small subunit ribosomal RNA (SSU rRNA) molecules.[18] Operational taxonomic units (OTUs) based on SSU rRNA sequence similarity have become indispensable proxies to the otherwise intangible concept of bacterial species. The sequence-based phylogenetic system, which is the current "gold standard" for profiling microbial communities, splits the prokaryotes in two domains—the bacteria and the archaea. A faster and cheaper strain dereplication tool has recently been described that compares the RNase P RNA gene (mpB) sequence.[19] The actual phylogenetic tree of life based on current molecular knowledge (SSU rRNA and additional molecular evidence) was described previously[20] (see also http://greengenes.lbl.gov).

Beside the astonishing macroscopic diversity of the oceans, real diversity is hidden in the sediments and in the sea water. One millilitre of sea water can contain up to one million microorganisms. The surface of plants and invertebrates are unique habitats to be colonised by partially endemic microbes. There exists a huge number of unknown actinomycetes with the potential to contain metabolic genes for new and highly unique bioactive metabolites.[21] Recently, two new genera, *Salinospora* and *Marinophilus*, have been described which are fundamentally different from the approximately 100 genera from terrestrial sources.[22]

A recent analysis of ~600 bacteria from US soil samples revealed that their metabolic machinery could subsist on single type antibiotics as carbon sources.[23] Does this open up a window into a metabolic niche of "uncultivable" bacteria?

So far, we have just scratched the surface of bacterial diversity, yet there are many unanswered questions which are intrinsically connected to the future success of natural product research. How many different microbes do exist? What is the real extent of this diversity? How does microbial diversity correlate to chemical diversity of the secondary metabolites produced? What does all that microbial diversity do? How do species evolve? Are there environmental factors that support horizontal gene transfer? Are there ecosystems and micro-habitats that have higher probabilities to attract or support microbes with biologically attractive secondary metabolism?

3 Fungi

A fungus differs from a bacterium at a number of fundamental levels, most importantly because it is eukaryotic rather than prokaryotic. The definition of a fungus is complex because the fungal kingdom covers a wide range of heterotrophic organisms with diverse cellular structures (nucleated and chitin cell wall generally), reproduction, physiology, biochemistry and secondary metabolism.[24]

As with bacteria, fungi are ubiquitous and have been isolated from every conceivable organic substrate examined regardless of where in the world it was collected. The number of known fungal species is approximately 80 000. In 2001, it was estimated that 74 000 species had been described in the literature; with a further disclosure rate of one thousand per year, this provides the current approximation—which may also be an underestimation.[25,26]

As with bacteria, the number of fungi isolated is a fraction of what is in nature: the total number of fungal species predicted to exist is between 500 000 species[27] and 9 900 000 species.[28] Estimates are based on a number of considerations, that of Hawkesworth[25] (in many ways the seminal prediction) being an extrapolation from the ratio of fungi to plants (6 : 1 was used) in a particular region (the native plants of Great Britain and Ireland, and the number of species in an alpine community) taken together with various allowances and considerations. Hawkesworth's initial 1990 estimates[25] were invaluable for a variety of reasons and provided encouragement to those interested in fungal metabolic diversity and especially to those engaged in the discovery of new compounds with medicinal or agricultural potential. Reviews and discussions of Hawkesworth's paper (and others) highlight a number of areas where the actual fungal diversity may be higher than that predicted, including that existing in the tropical and polar regions and insect-associated fungi, which alone have been estimated at 1 500 000 species.[29]

The extent of fungal novelty was also discussed by Hawkesworth, who stated in 2001: *"new species discovered . . . is an incontrovertible indicator of our ignorance"*.[26] Unlike the study of bacterial diversity, which may in part be limited by cultivation techniques, within the mycology world in many cases the extent of novelty can be demonstrated simply by studying a particular niche and cataloguing those organisms found; for example:

- new lichenicolous fungal species were observed at a rate of 92–96% of the total isolated;[30,31]
- new Mexican oak-pine macromycetes were observed at a rate of 64% (835 new species);[32]
- new marine species were found at a rate of 53% of total isolates.[33]

The question remains unanswered as to how this will lead to new chemical diversity.

As well as the potential from the biodiversity perspective, a similar level of undiscovered utility has been revealed by genomic studies. Taken together with physiological and biochemical studies addressing cultivation of these organisms to yield new metabolites, as with bacteria, there are staggering possibilities for novel chemistries.

4 Terrestrial and Marine Microorganisms

Recently, the marine environment has been targeted increasingly by microbiologists as a source for new organisms. On the other hand, soil is unquestionably the richest source for diverse microorganisms. This has been proven by the fact that 200 g of soil can contain up to 0.5 g of microbial cells. Numerous research groups are constantly demonstrating that new pretreatment and sophisticated isolation techniques furnish an impressive stream of new terrestrial microbes that have not been described before.[34] Gathering of terrestrial samples in times of high-tech marine submersibles seems unspectacular, but most new and novel natural products are still discovered from terrestrial microorganisms. This is not surprising because the overwhelming majority of soil-derived microbes have not been grown successfully under laboratory conditions (see Section 6). Furthermore, most of the marine microbes produce the same metabolites as comparable terrestrial microorganisms, probably because the shoreline offers a huge interface for horizontal gene exchange as well as environmental adaptation in both directions.

It remains to be seen if the expense of mining the marine environment will be compensated by commercial success stories as with terrestrial microbes.[35] The current interest in marine microorganisms might be related to the differing academic point of view or if the focus pinpoints more on new enzymes than on biologically active natural products. Given the examples described in several publications and the application of homologous and heterologous expression of biosynthetic gene clusters, we strongly believe that the exploration of marine endophytes or symbiotic microbes derived from macroalgae or invertebrates will furnish chemical breakthroughs in terms of therapeutic applications.[36–39]

5 Microbial Culture Collections

There are a number of larger established microbial culture collections such as:

- American Type Culture Collection in the USA (ATCC);
- Centraalbureau for Schimmelcultures in the Netherlands (CBS);
- German Collection of Microorganisms (DSMZ);
- Institute for Fermentation in Japan (IFO);
- US Department of Agriculture, Agricultural Research Service (NRRL).

As of 10 December 2008, 544 culture collections in 68 countries had been registered at WDCM (World Data Centre of Microorganisms; http://wdcm.nig.ac.jp/hpcc.html) containing 1 413 100 microbial cultures.

Interestingly, a number of smaller biotechnology companies have taken the place of the larger pharmaceutical companies in this area of research; for example:

- Albany Molecular (combined the microbial collection from former Panlabs and Eli Lilly);
- AnalytiCon (has access to ~10 000 stains of unknown heritage);
- Biofocus DPI (uses the collection from Labor Gruen-Wollny which consists of ~50 000 actinomycetes and 8000 fungi);
- InterMed Discovery (inherited ~40 000 strains of actinomycetes and fungi from the former Bayer collection);
- Nereus Pharmaceuticals (20 000 unique stains from marine habitats);
- MerLion Pharmaceuticals (strain collection of ~130 000 consisting of actinomycetes, fungi, eubacteria and myxomycetes partly based on the former GlaxoSmithKline collection);
- Mycosynthetix (owns a collection of approximately 55 000 fungi based on collections from OSI Pharmaceutical's MYCOsearch collection and that of Agrasol, and extended *via* its own collection programme).

In addition, there are two substantial academic institutes that are active in the field of natural product commercialisation:

- Hans-Knöll-Institut (HKI) has a collection of 30 000 actinomycetes, 5000 fungi and 3000 bacteria (the collection is based on the former collection of the Zentralinstitut für Mikrobiologie of the German Democratic Republic);
- Institut für Biotechnologie und Wirkstofforschung (IBWF) has 10 000 fungal strains equally distributed between basidiomycetes, ascomycetes, zygomycetes and imperfect fungi.

Of the classical pharma companies, only Novartis, Sanofi-Aventis and Wyeth remain active in the field and have similar size collections.

6 Evidence for "Uncultivable" Microbes

"For over 80 years it has been known that there is a large discrepancy between the number of bacterial colonies that form on solid media when soil is used as an inoculum and the total number of bacterial cells actually present in that same soil".[40]

It is estimated that only ~1% of the total microbial community has been cultured by standard isolation techniques. There is a growing conviction that this observation resulted from the selected media used, as well as incubation time and inoculum size. Additional stimulation factors for growth and sporulation (*e.g.* gamma-butyrolactones and indole-3-acetic acid) have been reported.[41,42] This implies that the majority of microbial biodiversity still remains to be discovered and exploited.

Until 1990, when Giovannoni and Ward applied 16S rRNA sequence analysis by extracting community DNA, microbiologists had to rely on plate counts of bacteria which could be cultured and isolated to obtain an idea of the number of bacteria in the sample.[43,44] This new approach allowed the study of the remaining 99% of bacteria which could not be isolated by nutrient media and liquid enrichment.

Genomic studies disclosed that many of those "uncultivable" microbes contain the potential to produce new secondary metabolites and there are a variety of routes to access this pool of novel natural products. For example, the community DNA gene clusters can be isolated and the genetic information for secondary metabolism expressed in heterologous hosts.[45–47] A more rewarding approach may be to focus on improved cultivation methods.[40,48]

In the meantime, numerous examples finally demonstrated that many of these so far unknown microbes do not resist cultivation. Special enrichment techniques in combination with various trap technologies furnish rare actinomycetes hitherto not isolated by standard methodologies.[49–51] This strategy needs to take several important factors into consideration: knowledge of the microbial ecology is needed, together with an understanding of species composition of a community, microbial physiology and the exploitation of neglected habitats. Development of unconventional pretreatment, enrichment and isolation conditions is essential. Selective agents for motile microorganisms[52] as well as microcapsules[53] and the chamber technique[54,55] are some of the new methods that have been applied successfully. The use of a soil-extract agar medium proved to be a successful method to access new species of rare actinomycetes.[56] A very useful strategy to generate microbial diversity is based on the combination of 16S rRNA sequencing, polyketide synthase (PKS) and non-ribosomal synthetase (NRPS) analysis, and the application of atypical isolation methodologies.[57,58]

Genomic studies have shown that the potential for producing secondary metabolites is not uniformly distributed across bacterial genera and species. *Streptomyces coelicolor*[59] and *S. avermitilis*[60,61] each possess more than 20 gene clusters responsible for the biosynthesis of secondary metabolites. The same applies for the so-called rare actinomycetes and the myxomycetes,[62] which contain multiple gene clusters for extensive secondary metabolism. A genetic comparison based on the presence of type I and II PKSs and NRPSs demonstrates that genome sizes of >5 MB (or better >8 MB) and filamentous growth of bacteria are prerequisites of substantial secondary metabolism, *i.e.* actinomycetes.[63]

The strong commitment of actinomycetes that devote 5–10% of their genetic capacity to secondary metabolism demonstrates why they are the unmatched leaders in the supply of novel antibiotics for over half a century. There is strong evidence that many hitherto undescribed actinomycetes taxa will be discovered among the "uncultivable" microbes. The successful laboratory cultivation of previously uncultured soil actinomycetes with a high degree of novelty demonstrates that even relatively simple isolation methods can still be extremely successful.[40,48,50,51]

7 Metagenomic Approach to Access Uncultivable Microbes

Despite the ongoing discussion of culturable and uncultivable microorganisms, a new technology has been developed that uses the cultivation of rare or slow-growing organisms based on the heterologous expression of community DNA which contain secondary metabolic gene clusters. First, one has to produce clone libraries in a cultivation-independent fashion, which is termed metagenomics.[64] The next steps consist of the insertion of large genomic clusters into suitable carriers, *e.g.* bacterial artificial chromosomes (BACs) or cosmids, followed by their evaluation in suitable heterologous hosts or fast-growing *Streptomyces*. This technology has been particularly successful for marine and terrestrial bacterial symbionts and the expression of their secondary metabolic potential.

The most used expression model was *Escherichia coli* and this system has been used for the successful production of 6-deoxyerythronolide B, epothilone and patellamide B. Other hosts include *Streptomyces* genera. In the species *S. albus*, the biosynthetic pathway for rebeccamycin was reconstituted and provided precursors without further genetic manipulation. In addition, the synthesis of a variety of indolocarbazoles, including staurosporine, demonstrated the impact of the strategy to generate biosynthetically based combinatorial libraries. On the other hand, the introduction of the biosynthetic gene cluster for fredericamycin A from *S. griseus* into *S. lividans* was only successful when the pathway-specific activator *fdmR* was overexpressed under the control of a strong constitutive promoter in order to generate comparable amounts of fredericamycin A as in the wild type strain.[65] One of the largest gene clusters (128 kb DNA) has been transferred from *S. roseosporus* into *S. lividans* to express successfully the antibiotic daptomycin (see Chapter 14, Section 3.1).[66]

Although, the true metagenomic approach has yet to prove its usefulness in discovering novel secondary metabolites,[67] there is still the opportunity to access the hitherto unculturable microbes and their array of important new chemical entities.

8 Culturing Techniques to Produce Secondary Metabolites

There exist countless publications about the general as well as specific culture conditions of bacteria and, in particular, actinomycetes used to produce secondary metabolites. One of the best introductions to the field of fermentation is provided by the *Manual of Industrial Microbiology and Biotechnology* published in 1999 with Demain and Davies as the editors in chief.[68] Countless scientific publications, *e.g. The Journal of Antibiotics*, give excellent additional information on the subject, but it should be noted that industrial processes for high-value products are usually not published in detail and are kept as company intellectual property or trade secrets.

Fungi can be cultured using a wide variety of techniques and large numbers of fungi are easily grown, as illustrated by the organisms that will quickly contaminate food. Providing the basic chemical components for growth is relatively straightforward and a medium comprising organic plant material will satisfy most of the fungal requirements. An energy source is also needed and this can be provided in the form of simple or complex carbohydrate. Since fungal nutrition falls on the periphery of this review, readers are encouraged to refer to other literature sources.[69]

The control of secondary metabolism is discussed earlier in this chapter. In summary, a number of factors have been identified which trigger and control secondary metabolism. Understanding these triggers can be useful during the design of the environmental conditions to be used to grow fungi.

The most practical approach to finding novel biologically active compounds produced by microorganisms is to culture the organism in a medium that provides all the essential ingredients incubated at an optimum temperature to produce growth and at the same time providing the correct environment, usually developed at the end of the growth phase, to stimulate and support the production of secondary metabolites. Conditions that produce the most abundant growth rate or cellular mass will not necessarily be the correct conditions to trigger and support secondary metabolism. Although there are serious quantitative aspects to the normal microbial metabolite screening programme, here we focus on the qualitative aspects of the growth and production phases.

Typical actinomycetes and fungal media are composed of organic plant material (complex media), although animal material is also used especially for the cultivation of bacteria and defined media are also employed for the cultivation of both fungi and bacteria. There may be incubation temperature differences as well as pH optima variations between bacteria and fungi. In general, actinomycetes are grown in highly aerobic environments created by vigorous shaking in liquid media, whereas fungi may be grown without vigorous shaking and also under aerobic conditions.

Fungi may be cultured in either submerged liquid fermentation or using a solid substrate containing some free water. Submerged liquid fermentations using industrial sized tanks containing thousands of litres of medium have enabled the efficient production of metabolites such as penicillin and citric acid.[70] In the case of penicillin G production, yields have been raised from initial levels of 5 mg/L to the current levels in the region of 70 g/L—an increase of 14 000 fold, which was achieved through a combination of strain improvement and fermentation optimisation based on submerged liquid fermentation.

Solid state fermentation approaches provide an alternative technology platform for the production of novel metabolites, with the growth environment presenting different physiological challenges to the fungus with corresponding differences in the organism's biochemistry.[71] Our own results and those of others have demonstrated the value of including simple solid state fermentations in microbial screening programmes and these have resulted in the discovery of novel compounds. Solid state fermentation techniques have been reviewed elsewhere[72] and we restrict our comments to two unusual variations of this approach.

One variation on the solid substrate theme employs inert membranes for the support of the mycelial growth matt as the solid part of the fermentation and this technique offers some unique advantages for culture manipulation. In a method developed by workers at Biodiversity Ltd in the UK, a miniature bioreactor (Reacsyn®) is employed to grow the filamentous organism on a membrane which draws nutrient medium from a reservoir to the growing organism. Upon developing biomass and at the point when the organism would normally be entering secondary metabolism, the fungus (or actinomycetes as the technique works equally well with any type of filamentous microorganism) is removed from the nutrient medium and placed in a simple solution containing, for example, carbohydrate. Secondary metabolism proceeds and metabolites are produced and excreted into a relatively simple environment free from most of the media components. Subsequent pharmacological and chemical analysis is unaffected by extraneous solutes. Natural product chemistry and dereplication are facilitated by the relatively simple composition of the broth; in a number of cases, higher productivity was observed (N. Porter, personal communication). Additional advantages include the possibility of directed fermentations by adding precursors and related analogues, as well as for biosynthetic studies using isotopically labelled substrates.

In a recent publication from the Wyeth natural products group,[73] the use of inert supports for fungal solid state fermentation resulted in the production of novel antibiotics, pyrrocidins A and B from *Cylindrocarpon* sp.[74] and acremonidins A–E from *Acremonium* sp.[75] In these experiments, the *Cylindrocarpon* strain was cultured on a polyester–cellulose support on malt extract agar wherein pyrrocidin, which contains an unusual 13-membered macrocyclic ring, was produced. In contrast, a simple liquid version of the same medium failed to support synthesis of the antibiotic. In the second case, an *Acremonium* strain produced the polyketides acremonidins A–E when cultured on a polyester–cellulosic support in malt extract medium at significantly elevated levels over those produced in cultures without the support.

These results and others, which illustrate the value of culturing microorganisms using inert supports, may in part be a reflection of the similarity to the conditions found in nature whereas homogenous nutritious media, either in solid or liquid form, are rarely encountered naturally.

9 Evidence for New Biosynthetic Pathways in Known Microbes

Modern molecular biology and genetic sequencing have revealed many interesting observations into the further potential of known microorganisms, including genetic engineering to produce new metabolites and the uncovering of silent gene clusters which have the potential of producing novel metabolites.

Many fungal and bacterial metabolites are derived from polyketide synthase (PKS) and non-ribosomal peptide synthase (NRPS) biosynthetic routes which involve the condensation of common primary metabolites in unique sequences or oxidation and reduction cycles which generate novel compounds with surprising elegance (see Chapter 10). Many different families of regulatory proteins are controlling the expression of secondary metabolites. The activation can be triggered by extracellular as well as intracellular eliciting or signalling molecules as described below.

The exploration of the silent secondary metabolic genes can be divided into two categories: the molecular biology-based techniques (see Section 10); and the cultivation-based strategies. Some well known signal transmitters, *e.g.* butyrolactones, PI-factor (2,3-diamino-2,3-bis(hydroxymethyl)-1,4-butanediol) and indole-3-acetic acid, are implicated in the activation of secondary metabolic pathways.[35,36] Furthermore, secondary metabolism and the activation of silent genes are also under epigenetic control of constitutive genes.[76] It was demonstrated that the treatment of 12 fungi with DNA methyltransferase and histone deacetylase inhibitors (5-azacytidine, suberoylanilide hydroxamic acid) furnished the production of new natural products and/or enhanced accumulation of constitutive secondary metabolites.

The Canadian company, Ecopia BioSciences, developed a high throughput genome scanning technology to detect secondary metabolite gene clusters with signature NRPS and PKS sequences and from these to predict the expected chemical structures. Fermentations in ~50 different media furnished not only compounds with the predicted physicochemical properties, but also the new antibacterial ECO-0501 (from *Amycolatopsis orientalis*) and antifungal ECO-02301 (from *Streptomyces aizunensis*) compounds.[77,78]

10 Genetic Pathway Engineering and Modulation of Post-translational Modification to Generate Novel Compounds

As briefly mentioned above, the accumulated knowledge and the discoveries connected with genome mining and accessing the metagenome of soil-borne microbes can be applied to the field of "recombineering" of large biosynthetic gene clusters. This field of synthetic biology gives rise to the ability to address intractable and cryptic genes, as well as to provide access to genes for the biosynthesis of biologically active secondary metabolites. The designed and cloned pathways can be expressed in homologous as well as heterologous hosts. The focus of application can be lead optimisation, structural diversification of natural product libraries and combinatorial biosynthetic engineering. Limitations of the approach result from the lack of understanding of the structure and interplay of the enzymes involved (see Chapter 10, Section 4). A milestone was the recently published X-ray structure of a non-ribosomal peptide synthase from *Bacillus subtilis*, which opens insights into the various catalytic domains.[79,80]

Numerous examples have demonstrated the proof-of-concept of metabolite engineering but also highlight its limitations.[81–86] An excellent introduction into the field has been given by Bode and Müller.[87]

Recently, the biosynthesis of PKS–NRPS hybrid metabolites has been described. The pathway-specific regulator was expressed under the control of an inducible promoter, which finally resulted in switching on of the silent pathway.[88] Although there are many successfully executed examples from academia, there is still a long way to go to transfer the technology into broadly applicable industrial processes.

Many biologically active and commercially important natural products are specifically glycosylated. The glycosylation pattern of the core scaffold is determined by highly selective glycosyltransferases (GTs). For drug development and structure–activity studies, it would be helpful to diversify the glycosylation pattern by expanding the promiscuity of the glycosyltransferases. Natural product targets of choice could include vancomycin, rebeccamycin, calicheamicin, novobiocin and avermectin, as well as the carbohydrate aminocyclitol antibiotics. Recent publications describe the present status of these techniques.[89,90]

11 Microbial Secondary Metabolites with Unique Biological Activity and Chemical Diversity

There have been excellent publications which demonstrate the biosynthetic and metabolic potential of microbes in general and specific genera which produce highly active compounds with substantial market perspectives,[91–96] and two good compilations from well-known academic research groups in Germany[97] and Japan.[98]

Despite enormous efforts in recent years to access new microbial sources, actinomycetes continue to be the largest group of microbial producers of biologically active natural products. Even though the genera of the *Streptomyces* family maintains the dominant position of secondary metabolite producers, rare actinomycetes from terrestrial, as well as marine habitats, are becoming excellent sources of new potential lead compounds.

Many of the natural compounds isolated and characterised in the past underwent a very limited biological profiling, *i.e.* they were usually tested only in antibacterial and/or antifungal assays. This applies particularly to the thousands of natural products isolated and characterised by academic groups. Testing of such compounds in biochemical target-based assays or in modern phenotypical cell-based assays should reveal new biological activities and many new applications.

Based on the belief that most of the genetically encoded natural products are waiting to be discovered, microorganisms could again prove to be a gold mine of useful compounds for medicinal, agrochemical and veterinary applications.

With improved microbe isolation and cultivation strategies, advances in dereplication and structural characterisation and the expected progress in triggering the silent genes of secondary metabolism, selective and sensitive target and cellular screens will furnish a plethora of new natural products with an excellent potential of being developed into drugs. This is also emphasised by the evidence that most natural compounds have been biosynthetically optimised in a cellular polypharmacological (*i.e.* multi-target) setting.

Looking at secondary metabolites from a biological perspective, their chemical diversity is the result of their diverse biological purposes, *i.e.* the requirement to interact with a large variety of target proteins. It is not surprising, therefore, that natural compounds can directly represent drug candidates without further modification.

Viewing secondary metabolites from a chemical perspective, the scaffold diversity and polyfunctional features offer significant potential to generate random libraries for screening purposes. The structural status of a compound on this level is, therefore, a "transition state" with the goal to become a hit or a lead for further development (Table 7.2). Various automatic chromatographic separation techniques and modern spectroscopic methodologies can be applied to generate diverse chemical compound libraries derived from microorganisms.[99,100]

The major biosynthetically derived compounds are related to the following structural classes:

- polyketides (from polyketide synthase type I–III);
- peptides and depsipeptides (from non-ribosomal peptide synthases);
- isoprenoids;
- aminoglycosides;
- others, *e.g.* chorismate derived, rare nucleosides and tetramic acid derivatives.

Covering numerous glycosylated secondary metabolites from various sources, Dembitsky has published seven excellent reviews about glycosides from fatty acids and alcohols,[101] polyether glycosidic ionophores and macrocyclic glycosides,[102] carotenoid glycosides and isoprenoid glycolipids,[103] acid amide glycosides and their analogues and derivatives,[104] biologically active glycosides of aromatic metabolites,[105] biologically active marine and terrestrial

Table 7.2 Secondary metabolites which have a proven biological/therapeutic activity.

Natural product	*Producing microorganism*	*Application, target*
Trierixin **23**[a]	*Streptomyces* sp.	Oncology, XBP1
Colletoic acid **24**	*Colletotrichum gloeosporioides*	Diabetes 2, 11β-HSD1
Staurosporine **17**	*Streptomyces* sp.	Oncology, FLT3
UCN-01 **25**	*Streptomyces* sp.	Oncology, CDK1
Calicheamycin γ_1^I **26**	*Micromonospora echinospora*	Oncology, DNA
Geldanamycin **27**	*Streptomyces hygroscopicus*	Oncology, HSP90
Heneicomycin **28**	*Streptomyces filipinensis*	Antibacterial, EF-Tu
Simocyclinone D8 **29**	*Streptomyces antibioticus*	Antibacterial, DNA-gyrase
Daunorubicin **30**	*Streptomyces peuceticus*	Oncology, Topo I
Lactacystin **31**	*Streptomyces* sp.	Oncology, 20I proteasome
Actinoplanic acid **32**	*Actinoplanes* sp.	Oncology, FPTase
Epothilone B **33**	*Sorangium cellulosum*	Oncology, tubulin
Acarbose **34**	*Actinoplanes* sp.	Diabetes 2, α-glucosidase
Pleuromutilin **35**	*Pleurotus mutilus*	Antibacterial, 50*S* subunit
Myriocin **36**	*Isaria sinclairii*	Immunology, SP-transferase
Rifamycin B **37**	*Amycolatopsis mediterranei*	Antibacterial, RNA polymerase
Bleomycin A_2 **38**	*Streptomyces verticillus*	Oncology, DNA
Avermectin $B1_a$ **39**	*Streptomyces avermitilis*	Anthelminth, Cl-channel
Nodulosporic acid A **40**	*Nodulosporium* sp.	Anthelminth, Cl-channel
Friulimicin B **41**	*Actinoplanes friuliensis*	Antibacterial, peptidiglycan synth.
Mannopeptimycin α **42**	*Streptomyces hygroscopicus*	Antibacterial, cell wall
Avilamycine A **43**	*Streptomyces viridochromogenes*	Antibacterial, 23*S* subunit
Daptomycin **44**	*Streptomyces roseosporus*	Antibacterial, cell wall
Platensimycin **45**	*Streptomyces platensis*	Antibacterial, FabF
Tylosin **46**	*Streptomyces fradiae*	Antibacterial, 50*S* subunit
Chlorofusin **47**	*Fusarium* sp.	Oncology, p53/MDM2
Lipstatin **48**	*Streptomyces toxytricini*	Metabolic syndrome, lipase
Pladienolide D **49**	*Straptomyces platensis*	Oncology, SF3b
Diazepinomycin **50**	*Micromonospora* sp.	Oncology, MAPK
L-783,281 **51**	*Pseudomassaria* sp.	Diabetes 2, IR tyrosine kinase
Piericidin A_1 **52**	*Streptomyces* sp.	Oncology, NADH-ubiquinone red.
Salinamide A **53**	*Streptomyces* sp.	Antibacterial, inflammation
Virginiamycin M_1 **54**	*Streptomyces virginiae*	Antibacterial, protein biosynthesis

[a]Numbers in bold refer to structures displayed at the end of the chapter.

Table 7.3 Microbial natural products with hit/lead potential.

Natural product	*Producing microorganism*	*Potential field*
Hormaomycin **1**[a]	*Streptomyces griseoflavus*	Open
Fredericamycin A **2**	*Streptomyces griseus*	Oncology
Nargenicin B_1 **3**	*Saccharopolyspora hirsute*	Infectology
Stachyflin **4**	*Stachybotrys* sp.	Virology
Borrelidin **5**	*Streptomyces rochi*	Oncology
Striatin A **6**	*Cyathus striatus*	Libraries
Abyssomycin C **7**	*Verrucosispora* sp.	Infectology
Fluostatin E **8**	*Streptomyces lavendulae*	Oncology
Aspochalamine A **9**	*Aspergillus niveus*	Oncology
Spirodionic acid **10**	*Streptomyces* sp.	Libraries
Pentalenolactone **11**	*Streptomyces arenae*	Libraries
Deflectin 1a **12**	*Aspergillus deflectus*	Libraries
Terrecyclic acid **13**	*Hymenoscyphus herbarum*	Libraries
Illudin S **14**	*Omphalotus olearius*	Libraries
Toyocamycin **15**	*Streptomyces toyocaensis*	Libraries
Sphaeropsidin A **16**	*Aspergillus chevalieri*	Libraries
Staurosporine **17**	*Streptomyces staurosporeus*	Oncology, libraries
Streptazolin **18**	*Streptomyces viridochromogenes*	Libraries
Lymphostin **19**	*Streptomyces* sp.	Oncology, libraries
Pyralomycin 1a **20**	*Microtetraspora spiralis*	Libraries
Psathyrellone B **21**	*Psathyrella* sp.	Libraries
Lachnumone **22**	*Lachnum papyraceum*	Libraries

[a]Numbers in bold refer to structures displayed at the end of the chapter.

alkaloid glycosides[106] and biologically active hemi- and monoterpenoid glycosides.[107]

In order to demonstrate the broad structural features, we compiled a list of various natural products that are not under development but might have the potential to generate lead compounds (Table 7.3).

Based on a survey by Ganesan[108] therapeutically relevant natural products can be subdivided into two major groups: half of them obey the "rule of five" and the other half does not. Obviously natural compounds are not only multi-target optimised[109] but, additionally, have the ability to use actively some of the genetically encoded 758 transport mechanisms in humans with structural features in common with microbial targets. It is surprisingly consistent, even for the so-called "parallel universe" of compounds that do not obey the "rule of five", that these natural compounds are compliant with regard to log P (Ganesan: "*log P is the lord of the rules*").

12 Microbial Secondary Metabolites with Unique Pharmacological Activity

From the very beginning of synthetic organic chemistry, studies have been intrinsically connected to natural products from varying sources. In contrast to plants that were used for medical treatments for thousands of years, microbial-derived drugs only relatively recently became an essential part of the physician's armamentarium. As pointed out in the first paragraph of this chapter, the breakthroughs were the discoveries of penicillin by Fleming *et al.* and various antibiotics by Waksman. After the "golden age of antibiotics", where most industrial and academic scientists connected microbial derived compounds with antibiotic activity, it took almost four decades until the statins and their hypocholsterolemic activity were described in 1979 by Endo.[110] Excellent articles have covered the subject for the last two decades.[111–119]

With a well proven track record for generating novel bioactive compounds, microorganisms remain a reliable source of innovative and therapeutically relevant products. In addition to the well-known advantages compared with randomly assembled synthetic libraries, a hitherto completely underestimated feature represents a major advantage: the polypharmacological status of natural products. What molecular pharmacologists describe as the newest trend in pre-clinical research has been successfully invented in microbial sources. Most and maybe all secondary metabolites address multiple targets that seem to be beneficial for their survival, but which has the positive effect that pathogens develop resistance to naturally occurring antibiotics at a slower rate compared with "mechanistically pure" drugs.[109] The multiple modes of action of numerous natural products, once called "dirty mechanism compounds", actually may be an advantage. Many examples are known for their mixed-mode-of-action, *e.g.* statins, borrelidin and gliotoxin—just to name a few.

Our compilation of biologically/therapeutically relevant microbial-derived natural products given in Table 7.2 contains compounds that are under development or have already been marketed. One important question for a successful future remains: what rationale could be used to guide the selection of specific microorganisms to furnish new chemical entities in a fast and economic fashion?

13 Conclusions

In this brief chapter we have attempted to highlight both the past and the present, and to hint at the future of the study of microbial compounds which we believe will continue to lead to the discovery and development of new and useful products. There is no question that in the past medicine and agriculture have benefited from the wide variety of potent microbial metabolites. The diversity of chemistry that microorganisms are able to exploit has led to large numbers of unique structures with three-dimensional characteristics giving them equally unexpected and unique biological activity; recent results have shown that these compounds frequently display multiple activities.

At this moment, as illustrated by this and other chapters in this book, there are many interesting and unusual microbial metabolites being studied as scientifically valuable probes and in development as either new pharmaceuticals or agrochemicals. Although the field of microbial metabolites has been investigated for the past three quarters of a century since the discovery of penicillin, the vast majority of bacteria and fungi have still not been examined and remain as potential sources for new compounds. In addition, from simple fermentation and cultivation approaches to the more sophisticated genetic and metabolic studies, new methods have demonstrated that much more can be obtained from even well-known and much studied microorganisms. Genetic engineering of new pathways and the use of surrogate organisms have also led to the production of new compounds and to higher yields more closely attuned to commercialisation.

For these reasons alone we propose that the future study of microbial metabolites will be an essential, worthwhile and profitable endeavour which, coupled with modern chemistry and pharmacology, will provide the means to address meaningful and complicated problems inaccessible using any other approach. Because of this, microbiology in all its forms should be an essential part of any pharmaceutical and agricultural discovery and development programme.

Structures Referred to in Tables 7.2 and 7.3

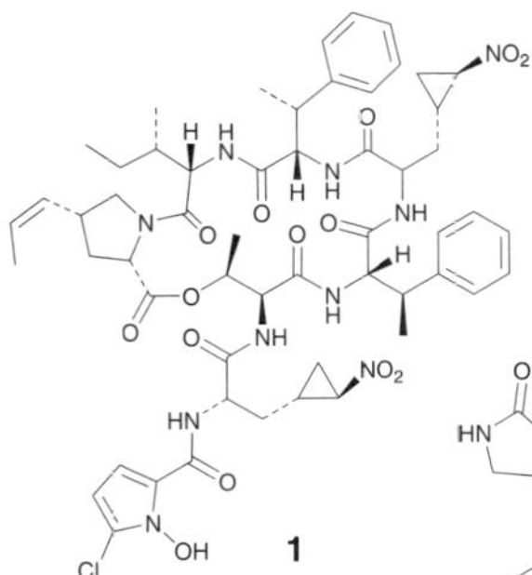

1

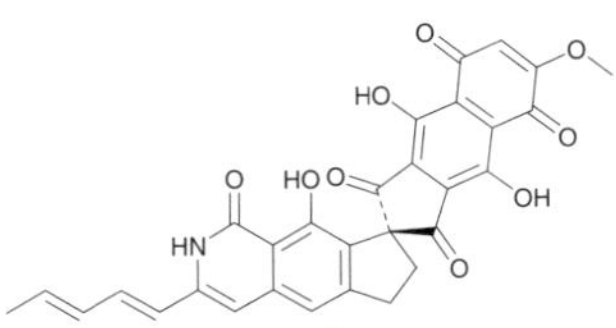

2

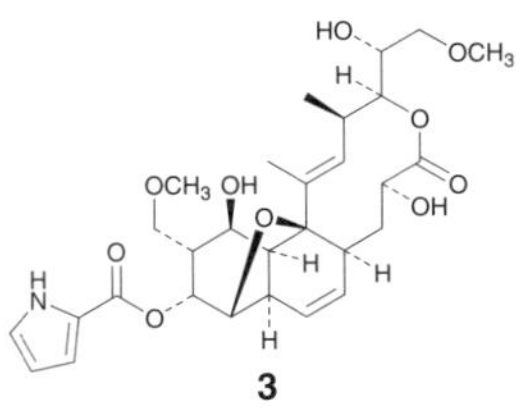

3

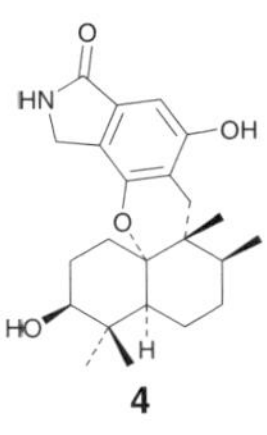

4

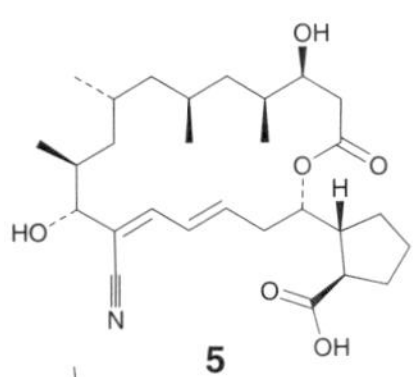

5

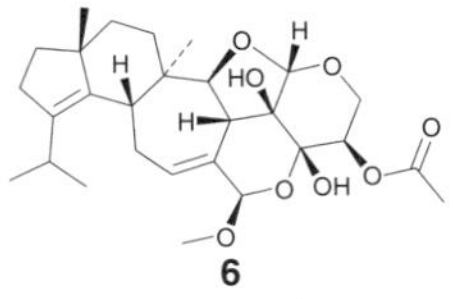

6

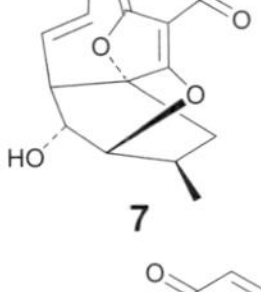

7

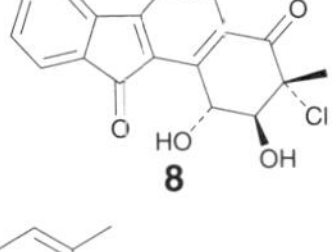

8

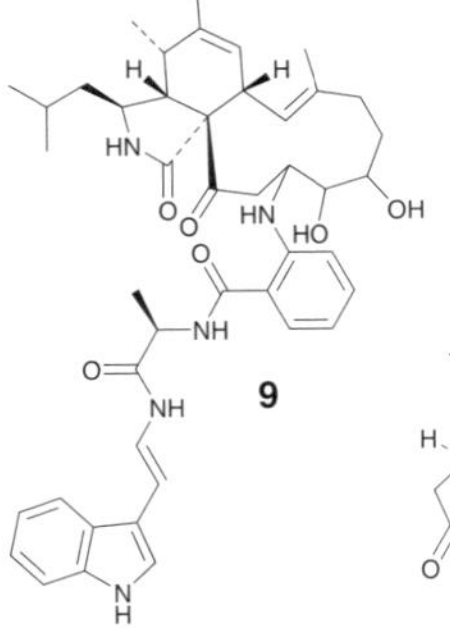

9

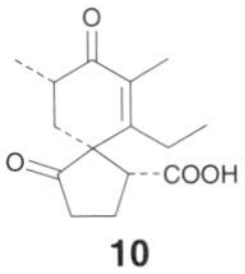

10

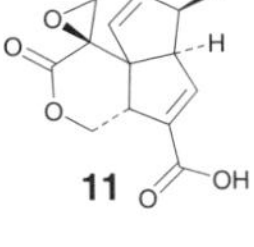

11

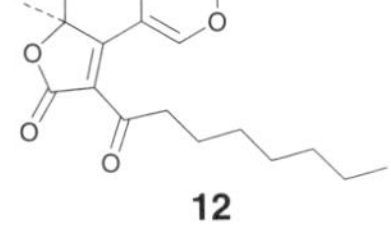

12

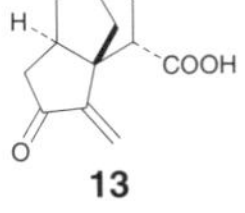

13

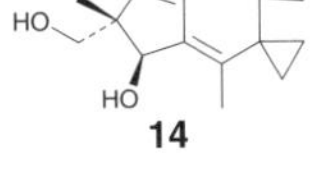

14

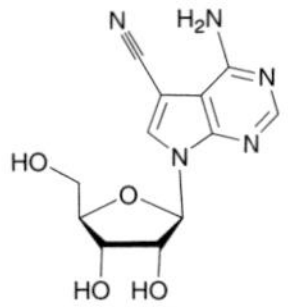
15

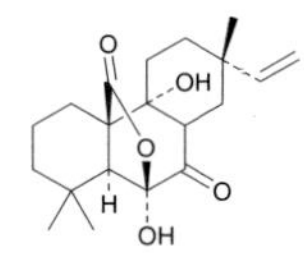
16

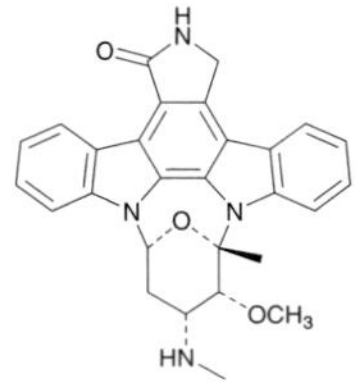
17

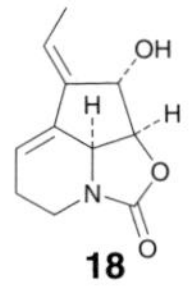
18

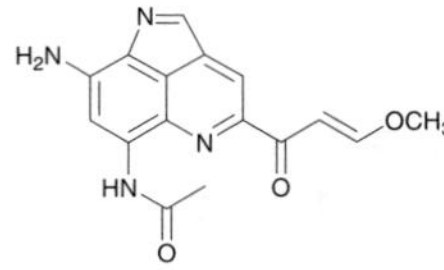
19

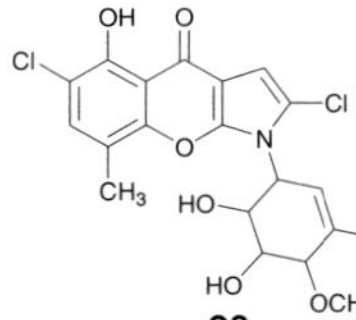
20

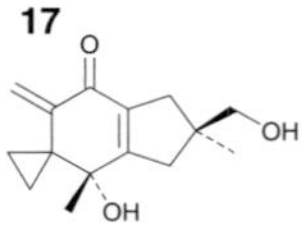
21

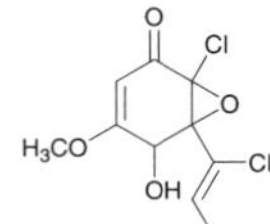
22

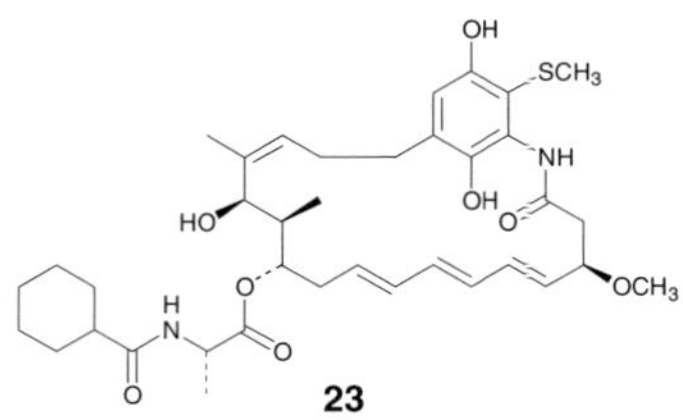
23

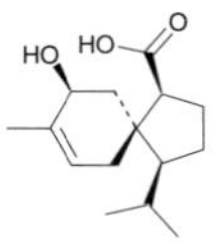
24

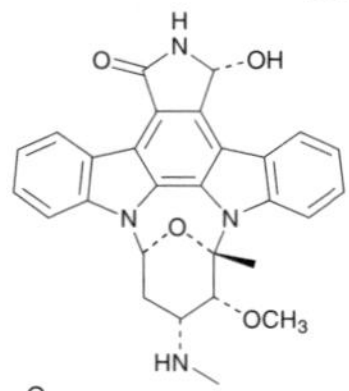
25

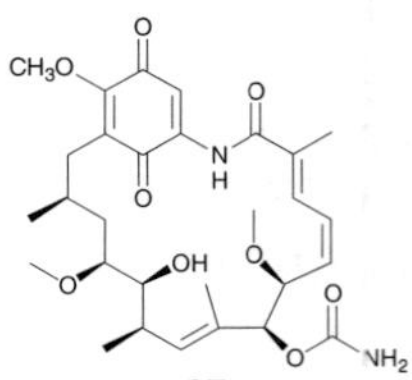
27

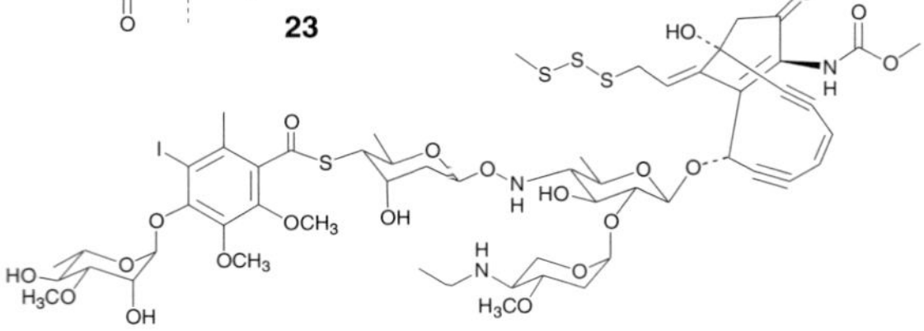
26

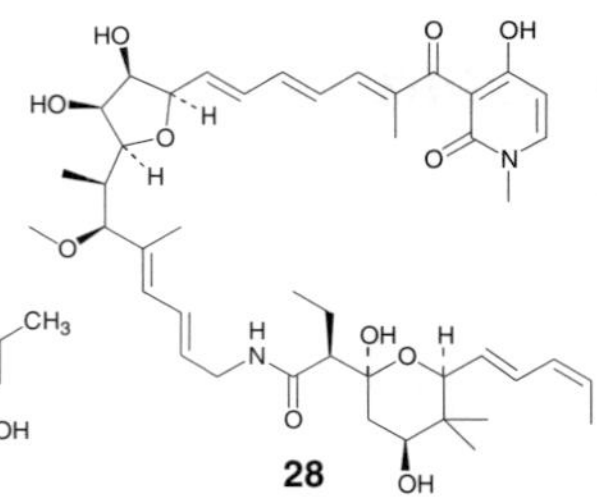
28

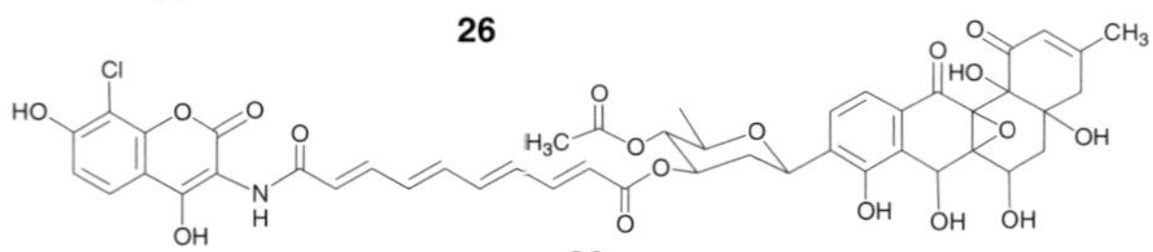
29

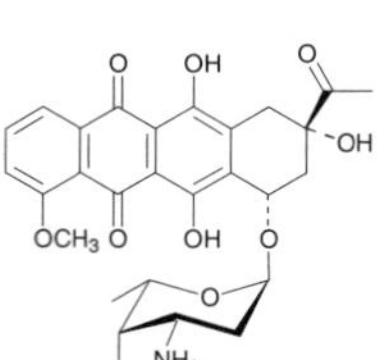
30

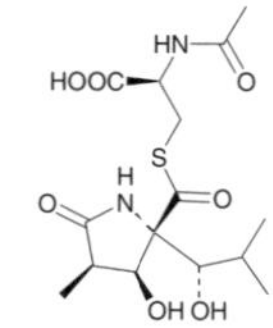
31

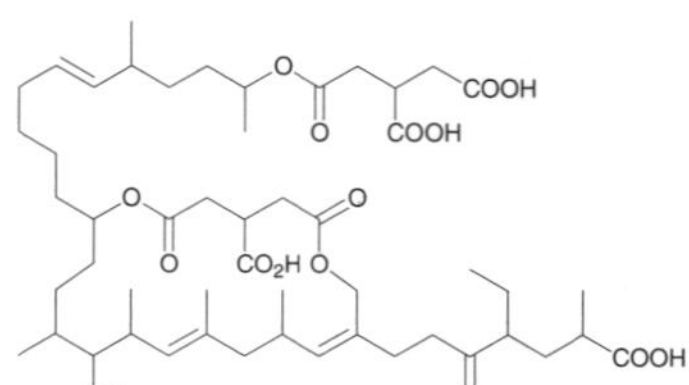
32

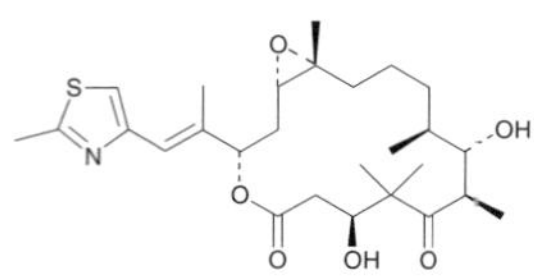
33

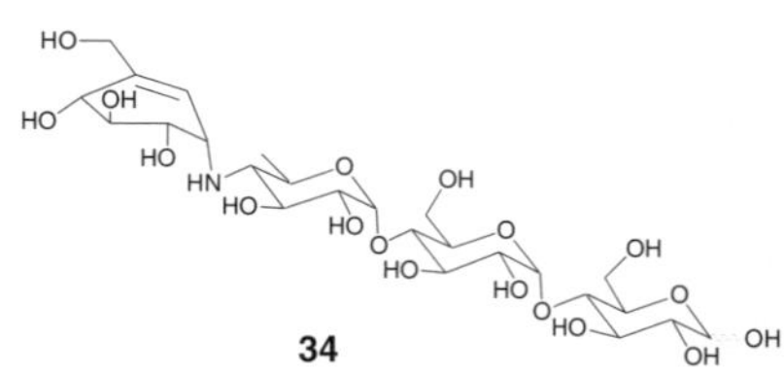
34

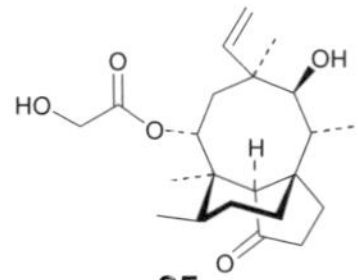
35

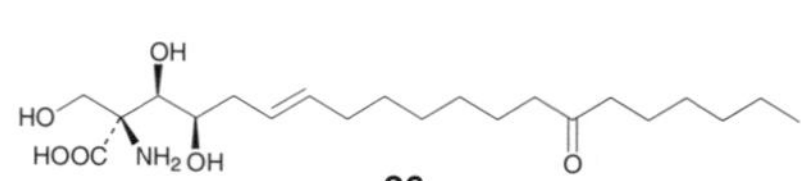
36

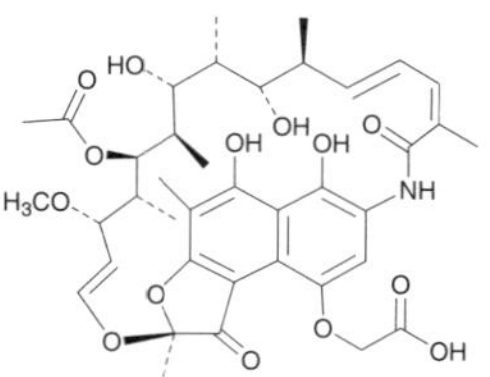

37

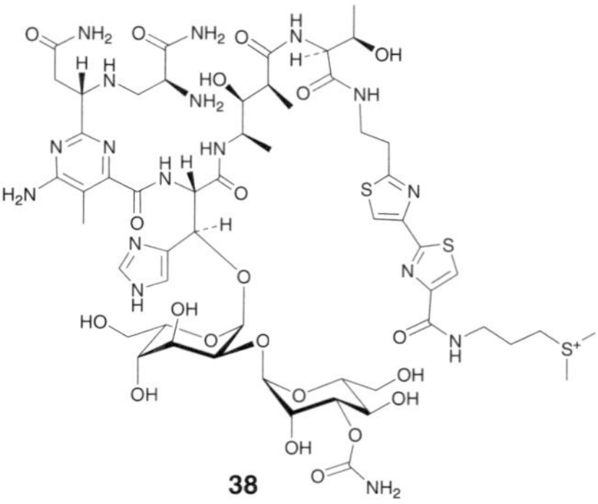

38

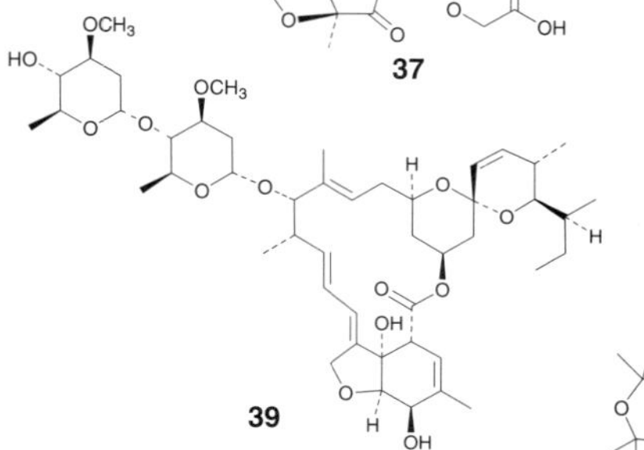

39

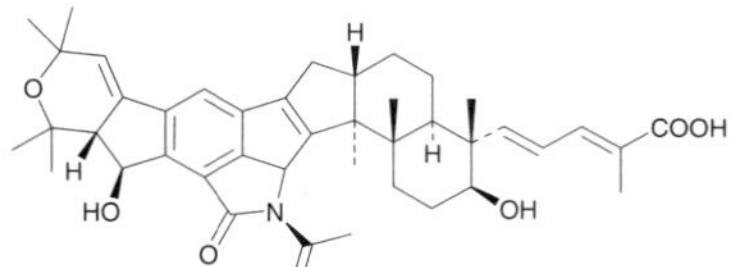

40

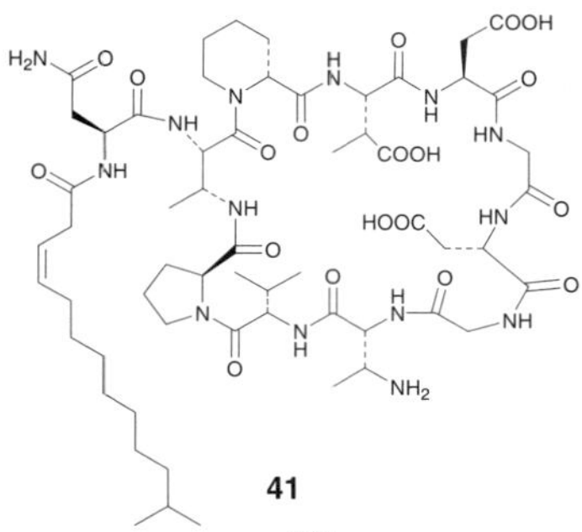

41

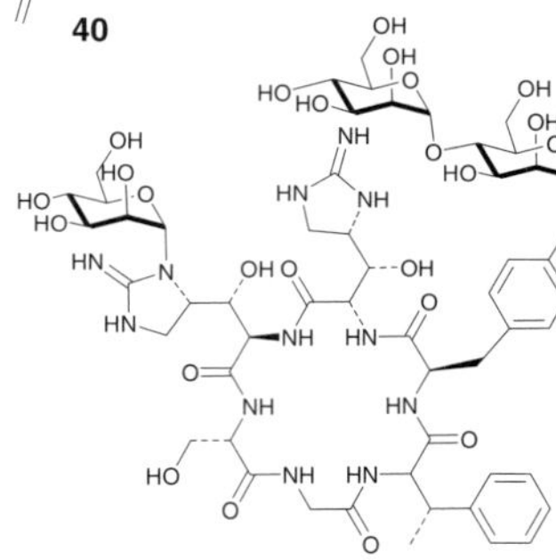

42

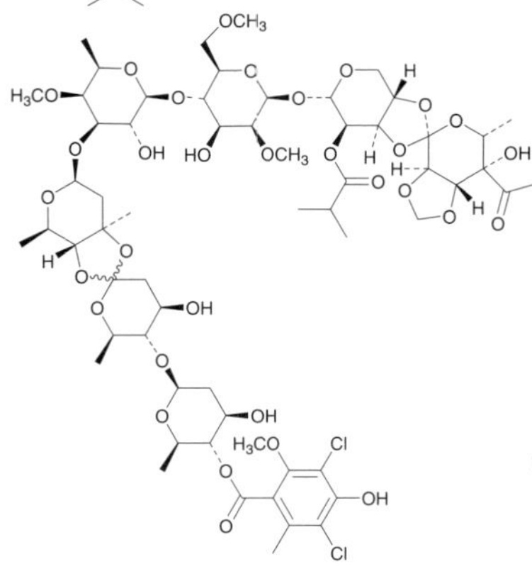

43

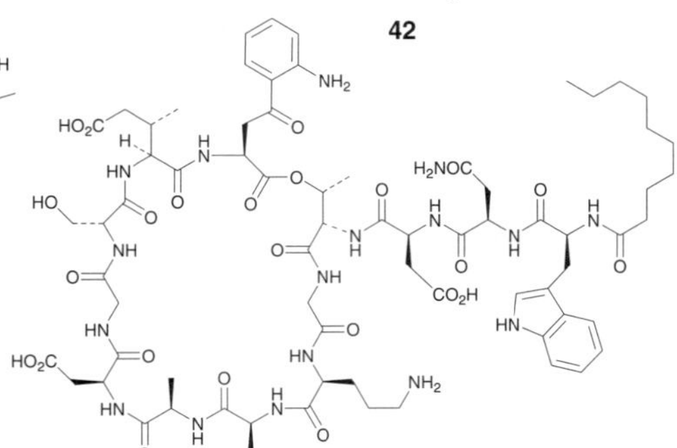

44

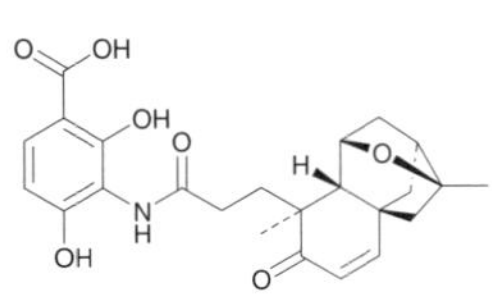

45

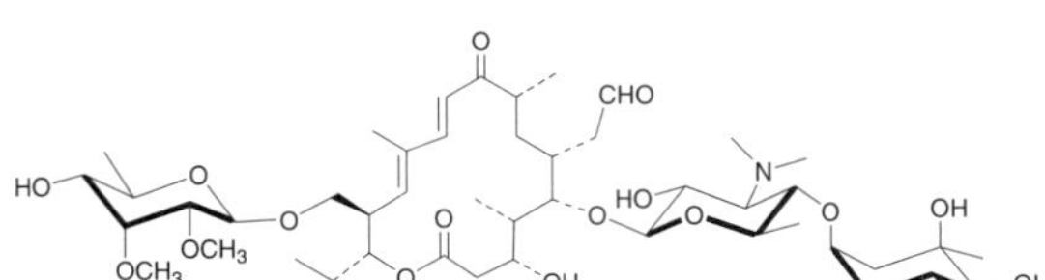

46

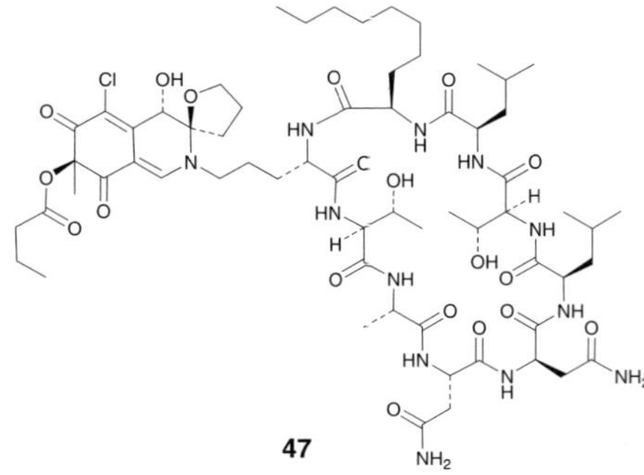

47

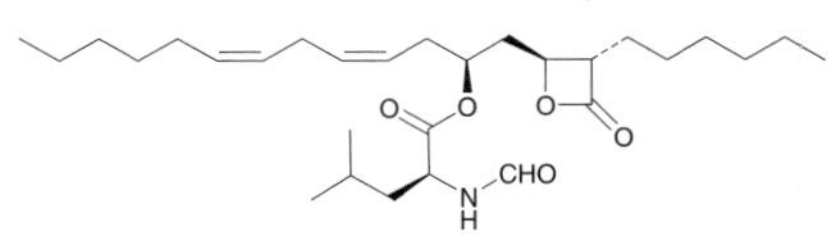

48

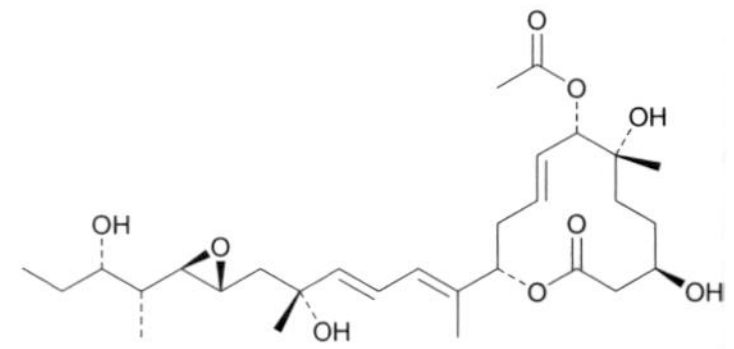

49

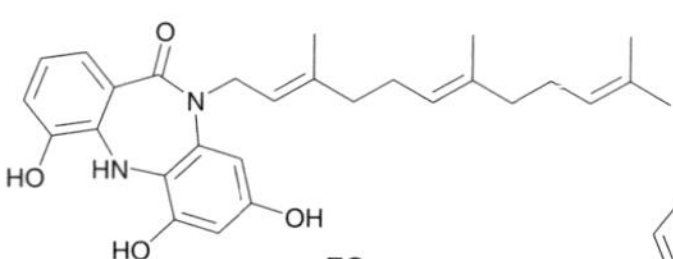

50

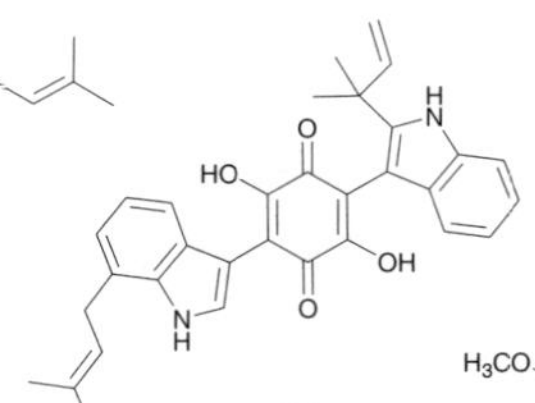

51

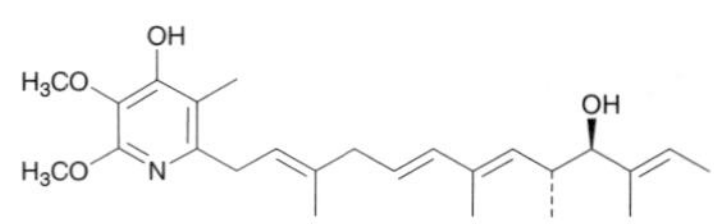

52

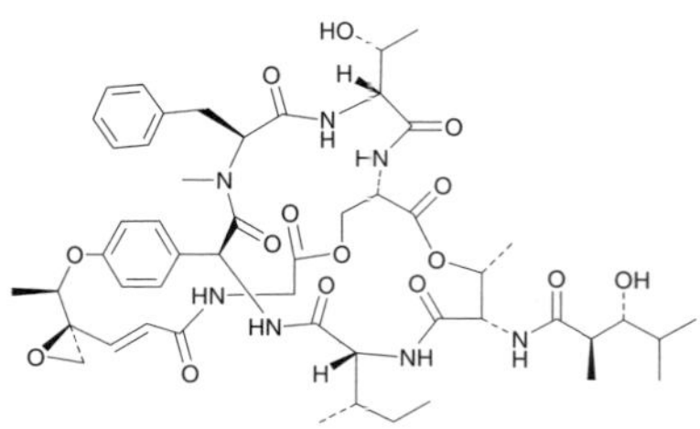

53

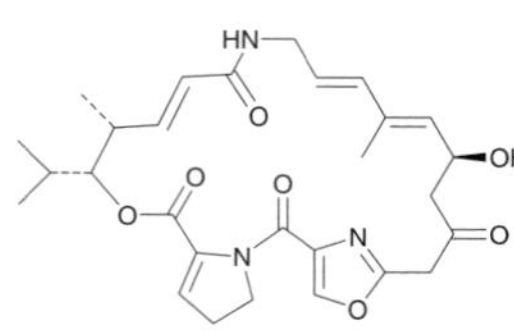

54

References

1. A. Fleming, *Br. J. Exp. Med.*, 1929, **10**, 226.
2. A. Schatz, E. Bugie and S. A. Waksman, *Proc. Soc. Exp. Biol. Med.*, 1944, **55**, 66.
3. S. B. Singh and F. Palaez, *Prog. Drug Res.*, 2008, **65**, 142.
4. (a) R. J. Cole and M. A. Schweikert, *Handbook of Secondary Fungal Metabolites*, **Vols I, II**, Academic Press, California, 2003; (b) R. J. Cole, M. A. Schweikert and B. B. Jarvis, *Handbook of Secondary Fungal Metabolites*, **Vol. III**, Academic Press, California, 2003; (c) C. Pearce, *Adv. Appl. Microbiol.*, 1997, **44**, 1.
5. J. Green and B. J. M. Bohannan, *Trends Ecol. Evol.*, 2006, **21**, 501.
6. L. G. M. Baas Becking, *Geobiologie of Inleiding to de Milieukunde*, van Stockum & Zoon, 1934.
7. R. de Wit and T. Bouvier, *Environ. Microbiol.*, 2006, **8**, 755.

8. A. T. Bull, M. Goodfellow and J. H. Slater, *Ann. Rev. Microbiol.*, 1992, **46**, 219.
9. N. Fierer and R. Jackson, *Proc. Natl. Acad. Sci. USA*, 2006, **103**, 626.
10. R. Conrad, 17th World Congress of Soil Science (WCSS), 14–21 August 2002, Thailand, Symposium No. 11, 2002, Paper No. 390.
11. H. J. Bach, J. Tomanova, M. Schloter and J. C. Munch, *J. Microbiol. Methods*, 2002, **49**, 235.
12. R. A. Olson and L. R. Bakken, *Microb. Ecol.*, 1987, **13**, 59.
13. R. I. Amann, W. Ludwig and K. H. Schleifer, *Microbiol. Rev.*, 1995, **59**, 143.
14. N. Boon, W. de Windt, W. Verstaete and E. M. Top, *FEMS Microbiol. Ecol.*, 2002, **39**, 101.
15. A. D. Kent, D. J. Smith, B. J. Benson and E. W. Triplett, *Appl. Environ. Microbiol.*, 2003, **69**, 6768.
16. C. Rösch and H. Bothe, *Appl. Environ. Microbiol.*, 2005, **71**, 2026.
17. J. O. Mc Inerney, M. Mullarkey, M. E. Warnecke and R. Powell, *Biodiversity*, 2002, **3**, 3.
18. C. R. Woese, *Microbiol. Rev.*, 1987, **51**, 221.
19. P. Monciardini, N. Montanini, M. Sosio and S. Donadio, *Lett. Appl. Microbiol.*, 2008, **46**, 87.
20. P. López-García and D. Moreira, *Res. Microbiol.*, 2007, **159**, 67.
21. P. R. Jensen, P. G. Williams, D.-C. Oh, L. Zeigler and W. Fenical, *Appl. Environ. Microbiol.*, 2006, **73**, 1146.
22. G. M. Zenova, N. V. Mikhailova and D. G. Zvyagintsev, *Microbiology*, 2000, **69**, 109.
23. G. Dantas, M. O. A. Sommer, R. D. Oluwasegun and G. M. Church, *Science*, 2008, **320**, 100.
24. D. L. Hawkesworth, B. C. Sutton and G. C. Ainsworth, *Ainsworth & Bisby's Dictionary of the Fungi*, Commonwealth Mycological Institute, Kew, UK, 1983.
25. D. L. Hawkesworth, *Mycol. Res.*, 1991, **95**, 641.
26. D. L. Hawkesworth, *Mycol. Res.*, 2001, **105**, 1422.
27. P. Hammond, in *Global Biodiversity: Status of the Earth's Living Resources*, ed. B. Groombridge, Cambridge University Press, Cambridge, 1992, pp. 17–19.
28. P. F. Cannon, *Biodiversity Conserv.*, 1997, **6**, 669.
29. N. L. Hywel-Jones, in *Aspects of Tropical Mycology*, ed. S. Isaac, J. C. Frankland, R. Watling and A. J. S. Whalley, Cambridge University Press, Cambridge, 1993, 300–301.
30. P. Diederich, *Bibl. Lichenol.*, 1995, **61**, 1.
31. D. L. Hawkesworth and A. Y. Rossman, *Phytopathology*, 1997, **87**, 888.
32. J. Cifuentes Blanco, M. Villegas Rios, J. L. Villarruel-Ordaz and S. Sierra Galván, in *Mycology in Sustainable Development: Expanding Concepts, Vanishing Borders*, ed. M. E. Palm and I. H. Chapela, Parkway Publishers, North Carolina, 1997, pp. 111–121.

33. K. D. Hyde, V. V. Sarma and E. B. G. Jones, in *Marine Mycology: A Practical Approach*, ed. K. D. Hyde and S. B. Pointing, Fungal Diversity Press, Hong Kong, 2000, 172–204.
34. Y. Takahashi, *Actinomycetologica*, 2004, **18**, 54.
35. G. Toledo, W. Green, R. A. Gonzalez, L. Christoffersen, M. Podar, H. W. Chang, T. Hemscheidt, H. G. Trapido-Rosenthal, J. M. Short, R. R. Bidigare and E. J. Mathur, *Oceanography*, 2006, **19**, 120.
36. W. Fenical and P. R. Jensen, *Nat. Chem. Biol.*, 2006, **2**, 666.
37. J. Kennedy, J. Marchesi and A. Dobson, *Appl. Microbiol. Biotechnol.*, 2007, **75**, 11.
38. J. Kennedy, J. R. Marchesi and A. D. W. Dobson, *Microb. Cell Fact.*, 2008, **7**, 27.
39. T. L. Simmons, R. C. Coates, B. R. Clark, N. Engene, D. Gonzalez, E. Esquenazi, P. C. Dorrestein and W. H. Gerwick, *Proc. Natl. Acad. Sci. USA*, 2008, **105**, 4587.
40. S. J. Joseph, P. Hugenholtz, P. Sangvan, C. A. Osborne and P. H. Janssen, *Appl. Environ. Microbiol.*, 2003, **69**, 7210.
41. E. Takano, *Curr. Opin. Microbiol.*, 2006, **9**, 287.
42. E. Matsukawa, Y. Nakagawa, Y. Iimura and M. Hayakawa, *Actinomycetologica*, 2007, **21**, 32.
43. S. J. Giovannini, T. B. Britschgi, C. L. Moyer and K. G. Field, *Nature*, 1990, **345**, 60.
44. D. M. Ward, R. Weller and M. M. Bateson, *Nature*, 1990, **345**, 63.
45. J. Handelsman, M. R. Rondon, S. F. Brady, J. H. Clardy and R. M. Goodman, *Chem. Biol.*, 1998, **5**, R245.
46. R. Sharma, R. Ranjan, R. K. Kapardar and A. Grover, *Curr. Sci.*, 2005, **89**, 72.
47. S. G. van Lauen and B. Shen, *Curr. Opin. Microbiol.*, 2006, **9**, 252.
48. M. Hayakawa, *Actinomycetologica*, 2008, **22**, 12.
49. A. Lazzarini, L. Cavaletti, G. Toppo and F. Marinelli, *Antonie van Leeuwenhoek*, 2000, **78**, 399.
50. D. I. Kurtböke, *Selective Isolation of Rare Actinomycetes*, Queensland Complete Printing Services, Nambour, Australia, 2003.
51. E. Matsukawa, Y. Nakagawa, Y. Iimura and M. Hayakawa, *Actinomyceteolgica*, 2007, **21**, 66.
52. M. Otoguro, M. Hayakawa, T. Yamazaki and Y. Iimura, *J. Appl. Microbiol.*, 2001, **91**, 118.
53. K. Zengler, M. Walcher, G. Clark, I. Haller, G. Toledo, T. Holland, E. J. Mathur, G. Woodnut, J. Short and M. Keller, *Methods Enzymol.*, 2005, **397**, 124.
54. T. Kaeberlein, K. Lewis and S. S. Epstein, *Science*, 2002, **296**, 1127.
55. E. Gavrish, A. Bollmann, S. Epstein and K. Lewis, *J. Microbiol. Methods*, 2008, **72**, 257.
56. T. Hamaki, M. Suzuki, R. Fudou, Y. Jojima, T. Kajima, A. Tabuchi, K. Sen and H. Shibai, *J. Biosci. Bioeng.*, 2005, **99**, 485.
57. M. Sosio and S. Donadio, *Microbiol. Aust.*, May 2004, 32.

58. D. Bentley, K. F. Chater, A.-M. Cerdeño-Tárraga, G. L. Challis, N. R. Thomson, K. D. James, D. E. Harris, M. A. Quail, H. Kieser, D. Harper, A. Bateman, S. Brown, G. Chandra, C. W. Chen, M. Collins, A. Cronin, A. Fraser, A. Goble, J. Hidalgo, T. Hornsby, S. Howarth, C.-H. Huang, T. Kieser, L. Larke, L. Murphy, K. Oliver, S. O'Neil, E. Rabbinowitsch, M.-A. Rajandream, K. Rutherford, S. Rutter, K. Seeger, D. Saunders, S. Sharp, R. Squares, S. Squares, K. Taylor, T. Warren, A. Wietzorrek, J. Woodward, B. G. Barrell, J. Parkhill and D. A. Hopwood, *Nature*, 2002, **417**, 141.
59. S. Ōmura, H. Ikeda, J. Ishikawa, A. Hanamoto, C. Takahashi, M. Shinose, Y. Takahashi, H. Horikawa, H. Nakazawa, T. Osonoe, H. Kikuchi, T. Shiba, Y. Sakaki and M. Hattori, *Proc. Natl. Acad. Sci. USA*, 2001, **98**, 12215.
60. H. Ikeda, J. Ishikawa, A. Hanamoto, M. Shinose, H. Shiba, Y. Sakaki, M. Hattori and S. Ōmura, *Nat. Biotechnol.*, 2003, **21**, 526.
61. M. Sosio, E. Bossi, A. Bianchi and S. Donadio, *Mol. Gen. Genet.*, 2000, **264**, 213.
62. K. Gerth, S. Pradella, O. Perlova, S. Beyer and R. Müller, *J. Biotechnol.*, 2003, **106**, 233.
63. R. H. Baltz, *Curr. Opin. Pharmacol.*, 2008, **8**, 557.
64. J. Handelsman and P. D. Schloss, *Genome Biol.*, 2005, **6**, 229.
65. E. Wendt-Pienkowski, Y. Huang, J. Zhang, B. Li, H. Jiang, H. Kwon, C. R. Hutchinson and B. Shen, *J. Am. Chem. Soc.*, 2005, **127**, 16442.
66. V. Miao, M.-F.e Coëffet-LeGal, P. Brian, R. Brost, J. Penn, A. Whiting, S. Martin, R. Ford, I. Parr, M. Bouchard, C. J. Silva, S. K. Wrigley and R. H. Baltz, *Microbiology*, 2005, **151**, 1507.
67. J. Handelsman, in *Microbial Diversity and Bioprospecting.*, ed. A. T. Bull, ASM Press, 2004, pp. 109–119.
68. A. L. Demain, J. E. Davies, R. M. Atlas, G. Cohen, C. Hershberger, W.-S. Hu, D. Sherman, R. Willson and J. D. Wu, *Manual of Industrial Microbiology and Biotechnology*, ASM Press, Washington, DC, 2nd edn, 1999.
69. D. H. Jennings, *The Physiology of Fungal Nutrition*, Cambridge University Press, Cambridge, 1995.
70. A. L. Demain, *J. Ind. Microbiol. Biotechnol.*, 2006, **33**, 486.
71. (a) C. Krishna, *Crit. Rev. Biotechnol.*, 2005, **25**, 1; (b) T. Robinson, D. Singh and P. Nigam, *Appl. Microbiol. Biotechnol.*, 2001, **55**, 284.
72. A. L. Demain, *Appl. Microbiol. Biotechnol.*, 1999, **52**, 455.
73. R. Bigelis, H. He, H. Y. Yang, L. Chang and M. Greenstein, *J. Ind. Microbiol. Biotechnol.*, 2006, **33**, 815.
74. H. He, H. Y. Yang, R. Bigelis, E. H. Solum, M. G. Greenstein and G. T. Carter, *Tetrahedron Lett.*, 2002, **43**, 1633.
75. H. He, R. Bigelis, E. H. Solum, M. G. Greenstein and G. T. Carter, *J. Antibiot. (Tokyo)*, 2003, **56**, 923.
76. R. B. Williams, J. C. Henrikson, A. R. Hoover, A. E. Lee and R. H. Cichewicz, *Org. Biomol. Chem.*, 2008, **6**, 1895.

77. A. H. Banskota, J. B. Mcalpine, D. Sørensen, A. Ibrahim, M. Aouidate, M. Piraee, A. M. Alarco, C. M. Farnet and E. Zazopoulos, *J. Antibiot. (Tokyo)*, 2006, **59**, 533.
78. J. B. McAlpine, *J. Nat. Prod.*, 2005, **68**, 493.
79. A. Tanovic, S. A. Samel, L.-O. Essen and M. A. Marahiel, *Science*, 2008, **321**, 6590.
80. S. Everts, *Chem. Eng. News*, 2008, **86**, 48.
81. H. Onaka, *Actinomycetologica*, 2006, **20**, 620.
82. C. Khosla and J. D. Kiesling, *Nat. Rev. Drug Discov.*, 2003, **2**, 1019.
83. K. Koketsu, H. Oguri, K. Watanabe and H. Oikawa, *Chem. Biol.*, 2008, **15**, 818.
84. B. Wilkinson and S. J. Moss, *Curr. Opin. Drug. Discov. Devel.*, 2005, **8**, 748.
85. E. S. Sattely, M. A. Fischbach and C. T. Walsh, *Nat. Prod. Rep.*, 2008, **25**, 757.
86. H. G. Menzella and C. D. Reeves, *Curr. Opin. Microbiol.*, 2007, **10**, 238.
87. H. B. Bode and R. Müller, *Angew. Chem., Int. Ed. Engl.*, 2005, **44**, 6828.
88. S. Bergmann, J. Schümann, K. Scherlach, C. Lange, A. A. Brakhage and C. Hertweck, *Nat. Chem. Biol.*, 2007, **3**, 213.
89. G. J. Williams, R. W. Gantt and J. S. Thorson, *Curr. Opin. Chem. Biol.*, 2008, **12**, 556.
90. G. J. Williams and J. S. Thorson, *Adv. Enzymol. Relat. Areas Mol. Biol.*, 2009, **76**, 55–119.
91. Ch. M. M. Franco and L. E. L. Coutinho, *Crit. Rev. Biotechnol.*, 1991, **11**, 193.
92. J. J. Sanglier, H. Haag, T. A. Huck and T. Fehr, *Res. Microbiol.*, 1993, **144**, 633.
93. J. J. Sanglier, H. Haag, T. A. Huck and T. Fehr, *Expert Opin. Investig. Drugs*, 1996, **5**, 207.
94. S. Grabley and R. Thiericke, *Drug Discovery from Nature*, Springer Verlag, Berlin, 1999.
95. L. Zhang and A. L. Demain, *Natural Products—Drug Discovery and Therapeutic Medicine*, Humana Press, NJ, 2005.
96. A. T. Bull, *Microbial Diversity and Bioprospecting*, ASM Press, Washington DC, 2004.
97. E. Fiedler and H.-P. Fiedler, *Secondary Metabolites Isolated by Hans Zähner and his Group 1954–1994*, Zeeb-Druck, Tübingen, 1994.
98. Research Center for Biological Function, The Kitasato Institute, *"Splendid Gifts from Microorganisms"—The Achievements of Satoshi Omura and Collaborators*, The Kitasato Institute, Tokyo, 1995.
99. U. Abel, C. Koch, M. Speitling and F. G. Hansske, *Curr. Opin. Chem. Biol.*, 2002, **6**, 453.
100. M. M. Wangenaar, *Molecules*, 2008, **13**, 1406.
101. V. M. Dembitsky, *Lipids*, 2004, **39**, 933.
102. V. M. Dembitsky, *Lipids*, 2005, **40**, 219.
103. V. M. Dembitsky, *Lipids*, 2005, **40**, 535.

104. V. M. Dembitsky, *Lipids*, 2005, **40**, 641.
105. V. M. Dembitsky, *Lipids*, 2005, **40**, 869.
106. V. M. Dembitsky, *Lipids*, 2005, **40**, 1081.
107. V. M. Dembitsky, *Lipids*, 2006, **41**, 1.
108. A. Ganesan, *Curr. Opin. Chem. Biol.*, 2008, **12**, 306.
109. H. Brötz-Oesterhelt and N. A. Brunner, *Curr. Opin. Pharm.*, 2008, **8**, 564.
110. A. Endo, M. Kuroda and Y. Tsujita, *J. Antibiot. (Tokyo)*, 1976, **29**, 1346.
111. Y.-Z. Shu, *J. Nat. Prod.*, 1998, **61**, 1053.
112. D. J. Newman, G. M. Cragg and K. M. Snader, *Nat. Prod. Rep.*, 2000, **17**, 215.
113. M. S. Butler, *J. Nat. Prod.*, 2004, **67**, 2141.
114. M. S. Butler, *Nat. Prod. Rep.*, 2005, **22**, 162.
115. J. Bérdy, *J. Antibiot. (Tokyo)*, 2005, **58**, 1.
116. D. J. Newman and G. M. Cragg, *J. Nat. Prod.*, 2007, **70**, 461.
117. D. D. Baker, M. Chu, U. Oza and V. Rajgarhia, *Nat. Prod. Rep.*, 2007, **24**, 1225.
118. M. S. Butler, *Nat. Prod. Rep.*, 2008, **25**, 475.
119. E. Stackebrandt and J. Ebers, *Microbiol. Today*, 2006, **33**, 152.

Section 3 Advances in Technology

CHAPTER 8

Advances in Biological Screening for Lead Discovery

CHRISTIAN N. PARKER*,[a] JOHANNES OTTL,[a] DANIELA GABRIEL[a] AND JI-HU ZHANG[b]

[a] Novartis Institutes for BioMedical Research, Center for Proteomic Chemistry, Forum 1, Novartis Campus, CH-4056, Basel, Switzerland;
[b] Novartis Institutes for BioMedical Research, Center for Proteomic Chemistry, 250 Massachusetts Avenue, Cambridge MA 02139, USA

1 Introduction

High throughput screening (HTS) has become one of the main methods for generating leads for drug discovery.[1] The capacity and throughput of screening have been growing steadily for the last two decades. HTS has transformed biological testing from a process using test tube and cuvette measurements to one using high density, low volume assay formats and screening systems. This transformation has been in response to rapid changes in biological target identification and validation, facilitated by advances in genomic and proteomic studies and compound synthetic methodologies. In particular, the development and utilisation of miniaturised, homogeneous assay technologies and various signal detection methodologies have made possible the screening of hundreds of thousands (even millions) of samples in a relatively short time period. Standardisation of assay formats, instrumentation, automation, assay miniaturisation and informatics tools for data analysis have all greatly facilitated the screening process by increasing reliability and reducing personnel workload.

RSC Biomolecular Sciences No. 18
Natural Product Chemistry for Drug Discovery
Edited by Antony D. Buss and Mark S. Butler

Published by the Royal Society of Chemistry, www.rsc.org

The involvement of HTS activities in different phases of drug discovery and development are shown in Figure 8.1.

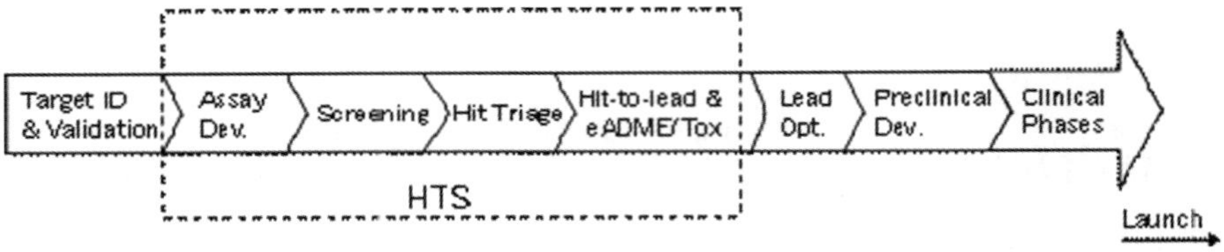

Figure 8.1 Stages of drug discovery and development and the expanded involvement of HTS activities. (Dev.=Development; Opt.=Optimisation; eADME/Tox=early Absorption, Distribution, Metabolism and Excretion/Toxicology)

From the range of different activities involved, HTS should be regarded as a multi-disciplinary science including activities as diverse as the creation of cell lines or recombinant proteins for screening, to computational and medicinal chemists designing and making compound screening libraries or from engineers, building suitable robots and instrumentation allowing the automation of biological assays, to statisticians and modellers analysing screening data to extract useful and predictive models of mode of interaction. Figure 8.2 illustrates the interconnection of scientific disciplines required for successful HTS and hit-to-lead optimisation processes.

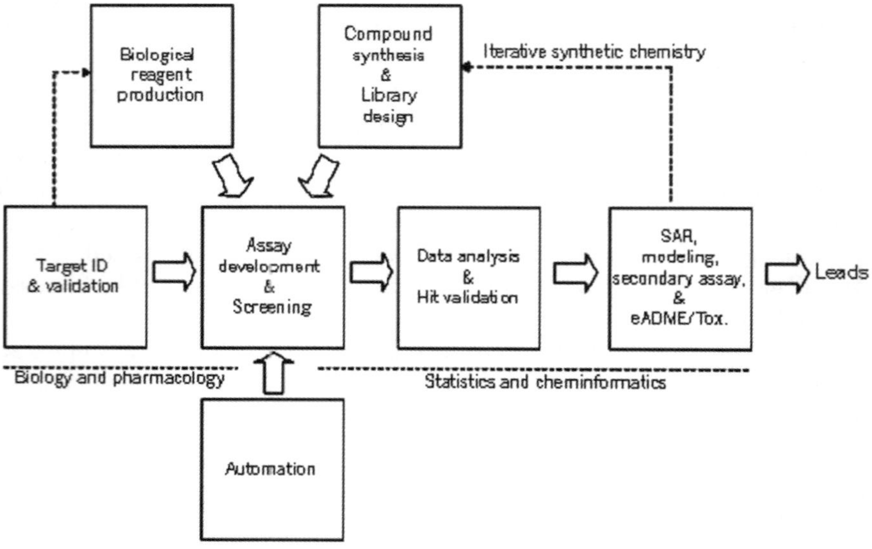

Figure 8.2 The interconnection of scientific disciplines and activities involved in HTS. (SAR=Structure–Activity Relationships).

The challenge facing HTS scientists is not only that an assay developed for a target must be biologically relevant and as specific as possible, but that it should also meet the limitations of practical constraints such as cost, scale and throughput. HTS must be properly designed and configured to maximise both screening efficiency and biological effectiveness. An improperly designed HTS assay may fail to detect the desired classes of hits even if it is trivial to execute. Therefore, HTS should be coupled to more pharmacological appropriate properties in order to make screening more relevant and predicative.

1.1 Natural Product Screening and the Development of HTS

At present, many of the large pharmaceutical companies have de-prioritised (or in all honesty, given up on) natural product research as a source of leads for drug discovery.[2] This is ironic considering that many HTS departments in pharmaceutical companies had their origins in natural product groups, for example, at Pfizer[3] and Upjohn/Pharmacia (C. Haber, personal communication). However, a number of different factors have led to the drift away from natural product research, including changes in work flow requiring that only pure compounds will be screened and taken on for project team evaluation.

In contrast, screening natural product extracts requires additional dereplication and validation stages adding time to the screening process. The presence of fluorescent or coloured compounds in natural product extracts is also an issue with the homogenous detection methods required for current HTS work processes. In addition, project team expectations are changing. The acceptance of the Lipinski "rule of five" and "rule of three" for fragment screening has led to the perception that natural products are too large, with too many functional groups to act as suitable lead compounds (even though a number of papers, including the original paper from Lipinski, have shown such assumptions to actually be false for natural products; see Chapter 2, Section 4). There is evidence to suggest that there will be a renewed interest in screening natural products.[4]

1.2 Chapter Objectives

The aim of this chapter is to review some of the assay technologies and strategies being used for high throughput screening. Finally, the chapter briefly reviews some of the emerging trends in the science of biomolecular screening.

2 Types of HTS Assays

Assays can be used to probe biological interactions in a continuum of complexity, ranging from simple binding events to effects at the whole organism level. High throughput screening assays can be classified into several different types. Broadly speaking, there are two major assay types: *in vitro* biochemical and cell-based assays. The biochemical assays can be further divided into molecular binding (affinity) and *in vitro* functional assays (*e.g.* enzymatic

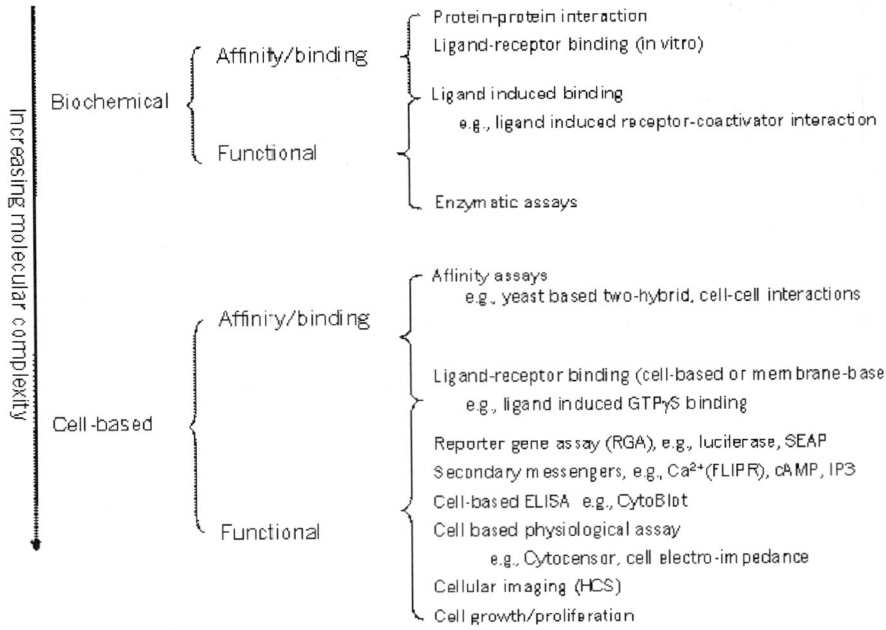

Figure 8.3 Hierarchy of assays monitoring biological processes of increasing molecular complexity.

reactions). There is sometimes, however, no clear distinction between the affinity binding and *in vitro* functional assays. For example, ligand-induced GTPγS binding to G proteins can be regarded as both a binding or a functional assay. Ligand-activated coactivator recruitment to the ligand-binding domain of a nuclear receptor is another such example. Cell-based assays can also be further divided into cell- or membrane-based binding assays and functional cell-based assays (Figure 8.3). This section provides an overview of the assay technologies commonly used for HTS.

2.1 *In vitro* Biochemical Assays

In vitro biochemical assays have been used extensively against various targets, *e.g.* enzymes, receptor–ligand and protein–protein interactions. This reductionist approach has been facilitated by various biochemical and genetic studies for delineating complicated metabolic, signal transduction pathways and biological systems involved in diseases. It offers the advantage of clear drug–target interactions, leading to clean mode of action and structure–activity relationships (SAR) during hit evaluation and hit-to-lead optimisation. The drawbacks of this approach include sometimes a lack of physiological context (*e.g.* cell permeability) compared with cell-based assays, which can lead to a lack of *in vivo* efficacy. A clear trend is to combine both approaches so as to gain the advantages of biochemical and cell-based screens while minimising the risks.

Targeting of specific enzymes and components in signal transduction, metabolic pathways, protein processing and inflammatory cascades has proven to be a valid approach for addressing disease processes. Many classes of therapeutically relevant enzymes have been actively pursued using HTS for lead generation. The majority of these enzymatic assays have been adapted to highly miniaturised microtitre plate formats, automated and used for compound screening.

Several of the most commonly used high throughput assay technologies based on direct molecular affinity binding events and enzymatic activities are highlighted below. Many of these assay and detection technologies can be used to measure both binding and enzymatic activities. Several reviews on HTS assays are available, for example, the books edited by Seethala and Fernandes[5] or Janzen[6] contain many excellent articles.

Fluorescence Based Assays

Fluorescence-based detection methods are the most commonly used readouts for HTS as these readouts are sensitive, usually homogeneous and can be readily miniaturised, even down to the single molecule level.[7,8] Fluorescent signals can be detected by methods such as fluorescence intensity (FI), fluorescence polarisation (FP) or anisotropy (FA), fluorescence resonance energy transfer (FRET), time-resolved fluorescence resonance energy transfer (TR-FRET) and fluorescence intensity life time (FLIM). Confocal single molecule techniques such as fluorescence correlation spectroscopy (FCS) and one- or two-dimensional fluorescence intensity distribution analysis (1D FIDA, 2D FIDA) have been reported but are not commonly used.

One disadvantage of fluorescence markers and dyes is that they can cause a loss of functional activity due to the presence of the, often hydrophobic, fluorescent dye. Intrinsic fluorescence and absorbance of test compounds, quench effects and turbidity of solutions can lead to false signals. In TR-FRET, the long lifetime of lanthanide metal fluorescence ($>$200–1000 ns) can be used in combination with a time-gated detection, eliminating prompt fluorescence disturbance from test compounds because of the typical fluorescence lifetime of organic molecule dyes ($<$20 ns).[9] TR-FRET is also useful for monitoring cell metabolites in cell-based assays.[10]

Whereas fluorescence is typically measured during non-separation, mix-and-read protocols offering high readout throughput, several approaches exist today that involve separation steps in front of the actual detection. In the Caliper Labchip™ technology, a multi-parallel microfluidic separation system coupled with fluorescence detection allows the monitoring of enzymatic reactions. Quantification of substrate and product of the enzymatic reaction, after separation from each other as well as from the test compound, minimises artefacts and offers ratiometric results, although with comparatively lower throughput. Typical applications of the system include kinase, phosphatase and protease assays.[11,12]

Enzyme-linked immunosorbent assays (ELISA) are still popular assays in the low to medium throughput screening field. A large number of biological assays have been developed using this basic assay technology, where an analyte of interest is first captured on a solid surface by specific or non-specific absorption. The amount of captured analyte is then monitored by an antibody that has been conjugated to an enzyme such as horse radish peroxidase or alkaline phosphatase. Such assay formats are widely used to monitor biological systems. However, such assays only have limited applicability due to the need for extensive plate handling and plate washing.

Fluorescent microvolume assay technology (FMAT™) is a bead-based or cell-based fluorescent technology for homogeneous ELISA-like assays. In FMAT, a laser beam is focused on the bottom of the assay well and the localised fluorescence intensity bound to beads (or cells) is detected as an area of intense fluorescence over the unbound and background fluorescence in solution. Different analytes can be detected with appropriate fluorophores and, by using different sized beads, the assay can be multiplexed to monitor multiple analytes.[13]

Luminex is also a bead-based, non-separation technology using the Luminex colour-coded beads and detection systems (Luminex 100 IS™ or Luminex HT™). The readers used for this assay format are based on the principle of flow cytometry. The system enables assays to be multiplexed, *i.e.* allowing different analytes to be monitored simultaneously. The Luminex HT™ system is compatible with 96- and 384-well microplates[14] but throughput of the reader is still a limiting factor for large-scale HTS.

Luminescence-Based Assays

Luminescence detection has the advantage of very low background compared with fluorescent technologies, so assay sensitivity can be extremely high.

The AlphaScreen™ assay format is another versatile, non-separation technology for HTS assays. The analyte being measured brings two beads into close proximity allowing a specific signal to be generated. AlphaScreen™ is a bead-based, chemiluminescent readout technology based on energy transfer by singlet state molecular oxygen, generating emitted light. The effective distance from donor to acceptor beads is $\sim$200 nm in aqueous solution, much larger than that in TR-FRET. Another feature of AlphaScreen™ is that the amplification of the signal is large, typically offering a greater sensitivity than TR-FRET. However, the amplified signal may show a higher variability compared with a TR-FRET detection format.[15] AlphaScreen™ is a powerful technique that is easily set up, but the readout can suffer from interference by test compounds altering the stability of the singlet-oxygen and other non-specific artefacts.

Electrochemiluminescence (ECL) is a homogeneous, bead-based technology using the Origen platform. ECL utilises ruthenium chelates conjugated to an antibody as a tracer, which reacts with trispropylamine (TPA) and emits luminescence upon application of a low voltage. Magnetic beads are utilised to align the beads to the surface of an electrode. This assay format has been mostly used in diagnostic bioassays but has also developed for HTS use.[16]

Chemiluminescence detection methods based on enzymes such as horseradish peroxidase have been used as improvements to standard ELISA-based assays. The number of different luminescence-based assays has been expanding. For example, the Kinase Glo™ assay format monitors the levels of ATP remaining in a reaction mix. Whilst this format was originally developed for assaying kinase targets, it could be easily adapted to monitor any ATP consuming (or generating) reaction.[17] Recently, additional luminescence assay formats have been developed using substrates that liberate luciferin upon action of the target enzyme and have been applied to proteases such as caspases[18] and cytochrome P450s.[19] As mentioned above, low background signals lead to very sensitive readout systems, but they can suffer from artefacts due to inhibition of the detection enzyme by the test compounds.

Colourimetric and Chromogenic Assays

Colourimetric assays are not used as frequently as fluorimetric assays for several reasons. First, colourimetric detection is usually less sensitive than the more widely used fluorimetric detection due to the dependency of the readout on the path length of light passing through the liquid sample (Beer's law), which limits the miniaturisation of such assays. However, clever methods to increase the apparent path length of light have been used to allow miniaturisation even into 1536-well plate formats with reaction volumes of only 10 μL.[20] The second constraint of colourimetric assays is that coloured compounds, which are present in most large compound libraries, can interfere with detection and lead to screening artefacts. This issue can be overcome by monitoring the enzyme reaction rate in kinetic mode instead of using a single endpoint readout. The change in reaction rate is less influenced by the presence of coloured compounds, as has been shown in screens using the relatively weak absorbance of NADPH[21] or nitrophenolate.[22]

Coupled Assays

Coupled assays have been used to monitor a variety of different enzymatic reactions and can be split into two types: chemically coupled and enzymatically coupled assays. In the former type, the product of the enzymatic reaction under study is detected by reaction with a reactive chemical to allow easy detection of the analyte of interest.[23] In enzymatically coupled reactions, the products or substrates of the reaction of interest are acted upon by a second enzyme, creating a tangible readout. One advantage of using such a standardised assay format is that the development of such a method allows the activity of a family of enzymes to be monitored by a single detection method.[24,25] The disadvantage of these coupled assay systems is that rigorous counterscreens are required to eliminate hits not acting directly upon the primary enzymatic reaction of interest.

One of the areas of assay development that has seen increased application is the use of enzyme complementation assays. While this has primarily been used in cell-based assays, a number of interesting applications of enzyme complementation for *in vitro* assays have also been reported.[26]

Radioisotope-based Assays

It has been anticipated for some time that biological assays based on radiolabelled substrates or ligands would be used less frequently for HTS.[27] However, the specificity with which substrates and ligands can be labelled with a radioisotope without perturbing their chemical properties and the high detection sensitivity of such assays ensures that radioactive tracers will remain an important tool for HTS. A number of well-developed technologies based on scintillation proximity, such as SPA beads or FlashPlates, allow complex biological assays to be conducted without the need for time-consuming separation steps. Technological advances such as the use of charged coupled device (CCD) based detection methods which monitor the activity of a whole assay plate at once have reduced the time required for conventional scintillation counting. These CCD-based plate readers, such as LeadSeeker™ and ViewLux™, also have the advantage that they allow the miniaturisation of assays to 1536-well formats. Other than the commonly used SPA and FlashPlates assay formats, traditional filter binding assays continue to be used for specific applications. There is a recent extensive review of this field by Glickman *et al.*[28]

Biophysical ("Label-free") Detection

All the technologies discussed so far rely on detection systems that are dependent on, or linked to, labels such as chromophores or radioisotopes. However, biophysics or "label-free" technologies are gaining importance in screening. Often their readout is more direct and relies on biophysical parameters. This offers several advantages that will complement existing assay technologies in the coming years. Methods are evolving such that they are already being used in productive screening as orthogonal or secondary readouts. Some of the assays are suitable for high or medium throughput screening.[29]

Recently, mass spectrometry coupled with liquid chromatography (LC/MS) based detection has demonstrated value as a label-free detection tool for HTS. By measuring the mass-to-charge (m/z) ratio of analytes, MS-based assays have the advantage of allowing generic, label-free detection of reaction substrates and products. The MS readout can focus on the substrate and product of the reaction and, therefore, gain high sensitivity even in complex reaction mixtures due to the chromatography step. A group at BioTrove has reported the use of LC/MS set ups for HTS.[30] In addition, Roddy *et al.*[31] have reported screening of several metabolic enzymes with LC/MS detection on a large scale. Desorption-based MS detection techniques for HTS applications are being developed and hold great promise.[32] MS-based HTS will play an increasingly important role in the coming years.

Chromatography-based assays that separate the bound and unbound ligands by affinity selection chromatography constitute a simple method to identify ligands binding a large biomolecule. Similar methods have been used previously for the detection and characterisation of small molecule binding to serum albumin.[33] This is similar to the SpeedScreen approach in which the target protein of interest is incubated with test compounds at high numbers.[34] The protein/compound complex is then separated from unbound compounds with a microtitre plate based chromatographic separation step. Even in the presence of the protein or other MS signals, the bound compounds can reliably be identified by focusing on the *m*/*z* signals from the known cocktail of test compounds with LC/MS.[34] Alternative approaches use a size-exclusion ultrafiltration step such that the protein–ligand complex remains trapped in the filter plate rather than flowing through,[35] as occurs with size-exclusion chromatography.[36] This screening method has been extended to allow screening of integral membrane proteins such as G-protein coupled receptors (GPCRs).[37] The major advantage of these methods is the obvious ease of set-up even in the absence of functional activity for the target, *e.g.* for orphan or new genomic targets. The disadvantage of these methods is that any binder, even promiscuous or non-productive, can be identified. Such artefacts have to be triaged out with other orthogonal readouts at a later stage.

Microcalorimetry has been used to monitor molecular interactions such as ligand binding by monitoring the enthalpic heat of interaction. Typically, isothermal titration calorimetry (ITC) and differential scanning calorimetry (DSC) have been used to measure the binding constants and thermostability of the protein–ligand complex. However, these traditional microcalorimetric methods have a very limited use in screening due to very low throughput and high protein consumption. Techniques to improve the throughput of calorimetric approaches include enthalpy arrays[38] and miniaturisation. A number of thermostability-based methods have been modified to allow screening of compounds with reasonable throughput. The ThermoFluor® affinity screening method is based on the fluorescence change seen upon binding of a dye to denatured *vs.* native protein. This assay technology has been shown to have a much higher throughput and lower protein consumption than other microcalorimetry methods.[39,40] Other approaches measure the change of fluorescence during the denaturation of a protein at a set temperature and the effect of ligands in altering the rate of denaturation,[41] or measure light scattering of these transitions without the need of an environmental sensitive fluorescent dye.[42,43]

NMR (nuclear magnetic resonance) is a versatile technology used in low or medium throughput screening. Different NMR readouts offer a variety of detection modes. The so-called ligand observation mode is used to monitor each compound (even in a mixture), assessing its true concentration (solubility) and integrity, as well as if the ligand is bound or not. In the protein observation mode, protein integrity (*e.g.* aggregation), concentration, the percentage of bound or unbound protein, and even which ligand the protein has bound to, can be determined. The binding state of the protein and each individual compound can be monitored directly in the mixture; even the binding site (with resonance assignments) and conformational changes of the protein can be assessed.[44,45] NMR can identify even extremely weak binder compounds, so it is often used for small molecular weight fragment screening.[46] The caveats of NMR detection are the relative low throughput at considerably high protein consumption. Depending on the detection mode, labelling of the protein with suitable isotopes may be necessary (*e.g.* ^{13}C or ^{19}F).

Immobilisation-based, label-free techniques utilise a variety of approaches and strategies; some where binding is subsequently detected by a labelled antibody or aptamer and some in which binding is detected due to changes in the surface properties of the immobilised target. The first types of immobilised assays include ELISA and ELISA-like methods. These methods require the availability of selective labelled binding partners such as antibodies or oligonucleotide aptamers for the protein of interest (for a review, see Nielsen and Geierstanger[47]). Another method for monitoring the effect of small molecule binding to proteins is frontal affinity chromatography coupled with MS detection (FAC-MS), first reported by Hage[48] and reviewed by Slon-Usakiewicz *et al.*[49] The drawback of this approach, as with all techniques requiring immobilisation, is that sufficient protein has to be immobilised onto the column and immobilised in a way that will not inhibit or influence the binding of small molecules. However, the use of MS to detect small molecule binding removes the need for labelled antibodies or aptamers.

One, relatively low throughput, embodiment of surface property-based detection methods is the Biacore instrumentation in which surface plasmon resonance (SPR) technology is used to detect molecular interactions with the immobilised target (reviewed by Lofas[50]). In the last few years, label-free detection systems have been developed and such assays have emerged as a new detection paradigm in HTS and related applications.[51,52] These include the development of evanescent wave technologies and interferometry; these have been adapted to allow the detection of binding at the surface of microtitre plates such as Corning's Epic™ system, the Bind™ system from SRU and Fortèbio's Octet™ system. Still other methods have been developed such as the use of acoustic sensing, the Resonant Acoustic Profiling (RAP™) system (Akubio) or wavelength-interrogated optical sensing (WIOS, CSEM Scientific). In these detection strategies, the change in the surface molecular properties detected is dependent on the size of the ligand interacting at the immobilisation surface.[53] Some of these technologies are already adapted to 384-well plate format suitable for medium to high throughput screening applications.

Whereas those technologies have been applied and optimised over the last decade to investigate binding events of relative large binding partners on typically surface immobilised ligands (*e.g.* protein–protein or antigen–antibody interactions), technologies are now being used to screen for specific binding of small molecular weight compounds (MW <400 Da) on large targets (*e.g.* proteins with MW >30 000 Da).

One key challenge is the lack of sufficiently sensitive analytical tools to probe the real-time relationship between structural change and molecular function or behaviour. The dual polarisation interferometry (DPI) from Farfield's AnaLight® technology studies measures optical interference patterns caused by changes in the structure and/or mass of immobilised molecules. Thus, DPI gives insights into the structural changes taking place in molecular systems as they function and interact.[54,55]

The necessity of immobilisation for many of the above-mentioned technologies can significantly limit their application. Recently free-solution, label-free molecular interactions were investigated with back-scattering interferometry (BSI) in an optical train composed of a helium–neon laser, a microfluidic channel and a position sensor.[56] Molecular binding interactions between proteins, ions and protein as well as small molecules and protein could be monitored without labelling or immobilising any of the interaction partners.

Even if most of these technologies do not become suitable for high throughput screening, they still have a very important role in drug discovery as secondary assays. With these detection methods, important binding parameters such as K_D, binding kinetics (compound on- and off-rate) and thermodynamics (ΔH, ΔG, ΔS) can be investigated and used to evaluate lead compounds (*e.g.* compounds with a slow off-rate may be favoured for long lasting compound effects).

2.2 Cell-based Assays

Cell-based assays have been important tools for finding drug candidates from the days of screening natural product extracts for antibacterial activity.[57] With the development of molecular biology and advances in cell culture, many more ways to monitor specific biological processes in cultured cells have become possible.

The use of cell-based assays in HTS for drug discovery has increased steadily over the past several years. Cell-based screening, in which the target activity is directly assessed in its cellular context, can improve the biological relevance of active compounds compared with screening isolated biochemical targets. Various cell-based assays such as reporter gene assays, secondary messenger assays, cell-based enzyme-linked immunosorbent assays, cell-based proximity assays and pathway screening assays are used for HTS. Other assays use visualisation tools to monitor the state and location of cellular constituents by using specific binding reagents, or use indirect technologies that monitor more general aspects of cell behaviour such as cellular growth, cell viability, shape or electrical conductivity.

Cell Growth

Advances in commonly used assays monitoring cell growth and proliferation now include systems to allow multi-target screens. For example, antimicrobial screens have been implemented by using the protection engendered by the active compounds to rescue mammalian cell growth.[58] Other similar systems have been developed using other model organisms, such as the co-culture systems developed by companies such as Athelas (now part of MerLion Pharmaceuticals) and others.[59,60] Such assay systems have the potential to identify compounds acting by specific inhibition of microbial growth. In addition, such assays could be used to identify compounds that act by interfering with bacterial or host functions essential for survival and pathogenesis of the pathogen.

Reporter Gene Assays

Reporter gene assays monitor changes in the expression levels of a particular gene. These methods use specific enzymes or fluorescent proteins whose expression level is controlled by a promoter or RNA sequence that regulates the expression of a gene of interest. Multiple reporter genes and cell lines have been used in antibiotic drug discovery,[61] as well as to help define the mechanism of action of compounds.[62] Technology development for reporter gene assays in mammalian cell systems has continued, with a wide range of technologies now being available. Some of the more commonly used reporter assay systems are highlighted below.

One recent improvement in the available reporter gene systems for mammalian cell assays was the development of the bacterial β-lactamase, bla, in conjunction with the CCF2-AM dye, *i.e.* the GeneBlazer™ system. This cephalosporin-based dye contains a set of acetate ester groups that are removed by intracellular esterases, resulting in intracellular accumulation of the dye. The dye molecule contains a FRET pair, allowing ratiometric estimates of bla activity. The system has been shown to work with a wide variety of different pharmaceutical targets including GPCRs,[63] or to allow detection of anti-viral compounds.[64]

Another reporter is based on a bacterial nitroreductase (NTR) and a cell permeable cyanine fluor, CytoCy5S, as its substrate. The nitro groups in the substrate can be reduced by the enzyme to generate a fluorescent product which is then retained within the cell.[65] One advantage of this reporter is the use of a red-shifted substrate, leading to less compound or cell autofluorescence interference. Also, the substrate CytoCy5S is reported to be photo-stable and soluble in water. The biggest difference of this system from the bla system may be that the product of the nitroreductase can be fixed in the cell. This can simplify scheduling issues during screening, allowing cells to be fixed and the NTR activity measured later. The ability to fix cells expressing this reporter can even make NTR compatible with high content imaging based assays.

Although secreted alkaline phosphatase (SEAP) does not allow the monitoring of intracellular location, there are a number of sensitive chemiluminescense-based assays for SEAP, allowing very low levels of gene expression to be monitored. The secretion of the reporter, as well as the stability of the SEAP enzyme, allows the accumulation of the reporter to be monitored over time. Recently, a SEAP reporter-based system has been developed for screening for effectors of the signal peptide translation and secretion process.[66]

Perhaps the most commonly used reporter gene systems are those based on luciferases. These assays have been widely accepted because of their sensitivity and large dynamic range. In addition, the availability of photon multiplier tube (PMT) based readers in most laboratories, or sensitive CCD cameras, means that such assays can be developed readily. Recently, there have been further developments in this area with the availability of the Chroma-Luc system in which two versions of the click beetle (*Pyrophorus plagiophalam*) luciferase have been cloned into expression vectors. These enzymes are essentially identical except for one amino acid, which causes one variant to emit a predominantly green light while the second mutant emits a predominantly blue light. By addition of only one detection reagent, it is possible to monitor both the specific target signal as well as the control signal. Luciferase systems can even use substrates that allow real-time, non-invasive and non-destructive monitoring of luciferase activity such as coelenterazine for the Renilla luciferase[67] or luciferin with firefly luciferase.[68] Luciferase can also be used to detect close proximity of protein partners using bioluminescence resonance energy transfer (BRET) (see below).

Enzyme complementation reporter assays can monitor translocation events causing both fragments of an enzyme to be present in the same cellular location, leading to formation of the active enzyme. This principle has been used for development of sensitive translocation assays with luciferase[69] or β-galactosidase. A recent development is based on the biomolecular fluorescence complementation by the formation of a fluorescent protein from two non-fluorescent fragments, *i.e.* yellow fluorescent protein (YFP) or variants of the green fluorescent protein (GFP).[70,71] This split GFP may also be useful for detecting protein–protein interactions *in vivo*, similar to the luciferase fragment complementation systems.

GFP has been widely used as fluorescent tracer and reporter for cellular imaging and for high content screening. GFP-fused to β-arrestin has been engineered to allow imaging of the translocation of GPCRs after activation and translocation into endosomes,[72] which can be monitored by sub-cellular imaging instruments (see below). Further translocation events can be monitored by GFP fusion to target proteins or transcription factors.[73,74] The advantage of translocation assays is the possibility to screen compounds on specific pathways in a functional environment.[75,76]

FRET and Bioluminescence Energy Transfer (BRET)

FRET can also be used for detection of cell metabolites, such as cAMP, with cell-based assays.[10]

BRET is based on the non-radiative transfer of energy between a bioluminescent donor protein (*e.g.* firefly or *Renilla reniformis* luciferase) and a fluorescent acceptor protein, most often a variant of GFP.[77,78] BRET is a technique to measure protein–protein interactions in as near a physiological environment as possible. GPCRs and β-arrestin were re-engineered to incorporate derivatives of *Renilla* luciferase (Rluc) and green fluorescent protein (GFP2), respectively. The interaction of the GFP2 arrestin fusion with an agonist-activated Rluc-GPCR fusion brings Rluc and GFP2 into close enough proximity to allow resonance energy transfer to occur.[79,80] The advantages of BRET include the possibility of live cell analysis.[77]

Cell-based ELISA

Cell-based ELISAs are a sensitive method for detecting and quantifying cellular proteins including post-translational modifications associated with cell activation (*e.g.* phosphorylation and degradation) without making extracts or performing electrophoresis and membrane blotting. Such assays can use colourimetric or chemiluminescent formats as with standard ELISA assays. These methods have the advantage of the wide availability of commercial antibodies to cell signalling events. However, the use of such assays in high throughput screening is still limited by the requirement of several washing steps. In fact, this assay format can be seen as the precursor to a number of newer assay technologies that seek to perform immuno detection of analytes from whole cells, such as the SureFire®[81] and LI-CQR® technologies.[82] The SureFire® assays are based on the AlphaScreen™ assay format, a homogeneous screening technology (see above). These newer methods have the advantage of being plate-based, homogeneous assay formats, but still require careful optimisation of the cell lysis and antibody binding conditions.

Kinetic Imaging Plate Reading

The FLIPR (Molecular Devices) and FDSS6000 (Hamamatsu) plate imagers enable the kinetic measurement of all wells by simultaneously imaging the whole plate at once. These readers are equipped with parallel non-contact dispensers able to inject reagents into the assay wells and are widely used to measure cytosolic Ca^{2+} flux elicited by activation of GPCRs or some ion channels. On-line injection enables real-time kinetic analysis with both adherent and non-adherent cells. The development of non-wash dyes (*e.g.* Calcium-4) and instrumentation such as FLIPR^{tetra} (Molecular Devices) and FDSS7000 have made it possible to miniaturise such assays to 384-well and 1536-well formats.[83,84] The use of voltage-sensitive indicator dyes also allows screening assays for a wider variety of ion channels.[85]

High throughput flash luminescence readers such as the FDS6000 and 7000, as well as the Lumax Flash HT, enable functional GPCR and calcium channel testing. A sensitive photon-counting CCD camera enables aequorin and luciferase activity to be measured by flash luminescence. Thus, these advances allow fluorescence-based assays to be replaced by luminescence-based assays, which have the advantages of greater sensitivity, lower background and less compound interference.

Automated Electrophysiology Assays

Recently developed technologies allowing automated electrophysiology, such as IonWorks Quattro™, PatchXpress™ (Molecular Devices) and QPatch (Sophion Inc.) have profoundly changed the ion channel research field.[86–88] These new, automated, electrophysiology platforms are able to fill the gap between FLIPR type assays and conventional patch clamp assays, not only by significantly increasing the throughput, but also by using more relevant testing conditions. Although the throughput for such technologies is not high enough for conventional HTS, it is feasible to run screens of focused compound libraries against ion channel targets with a direct electrophysiology readout.

High Throughput Flow Cytometry

High throughput flow cytometry has been implemented for HTS applications with the HyperCyt. The platform can be applied to general cell-based assays and particle-based multiplexed approaches.[89] The advantages this system offers are the ability to use the many pre-existing flow cytometry assays and the rich multidimensional data sets that have been developed for flow cytometers. The challenges that remain are the need for cells to be in suspension, the limited throughput of the system and the difficulty of sample preparation due to the washing steps when using antibody staining. In addition, the large number of cells needed for the assays due to liquid handling constraints and dead volumes may also be a limitation.

Sub-cellular Imaging and High Content Screening (HCS)

HCS instruments take microscope-based images of a limited region of each microtitre plate well with sub-cellular resolution. Proteins of interest can be detected by a fluorescent tag, such as GFP, or are identified by fluorescent antibodies. One of the characteristics of HCS assays is the ability to not only monitor the levels of different proteins within the cells but also their location, the morphology and shape of the cells by multiplexing several labels.[90–92]

Sub-cellular imaging instruments are automated (fluorescence) microscopes with integrated automated image analysis. Such instruments can be divided into two broad classes:[93]

- wide-field imagers, *e.g.* Cellomics ArrayScan (Thermo Scientific Cellomics);
- confocal imagers, *e.g.* the Opera (Perkin Elmer)

Such systems were originally being designed for secondary assays, but the increased application of sub-cellular imaging and the development of instruments capable of reading 1536-well plates, *e.g.* the Opera (Perkin Elmer), has enabled such assays to be used for primary screening. As the imaging systems have been adapted to allow HTS, plate preparation, involving washing steps, now limits the throughput of such assays.[94]

Instruments such as the Acumen (TTP Labtech)[95] or IsoCyte (Blueshift Biotechnologies) readers use an alternative approach for HCS. These instruments do not take microscopic images of the cells, but rather use lasers to scan the bottom of the plate, recording the peaks of fluorescence intensity along the scanning path. While it is possible to use these fluorescence peak intensities to create pseudo images, these are not actually required for analysis. The advantage of such methods is that the image recognition steps are no longer required.

One advantage of cell-based assays using sub-cellular imaging is the ability to select active molecules based on a cellular phenotype identified by image analysis of, for example, morphology changes.[96] Therefore, the screen does not require *a priori* knowledge of any specific biochemical target being regulated by the active compounds. The ability to perform an HTS campaign based on cellular phenotype allows the discovery of novel and unexpected compounds, but creates the issue of later defining the mechanism of action of such active compounds.[97] Image analysis, based on each cell, offers the possibility to identify sub-populations within a well that might be respondent to a compound dependent on its cell-cycle state. One of the limitations of these assays is the need for several washing steps, as well as appropriate labelling reagents in order to stain the cells for imaging. Alternatively, cell lines with fluorescence-tagged proteins have to be created.

Another emerging trend in HCS is multiplexed readout. With an appropriate image analysis algorithm, a combination of the most relevant parameters (often deduced after unbiased analysis of all the possible readouts using tools such as principal component analysis) can be used to better assess the action of each hit, eliminating false positive hits. One of the challenges for any multiplexed readout is the development of high throughput methods for analysis of such large data sets.[98]

Cell-based Label-free Readouts

Recently developed cell-based, label-free monitoring systems include electro impedance-based sensors, *e.g.* the CellKey system (MDS Sciex) based on cellular dielectric spectroscopy (CDS) or the RT-CES™ (real-time cell electronic sensing) System from ACEA Biosciences.[99] These sensor systems use impedance-based measurements to detect changes in the electronic properties or the passage of ions through cells. Such changes in the properties of cells can be brought about by several types of receptors including GPCRs and tyrosine kinase receptors. In a similar manner, the Epic (Corning), which measures refractive index changes, allows the detection of cellular changes at the surface of microtitre plates.[100,101]

For all these systems, cells are grown in the individual, sensor-containing wells of the microtitre plates. Changes to the biological status of the cells are measured automatically in real-time. Although these systems have been adapted to 96-well or even 384-well plate formats, they have only limited throughput at present. In addition, direct interpretation of the results in terms of the underlying molecular events is still not possible. These assays may be used in target validation and as orthogonal assays for evaluation of lead compounds because such methods can even use small numbers of primary non-engineered cells.

Multiplex mRNA Detection Assays

Another important set of multiplexed assays monitor mRNA transcript levels. The expression level of all the genes involved in a known signal transduction pathway or other selective genes can be monitored simultaneously as a way of following compound effects on a cell. The current technologies for multiple mRNA detection include quantitative reverse transcriptional PCR (qRT-PCR), qNPA (quantitative nuclease protection assays), mass array assay technologies and branched DNA detection on Luminex beads (Panomics). The applications of such multiplexed *in vitro* and cell-based detection systems should provide more predicative information in hit finding and lead characterisation.

2.3 Modelling to Identify False Positives and Negatives

Biological assays inherently generate false negatives as well as false positives. Significant effort has been invested in new technologies to identify and eliminate hits due to assay artefacts (false positives). Independently retesting samples is, in effect, the only way to identify false negatives. However, data mining offers an alternative route for rescuing false negatives. A number of approaches have been proposed using primary screening data to generate models for predicting the activity of compound samples. Such models are then used to predict the activity of individual compounds so that some of the hits missed by screening can be rescued by retesting. A number of different model building methods have been applied, including: logistic regression,[102] clustering[103] and Bayesian modelling.[104] In fact, the greater amount of screening data can allow the derivation of models as predictive as models derived using validated IC_{50} results.[105] Such methods can also find application in genetic screens of small interfering RNAs (siRNAs)[106] and the evaluation of gene chip expression profiling.[107] All of these methods, in a sense, make the assumption that, by "averaging" the activity of similar samples (*e.g.* characteristics or substructures depending on the type of descriptors), the potential activity of a compound can be predicted. Obviously, such methods can only be used when the structures of the compounds in a sample are known.

Prediction of Mechanism and Off-target Effects

The modelling and mining of HTS data is not limited to identifying active hits. Such methods have been used to predict the possible mechanism of action of active compounds, especially in cell-based assays where multiple targets may bring about the observed biological effect.[108,109] As with approaches to predict activity against a single target, these attempts to predict the mechanism of action of a compound can rely on different modelling methods. Essentially, these methods first create multiple models predicting the activity of compounds against a specific target or target class. The models in which the compound is predicted to be active are then noted as possible targets. Such methods have been used also to identify and explain the activity of compounds identified as frequent hitters in reporter gene assays[110] and thus, may help identify hits most suitable for lead optimisation.

3 Emerging Trends

The "hits" generated from HTS campaigns in the past are slowly but surely starting to have an impact on the compounds progressing through the clinic.[111] HTS capacity and throughput is now not perceived as a limiting factor in lead discovery. As a consequence, screening campaigns are no longer regarded as just an effort to screen as many compounds as possible. Instead "smart screening" approaches of testing focused or designed sub-libraries are being used more frequently. How to construct the most effectively diverse or target specific sub-libraries is still heavily debated. The structural diversity and well-annotated effects of natural products will continue to act as a source of samples for screening both for the discovery of novel lead compounds and as tool compounds to probe biology.

3.1 New HTS Approaches

HTS is facing new challenges and needs to adopt new approaches to improve its efficiency and predictability during early lead identification and the drug discovery process. Therefore, the quality of HTS assays needs to continually improve.

Improvement in Screening Libraries

The selection and quality of a screening library with drug-like and lead-like structures is a critical endeavour. The features of drug-like and lead-like structures continue to be better defined, at the same time as the diversity of drug-like and lead-like molecular space continues to be explored and categorised. Other areas of development focus on the discovery of small molecules suitable for modulating protein–protein interactions, with a greater focus on natural product-like compounds.

Combination of Potency Screening with Early ADME/Tox Testing

The hits from potency-based screens should be evaluated, as early as possible, with a set of profiling (or selectivity) assays and toxicology testing from early

ADME/Tox (adsorption, distribution, metabolism and excretion/toxicology). Such characterisation will allow prioritisation of HTS hits and thus, enhance the chance of focusing on the chemotypes that may have a lower attrition rate in the later stage of development.

Combining and Exploring Novel Screening Strategies

This may include evaluating strategies such as iterative screening in combination with strategies using lead modelling. Such iterative and combined screening strategies may reduce the apparent throughput and efficiency of screening (in terms of compounds screened per day) but may lead, overall, to a more effective hit-to-lead discovery process.

Applying HTS in New Target Areas

The needs of drug discovery will continue to drive the need for screening to be applied to new biological systems in order to support current drug discovery efforts. High throughput screening is a tool and an approach that has not been limited to its use in drug discovery but has also been applied in fields as diverse as genomic and proteomic discoveries, protein engineering and the selection of chemical catalysts.

Increase in Academia Outreach and Collaboration

HTS is no longer limited to industrial settings and is starting to play an increasingly important role in academic research.[112,113] With the new National Institutes of Health (NIH) initiatives in the USA to support a number of screening Molecular Libraries Probe Production Center Network (MLPCN) centres (http://ncgc.nih.gov/index.html) for academic research, it can be expected that the mutual impact of HTS and basic biology will increase.

Assay Miniaturisation

The pressure of cost constraints and throughput will continue to drive the need for assay miniaturisation. Plate-based assay technologies will continue to be the dominant assay format, but non-plate screening technologies may come to have more impact in future years. It is also possible that microfluidic assays and chip-based assay technologies will lead the miniaturisation of HTS assays into new dimensions.

Currently, liquids are most conveniently handled by pipette tips for aspiration and dispensing into microtitre plates. Recently, acoustic energy driven liquid dispensing devices have been available (*e.g.* the ECHO550 or EDC system), such that liquid can be transferred in small volume (nanolitre to μL) accurately and directly from the wells on one plate into wells on another plate.

With the continued drive for assay miniaturisation, non-contact liquid handling devices will play an increasingly important role. The advantage of these non-tip liquid transfer machines is to eliminate the need for tip washing and reduce the chance of liquid cross-contamination. So far, these devices have been used mostly for dispensing dimethyl sulfoxide (DMSO) compound solutions, making it possible to perform assays without the need for pre-dilution steps. Increasing the speed and throughput of such transfer will further increase the number of assays that can be miniaturised.

Advances in Multiplexed Readouts

Another emerging trend in HTS detection is to have increasingly multiplexed readouts. The traditional screening assay only monitors one or two signals of the test system. With some specialised assay detection technologies, multiple readouts can be obtained. For example, a multi-module detector can simultaneously detect the fluorescence intensity, fluorescent life-time, FP or FRET signals. Analysis of the multiplexed readouts can provide information to discriminate a hit as either a true inhibitor or as an artefact due to some interference with one of the detection modes. In a similar vein, high content imaging analysis-based screening (HCS) also provides multiple readouts depicting cellular changes. The expression level of all the genes involved in a known signal transduction pathway or relevant selective genes can be monitored simultaneously. The applications of such multiplexed *in vitro* and cell-based detection systems should provide more predicative information in hit and lead finding. One extension of this trend, that may have a profound impact, is the use of primary cells or stem cell systems to make the assay as physiologically relevant as possible.

Advances in Natural Product Screening

In addition to the new assay and detection technologies, there have also been a number of advances in natural product screening (many of which are reviewed in more detail in Chapter 9). These advances can be thought of as either technical (such as improvements in "hyphenated" technologies) or biological (such as new ways to identify sources of novel compounds or to express and modify compounds of interest). There is a constant improvement in chromatography techniques allowing an ever increasing improvement in the capacity and resolution of compounds isolated from complex mixtures, *e.g.* countercurrent chromatography.[114] Technology has advanced significantly over the last few decades so that, in combination with novel chromatography techniques, methods for structure elucidation have improved. It is now possible to define the structure of novel compounds with as little as 10–20 μg of compound with some of the methods, such as NMR, being non-destructive. In fact, it is possible to characterise compounds to a large extent even during purification.[115] This not only has an influence on determining the structure of compounds from natural product mixtures, but also on the ability of researchers to avoid the problem of re-isolation of compounds.

In addition, there have also been a number of efforts to find more efficient ways to categorise and enumerate compounds. One approach to defining the relationship of compounds to each other uses a chemical space defined by a set of ~500 compounds described with 60 molecular descriptors onto which novel compounds can then be mapped.[116] Another approach that has been described uses a rule-based classification scheme to define the common scaffold or core of compounds.[117] This approach has been taken further to develop a scoring function to allow the prioritisation of compound libraries for testing.[118]

Possibly one of the most interesting advances in biological assays for natural product screening has been the development of "chemical–genetic interaction" profiles using yeast.[119] This method profiles the sensitivity of a panel of viable yeast haploid strains to allow identification of the possible gene target of compounds. The exciting aspect of this approach is the fact that the target of the most potent compound present in a crude extract mixture can be identified, thus making it possible to profile extracts for interesting biological activity before fractionation and dereplication. This approach is conceptually similar to looking at the differential activity on cells in which a target gene has been depleted, thus making the cell line depleted in the target more sensitive to inhibition. This approach is being used with great success in screening for antibiotics.[120,121] There is even the potential this method could be applied to screening mammalian systems using siRNA to sensitise cells.

While each of these advances on their own may appear incremental, added together they constitute significant advances in biological screening. However, the main challenges facing HTS will remain the implementation of robust, sensitive assays that can be efficiently automated, but still accurately reflect the biology they seek to explore.

Acknowledgements

We would like to thank Dr S. Siehler for help in describing the reporter gene assays technologies, Dr Q. Lu for help in describing the ion channel assays and Dr G. Scheel for help in proof-reading the manuscript.

References

1. R. Macarron, *Drug Discov. Today*, 2006, **11**, 277.
2. K. S. Lam, *Trends Microbiol.*, 2007, **15**, 279.
3. D. A. Pereira and J. A. Williams, *Br. J. Pharm.*, 2007, **152**, 53.
4. D. J. Newman, *J. Med. Chem.*, 2008, **51**, 2589.
5. K. R. Oldenburg, I. Kariv, J.-H. Zhang, T. D. Y. Chung and S. Lin, in *Handbook of Drug Screening.*, ed. R. Seethala and P. Fernandes, Marcel Dekker, New York, 2001, pp. 525–562.

6. W. P. Janzen, *High Throughput Screening, Methods & Protocols*, Humana Press, Totowa, NJ, 2002, Methods in Molecular Biology Vol. 190.
7. A. J. Pope, U. M. Haupts and K. J. Moore, *Drug Discov. Today*, 1999, **8**, 350.
8. C. Eggeling, L. Brand, D. Ullmann and S. Jager, *Drug Discov. Today*, 2003, **8**, 632.
9. I. Hemmila and S. Webbs, *Drug Discov. Today*, 1997, **2**, 373.
10. D. Gabriel, M. Vernier, M. J. Pfeifer, B. Dasen, L. Tenaillon and R. Bouhelal, *Assay Drug Dev. Technol.*, 2003, **1**, 291.
11. J. Dunne, H. Reardon, V. Trinh, E. Li and J. Farinas, *Assay Drug Dev. Technol.*, 2004, **2**, 121.
12. D. Perrin, T. Martin, Y. Cambet, C. Fremaux and A. Scheer, *Assay Drug Dev. Technol.*, 2006, **4**, 185.
13. S. Miraglia, E. E. Swartzman, J. Mellentin-Michelotti, L. Evangelista, C. Smith, I. I. Gunawan, K. Lohman, E. M. Goldberg, B. Manian and P. M. Yuan, *J. Biomol. Screen.*, 1999, **4**, 193.
14. J. Seideman and D. Peritt, *J. Immunol. Methods*, 2002, **267**, 165.
15. X. Wu, M. A. Sills and J. H. Zhang, *J. Biomol. Screen.*, 2005, **10**, 581.
16. R. Golla and R. Seethala, *J. Biomol. Screen.*, 2004, **9**, 62.
17. M. Koresawa and T. Okabe, *Assay Drug Dev. Technol.*, 2004, **2**, 153.
18. J. J. Liu, W. Wang, D. T. Dicker and W. S. El-Deiry, *Cancer Biol. Ther*, 2005, **4**, 885.
19. M. F. Yueh, M. Kawahara and J. Raucy, *Drug Metab. Dispos.*, 2005, **33**, 38.
20. P. Zuck, G. T. O'Donnell, J. Cassaday, P. Chase, P. Hodderr, B. Strulovici and M. Ferrer, *Anal. Biochem.*, 2005, **342**, 254.
21. M. Zolli-Juran, J. D. Cechetto, R. Hartlen, D. M. Daigle and E. D. Brown, *Bioorg. Med. Chem. Lett.*, 2003, **13**, 2493.
22. R. Iyer, A. A Barrese III, S. Parakh, C. N. Parker and B. C. Tripp, *J. Biomol. Screen.*, 2006, **11**, 782.
23. D. Wahler and J. L. Reymond, *Agnew. Chem., Int. Ed. Engl.*, 2002, **41**, 1229.
24. C. Wang, S. Leffler, D. H. Thompson and C. A. Hrycyna, *Biochem. Biophys. Res. Comm.*, 2005, **331**, 351.
25. B. Zhang, D. Senator, C. J. Wilson and S. C. Ng, *Anal. Biochem.*, 2005, **345**, 326.
26. T. Naqvi, A. Lim, R. Rouhani, R. Singh and R. M. Eglen, *J. Biomol. Screen.*, 2004, **9**, 398.
27. L. Zemanova, A. Schenk, M. J. Valler, G. U. Nienhaus and R. Heilker, *Drug Discov. Today*, 2003, **8**, 1085.
28. J. F. Glickman, A. Schmid and S. Ferrand, *Assay Drug Dev. Technol.*, 2008, **6**, 433.
29. K. Shiau, M. Massari and C. C. Ozbal, *Comb. Chem. High Throughput Screen.*, 2008, **11**, 231.
30. C. C. Ozbal, W. A. LaMarr, J. R. Linton, D. F. Green, A. Katz, T. B. Morrison and C. J. Brenan, *Assay Drug Dev. Technol.*, 2004, **2**, 373.

31. T. P. Roddy, C. R. Horvath, S. J. Stout, K. L. Kenney, P. I. Ho, J. H. Zhang, C. Vickers, V. Kaushik, B. Hubbard and Y. K. Wang, *Anal. Chem.*, 2007, **79**, 8207.
32. Z. Shen, E. P. Go, A. Gamez, J. V. Apon, V. Fokin, M. Greig, M. Ventura, J. E. Crowell, O. Blixt, J. C. Paulson, R. C. Stevens, M. G. Finn and G. Siuzdak, *Chembiochem*, 2004, **5**, 921.
33. J. T. Wong and L. S. Porter, *Anal. Biochem.*, 1983, **130**, 491.
34. H. Zehender, F. Le Goff, N. Lehmann, I. Filipuzzi and L. M. Mayr, *J. Biomol. Screen.*, 2004, **9**, 498.
35. K. M. Comess, M. E. Schurdak, M. J. Voorbach, M. Coen, J. D. Trumbull, H. Yang, L. Gao, H. Tang, X. Cheng, C. G. Lerner, J. O. McCall, D. J. Burns and B. A. Beutel, *J. Biomol. Screen.*, 2006, **11**, 743.
36. D. A. Annis, N. Nazef, C.-C. Chuang, M. P. Scott and H. M. Nash, *J. Am. Chem. Soc.*, 2004, **126**, 15495.
37. C. E. Whitehurst, N. Nazeef, D. E. Annis, Y. Hou, D. M. Murphy, P. Spacciapoli, Z. Yao, M. R. Ziebell, C.-C. Cheng, G. W. Shipps, J. S. Felsch, D. Lau and H. M. Nash, *J. Biomol. Screen.*, 2006, **11**, 194.
38. F. E. Torres, P. Kuhn, D. De Bruyker, A. G. Bell, M. V. Wolkin, E. Peeters, J. R. Williamson, G. B. Anderson, G. P. Schmitz, M. I. Recht, S. Schweizer, L. G. Scott, J. H. Ho, S. A. Elrod, P. G. Schultz, R. A. Lerner and R. H. Bruce, *Proc. Natl. Acad. Sci. USA*, 2004, **101**, 9517.
39. M. D. Cummings, M. A. Farnum and M. I. Nelen, *J. Biomol. Screen.*, 2006, **11**, 854.
40. M. W. Pantoliano, E. C. Petrella, J. D. Kwasnoski, V. S. Lobanov, J. Myslik, E. Graf, T. Carver, E. Asel, B. A. Springer, P. Lane and F. R. Salemme, *J. Biomol. Screen.*, 2001, **6**, 429.
41. D. E Epps, R. W. Sarver, J. M. Roger, J. T. Herberg and P. K. Tomich, *Anal. Biochem.*, 2001, **292**, 40.
42. M. Vedadi, F. N. Niesen, A. Allali-Hassani, O. Y. Fedorov, P. J. Finerty, G. A. Wasney, R. Yeung, G. Arrowsmith, L. J. Ball, H. Berglund, R. Hui, B. D. Marsden, P. Nordlund, M. Sundstrom, J. Weigelt and A. L. Edwards, *Proc. Natl. Acad. Sci. USA*, 2006, **103**, 15835.
43. G. A. Senisterra, E. Markin, K. Yamazaki, R. Hui, M. Vedadi and D. E. Awrey, *J. Biomol. Screen.*, 2006, **11**, 940.
44. W. Jahnke and H. Widmer, *Cell. Mol. Life Sci.*, 2004, **61**, 580.
45. M. Pellecchia, I. Bertini, D. Cowburn, C. Dalvit, E. Giralt, W. Jahnke, T. L. James, S. W. Homans, H. Kessler, C. Luchinat, B. Meyer, H. Oschkinat, J. Peng, H. Schwalbe and G. Siegal, *Nat. Rev. Drug Discov.*, 2008, **7**, 738.
46. M. Schade, *IDrugs*, 2006, **9**, 110.
47. U. B. Nielsen and B. H. Geierstanger, *J. Immunol. Methods*, 2004, **290**, 107.
48. D. S. Hage, *J. Clin. Ligand Assay*, 1997, **20**, 293.
49. J. J. Slon-Usakiewicz, W. Ng, J. R. Dai, A. Pasternak and P. R. Redden, *Drug Discov. Today*, 2005, **10**, 409.

50. S. Lofas, *Assay Drug Dev. Technol.*, 2004, **2**, 407.
51. J. Comley, *Drug Discov. World*, 2004, Winter 2004/5, 63.
52. M. Cooper, *Anal. Bioanal. Chem.*, 2003, **377**, 834.
53. B. T. Cunningham and L. Laing, *Expert Rev. Proteomics*, 2006, **3**, 271.
54. G. Cross, A. Reeves, S. Brand, J. Popplewell, L. Peel, M. Swann and N. Freeman, *Biosen. Bioelectron.*, 2003, **19**, 383.
55. M. Swann, L. Peel, S. Carrington and N. Freeman, *Anal. Biochem.*, 2004, **329**, 190.
56. D. J. Bornhop, J. C. Latham, A. Kussrow, D. A. Markov, R. D. Jones and H. S. Sørensen, *Science*, 2007, **317**, 1732.
57. D. J. Newman, G. M. Cragg and K. M. Snader, *Nat. Prod. Rep.*, 2000, **17**, 215.
58. G. Kleymann and H. O. Werling, *J. Biomol. Screen.*, 2004, **9**, 578.
59. M. Benghezal, M. O. Fauvarque, R. Tournebize, R. Froquet, A. Marchetti, E. Bergeret, B. Lardy, G. Klein, P. Sansonetti, S. J. Charette and P. Cosson, *Cell Microbiol.*, 2006, **8**, 139.
60. T. I. Moy, A. R. Ball, Z. Anklesaria, G. Casadei, K. Lewis and F. M. Ausubel, *Proc. Natl. Acad. Sci. USA*, 2006, **103**, 10414.
61. H. P. Fischer, N. A. Brunner, B. Wieland, J. Paquette, L. Macko, K. Ziegelbauer and C. Freiberg, *Genome Res.*, 2004, **14**, 90.
62. E. B. Goh, G. Yim, W. Tsui, J. McClure, M. G. Surette and J. Davies, *Proc. Natl. Acad Sci. USA*, 2002, **99**, 17025.
63. P. Kunapuli, R. Ransom, K. L. Murphy, D. Pettibone, J. Kerby, S. Grimwood, P. Zuck, P. Hodder, R. Lacson, I. Hoffman, J. Inglese and B. Strulovici, *Anal. Biochem.*, 2003, **314**, 16.
64. P. Zuck, E. M. Murray, E. Stec, J. A. Grobler, A. J. Simon, B. Strulovici, J. Inglese, O. A. Flores and M. Ferrer, *Anal. Biochem.*, 2004, **334**, 344.
65. N. Thomas, *Drug Discov. World*, 2001, Winter 2001/2, 25.
66. J. Besemer, H. Harant, S. Wang, B. Oberhauser, K. Marquardt, C. A. Foster, E. P. Schreiner, J. E. de Vries, C. Dascher-Nadel and I. L. Lindley, *Nature*, 2005, **436**, 290.
67. S. Bhaumik and S. S. Gambhir, *Proc. Natl. Acad. Sci. USA.*, 2002, **99**, 377.
68. R. Paulmurugan and S. S. Gambhir, *Anal. Chem.*, 2005, **77**, 1295.
69. R. Paulmurugan and S. S. Gambhir, *Cancer Res.*, 2005, **65**, 7413.
70. S. Cabantous, T. C. Terwilliger and G. S. Waldo, *Nat. Biotechnol.*, 2005, **23**, 102.
71. T. K. Kerppola, *Nat. Rev. Mol. Cell Biol.*, 2006, **7**, 449.
72. D. A. Ross, S. Lee, V. Reiser, J. Xue, K. Alves, S. Vaidya, A. Kreamer, R. Mull, E. Hudak, T. Hare, P. A. Detmers, R. Lingham, M. Ferrer, B. Strulovici and F. Santini, *J. Biomol. Screen.*, 2008, **13**, 449.
73. B. K. Lundholt, V. Linde, F. Loechel, H. C. Pedersen, S. Møller, M. Præstegaard, I. Mikkelsen, K. Scudder, S. Petersen-Bjørn, M. Heide, P. O. G. Arkhammar, R. Terry and S. J. Nielsen, *J. Biomol. Screen.*, 2005, **10**, 20.

74. M. Simonen, Y. Ibig-Rehm, G. Hofmann, J. Zimmermann, G. Albrecht, M. Magnier, V. Heidinger and D. Gabriel, *J. Biomol. Screen.*, 2008, **13**, 456.
75. V. Unterreiner, Y. Ibig-Rehm, M. Simonen, H. Gubler and D. Gabriel, *J. Biomol. Screen.*, 2009, **14**, 59.
76. F. Zanella, A. Rosado, F. Blanco, B. R. Henderson, A. Carnero and W. Link, *Assay Drug Dev. Technol.*, 2007, **5**, 333.
77. N. Boute, R. Jockers and T. Issad, *Trends Pharmacol. Sci.*, 2002, **23**, 351.
78. K. D. Pfleger and K. A. Eidne, *Nat. Methods*, 2006, **3**, 165.
79. L. Bertrand, S. Parent, M. Caron, M. Legault, E. Joly, S. Angers, M. Bouvier, M. Brown, B. Houle and L. Menard, *J. Recept. Signal Transduct. Res.*, 2002, **22**, 533.
80. A. Prinz, M. Diskar and F. W. Herberg, *Chembiochem*, 2006, **7**, 1007.
81. H. J. Lee, H. C. Mun, N. C. Lewis, M. F. Crouch, E. L. Culverston, R. S. Mason and A. D. Conigrave, *Biochem. J.*, 2007, **404**, 141.
82. D. M. Olive, *Expert Rev. Proteomics*, 2004, **1**, 327.
83. I. C. Marshall, I. Boyfield and S. McNulty, *Methods Mol. Biol.*, 2006, **312**, 119.
84. P. Hodder, R. Mull, J. Cassaday, K. Berry and B. Strulovici, *J. Biomol. Screen.*, 2004, **9**, 417.
85. K. L. Whiteaker, S. M. Gopalakrishnan, D. Groebe, C. C. Shieh, U. Warrior, D. J. Burns, M. J. Coghlan, V. E. Scott and M. Gopalakrishnan, *J. Biomol. Screen.*, 2001, **6**, 305.
86. H. Zeng, J. R. Penniman, F. Kinose, D. Kim, E. S. Trepakova, M. G. Malik, S. J. Dech, B. Balasubramanian and J. J. Salata, *Assay Drug Dev. Technol.*, 2008, **6**, 235.
87. P. Moore, *Nature*, 2005, **438**, 699.
88. J. Kutchinsky, S. Friis, M. Asmild, R. Taboryski, S. Pedersen, R. K. Vestergaard, R. B. Jacobsen, K. Krzywkowski, R. L. Schroder, T. Ljungstrom, N. Helix, C. B. Sorensen, M. Bech and N. J. Willumsen, *Assay Drug Dev. Technol.*, 2003, **1**, 685.
89. S. M. Young, C. Bologa, E. R. Prossnitz, T. I. Oprea, L. A. Sklar and B. S. Edwards, *J. Biomol. Screen.*, 2005, **10**, 374.
90. D. L. Taylor, E. S. Woo and K. A. Guiliano, *Curr. Opin. Biotech.*, 2001, **12**, 75.
91. B. R. Conway, L. K. Minor, J. Z. Xu, J. W. Gunnet, R. DeBiasio, M. R. D'Andrea, R. Rubin, R. Biasio, K. Guiliano, L. Zhou and K. T. Demarest, *J. Biomol. Screen.*, 1999, **4**, 75.
92. D. L. Taylor and K. A. Guiliano, *Drug Discov. Today*, 2005, **2**, 149.
93. C. Smith and M. Eisenstein, *Nat. Methods*, 2005, **2**, 547.
94. E. A. Vaisberg, D. Lenzi, R. L. Hansen, B. H. Keon and J. T. Finer, *Methods Enzymol.*, 2006, **414**, 484.
95. W. P. Bowen and P. G. Wylie, *Assay Drug Dev. Technol.*, 2006, **4**, 209.
96. D. W. Young, A. Bender, J. Hoyt, E. McWhinnie, G. W. Chirn, C. Y. Tao, J. A. Tallarico, M. Labow, J. L. Jenkins, T. J. Mitchison and Y. Feng, *Nat. Chem. Biol.*, 2008, **4**, 59.

97. J. Low, S. Huang, W. Blosser, M. Dowless, J. Burch, B. Neubauer and L. Stancato, *Mol. Cancer Ther.*, 2008, **7**, 2455.
98. A. Kümmel, D. Gabriel, C. N. Parker and A. Bender, *Expert Opin. Drug Discov.*, 2009, **4**, 1.
99. K. Solly, X. Wang, X. Xu, B. Strulovici and W. Zheng, *Assay Drug Dev. Technol.*, 2004, **2**, 363.
100. K. Dodgson, P. Rawlings, I. Dale, M. Coldwell and S. Hill, Poster presentation at the Society for Biomolecular Sciences (SBS) Annual General Meeting, Seattle, 2006.
101. Y. Fang and A. Ferrie, *BMC Cell Biology*, 2007, **8**, 24.
102. M. F. Engels, L. Wouters, R. Verbeeck and G. Vanhoof, *J. Biomol. Screen.*, 2002, **7**, 341.
103. S. F. Yan, H. Asatryan, J. Li and Y. Zhou, *J. Chem. Inf. Model.*, 2005, **45**, 1784.
104. M. Glick, A. E. Klon, P. Acklin and J. W. Davies, *J. Biomol. Screen.*, 2004, **9**, 32.
105. M. Glick, J. L. Jenkins, J. H. Nettles, H. Hitchings and J. W. Davies, *J. Chem. Inf. Model.*, 2006, **46**, 193.
106. N. Friedman, M. Linial, I. Nachman and D. Peer, *J. Comput. Biol.*, 2000, **7**, 601.
107. Y. Xu and R. G. Brereton, *J. Chem. Inf. Model.*, 2005, **45**, 1392.
108. J. H. Nettles, J. L. Jenkins, A. Bender, Z. Deng, J. W. Davies and M. Glick, *J. Med. Chem.*, 2006, **49**, 6802.
109. S. F. Yan, F. J. King, Y. He, J. S. Caldwell and Y. Zhou, *J. Chem. Inf. Model.*, 2006, **46**, 2381.
110. T. J. Crisman, C. N. Parker, J. L. Jenkins, J. Scheiber, M. Thoma, Z. B. Kang, R. Kim, A. Bender, J. H. Nettles, J. W. Davies and M. Glick, *J. Chem. Inf. Model.*, 2007, **47**, 1319.
111. S. Fox, S. Farr-Jones, L. Sopchak, A. Boggs, H. W. Nicely, R. Khoury and M. Biros, *J. Biomol. Screen.*, 2006, **11**, 864.
112. S. M. Young, M. S. Curry, J. T. Ransom, J. A. Ballesteros, E. R. Prossnitz, L. A. Sklar and B. S. Edwards, *J. Biomol. Screen.*, 2004, **9**, 103.
113. N. Tolliday, P. A. Clemons, P. Ferraiolo, A. N. Koehler, T. A. Lewis, X. Li, S. L. Schreiber, D. S. Gerhard and S. Eliasof, *Cancer Res.*, 2006, **66**, 8935.
114. S. Wu, L. Yang, Y. Gao, X. Liu and F. Lui, *J. Chromatog. A*, 2007, **1180**, 99.
115. G. Bringmann, T. A. Gulder, M. Reichert and T. Gulder, *Chirality*, 2008, **20**, 628.
116. J. Larsson, J. Gottfries, L. Bohlin and A. Backlund, *J. Nat. Prod.*, 2005, **68**, 985.
117. A. Schuffenhauer, P. Ertl, S. Roggo, S. Wetzel, M. A. Koch and H. Waldmann, *J. Chem. Inf. Model.*, 2007, **47**, 47.
118. P. Ertl, S. Roggo and A. Schuffenhauer, *J. Chem. Inf. Model.*, 2008, **48**, 68.

119. A. B. Parsons, A. Lopez, I. E. Givoni, C. A Williams de Gray, J. Porter, G. Chua, R. Sopko, R. L. Brost, C. H. Ho, J. Wang, T. Ketela, C. Brenner, J. A. Brill, G. E. Fernandez, T. C. Lorenz, G. S. Payne, S. Ishihara, Y. Ohya, B. Andrews, T. R. Hughes, B. J. Frey, T. R. Graham, R. J. Andersen and C. Boone, *Cell*, 2006, **126**, 611.
120. J. G. Ondeyka, D. L. Zink, K. Young, R. Painter, S. Kodali, A. Galgoci, J. Collado, J. R. Tormo, A. Basilio, F. Vicente, J. Wang and S. B. Singh, *J. Nat. Prod.*, 2006, **69**, 377.
121. S. B. Singh, H. Jayasuriya, J. G. Ondeyka, K. B. Herath, C. Zhang, D. L. Zink, N. N. Tsou, R. G. Ball, A. Basilio, O. Genilloud, M. T. Diez, F. Vicente, F. Pelaez, K. Young and J. Wang, *J. Am. Chem. Soc.*, 2006, **128**, 11916.

CHAPTER 9

Advances in Instrumentation, Automation, Dereplication and Prefractionation

TIM S. BUGNI,* MARY KAY HARPER, MALCOLM W.B. MCCULLOCH AND EMILY L. WHITSON

University of Utah, Department of Medicinal Chemistry, 30 S. 2000 E. RM 307, Salt Lake City UT 84112, USA

1 Introduction

Many of the advances discussed in this chapter have been implemented to improve the quality of assay hits, shorten discovery timelines and increase the number of new chemical entities (NCEs); the number of NCEs recorded by the US Food and Drug Administration's Center for Drug Evaluation and Research (CDER) dropped from 53 in 1996 to just 17 in 2008.[1]

The demand for more efficient methods in natural product drug discovery stems from the fact that traditional methods relied on the time-consuming process known as bioassay-guided isolation. Although these methods have put many natural products into the clinic, today's timelines are shorter and the chance of encountering known compounds is higher. Therefore, researchers in the pharmaceutical industry, as well as academia, have embraced and developed methods that streamline the hit-to-lead process and improve the feasibility of natural products in high-throughput drug discovery programmes.

RSC Biomolecular Sciences No. 18
Natural Product Chemistry for Drug Discovery
Edited by Antony D. Buss and Mark S. Butler

Published by the Royal Society of Chemistry, www.rsc.org

High throughput screening (HTS) is notoriously sensitive with respect to the material being tested and, for the most part, has not been optimised for natural product mixtures.[2–7] Over the past decade, however, elegant examples have demonstrated that natural products can be integrated into high throughput screening campaigns. Examples have included developing high purity libraries,[8,9] partially fractionated libraries[5,10,11] and a few examples of HTS technologies that are compatible with mixtures of natural products.[12–14]

In many cases, technological advances in instrumentation have facilitated this change in natural product drug discovery. One cost-effective approach has been to develop microscale libraries for HTS.[8,15] Microscale libraries, compatible with HTS, can be generated on analytical scale to produce material that in the past would have been insufficient for complete characterisation by NMR (nuclear magnetic resonance). In particular, advances in NMR technology allow structures to be elucidated with only microgram quantities and have made it possible to generate large natural product screening libraries without the need to purify milligram quantities, thereby reducing time, cost and equipment needs. Improved high-performance liquid chromatography (HPLC) technology has resulted in efficient separations and has not only enhanced isolation and library generation, but has also made analytical analyses extremely rapid. The use of smaller particles in HPLC column packings and higher operating pressures has greatly improved resolution, speed and sensitivity. Advances in HPLC technology coupled with mass spectrometry (MS) have revolutionised sample throughput. Mass spectrometers have become widely available and improvements in accuracy and resolution have greatly reduced the time for dereplication and structure elucidation.

Integration of advances in NMR, MS and separation technologies has made dereplication a rapid, low-cost process and has greatly reduced the hit-to-lead timeline. Improvements in databases have also reduced the time for dereplication. For example, AntiMarin and MarinLit (see Section 2) have the option to search using a variety of parameters including functional groups. Since HSQC (heteronuclear single quantum coherence) and HMBC (heteronuclear multiple bond correlation) data can be acquired rapidly in NMR, many functional groups can be easily identified and used as parameters in database searches and can greatly enhance dereplication. Although still in its infancy, the ability to create searchable tandem mass spectrometry (MS/MS) fragment libraries provides the first potential avenue for automated dereplication.[16,17]

The main focus of this chapter is on technological advances that have impacted or have the potential to impact natural product discovery. Since there are a large number of technologies involved with natural product discovery, this chapter strives to cover areas that have not been broadly covered in recent reviews. In summary, advances that have the potential to streamline the hit-to-lead process, facilitate dereplication and support the high throughput paradigm are highlighted. In particular, this chapter looks primarily at separation technologies and advances in MS and NMR. Overall, we have highlighted technologies that can make major contributions within the context of a drug discovery programme.

2 Dereplication

The significance of efficient dereplication in optimising the hit-to-lead process in natural product drug discovery and development cannot be overestimated. The methodology is employed routinely to sort bioactive materials from HTS, to prioritise hits and to facilitate detection of previously encountered compounds whether published or recognised as unattractive (non-specific inhibitors and/or interfering nuisance compounds); early detection of replicates focuses efforts on novel, higher value chemical entities. Additionally, rapid dereplication is crucial for bioactive natural products to remain competitive with synthetic libraries in HTS campaigns.

Effective dereplication serves many other functions in streamlining the discovery process. This applies to natural products derived from extracts of plants, marine organisms or microbial fermentations, as well as complex neutraceuticals and dietary supplements. During the dereplication process, a complete structural assignment can often be made while avoiding unnecessary costly acquisition and detailed interpretation of spectroscopic data. Effective dereplication also affords comparison between samples to group like compound sources, to avoid redundancy and to discover new or alternative sources. Analyses of data across samples can aid in the identification of useful structural motifs, which in turn can facilitate analogue identification and prediction of drug-like utility.

Typically, dereplication is initiated with some analysis, chromatographic and/or spectroscopic, to recognise an active entity detected during HTS. Additional analyses are employed to rapidly establish the unambiguous identity of the compound. This fingerprint can then be used to search databases and reference libraries to link the structure to all chemical, spectral, bioactivity and pharmacokinetic data, as well as patent and publication information.

There are many tools available to fingerprint complex mixtures and define a unique signature for each component—including but not limited to infrared (IR), ultraviolet (UV), NMR and MS—often in conjunction with non-spectroscopic parameters and properties such as source organism taxonomy, chromatographic retention time and biological activity profiles.

Dereplication strategies and methods continuously evolve with advances in chromatography automation and spectroscopic technologies, both in terms of instrumentation and ease of automation. Automation of dereplication is often expedited by combining chromatography and spectroscopy, termed hyphenation. An inherent need for hyphenation is that unambiguous identification of a known natural product often requires high purity. Wolfender *et al.*[18] provided an exhaustive review of hyphenated techniques related to drug discovery from plant-derived natural products, but many of the discussed techniques are widely applicable to other source organisms.

Accelerated approaches to data mining strategies[19] involve sophisticated computer-driven data processing[20] such as the X-hitting algorithm[21] and cluster analyses.[22] Each facilitates the rapid identification of active compounds while providing a measure of chemical diversity and novelty, thus reducing the discovery timeline.

There are numerous natural product databases that are used in dereplication such as Chemical Abstracts Service (CAS), Berdy's Bioactive Natural Products Database, Dictionary of Natural Products, MarinLit[23] and AntiBase.[24] Integration of structure-based and chemical shift searches, such as NAPROC-13,[25] make these databases even more valuable. AntiMarin is a result of a recent merger between AntiBase and MarinLit;[26] MarinLit and AntiMarin are the first commercially available databases that contain structural information (functional groups) as searchable fields.[27]

Recent advances in chemoinformatics have greatly enhanced the utility of these resources and many are now accessible *via* the Internet.[28] CHEMnetBASE (www.chemnetbase.com) provides online access to a variety of databases including the Dictionary of Natural Products and the Dictionary of Marine Natural Products, although full access through CHEMnetBASE requires a subscription. The Chemical Structure Lookup Service (http://cactus.nci.nih.gov/lookup) is an open access database and incorporates information from more than 80 databases on over 27 million structures. PubChem (http://pubchem.ncbi.nlm.nih.gov/) is another open access database that also links bioassay data to each structure. Measures such as the US National Institutes of Health (NIH) Public Access Policy Mandate should increase the amount of openly accessible information available on the Internet and facilitate dissemination of information.

3 Extraction

Once a natural product source is selected or obtained, the first processing step is extract preparation. A detailed plan is required for each source organism since there are potentially interfering substances present in the source and the method must extract drug-like natural products. Thus, extraction methods that are designed for a thorough chemical analysis of a source may not be optimal for a drug discovery programme.

Methods to prepare crude extracts have evolved from simple solvent extractions to more sophisticated systems utilising pressure, ultrasound, supercritical fluids, heat and microwave energy. Developments in methods to generate crude extracts have been discussed recently in a detailed review.[29] Many modern commercial extraction systems offer benefits of faster extraction and reduced solvent consumption.[30]

In general, the extraction process must be streamlined to reduce solvent consumption since removal of solvents can represent a major bottleneck. Additionally, the extraction method chosen should support the first step of fractionation for library generation. Supercritical fluid extraction (SFE) represents an efficient extraction method in terms of low solvent consumption and extraction speed. Supercritical fluids exhibit high diffusivity with low viscosity and low surface tension; they can readily permeate biomass matrices and solvate molecules, including drug-like compounds, leading to efficient extractions. The addition of small amounts of organic co-solvents may enhance the extraction of small molecules.[31,32] However, the special equipment and associated capital costs required for SFE need to be considered before adopting this extraction methodology.

4 Prefractionation

In natural product drug discovery, separation technologies play important roles in three inter-related areas: prefractionation for library generation, preparative or large-scale isolation and analytical analysis. Although there is overlap among the technologies used in each area, there are also distinctions in new technologies and we discuss each separately.

Over the past decade, it has become clear that crude natural product extracts do not perform well in HTS programmes that have become the "mainstream" drug discovery route. Many crude extracts contain multiple components that have different pharmacological activities and can present problems even in low throughput assays where selectivity is a priority. Additionally, isolation of an active component from a crude extract is time-consuming and requires continued screening resources. Therefore, prefractionation strategies have been developed to generate natural product screening libraries that contain less complex mixtures; this limits interference and reduces isolation time.

Prefractionation strategies are controlled to a large extent by the cost of prefractionation, screening methodology and screening capacity. The importance of the prefractionation process cannot be overlooked. As pointed out in a recent paper by Koehn: *"high throughput screening of poorly designed or constructed libraries yields few viable hits"*.[7] Prefractionation should effectively remove components that interfere with assays and concentrate drug-like molecules. For example, a prefractionation strategy for plant extracts might remove tannins or concentrate alkaloids while a strategy for marine invertebrates might remove salts from the drug-like molecules. A prefractionation method should be simple and rapid, provide reasonable compound separation and concentrate organic constituents effectively.[10] Ideally, fractions should be generated using a minimum volume of solvent; as noted above, solvent removal can be a major bottleneck.

Prefractionation methods have evolved from simple solvent–solvent partition methods to fully automated chromatography. Liquid–solid techniques have been the most widely employed methods for automated prefractionation. Polymeric solid phase adsorbents have found wide application due to their high capacity and suitability for rapid prefractionation.[8,11,33] Countercurrent chromatography (CCC) and related liquid–liquid methods have also shown promise to generate prefractionated natural product libraries. However, liquid–liquid methods tend to be less amenable to automation and more difficult to employ in high-throughput drug discovery programmes.[34,35] Nonetheless, CCC has been used to prefractionate plant extracts successfully after removal of polyphenols.[36] To remove polyphenols prior to CCC, the plant extracts were treated with poly-*N*-vinylpyrrolidone. Subsequently, extracts were separated using a ternary solvent system, dichloromethane : methanol : water (5 : 6 : 4, v/v/v).[36] Using this methodology, interfering compounds such as tannins and lipids were resolved from the compounds of intermediate polarity, which increased the detection of minor bioactive compounds.

Prefractionation by liquid–solid chromatography may be performed by HPLC, flash chromatography, or solid phase extraction (SPE). In order to achieve sufficient resolution to produce high purity screening libraries (1–5 compounds per well), a combination of two opposing or orthogonal steps is typically required.

Macroporous resins (MPRs) are solid phase adsorbents that have been used successfully to fractionate natural products from plants,[37] microorganisms[38] and marine invertebrates.[8–11] Synthetic resins such as Diaion® HP20SS, ENV^+, D101 and Amberlite™ XAD™ are highly porous organic polymers (polystyrene, divinyl benzene) stable to both aqueous and organic solvents.[39] In the case of ENV^+, the styrene monomers have been hydroxylated. As well as acting as a reversed phase medium, these resins have a large surface area that provides high adsorption capacity, which reduces column and solvent volume. A range of bead sizes and types are available but, in general, smaller beads provide higher capacity and improved resolution. Average bead size is an important characteristic since a narrow range of bead sizes provides better reproducibility. MPRs provide an efficient means to produce drug-like natural product libraries by removing nuisance materials such as salts and proteins.[11,33]

The preparation of prefractionated natural product libraries for drug discovery can be made more efficient by automation. Automation reduces manpower requirements, improves efficiency and increases productivity. In addition, human errors are reduced and reproducibility is improved. Overall, automation increases throughput and can produce larger and more diverse screening libraries.

Early work describing an automated natural product fractionation system utilised a modified RapidTrace® workstation to perform automated multistep SPE in order to prepare fractionated natural product libraries for HTS.[40] As a proof of principle, microbial extracts were subjected to an automated two-step SPE procedure using different solvent systems on ENV^+ or by employing two separate solid phases, ENV^+ and C_8. The choice of the second step was dependant on the polarity of the first eluate. Since the process separated and concentrated metabolites effectively, the samples were better suited for HTS compared with the crude extracts. For one *Streptomyces* strain, the compound responsible for activity in a progesterone receptor assay was separated from the toxic component(s) present in the crude extract.[40] Without prefractionation, the original extract would have exhibited a high level of toxicity. An automated system that is completely compatible with 96-well formats, the CyBi™-XTract, has also been described for use in natural product fractionation to generate screening libraries.[41] In both these examples, a two-step SPE method provides effective separation, but the resolving power of SPE is much less than that of HPLC. Compared with HPLC, SPE systems generally lack a detector, but the high adsorption characteristics of ENV^+ allow separation of quantities that would usually require preparative HPLC.

Several pharmaceutical companies and academic groups have utilised HPLC directly to generate natural product libraries for HTS. For example, both MerLion Pharmaceuticals and Wyeth have prepared natural product libraries of microbial fermentation extracts using preparative reversed phase (RP) C_{18} HPLC systems. In both cases, the highly polar early eluting material containing mostly media components was eliminated from the libraries because these were unlikely to contain drug-like compounds. In addition, for microbial extracts, ~85% of the mass is in the solvent front and testing the polar fraction gives negligible advantage over screening a crude extract. Wyeth's approach generated ten fractions per extract[10] while MerLion prepared four fractions per extract and could generate up to 48 000 fractions per year.[5] In both cases, for approximately 80% of the HTS hits, biological activity was detected in the fractionated library that was not detected in the crude extracts. These examples clearly demonstrate the ability of a prefractionation method to concentrate minor bioactive natural products and facilitate detection of the active components.

A particularly powerful method for constructing screening libraries is the combination of automated flash chromatography with HPLC. The FlashMaster™ II (Biotage) is an automated system for flash chromatography that can be fitted with normal phase silica or RP C_{18} cartridges. The FlashMaster™ II was used successfully by Sequoia Sciences, Inc. to fractionate plant extracts.[9,42] Subsequently, the fractions generated with the FlashMaster™ II were separated using a high-throughput parallel four-channel preparative HPLC system to generate 160 fractions for each plant extract. Using this method, a library of 36 000 fractions was generated for drug discovery.[9]

5 Isolation and Purification

In natural product drug discovery, compound purification is an essential step in identifying a new chemical entity. In the modern HTS environment, speed is essential; for natural products to compete with synthetics, efficient purification is vital to facilitate rapid structural elucidation of hits. Additionally, preparative purification strategies need to provide large quantities for a lead natural product to move forward in the development process.

High level compound purity is essential for a hit to become a genuine lead. For example, the terpene, ursolic acid, has shown inconsistent biological activity across a number of diverse assays. Recently, purity–activity studies of its antimycobacterial activity were performed using ursolic acid samples of varying purity. Ursolic acid showed enhanced antimycobacterial activity when impure.[43]

Optimised isolation strategies aim to take crude extracts to pure target compounds in as few chromatographic steps as possible. The current trend is towards smaller scale "micro fractionations".[44] For mixtures, prioritisation of screening hits can be enhanced with efficient chromatography coupled with analytical techniques to determine the class(es) of compounds present.

A final scale-up isolation may be unnecessary with the microscale approach since modern spectroscopic techniques allow the solving of structures with microgram quantities.[8,44]

CCC is a useful alternative purification technique although, from a high throughput perspective, technical challenges associated with CCC make it a slower and less efficient technique than HPLC. However, CCC has high loading capacity and results in 100% recovery of sample since the stationary phase is liquid and is expelled from the column at the end of the separation. Additionally, a wide variety of stationary and mobile phases can be constructed for excellent specificity. A recent review on CCC and related techniques provides an update on technological advances and methods applicable to natural products.[45]

5.1 Automated Purification

Automated purification systems should have the capacity to take a crude sample or prefractionated extract through to a series of pure compounds with limited human input. A number of companies produce automated purification systems. Automated purification of a target compound can be configured quite readily on many instruments using, for example, a UV, MS, or evaporative light scattering (ELS) signal.

However, systems such as the SepBox® are designed to perform automated purification from an extract without prior knowledge of the natural product. The SepBox® combines the power of HPLC with SPE and can collect purified samples in microtitre plates. A collaborative effort between Aventis Pharma and AnalytiCon Discovery produced a library of 4000 pure (>80%) natural products for HTS within 18 months using a SepBox® light system; however, most extracts were enriched using automated flash chromatography, a Flash™-Biotage system, prior to SepBox® separation.[2] For large-scale production, the SepBox 2D-5000 provides automated purification equipment and can handle up to 5000 mg. Dionex and Waters both produce HPLC based auto-purification systems, but these systems have been used more for automated purification of combinatorial libraries. Nonetheless, any system that can trigger fraction collection based on an external signal (*e.g.* MS, UV, ELS) can serve to automatically purify a target compound.

6 HPLC Separation Technologies

For high-throughput drug discovery, HPLC is an indispensable tool. HPLC offers high resolving potential, is highly reproducible, can be scaled and is amenable to automation. Although most drug-like compounds can arguably be purified from prefractionated extracts using RP HPLC, optimised purification strategies depend on the characteristics of a particular sample. For example, acidic and basic compounds are often chromatographically "streaky" without appropriate mobile phase modifiers.

Contemporary HPLC columns are made from high purity silica with low metal content, which leads to reduced silanol activity and affords improved separations, especially for basic compounds. Other improvements for basic compounds include utilising a C_{12} bonded phase that allows greater bonded phase density and reduces active silanol sites.[46] While C_{18} columns are commonly used to purify natural products, a variety of stationary phases have been introduced in recent years, such as polar end-capped media and polar embedded media.[47] These modified media provide alternative selectivity and improved retention of polar compounds, *e.g.* highly glycosylated natural products. Depending on the manufacturer, the polar functional group varies (*e.g.* amide or ether) and provides different selectivity. As with any natural product purification, a rapid assessment of functional groups present facilitates selection of an optimal purification procedure.

A significant improvement for isolation of natural products has been the recent commercial availability of monolithic columns, which were first developed about 20 years ago.[48] Monolithic HPLC columns are comprised of a single porous silica backbone (monolith) in contrast to traditional particle packed columns. Monolithic columns have a macroporous structure that reduces back pressure and a mesoporous surface that provides an immense surface area for analytes to adsorb to. These physical characteristics result in reduced operating pressures, a higher loading capacity and increased separation efficiency compared with equivalently sized particle-packed columns.[48] Another benefit to using monolithic columns is that larger quantities of sample can be separated at lower flow rates and on smaller columns, thereby reducing solvent use, requiring less time and providing a significant cost benefit.

Since late-stage dereplication is costly, sensitive, rapid analytical analyses are necessary for prioritising hits and supporting dereplication. In many cases, analytical chromatographic methods differ from those designed for preparative isolation and library generation. Arguably, one of the most powerful tools for prioritisation and dereplication is LC-MS. An inherent difficulty with MS analysis of natural products is that ion suppression can occur for co-eluting compounds. Therefore, analyses of natural product mixtures by LC-MS require efficient chromatography. The current trend embraces small particle sizes ($<2\,\mu m$) and HPLC systems that can operate at pressures up to $\sim 15\,000$ psi.[49,50] Particle size is inversely proportional to efficiency and resolution. However, the flow rates required to maintain linear velocity with sub $2\,\mu m$ particles leads to pressures that exceed the capability of conventional HPLC systems. The Acquity UPLC™ developed by Waters Corporation combines sub $2\,\mu m$ particle-packed columns with pressures approaching 15 000 psi. The UPLC™ was made possible by developing small particles that are cross-linked with ethylene bridges within the silica backbone allowing them to withstand high pressures and maintain stability across a greater pH range.[49,50] The improved resolution obtained from UPLC™ was demonstrated by the analysis of an extract of the plant *Passiflora edulis* (Figure 9.1). Overall, the analysis time was reduced by a factor of five. Additionally, the increased peak height improved MS detection by increasing the ion intensity, which

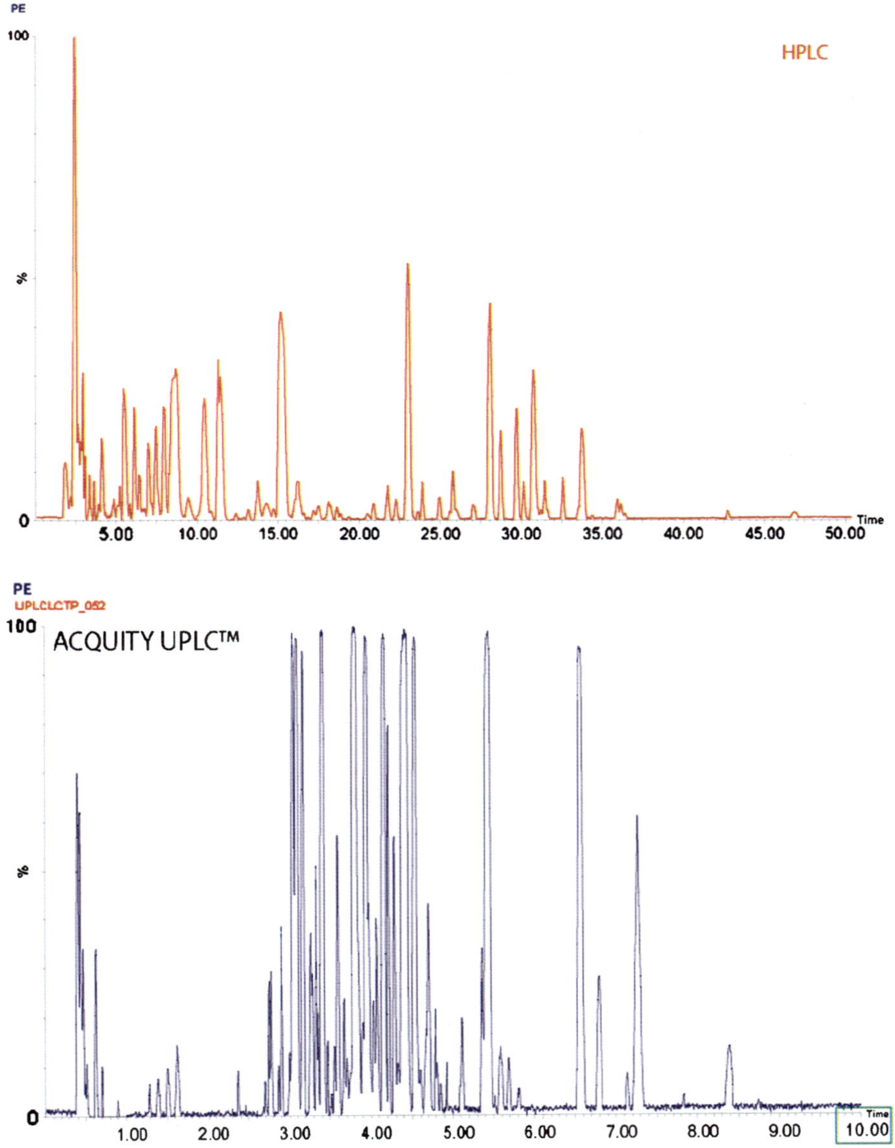

Figure 9.1 Analysis of *Passiflora edulis*: comparison of HPLC *vs*. UPLC (kindly provided by Waters Corporation).

greatly improved acquisition of MS/MS spectra. Other ultra high-pressure LC systems include Jasco's X-LC and Thermo's Accela high speed LC. Other LC systems utilising sub 2 μm particles include Shimadzu's ultra fast liquid chromatography (UFLC) and Agilent's Rapid Resolution System. Phenomenex's HST (High Speed Technology) columns have 2.5 μm particles that can withstand higher pressures (5800 psi) than conventional HPLC columns, but can still be used on conventional HPLC systems. This technology allows rapid analyses with improved resolution, but without the need for a new HPLC instrument.

Advances in LC technologies assist the drug discovery process by speeding up analysis and by generating more information content about a sample (*e.g.* higher chromatographic resolution improves sensitivity and detection of minor components). The key developments in LC technologies include improved column packings, smaller column sizes and developments in elution processes (*e.g.* supercritical fluid chromatography[51]).

7 Mass Spectrometry

Mass spectrometry has become increasingly important in natural product drug discovery programmes. MS and LC-MS offer high sensitivity and rapid throughput for analysis of natural products and can provide a great deal of structural information. MS technologies play major roles at nearly every stage of the lead generation process in natural product discovery programmes. The ease of automating LC-MS combined with faster separation techniques provides a platform for high-throughput analysis of natural products. New ionisation techniques coupled to new mass analysers with improved resolution and mass accuracy have greatly facilitated natural product dereplication, structure elucidation and shortened the hit-to-lead generation timeline. For example, ultra-high resolution Fourier transform mass spectrometry (FTMS) has provided unparalleled accuracy and has increased the level of confidence in identifying a molecular formula. Since many modern mass spectrometers have provided options for MS^n (where n = number of stages), detailed analyses of fragmentation pathways have expedited *de novo* structure elucidation. Overall, mass spectrometry provides the fastest most sensitive analytical tool to uncover novel natural products.

Several major goals can be achieved by MS analyses in the early phases of natural product drug discovery. A molecular formula can often be identified using HPLC coupled to a time-of-flight (TOF) MS since most modern TOF instruments (*e.g.* Waters i-Fit) are capable of accurate mass measurements with the aid of predicted isotope peak matching. Since the overarching goal of any natural product drug discovery programme is to identify novel entities with potent biological effects, identifying a molecular formula in the early phases can help prioritise hits toward novel chemotypes. Although a molecular formula alone cannot distinguish among possible isomers, MS/MS can provide additional structural information that is capable of distinguishing, for example, constitutional isomers. Additionally, accurate MS/MS data can help find a unique elemental composition[52–54] of the parent ion since a unique elemental composition can be assigned to small fragments; furthermore, these MS/MS acquisitions can be obtained in a data-dependant automated fashion on modern mass spectrometers. Recent reports indicate the feasibility of MS/MS searchable libraries[16,17] and the use of MS/MS for metabolite identification[54] and structural characterisation[53–55] can help with the otherwise time-consuming task of data interpretation. Future informatics systems that can take advantage of the information produced by LC-MS and LC-MS/MS would have the ability to automatically dereplicate many natural product samples.

Historically, common ionisation techniques such as electron impact (EI) and chemical ionisation (CI) often did not ionise highly functionalised or high molecular weight natural products efficiently. Other ionisation methods, such as fast atom bombardment (FAB), provided better ionisation of larger, highly functionalised natural products. The advent of electrospray ionisation (ESI) and matrix-assisted laser desorption/ionisation (MALDI) greatly improved the ionisation of most natural products, while atmospheric pressure chemical ionisation (APCI) greatly improved ionisation and the analysis of non-polar natural products. Newer methods of ionisation include atmospheric pressure photo ionisation (APPI),[56] direct analysis in real time (DART),[57,58] desorption atmospheric pressure chemical ionisation (DAPCI),[59] desorption electrospray ionisation (DESI),[60] atmospheric pressure solids analysis probe (ASAP)[61] and ESC® which combined the benefits of ESI and APCI in one source.[62] DART, DESI and ASAP require no sample preparation and allow mass spectrometric analysis directly from solid surfaces. For example, DART MS was used to chemically analyse live male and female flies and detect spatial distribution of hydrocarbon pheromones.[63]

Although MS has been used traditionally for the most part to obtain structural information (molecular formula), MS techniques are finding more uses in natural products research such as HTS and imaging. In terms of screening, two notable examples can be highlighted where natural product mixtures were screened for interactions with both RNA and protein targets. They demonstrate clearly that ESI-MS can be used to investigate non-covalent interactions[64] and has great potential for high throughput screening of natural product mixtures. Hofstadler and co-workers from Ibis Therapeutics demonstrated that electrospray ionisation Fourier transform ion cyclotron resonance mass spectrometry (ESI-FTICR-MS) could be used to identify known and unknown natural products that could bind RNA targets in a high throughput fashion.[12,65] The methodology was termed multi-target affinity/specificity screening (MASS) and the technique was capable of screening 67 000 putative ligand-substrate pairs in 24 hours.[65] More recently, ESI-FTICR-MS was used to identify inhibitors of bovine carbonic anhydrase II from ten alkaloid-enriched plant extracts and eight desalted marine extracts.[14] Once the mass was determined from the screening procedure, mass-directed purification provided enough sample for NMR studies that led to the identification of 6-(1*S*-hydroxy-3-methyl-butyl)-7-methoxy-2*H*-chromen-2-one as the active component. These screening techniques are beneficial because they can be utilised to screen natural product mixtures that have not performed well in typical HTS campaigns.

In terms of HTS campaigns, there has been a move toward screening purified natural products libraries.[6,7] Although more chemical diversity can be sampled from crude extract libraries or partially purified natural product libraries, pure natural product libraries have the advantage of being better suited for HTS while allowing rapid identification of hits.[6] Additionally, purified libraries have the potential to negate the use of bioassay-guided isolation. Utilising MS in parallel with generating natural product libraries for HTS has an obvious advantage. For the most part, dereplication can be performed immediately and novel compounds can be identified at the hit stage. Analysis of large natural product libraries requires high-throughput techniques in order to characterise the library. To make analysis of a large plant natural product library feasible, high-throughput eight-channel parallel LC-MS in conjunction with evaporative light scattering detectors (ELSD) was utilised to characterise a library that was created by Sequoia Sciences containing 36 000 partially purified fractions.[66] More recently, large marine natural product libraries were generated while obtaining accurate mass measurements in parallel with library generation by splitting the HPLC flow between a mass spectrometer and fraction collector.[8] This approach has the advantage that data are obtained up-front and does not require a separate analysis, which can serve as a quality control measure.

Recently, imaging mass spectrometry has been applied to natural products research and might have potential in natural product drug discovery. In particular, Gerwick, Dorrestein and co-workers were able to detect new natural products while imaging assemblages of cyanobacteria, individual cyanobacteria and sponges using MALDI.[67] This technology, especially in conjunction with ambient ionisation techniques, could provide natural product chemists with tools to map metabolites of interest directly from field collections. In another example, positive ion DESI was used to analyse the distribution of γ-coniceine across the stem of a sample of *Conium maculatum*.[68] Subsequently, DESI was used to directly analyse alkaloids from freshly cut plant tissues and could detect differences in γ-coniceine, *N*-methyl coniine and conhydrine among different plant parts.[69] Analysis of *Datura stramonium* root using DESI-MS allowed 15 out of 19 alkaloids to be identified and confirmed by MS/MS experiments.[69] These examples clearly demonstrate the potential of DESI and ambient ionisation techniques for direct analysis of natural product sources with no sample preparation and could be used for a number of applications including quality control in the herbal and nutraceutical industry.

The availability of ultra-high resolution FTMS has greatly facilitated the structure elucidation of novel natural products in drug discovery programmes. FTMS provides, in many cases, sub-parts per million (ppm) mass accuracy, which greatly limits the number of possible formulae. Additionally, the resolution provided by FTMS instruments allows quantitation of sulfur, which further limits the number of possible formulae for a novel natural product. The potential of FTMS has been clearly demonstrated by the natural products research group at Wyeth.[53] In particular, the use of a multi-CHEF (correlated harmonic excitation field) waveform offers a method to provide reference ions in MS/MS spectra, which greatly improves the accuracy obtained for fragment ions.[52] However, the cost and difficulty of operation makes FTMS more difficult to access.

8 NMR

8.1 Probe Technology

NMR is an indispensable tool for natural product structure elucidation even though it is notoriously insensitive compared with MS. A common problem in natural product discovery is that biologically relevant natural products are often present in limited amounts, creating a challenge for NMR sample preparation and detection. These challenges are being met to a larger degree with improvements in NMR probe technology rather than increasing field strength. Capillary NMR (CapNMR) and cryogenic[70] probes have reduced sample requirements significantly and made several natural product studies feasible.[71–76] For example, a number of novel steroids were identified from a mass-limited collection of a rare firefly, *Lucidota atra*, using capillary probe technology.[71] Since obtaining large numbers of *L. atra* was not feasible, the study was not tractable without the use of a CapNMR probe.

CapNMR has emerged as a cost-effective method for high sensitivity NMR spectroscopic analysis and has allowed NMR to be combined with other analytical techniques such as HPLC and MS. CapNMR applications for characterisation of mass-limited small molecules, rapid screening of small molecule libraries and hyphenated techniques have been recently reviewed.[77,78] Advances in CapNMR probe technology include a microsolenoid[77] and a multi-coil microsolenoid probe[79] that provide high quality spectra for mass-limited samples.

Compared with conventional NMR probes, CapNMR probes reduce the required sample volume by a factor of 20–100.[71] The reduction in volume provides an effective concentration increase within the receiver coil and reduces the amount of sample required to obtain quality spectra. Therefore, the mass sensitivity of a given coil geometry is inversely proportional to coil diameter.[80] CapNMR is amenable to automation and can reduce ^{1}H acquisition times to one minute, enabling a complete automated analysis of a 96-well plate in an overnight run.[81] CapNMR probes are less expensive than cryoprobes and smaller volumes reduce expensive deuterated solvent costs. With a CapNMR probe, sample handling can prove difficult when dealing with small volumes of volatile solvents, but can be overcome by using automated sample handlers.

A further milestone in NMR sensitivity was the introduction of probes where the radiofrequency receiver coil and the preamplifiers are cooled cryogenically, which almost eliminates thermal noise and provides an 4–5 fold increase in sensitivity.[70] In 2005, Lewis and co-workers compared the sensitivity and solvent suppression of various NMR probes (including a capillary probe, a standard analytical flow probe, a 5 mm inverse probe, a 3 mm inverse probe and a 5 mm dual $^{13}C/^{1}H$ cryoprobe).[82] According to the results, a combination of a 3 mm Shigemi tube in a 5 mm dual $^{13}C/^{1}H$ cryoprobe afforded the best sensitivity, while suppression of residual solvent signals was much higher with the capillary probe.[82] Since then, cryoprobes have been improved by reducing the volume of the detection coil. The high-temperature 1 mm superconducting probe has been reported to have ~25 times greater sensitivity than a conventional probe.[83] The capabilities of the 1 mm superconducting probe were clearly demonstrated in a study where defensive secretions from a single walking stick insect were analysed.[84]

Commercially available cryogenically cooled probes include the 1.7 mm MicroCryoProbe™ from Bruker and the 5 mm XSens™ from Varian. Both probes have a cryogenically cooled ^{13}C channel, which results in large sensitivity gains for ^{13}C acquisition. The MicroCryoProbe™ also benefits from the small volume required (30 μL) and provides a 14-fold gain in ^{1}H sensitivity compared with a conventional 5 mm probe. The 5 mm Varian probe has been optimised for ^{13}C detection and provides about a ten-fold gain for ^{13}C-detected experiments (see Figure 9.2) compared with a conventional 5 mm probe. Since the 1D ^{13}C experiment is one of the least sensitive experiments for a natural product, this probe should be beneficial for structure elucidation. Additionally, the potential to use this probe to acquire INADEQUATE (incredible natural abundance double quantum transfer experiment) spectra has been demonstrated recently in conjunction with a new pulse sequence that allows a complete structure from one NMR experiment through parallel acquisition.

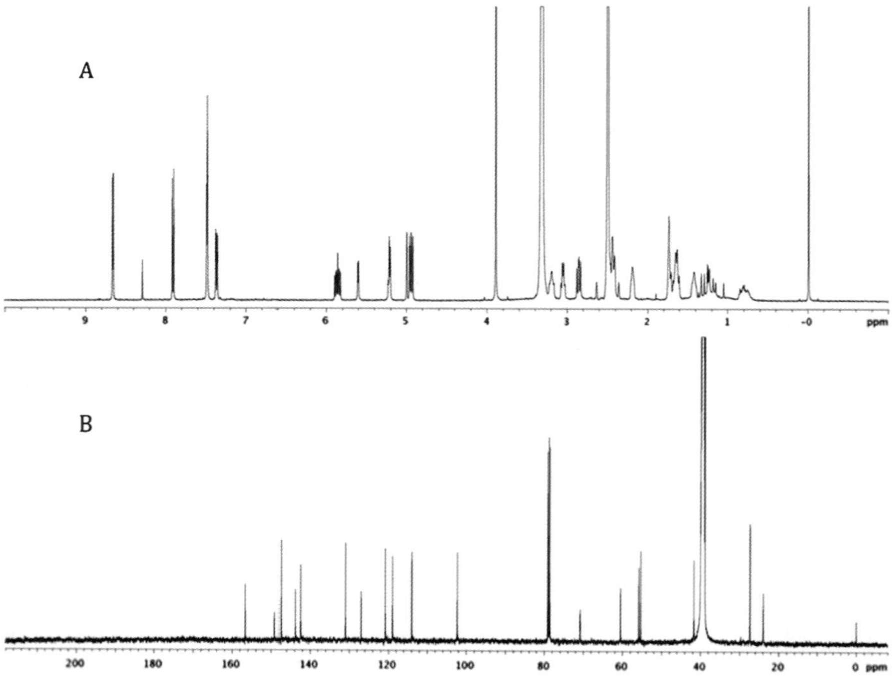

Figure 9.2 Spectra obtained using the Varian 5 mm XSens™ dual cold probe. (A) 500 MHz ^{1}H spectrum, 50 μg quinine, 3 mm NMR tube, 1 scan acquisition. (B) 125 MHz ^{13}C spectrum, 50 μg quinine, 3 mm NMR tube, 2.5 hour acquisition (data kindly provided by Varian Inc.).

This new structure determination scheme has been termed PANACEA (parallel acquisition NMR, an all-in-one combination of experimental applications) and is outlined in Section 8.3.

Advances in NMR technology have allowed researchers to identify molecules using only very small quantities of material. However, working with minuscule amounts of sample presents additional challenges. For 5 mm cryoprobes, the volume of solvent can be critical and typical deuterated solvents result in large solvent peaks that can negatively affect two-dimensional (2D) spectra.

The sensitivity of new NMR equipment allows analysis of samples that cannot be accurately weighed, but quantification is necessary for accurate biological testing. Recently, Quinn and co-workers developed an NMR protocol that uses the residual proton signal from DMSO-d_6 as an internal standard to determine the molar concentration of a compound in the absence of a molecular weight.[85] With this methodology, estimates of the weight of a compound can be obtained and an accurate measure of biological potency can be determined. In this regard, microscale purification and screening can be implemented and negate large-scale isolation.

8.2 Structure Elucidation

A 2002 review by Reynolds and Enriquez describes the most effective pulse sequences for natural product structure elucidation.[86] For natural product chemists, the review recommends HSQC over HMQC, T-ROESY (transverse rotating-frame Overhauser enhancement) in place of NOESY (nuclear Overhauser enhancement spectroscopy) and CIGAR (constant time inverse-detected gradient accordion rescaled) or constant time HMBC over HMBC. HSQC spectra provide better line shapes than HMQC spectra, but are more demanding on spectrometer hardware. The T-ROESY or transverse ROESY provides better signal to noise for most small molecules compared with a NOESY and limits scalar coupling artefacts. In small-molecule NMR at natural abundance, the 2D HMBC or variants experiment stands out as one of the key NMR experiments for structure elucidation. HMBC spectra provide correlations over multiple bonds and, while this is desirable, it poses the problem of distinguishing between two- and three-bond correlations.

H2BC (heteronuclear 2-bond correlations) affords an HMBC-type spectrum that shows two-bond correlations almost exclusively.[87] Multiplicity-editing, where $[CH+CH_3]$ and CH_2 are phased differently, adds to the wealth of information that the H2BC provides.[88] One limitation of the H2BC experiment is that two-bond correlations are weak or absent in the spectra if the $^3J_{HH}$ is small or vanishes. HAT-HMBC (homonuclear J attenuated heteronuclear multiple bond correlations),[89] a hybrid of H2BC and HMBC, can be used in addition to H2BC and HMBC; small J_{HH} coupling constants are the most attenuated in a HAT-HMBC spectrum in comparison with a regular HMBC spectrum.

Hardware plays a major role in what pulse sequences will perform best. For example, traditional HMBC, which was a simple variant of HMQC, does not contain any 180° pulses on X-nuclei and performs well on most spectrometers, while CIGAR contains 180° inversion pulses and may lead to reduced intensity near the edges of a large ^{13}C sweep width depending on the length of the pulse. Typically, this is not an issue on micro probes (≤3 mm). However, on 5 mm inverse probes, especially cryogenically cooled probes, this can become a serious problem. A large number of new pulse sequences have been published to address this problem. For example, in C2HSQC (doubly compensated heteronuclear single quantum coherence), the conventional proton 180° pulses of a multiplicity-edited ^{1}H–^{13}C HSQC pulse sequence are replaced with broadband inversion pulses (BIPs). This sensitivity gain is mainly the result of improved tolerance to radiofrequency (rf) inhomogeneity of the BIPs relative to conventional 180° pulses.[90] Considering the typical sweep width for an HSQC type experiment is much narrower than a CIGAR type experiment, long-range correlation experiments will benefit substantially from BIPs.

8.3 Methods for Fast NMR

Although increasing sensitivity can drastically reduce NMR acquisition time, new methods have evolved that can have a major impact on acquisition and structure determination. Three areas will be briefly presented here that offer reduction in acquisition time, either directly by reducing the time that a sample remains in the NMR or indirectly through processing.

Ultrafast 2D acquisition provides 2D NMR data sets in single scan experiments, but suffers from low sensitivity.[91] As the absolute sensitivity of NMR probes increases, these methods should find more application in natural products research. Traditional 2D NMR experiments are acquired by incrementing an evolution delay (t_1) and require an entire acquisition for each t_1 to generate 2D information. Incrementing the evolution delay provides a route to sample the evolution of all frequencies in the indirect dimension. Alternatively, ultrafast 2D acquisition utilises a spatial encoding strategy that allows frequencies in the indirect dimension to evolve throughout the sample, for example by applying a z-axis gradient in conjunction with frequency-modulated pulses. Please see ref. 92 for a description of spatial encoding strategies. The end result is that the indirect frequencies can be sampled without the need for incrementing t_1 and allow 2D acquisition in a single scan.

Other strategies that show great promise in reducing NMR acquisition time utilise methods to obtain multiple sets of data from one experiment through a concept known as time-shared evolution. An example of this process that should find utility in natural products elucidation was demonstrated by a pulse sequence called CN-HMBC.[93] Traditionally, a separate ^{13}C-HMBC and ^{15}N-HMBC were acquired independently. However, the CN-HMBC allows both ^{13}C- and ^{15}N-HMBC spectra to be obtained simultaneously. By acquiring both data sets simultaneously, an effective 50% time reduction can be achieved.[93] This approach has also been demonstrated for a sensitivity-enhanced 2D HSQC-TOCSY (heteronuclear multiple bond correlation total correlation spectroscopy) and HSQMBC (heteronuclear single quantum multiple bond correlation).[94] For peptides, obtaining ^{13}C-HSQC-TOCSY and ^{15}N-HSQC-TOCSY spectra simultaneously should greatly facilitate structure elucidation. However, the HSQC-TOCSY experiment is inherently insensitive and observation of the ^{15}N resonances may require long acquisition times.

Parallel acquisition NMR spectroscopy (PANSY) has emerged as a method to provide great time savings for NMR data acquisition and structure elucidation.[95] In contrast to the concept of time-sharing evolution, PANSY experiments use a separate receiver for each nucleus. As a proof of concept, Kupce and co-workers demonstrated the potential of PANSY by simultaneously acquiring ^{1}H–^{1}H correlation and ^{1}H–^{13}C correlation spectra, both direct and long range.[95] Using four transients, the experiment was complete in 40 minutes. Additionally, a similar data set was obtained in 22 seconds on a sample of inosine using Hadamard encoding in conjunction with parallel acquisition. In a more recent application of parallel acquisition, a structure determination scheme termed PANACEA (parallel acquisition NMR, an all-in-one combination of experimental applications) was described that allows for small molecule structure determination from a single NMR experiment.[96] Two-dimensional ^{13}C–^{13}C correlations, single- and multi-bond ^{13}C–^{1}H correlations and the conventional ^{13}C spectrum were all recorded in parallel, making use of separate receiver channels for acquisition of ^{13}C and ^{1}H signals. A high-sensitivity cryogenically cooled probe, optimised for ^{13}C detection, markedly improved the feasibility of natural-abundance ^{13}C–^{13}C correlation experiments such as INADEQUATE. This procedure was also extended with three parallel receivers to include natural abundance ^{15}N–^{1}H long-range correlations. PANACEA represents the first approach toward obtaining a structure from one NMR experiment. However, the inherently low sensitivity of INADEQUATE, in most cases, would require a ^{13}C cryogenically cooled probe.

A third area of development that has affected the speed of obtaining molecular connectivity information from NMR takes advantage of the information inherently present in two separate experiments. Traditionally, an analyst would use the information from a group of separate experiments to draw conclusions about molecular connectivity. In recent years, the projection–reconstruction technique[97,98] and indirect covariance NMR[99] have allowed information from two separately acquired experiments to be correlated into an additional experiment. Both techniques can increase the dimensionality of NMR data providing information that would otherwise require time-consuming acquisitions.

Indirect covariance spectroscopy quickly found applications and demonstrated that ^{13}C–^{13}C connectivity could be obtained from HSQC-TOCSY. Some artefacts were observed, but unsymmetrical indirect covariance processing was developed and overcame the problems observed in early processing methods. The result was the ability to co-process a pair of independently acquired 2D NMR spectra to afford the equivalent of hyphenated 2D NMR spectra.[100,101] Generally, hyphenated pulse sequences tend to give low signal-to-noise even though acquisition of high quality data for each independent experiment is easily attained. Of greatest applicability to small molecule spectroscopy are the calculation of a ^{13}C–^{13}C homonuclear correlation spectrum derived from an HSQC-TOCSY spectrum[102] and an *m,n*-ADEQUATE generated from independently acquired GHSQC (gradient heteronuclear single quantum coherence) and GHMBC (gradient heteronuclear multiple bond correlation) spectra.[100] Co-processing provides a way to obtain hyphenated data without having to run time-consuming insensitive hyphenated NMR experiments. However, the applicability of this technique to complex structures is limited as signals must be well-resolved to obtain high-quality co-processed data. Currently, NMR software from ACD Labs can be used for indirect covariance processing.[103]

The projection–reconstruction approach is a technique unrelated to covariance processing which can provide data typically inaccessible to the natural product chemist. For example, ^{13}C–^{15}N correlation spectra were obtained for vitamin B12 at natural abundance.[104] Compared with a conventional three-dimensional ^{13}C–^{15}N correlation experiment, the projection–reconstruction method provides a sensitivity enhancement of two orders of magnitude. The final ^{13}C–^{15}N spectrum was reconstructed from data obtained from ^{1}H–^{15}N and ^{1}H–^{13}C correlations acquired using a time-shared evolution pulse sequence that allowed all the information to be obtained in one experiment.[104] Martin and co-workers also demonstrated the ability to generate ^{13}C–^{15}N correlation spectra using unsymmetrical indirect covariance NMR with vinblastine as an example.[105] In the latter case, ^{13}C–^{15}N correlation spectra were obtained from ^{1}H–^{13}C HSQC data and ^{1}H–^{15}N HMBC data that were acquired separately. Both methods provide access to correlations that would be inaccessible for most natural products at natural abundance.

8.4 Automated Structure Elucidation

A number of NMR spectral databases exist to aid the natural product chemist in structure elucidation. SpecInfo currently contains 359 000 ^{13}C NMR spectra and 130 000 ^{1}H NMR assigned spectra.[106] CSearch is another repository with a number of data sets.[107] Both SpecInfo and CSearch provide structure prediction based on the database content. NMRShiftDB is an open access, open submission NMR web database for structures and their NMR spectra. It allows users to predict spectra and search for spectra and structures.[108,109] NMRPredict is offered with MestReNova and predicts ^{1}H and ^{13}C spectra from a structure.[110] The Madison Metabolomics Consortium Database (MMCD; http://mmcd.nmrfam.wisc.edu/) is a web-based bioinformatics resource that contains experimental NMR data on 447 compounds.[111] Additionally, the system contains information on more than 20 000 small molecules and can be queried using text, structure, NMR, mass and miscellanea.[111,112] ChemGate allows users to search for NMR data by structures or substructures and also predicts NMR spectra.[113]

A 2004 review outlines the developments in computer-assisted structure elucidation (CASE) between 1999 and 2004.[114] Elyashberg *et al.* have recently reviewed the current state-of-the-art in the field of CASE and structure verification.[115] This thorough review presents a number of examples of the automated elucidation of complex natural product structures that define the capabilities of each system. NMR Analyst[116] and ACD/Structure Elucidator[117–119] are two of the major programs that are capable of providing automated structure elucidation. Protasis Corporation[81] recently released its new Discovery High Throughput Structure Confirmation (HTSC) NMR Workstation for use with Bruker or Varian spectrometers and structure elucidation software such as ACD Labs. With the assistance of these CASE programs, rapid dereplication is possible and reduces the number of structures that need to be fully elucidated. CASE systems have the potential to contribute to higher throughput in laboratories and help solve complex structure elucidation problems.

8.5 Configuration by NMR

A recent review by Bifulco *et al.* covers the methods for determining relative configuration by NMR with the aid of computational methods, including *J*-based analysis, the Universal NMR Database and the quantum mechanical calculation of NMR parameters.[120]

Long-range heteronuclear coupling constants have been used extensively to determine the configuration of natural products.[121] As a result, a number of methods have been developed in an effort to provide more accurate measurements of long-range heteronuclear coupling constants; these have been outlined previously by Marquez *et al.*[122] Recent improvements include Carr–Purcell–Meiboom–Gill (CPMG) HSQMBC, which suppresses the evolution of homonuclear couplings and reduces the multiplet distortion often observed in the HSQMBC for proton signals with large homonuclear coupling constants.[123] Modifications to an HSQC-TOCSY have also been reported.[124] In well resolved regions of a spectrum, coupling constants can be extracted from exclusive correlation spectroscopy (E.COSY) type patterns. The modifications to the spin-edited sensitivity-enhanced HSQC-TOCSY allow for the determination of the sign and size of heteronuclear and homonuclear coupling constants and are useful for compounds with a great deal of spectral overlap. However, correlations to non-protonated centres cannot be observed in an HSQC-TOCSY spectrum.

A 2004 review focuses on assignment of absolute configuration for alcohols, amines, carboxylic acids and sulfoxides using NMR and chiral derivatising reagents.[125] There have been many developments in the use of chiral auxiliary reagents since this review, which are summarised below. Silyl ether reagents were developed for determining the enantiomeric purity and absolute configuration of secondary alcohols.[126] These *R*- and *S*-α-(trifluoromethyl)benzyl silyl derivatives can be easily synthesised and the original chiral alcohol can be easily recovered. More recently, procedures that require the synthesis of only one derivative or simplify the preparation have emerged and require less of the parent natural product.[127–129] New methodologies for determining absolute configuration of 1,2-primary/secondary diols,[127,130] secondary/secondary diols,[131,132] 1,2-amino alcohols[129,133,134] and 1,2,3-primary/secondary/secondary triols[135] have been reported.

8.6 Residual Dipolar Couplings

Residual dipolar couplings (RDCs) provide a wealth of structural information and have shown great potential to determine the configuration of stereocentres in natural products, as well as confirm molecular structure.[136] Dipolar couplings are normally large and, for that reason, solid state NMR requires magic angle spinning. In isotropic solutions, small molecules tumble rapidly and experience all possible angles between nuclei and the dipolar couplings average to zero. However, when a molecule is partially aligned, the dipolar couplings do not average and the RDCs can be measured. The RDCs can be extracted from NMR spectra that have coupling information since the RDCs add to the scalar couplings ($J + D$). The magnitude of the coupling is dependant on the angle and the distance (r^{-3}) between two nuclei. Therefore, as long as relevant models can provide interatomic distances, detailed information about the angle between two nuclei can be obtained. In theory, the angular information will allow the configuration to be determined.

RDCs have not been applied extensively to natural products because most partial alignment media have been optimised for aqueous solutions for large biomolecules.[136] More recently, alignment media that are compatible with organic solvents have become available. The most commonly employed approach has been to swell a polymeric gel in an NMR tube. Upon swelling, partial alignment of small molecules is observed. Gels compatible with organic solvents include polystyrene,[137] polyacrylamide,[138] polyvinyl acetate[139] and, more recently, polyacrylonitrile.[140]

The difficulty of using RDCs to establish configuration is that representative molecular models for each possible stereoisomer are required; because RDCs are calculated using those models and compared with the experimentally measured RDCs. Nonetheless, RDCs were used to distinguish between two stereoisomers of sagittamide A, which is a flexible molecule and for which, at first glance, it would be difficult to develop appropriate models.[141] Prior to sagittamide A, most work had been performed on fairly rigid cyclic systems. This work clearly demonstrates the potential of RDCs in natural products.

9 Conclusions

Technological advances have provided a path to integrate the chemical diversity found in natural products in high-throughput drug discovery programmes. Many of the previous bottlenecks that made natural product discovery a slow laborious process have been effectively removed. Automation in NMR and MS has provided a path toward automated isolation of active molecules. Sensitivity improvements in NMR probe technology have allowed structures to be determined with very small quantities of material. On the other hand, these sensitivity improvements have been combined with pulse programs that, in theory, could lead to automated structure elucidation from an overnight experiment (*e.g.* PANACEA). These improvements in technology have also provided an opportunity for researchers to investigate the roles of secondary metabolites in nature. Understanding this fundamental aspect could provide an immense opportunity for the discovery and development of therapeutics as well as agricultural products.

In addition, technological advances have led to an increasing amount of analytical data, which requires proper data management. As more processes become automated, data management will become increasingly more difficult and informatics tools that can be used to mine the data will be of great value. In terms of structure elucidation, automation can decrease the amount of human input, but in the end, a well-trained expert will be required to interpret and validate the outcome. In academic labs, it will become increasingly important to balance automation with traditional training models to ensure that knowledge is effectively passed on.

The improvements in technology have made it feasible for natural products to be integrated into an HTS platform, which has allowed, in many cases, natural products to complement synthetic libraries in high-throughput drug discovery programmes. The ability to operate on a microscale has also facilitated the production of large and diverse natural product libraries without the need for industrial scale processes such as HPLC. Production of natural product libraries in academic labs provides a path toward complementing libraries of industrial screening partners. Overall, the current trend suggests that natural product drug discovery has been reinvigorated and will continue to be an important source of therapeutic leads.

References

1. www.fda.gov/CDER/rdmt/default.htm, accessed 29 April 2009.
2. K. U. Bindseil, J. Jakupovic, D. Wolf, J. Lavayre, J. Leboul and D. van der Pyl, *Drug Discov. Today*, 2001, **6**, 840.
3. B. Y. Feng and B. K. Shoichet, *J. Med. Chem.*, 2006, **49**, 2151.
4. A. L. Harvey, *Curr. Opin. Chem. Biol.*, 2007, **11**, 480.
5. D. R. Appleton, A. D. Buss and M. S. Butler, *Chimia*, 2007, **61**, 327.
6. F. E. Koehn and G. T. Carter, *Nat. Rev. Drug Discov.*, 2005, **4**, 206.
7. F. E. Koehn, *Prog. Drug Res.*, 2008, **65**, 175.
8. T. S. Bugni, B. Richards, L. Bhoite, D. Cimbora, M. K. Harper and C. M. Ireland, *J. Nat. Prod.*, 2008, **71**, 1095.
9. G. R. Eldridge, H. C. Vervoort, C. M. Lee, P. A. Cremin, C. T. Williams, S. M. Hart, M. G. Goering, M. O'Neil-Johnson and L. Zeng, *Anal. Chem.*, 2002, **74**, 3963.
10. M. M. Wagenaar, *Molecules*, 2008, **13**, 1406.
11. T. S. Bugni, M. K. Harper, M. W. B. McCulloch, J. Reppart and C. M. Ireland, *Molecules*, 2008, **13**, 1372.

12. L. L. Cummins, S. Chen, L. B. Blyn, K. A. Sannes-Lowery, J. J. Drader, R. H. Griffey and S. A. Hofstadler, *J. Nat. Prod.*, 2003, **66**, 1186.
13. H. Zou, Q. Zhang, Z. Guo, B. Guo, Q. Zhang and X. Chen, *Angew. Chem., Int. Ed. Engl.*, 2002, **41**, 646.
14. H. Vu, N. B. Pham and R. J. Quinn, *J. Biomol. Screen.*, 2008, **13**, 265.
15. T. Wennberg, K. Kreander, M. Laehdevuori, H. Vuorela and P. Vuorela, *J. Liq. Chromatogr. Relat. Technol.*, 2004, **27**, 2573.
16. K. Volna, M. Holcapek, L. Kolarova, K. Lemr, J. Caslavsky, P. Kacer, J. Poustka and M. Hubalek, *Rapid Commun. Mass Spectrom.*, 2008, **22**, 101.
17. C. Hopley, T. Bristow, A. Lubben, A. Simpson, E. Bull, K. Klagkou, J. Herniman and J. Langley, *Rapid Commun. Mass Spectrom.*, 2008, **22**, 1779.
18. J.-L. Wolfender, E. F. Queiroz and K. Hostettmann, *Exp. Opin. Drug Discov.*, 2006, **1**, 237.
19. D. Wolf and K. Siems, *Chimia*, 2007, **61**, 339.
20. B. Schmidt, J. W. Jaroszewski, R. Bro and M. Witt, *Anal. Chem.*, 2008, **80**, 1978.
21. M. E. Hansen, J. Smedsgaard and T. O. Larsen, *Anal. Chem.*, 2005, **77**, 6805.
22. K. Boroczky, H. Laatsch, I. Wangner-Dobler, K. Stritzke and S. Schulz, *Chem. Biodivers.*, 2006, **3**, 622.
23. *MarinLit database*; Department of Chemistry, University of Canterbury, Christchurch, New Zealand; www.chem.canterbury.ac.nz/marinlit/marinlit.shtml, accessed 29 April 2009.
24. H. Laatsch, *AntiBase 2009: The Natural Compound Identifier*, Wiley-VCH, Weinheim, Germany, 2009, www.wiley-vch.de/stmdata/antibase.php, accessed 29 April 2009.
25. J. L. Lopez-Perez, R. Theron, E. del Olmo and D. Diaz, *Bioinformatics*, 2007, **23**, 3256.
26. *AntiMarin* database; J. W. Blunt, M. H. G. Munro and H. Laatsch, University of Canterbury, Christchurch, New Zealand and University of Gottenberg, Gottenberg, Germany, 2007.
27. G. Lang, N. A. Mayhudin, M. I. Mitova, L. Sun, S. van der Sar, J. W. Blunt, A. L. J. Cole, G. Ellis, H. Laatsch and M. H. G. Munro, *J. Nat. Prod.*, 2008, **71**, 1595.
28. M. Sitzmann, I. V. Filippov and M. C. Nicklaus, *SAR QSAR Environ. Res.*, 2008, **19**, 1.
29. O. Sticher, *Nat. Prod. Rep.*, 2008, **25**, 517.
30. P. Ralifo, K. Tenney, F. A. Valeriote and P. Crews, *J. Nat. Prod.*, 2007, **70**, 33.
31. K. Khosravi-Darani and E. Vasheghani-Farahani, *Crit. Rev. Biotechnol.*, 2005, **25**, 231.
32. M. C. Henry and C. R. Yonker, *Anal. Chem.*, 2006, **78**, 3909.
33. L. Liu, Y.-F. Li and Y.-Y. Cheng, *J. Chromatogr. B: Biomed. Appl.*, 2008, **862**, 196.
34. A. Marston and K. Hostettmann, *J. Chromatogr. A*, 2006, **1112**, 181.

35. R. J. Case, Y. Wang, S. G. Franzblau, D. D. Soejarto, L. Matainaho, P. Piskaut and G. F. Pauli, *J. Chromatogr. A*, 2007, **1151**, 169.
36. J. A. Armbruster, R. P. Borris, Q. Jimenez, N. Zamora, G. Tamayo-Castillo and G. H. Harris, *J. Liq. Chromatogr. Relat. Technol.*, 2001, **24**, 1827.
37. H. Kikuchi, T. Ohtsuki, T. Koyano, T. Kowithayakorn, T. Sakai and M. Ishibashi, *J. Nat. Prod.*, 2007, **70**, 1910.
38. S. Ayers, D. L. Zink, K. Mohn, J. S. Powell, C. M. Brown, T. Murphy, A. Grund, O. Genilloud, O. Salazar, D. Thompson and S. B. Singh, *J. Nat. Prod.*, 2007, **70**, 1371.
39. I. M. Abrams and J. R. Millar, *React. Funct. Polym.*, 1997, **35**, 7.
40. I. Schmid, I. Sattler, S. Grabley and R. Thiericke, *J. Biomol. Screen.*, 1999, **4**, 15.
41. R. Thiericke, I. Schmid, T. Moore and G. Ebert, *Am. Biotech. Lab.*, 2000, **18**, 66.
42. J.-F. Hu, E. Garo, G. W. Hough, M. G. Goering, H.-D. Yoo, M. O'Neil-Johnson and G. R. Eldridge, *J. Chem. Res.*, 2007, **296**.
43. B. U. Jaki, S. G. Franzblau, L. R. Chadwick, D. C. Lankin, F. Zhang, Y. Wang and G. F. Pauli, *J. Nat. Prod.*, 2008, **71**, 1742.
44. G. Lang, M. I. Mitova, G. Ellis, S. vanderSar, R. K. Phipps, J. W. Blunt, N. J. Cummings, A. L. J. Cole and M. H. G. Munro, *J. Nat. Prod.*, 2006, **69**, 621.
45. G. F. Pauli, S. M. Pro and J. B. Friesen, *J. Nat. Prod.*, 2008, **71**, 1489.
46. www.phenomenex.com, accessed 29 April 2009.
47. J. W. Coym, *J. Sep. Sci.*, 2008, **31**, 1712.
48. F. Svec and C. G. Huber, *Anal. Chem.*, 2006, **78**, 2100.
49. J. R. Mazzeo, U. D. Neue, M. Kele and R. S. Plumb, *Anal. Chem.*, 2005, **77**, 460A.
50. M. E. Swartz, *J. Liq. Chrom. Rel. Tech.*, 2005, **28**, 1253.
51. L. T. Taylor, *Anal. Chem.*, 2008, **80**, 4285.
52. L. A. McDonald, L. R. Barbieri, G. T. Carter, G. Kruppa, X. Feng, J. A. Lotvin and M. M. Siegel, *Anal. Chem.*, 2003, **75**, 2730.
53. X. Feng and M. M. Siegel, *Anal. Bioanal. Chem.*, 2007, **389**, 1341.
54. Y. Konishi, T. Kiyota, C. Draghici, J. M. Gao, F. Yeboah, S. Acoca, S. Jarussophon and E. Purisima, *Anal. Chem.*, 2007, **79**, 1187.
55. H. P. Benton, D. M. Wong, S. A. Trauger and G. Siuzdak, *Anal. Chem.*, 2008, **80**, 6382.
56. A. Raffaelli and A. Saba, *Mass Spectrom. Rev.*, 2003, **22**, 318.
57. R. B. Cody, J. A. Laramee and H. D. Durst, *Anal. Chem.*, 2005, **77**, 2297.
58. R. B. Cody, J. A. Laramée, J. M. Nilles, H. D. Durst and G. C. Inc, *JEOL News*, 2005.
59. L. Charles, L. S. Riter and R. G. Cooks, *Anal. Chem.*, 2001, **73**, 5061.
60. Z. Takáts, J. M. Wiseman and R. G. Cooks, *J. Mass Spectrom.*, 2005, **40**, 1261.
61. C. N. McEwen, R. G. McKay and B. S. Larsen, *Anal. Chem.*, 2005, **77**, 7826.

62. www.waters.com.
63. J. Y. Yew, R. B. Cody and E. A. Kravitz, *Proc. Natl. Acad. Sci. USA*, 2008, **105**, 7135.
64. S. A. Hofstadler and K. A. Sannes-Lowery, *Nat. Rev. Drug Discov.*, 2006, **5**, 585.
65. K. A. Sannes-Lowery, R. H. Griffey and S. A. Hofstadler, *Anal. Biochem.*, 2000, **280**, 264.
66. P. A. Cremin and L. Zeng, *Anal. Chem.*, 2002, **74**, 5492.
67. E. Esquenazi, C. Coates, L. Simmons, D. Gonzalez, W. H. Gerwick and P. C. Dorrestein, *Mol. Biosyst.*, 2008, **4**, 562.
68. Z. Takats, J. M. Wiseman, B. Gologan and R. G. Cooks, *Science*, 2004, **306**, 471.
69. N. Talaty, Z. Takats and R. G. Cooks, *Analyst*, 2005, **130**, 1624.
70. H. Kovacs, D. Moskau and M. Spraul, *Prog. Nucl. Magn. Reson. Spectrosc.*, 2005, **46**.
71. M. Gronquist, J. Meinwald, T. Eisner and F. C. Schroeder, *J. Am. Chem. Soc.*, 2005, **127**, 10810.
72. A. M. Wolters, D. A. Jayawickrama and J. V. Sweedler, *J. Nat. Prod.*, 2005, **68**, 162.
73. K. L. McPhail, D. France, S. Cornell-Kennon and W. H. Gerwick, *J. Nat. Prod.*, 2004, **67**, 1010.
74. D. J. Russell, C. E. Hadden, G. E. Martin, A. A. Gibson, A. P. Zens and J. L. Carolan, *J. Nat. Prod.*, 2000, **63**, 1047.
75. D. Saman, J. Cvacka, A. Svatos, E. A. Bouman and B. Kalinova, *J. Nat. Prod.*, 2006, **69**, 1203.
76. E. W. Rogers and T. F. Molinski, *J. Nat. Prod.*, 2005, **68**, 450.
77. F. C. Schroeder and M. Gronquist, *Angew. Chem., Int. Ed. Engl.*, 2006, **45**, 7122.
78. A. G. Webb, *J. Pharm. Biomed. Anal.*, 2005, **38**, 892.
79. H. Wang, L. Ciobanu, A. S. Edison and A. G. Webb, *J. Magn. Reson.*, 2004, **170**, 206.
80. A. Webb, *Anal. Bioanal. Chem.*, 2007, **388**, 525.
81. www.protasis.com, accessed 29 April 2009.
82. R. J. Lewis, M. A. Bernstein, S. J. Duncan and C. J. Sleigh, *Magn. Reson. Chem.*, 2005, **43**, 783.
83. W. W. Brey, A. S. Edison, R. E. Nast, J. R. Rocca, S. Saha and R. S. Withers, *J. Magn. Reson.*, 2006, **179**, 290.
84. A. T. Dossey, S. S. Walse, J. R. Rocca and A. S. Edison, *ACS Chem. Biol.*, 2006, **1**, 511.
85. G. K. Pierens, A. R. Carroll, R. A. Davis, M. E. Palframan and R. J. Quinn, *J. Nat. Prod.*, 2008, **71**, 810.
86. W. F. Reynolds and R. G. Enriquez, *J. Nat. Prod.*, 2002, **65**, 221.
87. N. T. Nyberg, J. O. Duus and O. W. Sorensen, *J. Am. Chem. Soc.*, 2005, **127**, 6154.
88. N. T. Nyberg, J. O. Duus and O. W. Sorensen, *Magn. Reson. Chem.*, 2005, **43**, 971.

89. A. J. Benie and O. W. Sorensen, *J. Magn. Reson.*, 2007, **184**, 315.
90. H. Hu and K. Krishnamurthy, *Magn. Reson. Chem.*, 2008, **46**, 683.
91. L. Frydman, A. Lupulescu and T. Scherf, *J. Am. Chem. Soc.*, 2003, **125**, 9204.
92. Y. Shrot and L. Frydman, *J. Chem. Phys.*, 2008, **128**, 052209.
93. M. Perez-Trujillo, P. Nolis and T. Parella, *Org. Lett.*, 2007, **9**, 29.
94. P. Nolis, M. Perez and T. Parella, *Magn. Res. Chem.*, 2006, **44**, 1031.
95. E. Kupce, R. Freeman and B. K. John, *J. Am. Chem. Soc.*, 2006, **128**, 9606.
96. E. Kupce and R. Freeman, *J. Am. Chem. Soc.*, 2008, **130**, 10788.
97. R. Freeman and E. Kupce, *J. Biomol. NMR*, 2003, **27**, 101.
98. E. Kupce and R. Freeman, *J. Am. Chem. Soc.*, 2006, **128**, 6020.
99. F. Zhang and R. Bruschweiler, *J. Am. Chem. Soc.*, 2004, **126**, 13180.
100. K. A. Blinov, N. I. Larin, M. P. Kvasha, A. Moser, A. J. Williams and G. E. Martin, *Magn. Reson. Chem.*, 2005, **43**, 999.
101. K. A. Blinov, N. I. Larin, A. J. Williams, M. Zell and G. E. Martin, *Magn. Reson. Chem.*, 2006, **44**, 107.
102. F. Zhang and R. Bruschweiler, *J. Am. Chem. Soc.*, 2004, **126**, 13180.
103. www.ACDLabs.com, accessed 29 April 2009.
104. E. Kupce and R. Freeman, *Magn. Reson. Chem.*, 2007, **45**, 103.
105. G. E. Martin, P. A. Irish, B. D. Hilton, K. A. Blinov and A. J. Williams, *Magn. Reson. Chem.*, 2007, **45**, 624.
106. R. Neudert and M. Penk, *J. Chem. Inf. Comput. Sci.*, 1996, **36**, 244.
107. V. Schutz, V. Purtuc, S. Felsinger and W. Robien, *Fresenius' J. Anal. Chem.*, 1997, **359**, 33.
108. C. Steinbeck and S. Kuhn, *Phytochemistry*, 2004, **65**, 2711.
109. C. Steinbeck, S. Krause and S. Kuhn, *J. Chem. Inf. Comput. Sci.*, 2003, **43**, 1733.
110. www.mestrec.com, accessed 29 April 2009.
111. J. L. Markley, M. E. Anderson, Q. Cui, H. R. Eghbalnia, I. A. Lewis, A. D. Hegeman, J. Li, C. F. Schulte, M. R. Sussman, W. M. Westler, E. L. Ulrich and Z. Zolnai, *Pac. Symp. Biocomput.*, 2007, **12**, 157.
112. Q. Cui, I. A. Lewis, A. D. Hegeman, M. E. Anderson, J. Li, C. F. Schulte, W. M. Westler, H. R. Eghbalnia, M. R. Sussman and J. L. Markley, *Nat. Biotechnol.*, 2008, **26**, 162.
113. http://chemgate.emolecules.com/index.php, accessed 29 April 2009.
114. C. Steinbeck, *Nat. Prod. Rep.*, 2004, **21**, 512.
115. M. E. Elyashberg, A. J. Williams and G. E. Martin, *Prog. Nucl. Magn. Reson. Spectrosc.*, 2008, **53**, 1.
116. R. Dunkel and X. Wu, *J. Magn. Reson.*, 2007, **188**, 97.
117. M. E. Elyashberg, K. A. Blinov, A. J. Williams, S. G. Molodtsov, G. E. Martin and E. R. Martirosian, *J. Chem. Inf. Comput. Sci.*, 2004, **44**, 771.
118. S. G. Molodtsov, M. E. Elyashberg, K. A. Blinov, A. J. Williams, E. E. Martirosian, G. E. Martin and B. Lefebvre, *J. Chem. Inf. Comput. Sci.*, 2004, **44**, 1737.

119. M. E. Elyashberg, K. A. Blinov, A. J. Williams, E. R. Martirosian and S. G. Molodtsov, *J. Nat. Prod.*, 2002, **65**, 693.
120. G. Bifulco, P. Dambruoso, L. Gomez-Paloma and R. Riccio, *Chem. Rev.*, 2007, **107**, 3744.
121. N. Matsumori, D. Kaneno, M. Murata, H. Nakamura and K. Tachibana, *J. Org. Chem.*, 1999, **64**, 866.
122. B. L. Marquez, W. H. Gerwick and R. T. Williamson, *Magn. Reson. Chem.*, 2001, **39**.
123. K. E. Kover, G. Batta and K. Feher, *J. Magn. Reson.*, 2006, **181**, 89.
124. P. Nolis and T. Parella, *J. Magn. Reson.*, 2005, **176**, 15.
125. J. M. Seco, E. Quinoa and R. Riguera, *Chem. Rev.*, 2004, **104**, 17.
126. R. T. Williamson, A. C. Barrios Sosa, A. Mitra, P. J. Seaton, D. B. Weibel, F. C. Schroeder, J. Meinwald and F. E. Koehn, *Org. Lett.*, 2003, **5**, 1745.
127. F. Freire, J. M. Seco, E. Quinoa and R. Riguera, *Org. Lett.*, 2005, **7**, 4855.
128. F. Freire, J. M. Seco, E. Quinoa and R. Riguera, *Chem. Commun.*, 2007, 1456.
129. V. Leiro, J. M. Seco, E. Quinoa and R. Riguera, *Org. Lett.*, 2008, **10**, 2733.
130. F. Freire, J. M. Seco, E. Quinoa and R. Riguera, *Chemistry*, 2005, **11**, 5509.
131. F. Freire, J. M. Seco, E. Quinoa and R. Riguera, *J. Org. Chem.*, 2005, **70**, 3778.
132. F. Freire, F. Calderon, J. M. Seco, A. Fernandez-Mayoralas, E. Quinoa and R. Riguera, *J. Org. Chem.*, 2007, **72**, 2297.
133. V. Leiro, F. Freire, E. Quinoa and R. Riguera, *Chem. Commun.*, 2005, 5554.
134. V. Leiro, J. M. Seco, E. Quinoa and R. Riguera, *Org. Lett.*, 2008, **10**, 2229.
135. E. Lallana, F. Freire, J. M. Seco, E. Quinoa and R. Riguera, *Org. Lett.*, 2006, **8**, 4449.
136. R. M. Gschwind, *Angew. Chem., Int. Ed. Engl.*, 2005, **44**, 4666.
137. B. Luy, K. Kobzar and H. Kessler, *Angew. Chem., Int. Ed. Engl.*, 2004, **43**, 1092.
138. P. Haberz, J. Farjon and C. Griesinger, *Angew. Chem., Int. Ed. Engl.*, 2005, **44**, 427.
139. J. C. Freudenberger, S. Knor, K. Kobzar, D. Heckmann, T. Paululat, H. Kessler and B. Luy, *Angew. Chem., Int. Ed. Engl.*, 2005, **44**, 423.
140. G. Kummerlowe, J. Auernheimer, A. Lendlein and B. Luy, *J. Am. Chem. Soc.*, 2007, **129**, 6080.
141. A. Schuetz, J. Junker, A. Leonov, O. F. Lange, T. F. Molinski and C. Griesinger, *J. Am. Chem. Soc.*, 2007, **129**, 15114.

CHAPTER 10

Natural Product Combinatorial Biosynthesis: Promises and Realities

DANIEL W. UDWARY

University of Rhode Island, 41 Lower College Drive, Kingston RI 02881, USA

1 Introduction

There is no shortage of interesting molecular structures with great medicinal value described in this book. How organisms are able to biosynthesise such varied structures with exquisite regio- and stereochemistry is of further importance. Acquiring a deep knowledge of natural product chemistry requires a cross-disciplinary approach and the most prominent biosynthesis researchers are also experts in some combination of organic chemistry, enzymology, microbiology, molecular biology and, increasingly, genomics and bioinformatics. In many ways, the study of natural product biosynthesis has been a microcosm of the biotechnological revolution of the last two decades. While biosynthesis researchers are perhaps not always leaders in generating novel advances in biotechnology and molecular methods, they are often "early adopters" of new technology, finding novel applications for recent advances to promote the understanding and manipulation of these simultaneously specific and unlimited biological systems.

The ultimate application of that information varies widely. In perusing the literature in this field, the stated goals of biosynthesis research ranges from

RSC Biomolecular Sciences No. 18
Natural Product Chemistry for Drug Discovery
Edited by Antony D. Buss and Mark S. Butler

Published by the Royal Society of Chemistry, www.rsc.org

direct drug discovery purposes,[1] to probing unique and unusual enzymology,[2,3] to exploring molecular evolution of biochemical pathways,[4] to bioprospecting for new medicines in novel environments[5] and to metabolic engineering for the purpose of building "non-natural" or "designer" natural products. This chapter explores a brief history of the technology that has been used to explore and increase our understanding of the details of natural product biosynthesis, particularly in the systems of most active research, the modular polyketide synthases (PKS) and non-ribosomal peptide synthetases (NRPS). Without belabouring the chemical mechanisms of biosynthesis (which are well documented elsewhere), the chapter examines the history and promise of harnessing biosynthetic systems for the production of compounds and the pitfalls and technical difficulties that have been the hallmark of this field in the last decade. Finally, as scientific research enters the post-genomic era, the new technologies on the horizon that may help us to solve previously intractable problems are explored.

2 A Brief History of Natural Product Biosynthesis

The origins of biosynthetic studies began in the late 19th century, when it became practical to determine the chemical structures of the most common biologically derived small molecules and, of course, to ponder where they came from and how they were constructed by living cells. In 1907, Collie[6] postulated that the lichen metabolite orsellinic acid might be constructed as a polymer of ketene ($CH_2{=}C{=}O$) and called this and related compounds "polyketides". Many decades later, Birch showed *via* doubly labelled acetate incorporation studies and sound reasoning that polyketides are in fact polymers of acetate, which are then often aromatised. Birch wrote in his paper[7] that now the origin and fates of the individual carbon atoms of the polyketide backbone had been determined, there was still much work to be done to determine the sequence of biosynthetic intermediates and details of the enzymology, but that *"as organic chemists, we have passed on to other areas after a most enjoyable excursion"*.

Understanding all the details of that enzymology is, of course, still an active area of research. Because many polyketide biosynthetic pathways have less straightforward mechanisms and may occasionally incorporate more unusual carbon units, the specific origins and fates of the carbon atoms of many natural products were not always easily determined by analogy to orsellinic acid and so labelling studies remained the most straightforward means of studying biosynthesis for many decades. Thanks to the subsequent efforts of Staunton, Simpson and many others working with stable isotopes and NMR, many of the chemical details of those pathways were well-established by the late 1970s, long before the enzymes were identified.[8,9]

The modern, genetics-guided study of natural products biosynthesis and the recognition of the promises it could deliver was initiated in 1984 by David Hopwood's group with their publication in *Nature* on the cloning and heterologous expression of an entire biosynthetic pathway[10] for the antibiotic actinorhodin from *Streptomyces coelicolor*, which has become the model organism for *Streptomyces* genetics. Previous extensive mutagenesis work had suggested that *S. coelicolor* mutations that destroyed actinorhodin (act) production mapped to the same region of the chromosome,[11] indicating that the genes encoding the biosynthetic pathway are clustered together, which would be a common theme and major advantage for microbial natural product biosynthesis studies. They then isolated this stretch of coding DNA and found that, after transforming their mutants, they could restore the pathway to full activity. Furthermore, if they introduced the DNA into another strain of streptomycete that was not an actinorhodin producer, they found that the transformed strain would produce it.

But Hopwood's work with these so-called "aromatic" PKSs did not stop there and, in 1985, the field of engineered biosynthesis began. Hopwood transformed portions of the *S. coelicolor* act cluster into the granaticin producer *Streptomyces* AM-7161 and the medermycin producer *S. violaceoruber*, or non-producing mutants of them.[12] Hopwood's group found that they could then produce hybrid aromatic polyketides, to which they gave hybrid names like mederrhodin or dihydrogranatirhodin. This work established two important points. First, that genetic manipulation or modification was a viable route to the alteration of a microorganisms' secondary metabolism. Secondly, it showed that PKS systems were, on some level, fundamentally compatible with one another and strongly suggested that parts could be swapped in a rational way (if all of the parts could be sufficiently well understood) in order to construct novel products using the biosynthetic enzymes.

Subsequent sequencing work established that the act cluster was composed of genes with some homology to fatty acid synthases (FAS), including, apparently, two poly-β-ketoacyl synthase (KS) homologues, as well as acyl carrier protein (ACP) and ketoreductase (KR) homologues. Additional genes were present that were specific for aromatisation of the polyketide chain. Thus, the FAS biosynthetic analogy was confirmed genetically. Conveniently, the act genes were distant enough from their FAS forebears that they could be used as probes to locate and identify other, similar aromatic PKS clusters.[13]

Identification and genetic manipulations of aromatic polyketide clusters proceeded apace for the next few years, until the polyketide field was given another major stimulus with the publication of a sequenced portion of the PKS for the macrolide aglycone core of the antibiotic erythromycin from *Saccharopolyspora erythraea* in 1990. Leadlay and co-workers[14] found that, unlike the average-sized, discrete enzymes which made up the act cluster, at least one gene of the erythromycin cluster was very large (over 9 kb) and that gene contained regions within it that were very similar to FAS or PKS genes. The sequence showed a repetitive, modular structure that was clearly multifunctional. Analogous to experiments that, at the time, suggested a head-to-tail dimerisation of the multifunctional vertebrate FAS, Leadlay speculated that the erythromycin PKS may be similarly arranged. However, subsequent sequencing and analysis showed that two additional very large genes were adjacent upstream; their modular structure was consistent with the idea that each module performed a condensation reaction, with specialised modifications to the resultant carbon chain being dependant on the presence or absence of KR, enoyl reductase (ER) or dehydratase (DH) domains within a given module. Additionally, the N-terminus of the first open reading frame comprised a module containing only acyltransferase (AT) and ACP domains, strongly suggesting "loading" of the propionyl-CoA starter unit. The domain found at the C-terminus of the third open reading frame bore homology to thioesterases (TEs) and was later confirmed to catalyse macrocyclisation to the erythromycin aglycon 6-deoxyerythronolide B (DEB). The entire arrangement of the DEB synthase (DEBS, Figure 10.1) immediately suggested a tantalising assembly line mechanism[15]—one ripe for rational genetic modification on a level conceptually impossible, even today, for the aromatic PKS systems.

Intriguingly, the modular PKS organisation was quite analogous to non-ribosomal peptide synthetase (NRPS) genes that were reported in the same year (Figure 10.2).[16–18] Like modular PKSs, the NRPSs are comprised of often very large, multifunctional enzymes with domains that control extender unit (in this case, an amino acid) specificity by chemically activating them *via* ATP-dependent adenylation (A domain), condensation of extender units (C domain) and occasionally contain modifying or tailoring domains which may catalyse methylations, or often epimerisation of the attached amino acid. Intriguingly, A domains are not strictly bound to the activation of the proteinogenic amino acids and many have adapted to activation of very specialised carboxylic acid containing molecules. Similarly, NRPSs contain homologous ACP-like domains—often called peptidyl carrier proteins (PCP) or thiolation (T) domains—to shuttle potentially active intermediates from module to module, and TE domains to hydrolyse or intermolecularly cyclise the completed product. However, C and A domains appear to bear no relationship to KS or AT domains, and NRPS and PKS modularity appears to be an astounding example of convergent evolution. Furthermore, there are now numerous examples of hybrid modular PKS–NRPS systems, examples of which include biosynthetic clusters for the microcystin[19,20] and curacin[21] series of cyanobacterial toxins and the myxobacterium-derived electron transport inhibitor, myxothiazol.[22]

These discoveries coincided with rapid advances in DNA sequencing technology. The switch from radiolabelled Sanger dideoxy-terminator technology to capillary electrophoresis and dye terminators made sequencing affordable, practical and fast. PKS and NRPS researchers leapt aboard and used this new sequencing technology to identify and analyse dozens and eventually hundreds of entire biosynthetic gene clusters, with the rationale that nearly every new cluster was leading to the discovery of some new, interesting chemoenzymatic mechanism that could go into the metabolic engineer's toolbox. At the same time, pioneering advances in *Streptomyces* genetics by Hopwood's laboratory and others from the previous decades allowed for very precise manipulations of these newly discovered clusters. So much so that by 2001, a review article stated that there were simply too many published examples of metabolic engineering of natural product systems to effectively discuss.[23] Khosla, Leadlay, Staunton,

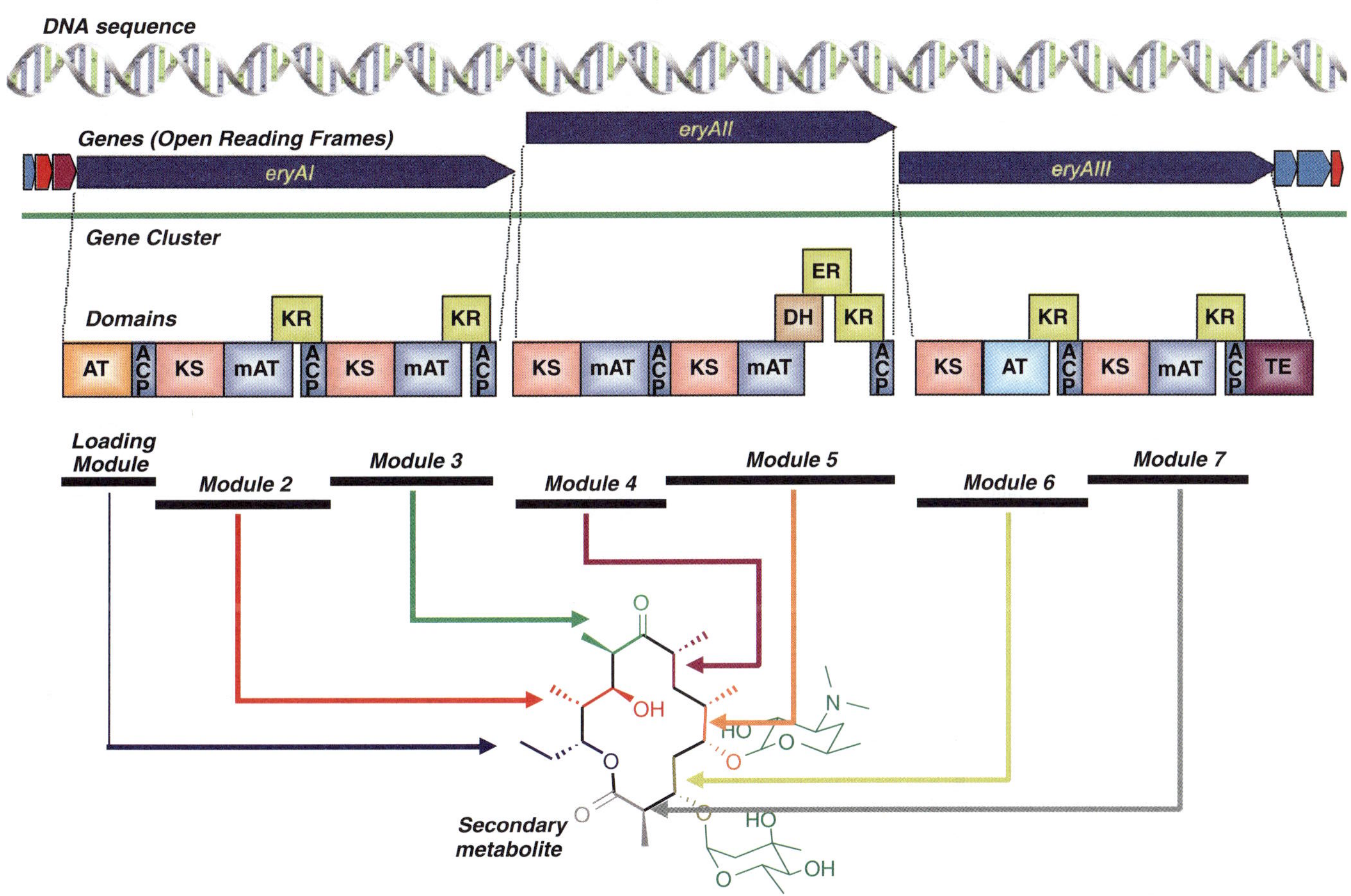

Figure 10.1 Domain and modular structure of 6-deoxyerythronolide B synthase (DEBS), the model modular PKS. ACP = acyl carrier protein; AT = acyl transferase; DH = dehydratase; ER = enoyl reductase; KR = ketoreductase; KS = ketosynthase; mAT = methylmalonyl-specific acyl transferase.

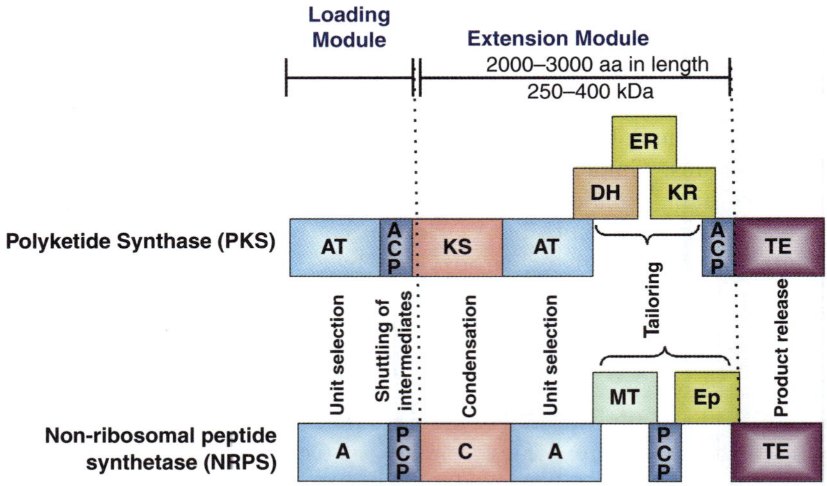

Figure 10.2 The PKS/NRPS biosynthetic paradigm, showing the most common domains and their relative positions within a modular PKS/NRPS enzyme. A = adenylation; AT = acyl transferase; C = condensation; DH = dehydratase; Ep = epimerase; ER = enoyl reductase; KR = ketoreductase; KS = ketosynthase; MT = methyltransferase; PCP = peptidyl carrier protein; TE = thioesterase.

Hopwood, Marahiel, Cane, Hutchinson and many others led the way in this exciting new endeavour. Manipulations included examples of loading molecule alteration, control of extender units, deletion or additions of reductive domains and altered placement of the TE domain to "short circuit" synthesis and create smaller molecules.[24] Attempts to "mix and match" modules from different systems in a truly combinatorial fashion have met with considerably less success. However, there was great optimism during most of the 1990s, as it seemed biosynthetic organic chemistry with unlimited potential was just around the corner and numerous biotechnology companies, mostly founded by academic leaders in the field, sprang up to capitalise on the coming new technologies. These most notably include Kosan Biosciences, Diversa, Ecopia and Biotica, of which only the latter exists today, the others having been absorbed by larger entities.

3 Promises

Conceptually, the modular biosynthetic systems with their assembly line structure are, perhaps, the most appealing for the purpose of rational biosynthetic engineering. If they could be properly harnessed, one can reasonably envision the construction of large, macrocyclic structures comprised of hydrocarbons of varying length, branches and oxidation *via* PKS modules, as well as nearly any amino acid imaginable *via* NRPS modules. Indeed, so many modular PKSs and NRPSs have been sequenced and probed biochemically that it is now often possible to predict the products of newly identified modular systems by carefully analysing the sequence alone. Outside the obvious steps of counting modules and noting type, much more information can be gleaned simply by examining binding motifs and/or properties of active site residues in specific domains. Guidelines exist to predict the degree of oxidation of the polyketide chain,[24] malonyl-CoA *vs.* methylmalonyl-CoA *vs.* ethylmalonyl-CoA and other specificities of AT domains,[25] to predict activation of over 100 different amino acids or carboxylates by A domains,[26,27] predict stereo-specificity of KR domains[28] and thiazole or oxazole ring formation during NRPS condensation reactions in C domains.[29,30] It stands to reason that if such chemistry can be predicted, it should be reasonably possible to introduce it into an engineered system.

Although the natural products of aromatic PKSs can be much more challenging to predict, their enzymes are much smaller and are often considered to be more tractable to routine heterologous expression, genetic modification and protein structure determination. Indeed, because large modular systems are so much more difficult to work with *in vitro*, much of what we now know about modular systems has been inferred from direct analogy to biochemical studies of aromatic systems. Aromatic PKSs can now be classified into several specialised families based upon both chemical product type and domain structure. Notably, NRPS analogues of the aromatic PKSs have not been observed.

Type I PKSs are considered to be any PKS in which multiple catalytic units, or domains, are found within a single peptide chain. This classification is then subdivided into the Type I modular PKSs (discussed above), of which DEBS is the classical model and the Type I iterative PKSs, which includes the fungal norsolorinic acid[31] and lovastatin synthases,[32] 6-methyl salicylic acid synthase (6-MSAS)[33] and the enediyne-involved family of polyketide synthases.[34] Their exact domain structure can vary, but these enzymes typically contain at least a single KS, AT, ACP and TE domain on a single peptide chain and may additionally carry methyl transferase, loading, KR, DH, ER and other, more cryptic domains, depending upon the product they produce. Unlike the modular systems, the primary active sites (KS, AT, ACP) are used many times to construct long polyketide chains, while tailoring domains will be very specific, only acting at a very specific time point in the reaction. It is currently not well understood how iterative PKSs (including the Type II and III PKSs discussed below) are able to so carefully control and aromatise their highly reactive poly-β-keto intermediates, although it is thought that there must be some degree of binding and stabilisation within the enzyme itself.

The Type II PKSs, of which the actinorhodin cluster is the model, consist minimally of an ACP and a heterodimeric ketosynthase complex and typically construct 2–4 ring aromatic molecules. Unlike modular systems, Type II PKSs are found as discrete proteins which probably associate as a complex and their reaction mechanism is iterative. Therefore, predicting the resulting size and cyclisation pattern of a Type II PKS product can be problematic, though enough systems have been studied that phylogenetic analysis is often helpful.

The primary KS, sometimes referred to as KS_{α}, is the site at which decarboxylative condensation occurs, while its partner, the KS_{β} or chain length factor (CLF) which is structurally homologous but chemically inactive, contains a long, narrow, charged channel that stabilises and holds in place the highly reactive, growing poly-β-keto intermediate. Comparison with X-ray crystal structures of the act KS-CLF heterodimer[35] allow one to predict the approximate product chain lengths of similar KS-CLFs by considering properties of residues in the CLF channel. Numerous additional genes may be present and clustered with a Type II PKS, including specialised ATs, which allow uptake of starter or extender units other than malonyl-CoA, aromatases and cyclases that control regiospecificity of the polyketide chain aromatisation and KRs which selectively reduce the polyketide chain at specific locations (and may also, as a result, play a role in aromatisation specificity).

Type III PKSs are small, homodimeric enzymes which can be found in both plants and a wide array of microbes. They generally construct fairly low molecular weight 1–2 ring aromatic molecules.[36] Examples include plant chalcone and stilbene synthases and the *S. coelicolor* gene tetrahydroxy naphthalene synthase (THNS), each of which has been structurally characterised by X-ray diffraction. These enzymes are often easy to express heterologously in *E. coli* and much work has been done to understand their complicated enzymology as a model for the larger, more complex Type I and II systems.[2]

Other non-PKS/NRPS biosynthetic pathways, of course, exist and varying progress has been made to their understanding and manipulation. The enzyme systems for glycosyl transferases have proven remarkably susceptible to metabolic engineering.[37] Terpene synthases, when handled *in vitro*, demonstrate enormous product flexibility, synthesising varying molecules in numbers comparable with combinatorial methods.[38] Certain alkaloid pathways, such as those encoded by the rebeccamycin and staurosporine clusters, show significant product flexibility when the cluster genes are recombined.[39]

No area of science promises to revolutionise biology as much as the rapid advances in DNA sequencing technology and bioinformatics analysis tools, which allow for the assembly and annotation of entire genomes. Analysis of microbial DNA sequence information has been the primary route toward investigation of the very large modular enzyme systems, as difficulties regarding cloning and expression prevent many of the classical biochemical experiments available to researchers in other areas of enzymology. Cloning of microbial DNA fragments into cosmid libraries and sequencing of the chromosomal regions of interest (*i.e.* regions containing natural product biosynthetic gene clusters) was a staple of natural products investigation at the turn of the Millennium. At the time, sequencing of an entire genome was a luxury no individual laboratory could afford to fund, let alone manage and interpret the flood of data that would result.

The first genome sequences of natural product producers came in 2001 with the completion of the model actinomycete *S. coelicolor* A3(2) led by Hopwood,[40] and, nearly simultaneously, *S. avermitilis*, completed by researchers in Japan.[41] Astoundingly, despite the decades of culturing and chemical extraction work that had gone into both organisms, each genome contained several biosynthetic clusters for molecules that have still not been identified today. Perhaps more amazing, and occasionally overlooked, was that very few of the biosynthetic clusters were shared between these closely related species. This has been a trend among sequenced natural product producers. Perhaps this is because genome sequencing has been relatively rare, thus far, for such organisms and only strains demonstrating unique chemistry have been sequenced, or perhaps it is because biosynthetic gene clusters evolve at a rate that far exceeds speciation. Or, perhaps it is both. As evidence, *S. coelicolor* and *S. avermitilis* each have a linear genome, approximately 9 Mb in size and the ends of their chromosomes, where the DNA is less stable and prone to mutation or recombination, contain many clearly inactive biosynthetic gene or cluster fragments. It has since been hypothesised that the linear chromosomal ends act as the "R&D" centre of the cell, with biosynthetic gene clusters merging, splitting and mutating until a pathway is created that synthesises some molecule conferring a selective advantage to the organism. These phenomena are similarly observed as more natural product producer organisms have been sequenced (Table 10.1) and indicates that there is a depth of biosynthetic potential that is not being accessed by conventional culturing and chemical screening methods alone. Large-scale sequencing coupled with bioinformatics-guided natural product isolations could prove to be an appealing route to the discovery of novel, medicinally useful compounds. So-called next-generation sequencers, which can provide many Mb of sequence data for only a few thousand dollars, have opened the floodgates to researchers who wish to look at all of the biosynthetic clusters in their organisms of interest. We can assume that there are few major natural products biosynthesis-focused labs that have not considered or already embarked upon "boutique" microbial genome sequencing.

4 Realities

There are many bright spots and promising routes to drug discovery by exploration of natural product biosynthetic gene clusters. Why then has there not been greater progress and greater interest in industry for this exploration? Why doesn't every large pharmaceutical company have a biosynthesis division, which uses genetically engineered microorganisms to churn out unique, complicated drug-like molecules ripe for assay? As it turns out, there are many technical hurdles even today to designing, expressing and otherwise working with biosynthetic gene clusters.

The primary technical hurdle lies with limitations of cloning and DNA manipulation technologies when working with large gene clusters and proteins. The typical modular PKS cluster may contain several 2–4 module open reading frames, with each module averaging 2–2.5 kb. One must also consider what can be dozens of tailoring genes, adding perhaps several kb more. There is also

Table 10.1 Natural product biosynthetic gene cluster information derived from selected genomes.

Organism	*Streptomyces coelicolor A3(2)*[40]	*S. griseus IFO 13350*[42]	*Salinispora tropica CNB-440*[5]	*Mycobacterium tuberculosis H37Rv*[43]	*Frankia* sp *CcI3*[44]	*N. farcinica IFM 10152*[45]
Size	8.72 Mb	8.5 Mb	5.18 Mb	4.41 Mb	5.43 Mb	6.01 Mb
Chromosome organisation	linear	linear	circular	circular	circular	Circular
%G + C content	72%	72%	69.5%	67%	70%	70%
Major NP clusters						
Modular Type I PKS	2	3	1	7	4	4
Enediyne PKS	–	1	2	–	–	–
Type II PKS	2	1	2	1	2	1
Type III PKS	3	2	1	3*	–	1
Mixed PKS/NRPS	–	4	4	–	1	1
NRPS	3	7	3	2	3	7
non-NRPS siderophore	1	1	1	–	1	–

often a regulatory system encoded within the cluster, as well as genes necessary to confer self-resistance, as many PKS and NRPS products act as antibiotics in their natural environment. Thus, if one wishes to clone a complete cluster, it typically requires the accurate cloning of about 20 kb of DNA and there are actinomycete clusters as large as 100 kb or more. Under normal circumstances, DNA amplification by PCR is limited to 10–12 kb and so it is typically necessary to prepare cosmid libraries (of which each cosmid vector can normally hold only about 40 kb of DNA) and screen for a sequence of interest, probing the ends to be sure that the entirety is contained within one clone (if possible). In addition, modular PKSs and NRPSs are fairly repetitive sequences, with modules often bearing >90% sequence identity, likely because they have originated by gene duplication events. Therefore, the long stretches of PKS or NRPS DNA are frequently unstable and prone to recombination or copy errors once clones have been introduced into a host. Despite these serious problems, this approach has worked reasonably well over the years, resulting in hundreds of published biosynthetic gene cluster sequences.

Assuming that it is possible to cross the sometimes considerable hurdle of cloning an intact biosynthetic gene cluster, the next idealised step would be to put that DNA into an expression host and observe the biosynthetic pathway in a more conveniently managed organism. However, this is rarely achieved for many technical reasons. First and foremost, the most studied natural product producers have been high G+C, Gram-positive actinomycetes. However, the most common and widely used expression host is the low G+C, Gram-negative *Escherichia coli*. The two organisms have differences in codon usage and introducing such "foreign" DNA means that expression of individual actinomycete genes in *E. coli* can be problematic, let alone lead to expression of entire operons. Another major problem is that many actinomycete compounds are antibiotics that may kill the heterologous host if self-resistance mechanisms are not present or are not functional. *E. coli*, apparently, has dramatically different metabolite pools compared with actinomycetes, which will need to be manipulated—either through culture conditions or additional genetic modifications—to provide appropriate substrates for reaction with the PKS or NRPS. Acyl carrier proteins in PKSs and peptidyl carrier proteins in NRPSs also require post-translational phosphopantetheinylation. While *E. coli* does have phosphopantetheinyl transferase (PPTase) genes required for FAS post-translational modification, it is apparently incompatible with streptomycete and other foreign ACPs and so a broad-specificity PPTase (such as the *Streptomyces verticillus* Svp gene[46]) must also be provided in the heterologous host. Furthermore, regulatory controls between any two microbial families or orders will almost always prove incompatible and so expression of more than one gene will require significant alteration of operon structure. For all of these reasons, use of an *E. coli*, yeast, mammalian or any other commonly used expression host is generally a non-starter in terms of heterologous pathway expression. The most common heterologous expression methods make use of a combination of integrated plasmids and a non-producing mutant streptomycete such as *Streptomyces* sp. CH999,[47] although, in this case, one must contend with much more troublesome genetics methods that are described below. A few examples of biosynthetic cluster expression do exist, the most prominent being that of a highly modified DEBS system expressed in a highly modified *E. coli* that has provided an important lesson on all of the potential associated problems presented here.[48] However, such examples are, to date, exceedingly rare.

As a result, rather than express a cluster or transport it to a more desirable organism, most researchers choose to directly manipulate the cluster in its native host. The modifications conducted most frequently are deletions of genes or domains to attempt to alter products of PKSs or NRPSs. In-frame insertions are also equally possible with genetic methods, but more rarely produce active enzymes, for reasons discussed later. General genetics systems for *Streptomyces* species are now fairly well established, although they require protoplasting, chromosomal integration, slow growth conditions and extensive screening techniques more akin to fungal genetics methods than typical bacterial techniques.[49] As a consequence, a single genetic manipulation can take a month or more from start to finish even in the hands of very experienced personnel. Thus, relatively few research groups are eager or able to tackle direct genetic alteration of natural product systems.

Assuming then that one has overcome problems regarding cloning and has achieved the technical ability to make specific modifications to genes in a given biosynthetic pathway, how does one know what alterations to make? Site specific mutagenesis to modulate activity in the absence of a thorough understanding of all of the dynamics of the reaction of an enzyme is often problematic and any mutation which alters activity can also result in severe losses in catalytic efficiency. For example, in the case of NRPS A domains, it is well-established that the nature of ten residues in the binding site of the domain can often allow prediction of the specific amino acid that will be bound and adenylated by that domain.[27] Clearly then, it should be conceptually straightforward to imagine making selective alterations to active site residues to alter the domain's substrate specificity. However, the lack of reported successes in the literature suggests that there is a body of unreported data showing that this is not the case. Iterative systems are even more problematic for mutagenesis, simply because so little is understood about the enzymology of the iterative process on an atomic level. Instead, the most effective means of altering enzyme activities and specificities has been by directed evolution, at its simplest a repeated process of random mutagenesis and selection of enzymes with desirable traits. However, directed evolution is best for selectively altering a single gene at a time and is simply not practical for pathway manipulation. Nor are screening methods, especially in the absence of heterologous expression. That said, destruction of activity of a given domain within a module is sometimes desirable and so this is the avenue where site-directed mutagenesis has most often been applied.

This lack of dynamic structure is one, but not the only barrier to swapping of modules. As stated at the outset, it is conceptually simple to imagine swapping a module or modules of one PKS or NRPS with another, resulting in a chimeric product structure. However, it seems that modules are much more than simple parts in an assembly line and instead these modular systems are complicated machines that have had perhaps millions of years to evolve into a tightly controlled, highly specialised and remarkably efficient biosynthetic apparatus. Experiments suggest that in most modular PKS and NRPS systems that, rather than a blind assembly line, there is downstream recognition of either carrier proteins or intermediates by KS or C domains[50] and there may be unknown checks and error-correcting mechanisms present throughout the systems. Structural characterisation of TE domains also shows some strong evidence for substrate recognition.[51] Thus, it would appear that there are bound to be fundamental incompatibilities between PKS and NRPS modules of different systems and, as discussed above, selectively overcoming those incompatibilities is not yet feasible until a greater understanding of the enzyme dynamics on an atomic level is achieved.

Assuming that all of the technical concerns described above can be overcome, there remains one final hurdle to drug discovery by metabolic engineering that is largely conceptual. In most every PKS or NRPS gene cluster, there are typically several so-called tailoring enzymes. These often catalyse simple reactions such as methylations, prenylations, amino group transfers, simple oxidative or reduction reactions or more complicated alterations (*e.g.* extensive oxidative cleavage or rearrangement to the PKS or NRPS chemical skeleton). These alterations are nearly always essential to modulate the binding ability of the potential drug. Therefore, by fundamentally altering a PKS or NRPS product, one could also inevitably lose all of the additional modulation reactions that confer bioactivity to the final product. Nature, with its millions of years of evolution, representing perhaps billions of generations of microbes, has often found the most biologically effective molecule for its purposes. That should not necessarily, of course, preclude pharmaceutical researchers from seeking to alter the bioactivity of a natural product when attempting to find a molecule more suited to human medical needs. However, there is perhaps a strong argument to be made regarding whether one may achieve better results in finding medically useful molecules by altering a biosynthetic pathway's tailoring enzymes, rather than the sometimes more simple modifications to a natural product's carbon skeleton. Unfortunately, there are hundreds, if not thousands, of specialised tailoring enzymes and engineering must be applied on a case-by-case basis without the clear conceptual framework and potential ruleset of PKS/NRPS engineering.

Compounding all these technical issues, one of the primary drivers away from studying biosynthesis was a general historical trend in the 1990s by the pharmaceutical industry to stop natural product discovery altogether. After decades of successful discovery programmes, big pharma companies were slowly seeing their natural product-derived drug pipelines drying up. This was partially because many of the common, soil-dwelling microbial species favoured by industry had been wrung dry of active compounds under standard laboratory culture conditions. In addition, the majority of compounds discovered in these organisms demonstrated antibiotic activity, which was no longer considered a pressing (nor as profitable) discovery need at the time. This should, perhaps, have been obvious with hindsight. Arguably, most PKS or NRPS products are produced by microbes because they have antimicrobial activity and are used in chemical warfare to keep other microbes that would encroach upon their territory away. Other non-anti-infective activities beneficial to human medicine are more likely to be happy accidents. At the same time, the concepts and practice of synthetic combinatorial chemistry[52] gained popularity and the ability to rapidly provide thousands, or even millions, of unique compounds, was seen as a better, more cost-efficient route to providing novel molecules for screening. As a result, most pharma companies gutted their natural product divisions entirely and metabolic engineering of natural product biosynthetic systems was never taken on as a major initiative by the wider industrial community. Unfortunately, combinatorial chemistry has largely been ineffective as a drug discovery method, perhaps because the libraries of molecules constructed typically have lacked the structural rigidity or stereocentres commonly found in natural products that allow for very specific receptor-site binding.

With the realisation of the dangers of methicillin-resistant *Staphylococcus aureus*, various Gram-negative bacteria, multidrug-resistant tuberculosis and numerous other dangerous antibiotic resistant pathogens, there is an urgent need for new antibiotics, as well as some recognition that new environments can be utilised to discover new microbial species and the novel bioactive products that they often make. Yet, pharma companies are still reluctant to reinvest in natural products discovery, despite the failure of combinatorial chemistry to produce effective new medicines and are, instead, largely content to capitalise upon or purchase intellectual property rights from smaller start-up companies that often arise from academic discoveries.

5 Future Biotechnological Promises

If the previous section presents a bleak prospect for rational engineering of natural product pathways, it also hopefully serves as a guideline for the kinds of technological developments necessary to get us out of the darkness. One can expect genomics to play a major role in the development of natural product research, as it already has become in every biological discipline. The traditional pipeline of DNA preparation, library construction, probing and sequencing of selected cosmids has become, by comparison, relatively costly. Next generation sequencers will easily enable high-coverage draft genome sequences for only a few thousand US dollars and it is expected that many researchers still searching for interesting biosynthetic mechanisms and natural products will make use of the opportunities that whole genome analysis provides. Already there are microbial genome sequences deposited in public databases containing many biosynthetic gene clusters that have been overlooked because the researchers who sequenced the organism were interested in some other aspect of its biology. Whereas in the past, a researcher might only go to the expense of sequencing a cosmid clone that contained a novel pathway, by sequencing an entire genome we can expect to see "accidental" sequencing of many more pathways also found in other organisms. This provides a tremendous opportunity to study mechanisms of horizontal gene transfer and to make comparative assessments of whole pathways and enzymes, which can only lead to a greater understanding of natural product biochemistry and how evolutionary mechanisms affect biosynthetic pathways and product formation.

As genomics has rapidly become the routine approach to identify natural product clusters, metagenomics will be the next step. Disregarding considerations of lineage, the most productive natural product producers with the richest biosynthetic genetic diversity are species that grow in organised colonies or filamentous forms. This is entirely logical as, in the absence of motility, these species must synthesise chemical defences to defend their environment against encroachment by other microbial predators. Thus, microbial natural products work has, for many years, focused on the rich diversity to be found in sponges, soil and marine sediments where these organisms are easily found. However, culturing from some environments has been met with only limited success, as in many cases removing these species from their natural habitats does not allow them to grow under laboratory culture conditions. Metagenomics, the direct sequencing of DNA isolated from environmental samples,[53] may be the answer to accessing the desirable genetic material to be found in these environments. At present, natural product applications may be limited as assembly of gene sequences deriving from what may be hundreds or thousands of species can be enormously problematic and it can thus be difficult to retrieve intact gene clusters. Per-organism sequence reads will also be dependent upon the populations of species within the sample. Therefore, if a sample is overpopulated with natural product non-producers, biosynthetic gene clusters will not be observed. In theory, all of this can be overcome by improvements in the methodology to "filter" environmental samples to enrich for desired organism types, as well as increased depth of sequencing coverage—which is only bounded by cost.

There is an important need for greater use of information technology to catalogue and understand all of this newly arriving data, primarily *via* more extensive bioinformatics analysis. There are already many reported methods to facilitate sequence analysis of PKS and NRPS genes, but the few PKS/NRPS automated analysis tools that do exist online focus primarily on BLAST (Basic Local Alignment Search Tool) analysis to determine the rough domain structure of input gene sequences. Even these simple tools are not in use to analyse entries submitted to either the US National Center for Biotechnology Information (NCBI) or the European Bioinformatics Institute (EBI) databases, and so most PKS or NRPS genes in natural product gene clusters end up cited, respectively, as "poly-β-ketoacyl synthase homologue" or "AMP-dependent ligase" at best, or "unknown" at worst. A much more comprehensive approach to labelling and identifying biosynthetic gene clusters will be necessary for the field to progress and should be quite feasible with current bioinformatic and programming tools.

What, then, to do with all of these new sequences identified over the last few decades and expect to continue to identify in the coming years? The burgeoning field of synthetic biology promises to offer a better route to expression of pathways and to lend greater insight into rational engineering of biological systems. There have been many recent advances in DNA synthesis and it is now possible to synthesise tens of kilobases of sequence with a speed and cost that was impossible only a few years ago. Synthetic biologists seek to use this technology to design biological systems from the ground up and their efforts have resulted, as one prominent example, in the construction of a wholly synthetic minimal bacterial genome.[54] One can imagine adapting the DNA sequence of a biosynthetic gene cluster to match a host's G+C bias, codon usage and regulatory mechanisms to enable far more straightforward heterologous expression. Much of current synthetic biology research focuses on the construction of useful biological circuits and the synthesis of very large biosynthetic gene clusters is probably outside the price range for most academic researchers. However, with more recognition of overlapping goals between natural product biosynthesis engineers and synthetic biologists, increased communication in the coming years could lead to fruitful collaboration.

From the very beginnings of its study, there has been tremendous excitement for natural product biosynthesis, the promises it holds for synthetic organic chemistry and its potential for controlled, combinatorial drug or complex molecule production. Though that promise has clearly not yet been realised, we can expect researchers in this ever-changing field to continue to utilise and generate novel advances in genomics, protein and operon expression, synthetic biology and biochemistry to advance steadily toward this goal. Perhaps, then, the greatest promise of natural products biosynthesis will be in the exciting highly cross-disciplinary research and researchers that its study produces.

References

1. J. B. McAlpine, B. O. Bachmann, M. Piraee, S. Tremblay, A. M. Alarco, E. Zazopoulos and C. M. Farnet, *J. Nat. Prod.*, 2005, **68**, 493.
2. M. B. Austin, P. E. O'Maille and J. P. Noel, *Nat. Chem. Biol.*, 2008, **4**, 217.
3. D. W. Udwary, L. K. Casillas and C. A. Townsend, *J. Am. Chem. Soc.*, 2002, **124**, 5294.
4. C. P. Ridley, H. Y. Lee and C. Khosla, *Proc. Natl. Acad. Sci. USA*, 2008, **105**, 4595.
5. D. W. Udwary, L. Zeigler, R. N. Asolkar, V. Singan, A. Lapidus, W. Fenical, P. R. Jensen and B. S. Moore, *Proc. Natl. Acad. Sci. USA*, 2007, **104**, 10376.
6. J. N. Collie, *J. Chem. Soc.*, 1907, **91**, 1806.
7. A. J. Birch, *Science*, 1967, **156**, 202.
8. T. J. Simpson, *J. Chem. Soc.*, 1975, **4**, 497.
9. M. J. Garson and J. Staunton, *Chem. Soc. Rev.*, 1979, **8**, 539.
10. F. Malpartida and D. A. Hopwood, *Nature*, 1984, **309**, 462.

11. B. A. Rudd and D. A. Hopwood, *J. Gen. Microbiol.*, 1979, **114**, 35.
12. D. A. Hopwood, F. Malpartida, H. M. Kieser, H. Ikeda, J. Duncan, I. Fujii, B. A. Rudd, H. G. Floss and S. Ōmura, *Nature*, 1985, **314**, 642.
13. F. Malpartida, S. E. Hallam, H. M. Kieser, H. Motamedi, C. R. Hutchinson, M. J. Butler, D. A. Sugden, M. Warren, C. McKillop, C. R. Bailey, G. O. Humphreys and D. A. Hopwood, *Nature*, 1987, **325**, 818.
14. J. Cortes, S. F. Haydock, G. A. Roberts, D. J. Bevitt and P. F. Leadlay, *Nature*, 1990, **348**, 176.
15. N. Wu, F. Kudo, D. E. Cane and C. Khosla, *J. Am. Chem. Soc.*, 2000, **122**, 4847.
16. A. Lawen and R. Zocher, *J. Biol. Chem.*, 1990, **265**, 11355.
17. H. J. Skarpeid, T. L. Zimmer, B. Shen and H. von Döhren, *Eur. J. Biochem.*, 1990, **187**, 627.
18. H. J. Skarpeid, T. L. Zimmer and H. von Döhren, *Eur. J. Biochem.*, 1990, **189**, 517.
19. T. Nishizawa, A. Ueda, M. Asayama, K. Fujii, K. Harada, K. Ochi and M. Shirai, *J. Biochem.*, 2000, **127**, 779.
20. D. Tillett, E. Dittmann, M. Erhard, H. von Dohren, T. Borner and B. A. Neilan, *Chem. Biol.*, 2000, **7**, 753.
21. Z. Chang, N. Sitachitta, J. V. Rossi, M. A. Roberts, P. M. Flatt, J. Jia, D. H. Sherman and W. H. Gerwick, *J. Nat. Prod.*, 2004, **67**, 1356.
22. B. Silakowski, B. Kunze and R. Muller, *Gene*, 2001, **275**, 233.
23. W. R. Strohl, *Metab. Eng.*, 2001, **3**, 4.
24. J. Staunton and K. J. Weissman, *Nat. Prod. Rep.*, 2001, **18**, 380.
25. C. D. Reeves, S. Murli, G. W. Ashley, M. Piagentini, C. R. Hutchinson and R. McDaniel, *Biochemistry*, 2001, **40**, 15464–15470.
26. T. Stachelhaus, H. D. Mootz and M. A. Marahiel, *Chem. Biol.*, 1999, **6**, 493.
27. G. L. Challis, J. Ravel and C. A. Townsend, *Chem. Biol.*, 2000, **7**, 211.
28. A. P. Siskos, A. Baerga-Ortiz, S. Bali, V. Stein, H. Mamdani, D. Spiteller, B. Popovic, J. B. Spencer, J. Staunton, K. J. Weissman and P. F. Leadlay, *Chem. Biol.*, 2005, **12**, 1145.
29. T. A. Keating, D. A. Miller and C. T. Walsh, *Biochemistry*, 2000, **39**, 4729.
30. C. G. Marshall, N. J. Hillson and C. T. Walsh, *Biochemistry*, 2002, **41**, 244.
31. C. M. Watanabe and C. A. Townsend, *Chem. Biol.*, 2002, **9**, 981.
32. J. Kennedy, K. Auclair, S. G. Kendrew, C. Park, J. C. Vederas and C. R. Hutchinson, *Science*, 1999, **284**, 1368.
33. J. B. Spencer and P. M. Jordan, *Biochem. J*, 1992, **288**(Pt 3), 839.
34. J. Ahlert, E. Shepard, N. Lomovskaya, E. Zazopoulos, A. Staffa, B. O. Bachmann, K. Huang, L. Fonstein, A. Czisny, R. E. Whitwam, C. M. Farnet and J. S. Thorson, *Science*, 2002, **297**, 1173.
35. A. T. Keatinge-Clay, D. A. Maltby, K. F. Medzihradszky, C. Khosla and R. M. Stroud, *Nat. Struct. Mol. Biol.*, 2004, **11**, 888.
36. M. B. Austin and J. P. Noel, *Nat. Prod. Rep.*, 2003, **20**, 79.
37. S. Blanchard and J. S. Thorson, *Curr. Opin. Chem. Biol.*, 2006, **10**, 263.

38. B. T. Greenhagen, P. E. O'Maille, J. P. Noel and J. Chappell, *Proc. Natl. Acad. Sci. USA*, 2006, **103**, 9826.
39. C. Sanchez, L. Zhu, A. F. Brana, A. P. Salas, J. Rohr, C. Mendez and J. A. Salas, *Proc. Natl. Acad. Sci. USA*, 2005, **102**, 461.
40. S. D. Bentley, K. F. Chater, A. M. Cerdeno-Tarraga, G. L. Challis, N. R. Thomson, K. D. James, D. E. Harris, M. A. Quail, H. Kieser, D. Harper, A. Bateman, S. Brown, G. Chandra, C. W. Chen, M. Collins, A. Cronin, A. Fraser, A. Goble, J. Hidalgo, T. Hornsby, S. Howarth, C. H. Huang, T. Kieser, L. Larke, L. Murphy, K. Oliver, S. O'Neil, E. Rabbinowitsch, M. A. Rajandream, K. Rutherford, S. Rutter, K. Seeger, D. Saunders, S. Sharp, R. Squares, S. Squares, K. Taylor, T. Warren, A. Wietzorrek, J. Woodward, B. G. Barrell, J. Parkhill and D. A. Hopwood, *Nature*, 2002, **417**, 141.
41. H. Ikeda, J. Ishikawa, A. Hanamoto, M. Shinose, H. Kikuchi, T. Shiba, Y. Sakaki, M. Hattori and S. Ōmura, *Nat. Biotechnol.*, 2003, **21**, 526.
42. Y. Ohnishi, J. Ishikawa, H. Hara, H. Suzuki, M. Ikenoya, H. Ikeda, A. Yamashita, M. Hattori and S. Horinouchi, *J. Bacteriol.*, 2008, **190**, 4050.
43. S. T. Cole, R. Brosch, J. Parkhill, T. Garnier, C. Churcher, D. Harris, S. V. Gordon, K. Eiglmeier, S. Gas, C. E. Barry 3rd, F. Tekaia, K. Badcock, D. Basham, D. Brown, T. Chillingworth, R. Connor, R. Davies, K. Devlin, T. Feltwell, S. Gentles, N. Hamlin, S. Holroyd, T. Hornsby, K. Jagels, A. Krogh, J. McLean, S. Moule, L. Murphy, K. Oliver, J. Osborne, M. A. Quail, M. A. Rajandream, J. Rogers, S. Rutter, K. Seeger, J. Skelton, R. Squares, S. Squares, J. E. Sulston, K. Taylor, S. Whitehead and B. G. Barrell, *Nature*, 1998, **393**, 537.
44. P. Normand, P. Lapierre, L. S. Tisa, J. P. Gogarten, N. Alloisio, E. Bagnarol, C. A. Bassi, A. M. Berry, D. M. Bickhart, N. Choisne, A. Couloux, B. Cournoyer, S. Cruveiller, V. Daubin, N. Demange, M. P. Francino, E. Goltsman, Y. Huang, O. R. Kopp, L. Labarre, A. Lapidus, C. Lavire, J. Marechal, M. Martinez, J. E. Mastronunzio, B. C. Mullin, J. Niemann, P. Pujic, T. Rawnsley, Z. Rouy, C. Schenowitz, A. Sellstedt, F. Tavares, J. P. Tomkins, D. Vallenet, C. Valverde, L. G. Wall, Y. Wang, C. Medigue and D. R. Benson, *Genome Res.*, 2007, **17**, 7.
45. J. Ishikawa, A. Yamashita, Y. Mikami, Y. Hoshino, H. Kurita, K. Hotta, T. Shiba and M. Hattori, *Proc. Natl. Acad. Sci. USA*, 2004, **101**, 14925.
46. C. Sanchez, L. Du, D. J. Edwards, M. D. Toney and B. Shen, *Chem. Biol.*, 2001, **8**, 725.
47. Z. Hu, D. A. Hopwood and C. R. Hutchinson, *J. Ind. Microbiol. Biotechnol.*, 2003, **30**, 516.
48. B. A. Pfeifer, S. J. Admiraal, H. Gramajo, D. E. Cane and C. Khosla, *Science*, 2001, **291**, 1790.
49. D. Hopwood, T. Kieser, M. Bibb, M. Buttner and K. Chater, *Practical Streptomyces Genetics*, John Innes Foundation, Norwich, 2000.
50. J. R. Lai, A. Koglin and C. T. Walsh, *Biochemistry*, 2006, **45**, 14869.
51. S. D. Bruner, T. Weber, R. M. Kohli, D. Schwarzer, M. A. Marahiel, C. T. Walsh and M. T. Stubbs, *Structure*, 2002, **10**, 301.

52. F. Guillier, D. Orain and M. Bradley, *Chem. Rev.*, 2000, **100**, 2091.
53. J. Handelsman, *Microbiol. Mol. Biol. Rev.*, 2004, **68**, 669.
54. D. G. Gibson, G. A. Benders, C. Andrews-Pfannkoch, E. A. Denisova, H. Baden-Tillson, J. Zaveri, T. B. Stockwell, A. Brownley, D. W. Thomas, M. A. Algire, C. Merryman, L. Young, V. N. Noskov, J. I. Glass, J. C. Venter, C. A. Hutchison III and H. O. Smith, *Science*, 2008, **319**, 1215.

Section 4 Natural Products in Clinical Development

CHAPTER 11

A Snapshot of Natural Product-Derived Compounds in Late Stage Clinical Development at the End of 2008

MARK S. BUTLER[a, b]

[a] MerLion Pharmaceuticals, 1 Science Park Road, The Capricorn #05-01, Singapore Science Park II, Singapore, 117528; [b] Department of Chemistry, National University of Singapore, Science Drive 3, Singapore, 117543

1 Introduction

Natural products (NPs) have played a pivotal role in drug discovery with over 50% of today's drugs being derived from NPs,[1–14] including the "statin" family of hypolipidemics, the anticancer drugs temsirolimus, trabectedin and ixabepilone, the immunosuppressant everolimus and the antimicrobials, daptomycin, tigecycline, doripenem and anidulafungin (Table 11.1). This chapter is a snapshot of NP-derived drug development at the end of 2008 with NP-derived drugs launched since 2003 detailed in Section 2 and NP-derived compounds that are undergoing late stage clinical evaluation in Section 3.

This chapter is an update of the *Natural Product Reports* reviews, *Natural products to drugs: natural product-derived compounds in clinical trials*, published in 2005[1] and 2008[2] except that compounds in development for oncology are discussed in detail. Compounds in this chapter are classified into three groups: NPs,

RSC Biomolecular Sciences No. 18
Natural Product Chemistry for Drug Discovery
Edited by Antony D. Buss and Mark S. Butler

Published by the Royal Society of Chemistry, www.rsc.org

Table 11.1 Natural product derived drugs launched since 2003 with reference to their classification, lead compound and disease area.

Year	*Generic name (trade name)*	*Classification*	*Lead compound (source, year reported)*[a]	*Disease area*
2003	miglustat (Zavesca®) **1**	semi-synthetic NP	1-deoxynojirimycin **2** (plant, 1966)[15]	type 1 Gaucher disease
2003	mycophenolate sodium (Myfortic®)	NP	mycophenolic acid **3** (fungus, 1952)[16]	immunosuppression
2003	rosuvastatin **4** (Crestor®)	NP derived	mevastatin **5** (fungus, 1976)[17]	dyslipidemia
2003	pitavastatin **6** (Livalo®)	NP derived	mevastatin **5** (fungus, 1976)[17]	dyslipidemia
2003	daptomycin **7** (Cubicin™)	NP	daptomycin **7** (actinomycetes, 1987)[18]	antibacterial
2004	everolimus **8** (Certican™)[b]	semi-synthetic NP	sirolimus **9** (actinomycetes, 1975)[19]	immunosuppression
2005	fumagillin **10** (Flisint®)	NP	fumagillin **10** (fungus, 1960)[20]	antiparasitic
2005	dronabinol **11**/ cannabidiol **12** mixture (Sativex®)	NPs	dronabinol **11** (plant, 1964)[21] cannabidiol **12** (plant, 1963)[22]	Pain
2005	doripenem **13** (Finibax®/ Doribax™)	NP derived	thienamycin **14** (actinomycetes, 1979)[23]	antibacterial
2005	tigecycline **15** (Tygacil®)	semi-synthetic NP	tetracycline **16** (actinomycetes, 1952)[24]	antibacterial

2005	ziconotide **17** (Prialt®)	NP	ziconotide **17** (cone shell, 1987)[25]	pain
2005	zotarolimus **18** (Endeavor™ stent)	semi-synthetic NP	sirolimus **9** (actinomycetes, 1975)[19]	cardiovascular surgery
2006	anidulafungin **19** (Eraxis™/ Ecalta™)	semi-synthetic NP	echinocandin B **20** (fungus, 1974)[26]	antifungal
2006	exenatide **21** (Byetta™)	NP	exenatide-4 **21** (lizard, 1992)[27]	diabetes
2007	lisdexamfetamine **22** (Vyvanse™)	NP derived	ephedrine **23** (plant, 1925)[28]	ADHD[c]
2007	retapamulin **24** (Altabax™/ Altargo™)	semi-synthetic NP	pleuromutilin **25** (fungus, 1962)[29]	antibacterial (topical)
2007	temsirolimus **26** (Torisel™)	semi-synthetic NP	sirolimus **9** (actinomycetes, 1975)[17]	oncology
2007	trabectedin **27** (Yondelis™)	NP	trabectedin **27** (ascidian, 1992)[30]	oncology
2007	ixabepilone **28** (Ixempra™)	semi-synthetic NP	epothilone B **29** (myxobacteria, 1995)[31]	oncology
2008	methylnaltrexone **30** (Relistor™)	NP derived	morphine **31** (plant, 1923)[32]	opioid-induced constipation
2008	ceftobiprole medocaril **32** (Zeftera™)	semi-synthetic NP	cephalosporin **33** (fungus, 1961)[33]	antibacterial

[a]This is the year that the lead compound's correct structure was reported in a journal. Please note that some of the compounds (*e.g.* **3**, **23** and **31**), were isolated many years before a definitive structure was determined, while the structure of some of the compounds were disclosed in earlier patents.

[b]Everolimus **8** is also in late stage clinical development by Novartis as RAD001 for the treatment of various cancers.

[c]attention deficit and hyperactivity disorder (ADHD).

semi-synthetic NPs and NP-derived. NPs are still classified as NPs even if the compound is produced synthetically for clinical studies or for the market. Semi-synthetic NPs are compounds that were derived from a NP template using semi-synthesis, while NP-derived compounds are synthetically derived or inspired from a NP template. Compounds derived from primary metabolites, vitamins, hormones, protein fragments, herbal mixtures and new uses of existing drugs are not discussed.

The development status of compounds undergoing clinical investigation can change rapidly and readers are encouraged to consult the recent literature, company web pages and clinical trial registers (*e.g.* www.clinicaltrials.gov) for the latest information.

2 NP-derived Drugs Launched in the Last Five Years

Twenty one NP-derived drugs have been launched since 2003 (Table 11.1), of which seven are the original NPs with no structure modification, nine are semi-synthetic NPs and five are NP-derived. The drugs span across many therapeutic areas with anti-infectives (five antibacterial, one antifungal and one antiparasitic) and oncology (three compounds) well-represented. This trend is likely to continue with a predominance of anticancer and antibacterial compounds in late stage development (Section 3). The newly launched drugs of 2008, methylnaltrexone **30** and ceftobiprole medocaril **32**, are described in this section in detail.

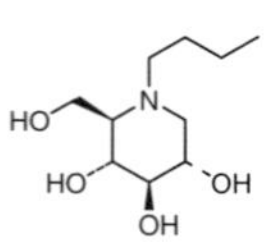
1

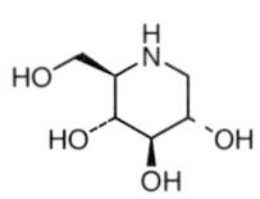
2

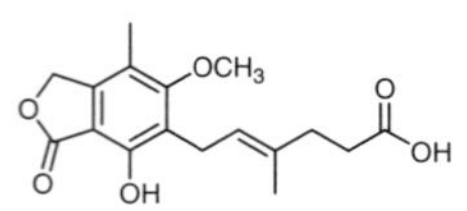
3

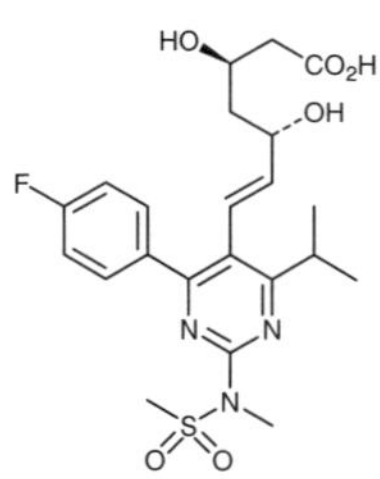
4

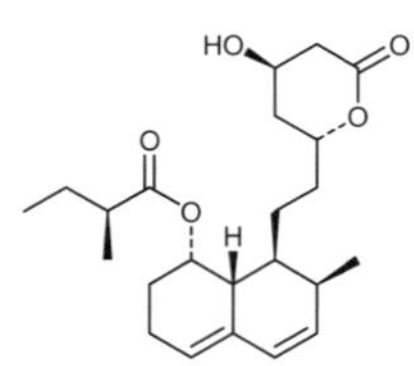
5

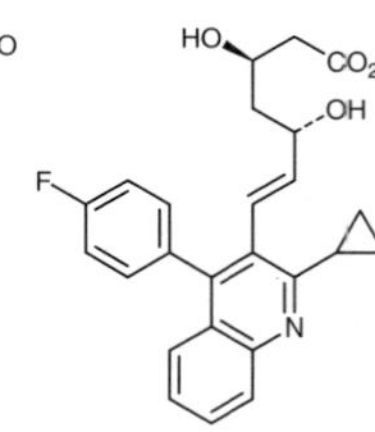
6

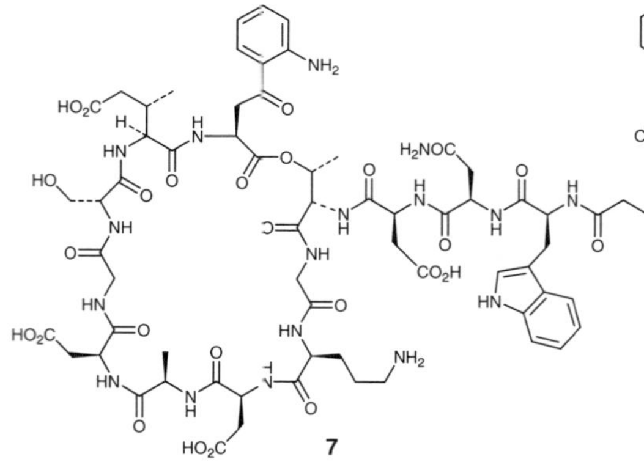
7

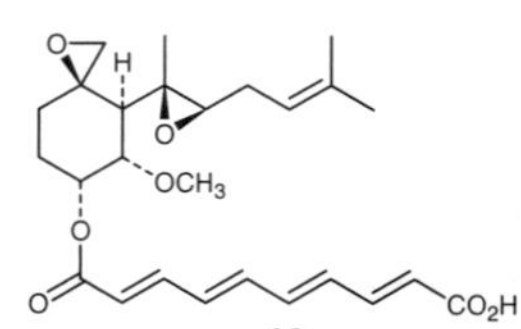
10

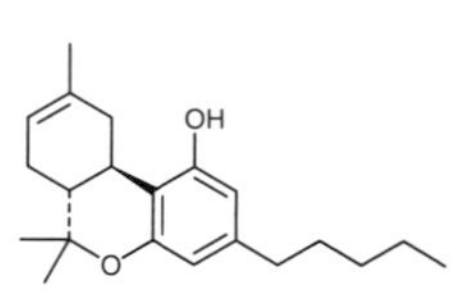
11

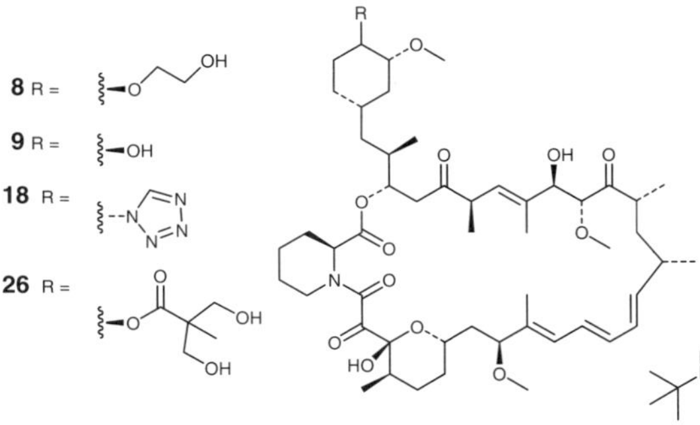

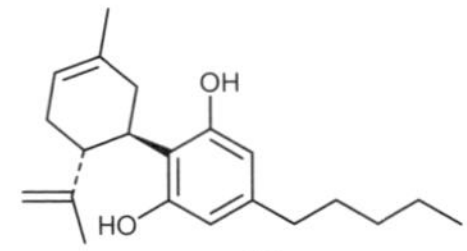

12

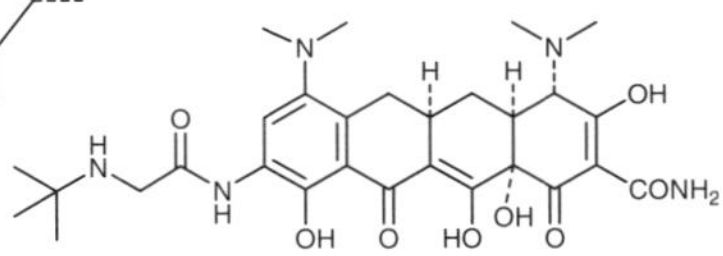

15

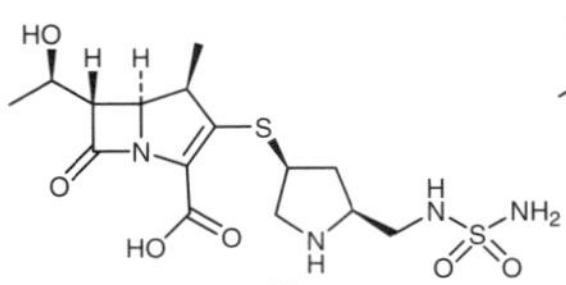

13

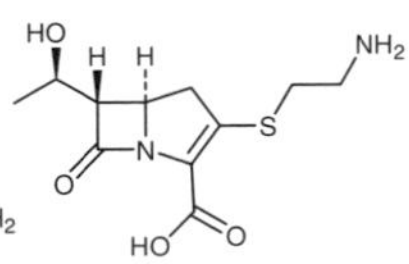

14

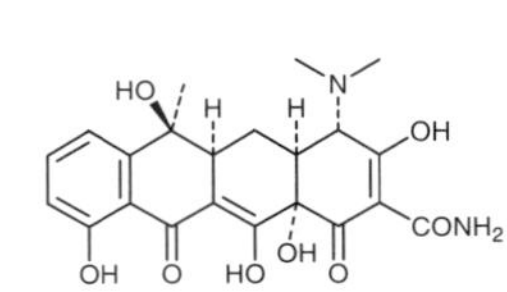

16

H_2N-CKGKGAKCSRLMYDCCTGSCRSGKC-$CONH_2$

17

HGEGTFTSDLSKQMEEEAVRLFIEWLKNGGPSSGAPPPS

21

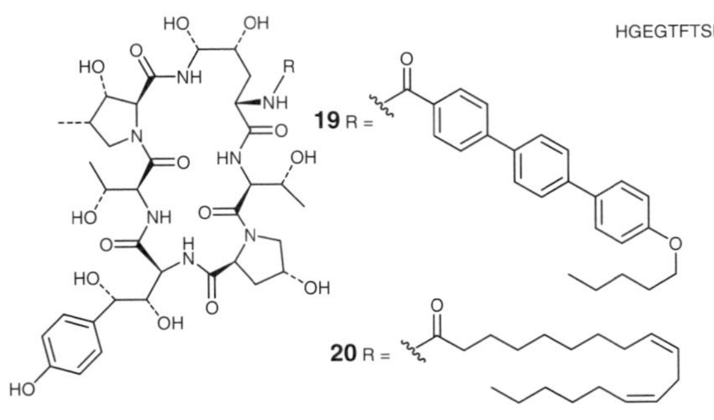

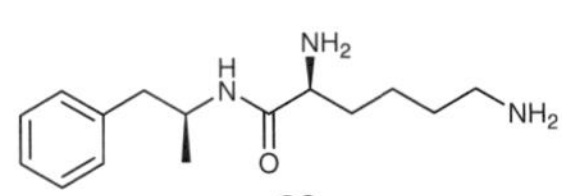

22

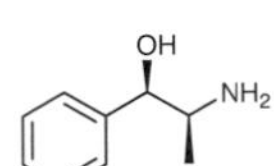

23

24 R =

25 R = OH

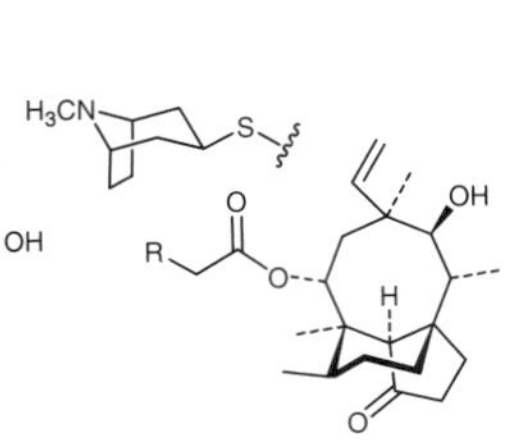

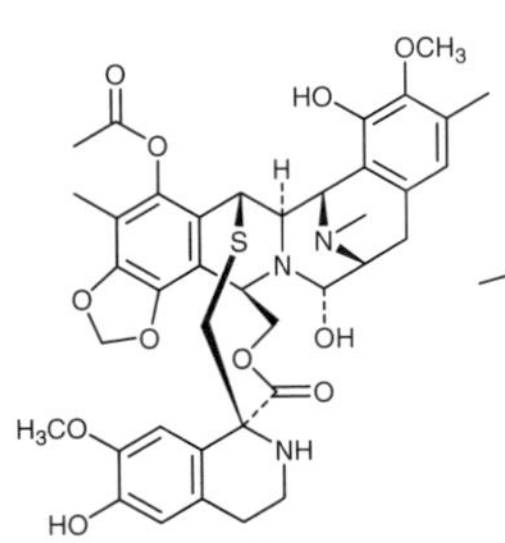

27

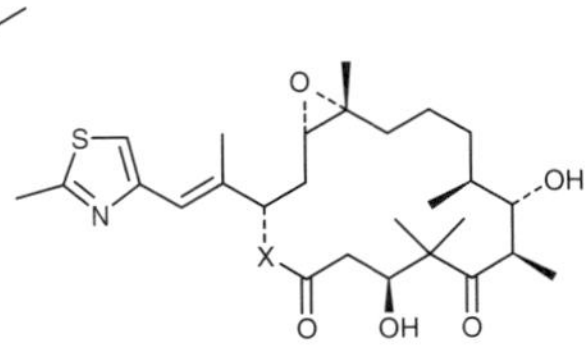

28 X = NH
29 X = O

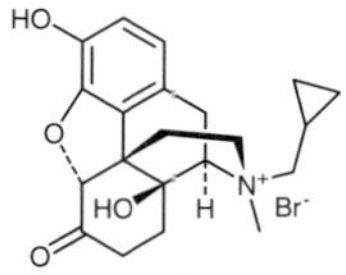

30

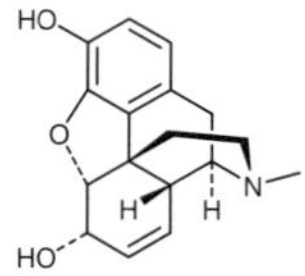

31

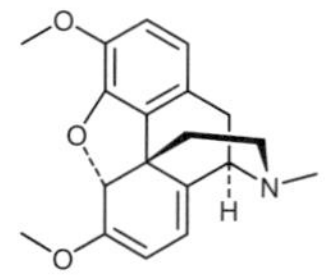

34

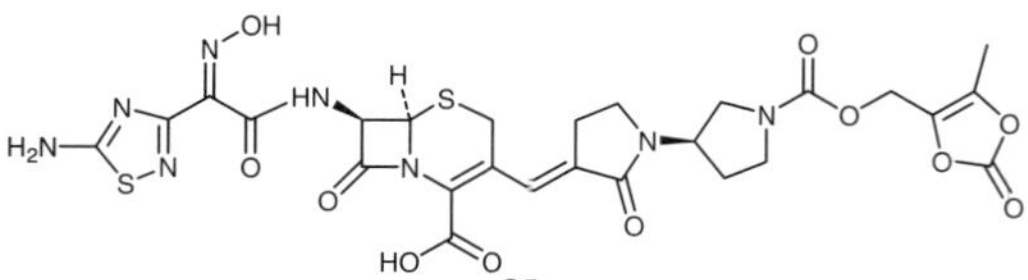

32

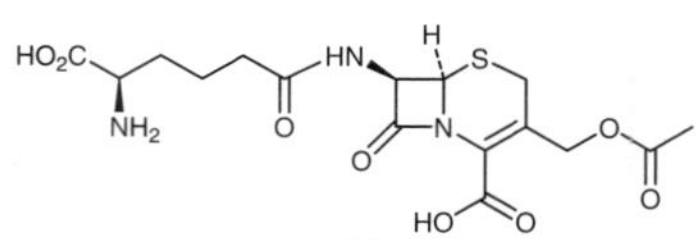

33

Methylnaltrexone bromide (MOA-728) **30** (Wyeth and Progenics) was first approved in Canada in April 2008 for the treatment of opioid-induced constipation using subcutaneous dosing.[34] Approval was obtained in the USA later that April and in Europe in July 2008.[34] Methylnaltrexone **30** is also being evaluated in Phase II trials for the treatment of opioid-induced constipation with oral dosing and Phase III trials for post-operative ileus with intravenous dosing. Methylnaltrexone **30** works by blocking peripheral opioid receptors activated by other opioids administered for pain relief that cause side effects such as constipation, urinary retention and severe itching. Methylnaltrexone **30** has minimal pain-relieving properties as it does not cross the human blood–brain barrier.[35,36] Methylnaltrexone **30** is the *N*-methyl derivative of naltrexone, which is an approved drug used in the management of alcohol and opioid dependence and is made semi-synthetically from thebaine **34**. The alkaloids morphine **31** and thebaine **34** are both produced by the opium poppy, *Papaver somniferum*.

Ceftobiprole medocaril (BAL-5788) (Basilea Pharmaceutica and Johnson & Johnson) **32** was first approved in Canada (June 2008) and later in Switzerland (November 2008) for the treatment of complicated skin and soft structure infections (cSSSIs).[37] Ceftobiprole medocaril **32** is also undergoing various Phase III trials for the treatment of hospital- and community-acquired pneumonia (HAP/CAP). Ceftobiprole medocaril **32** is a fourth-generation semi-synthetic cephalosporin derivative that has potent bactericidal activity against methicillin-resistant *Staphylococcus aureus* (MRSA) and penicillin-resistant *Streptococcus pneumoniae*, as well as *Enterobacter cloacae*, *Escherichia coli*, *Klebsiella pneumoniae*, *Proteus mirabilis* and *Streptococcus pyogenes*.[38–41] The structure of the first cephalosporin, cephalosporin C **33**, was reported from the fungus *Acremonium chrysogenum* (previously *Cephalosporium acremonium*) by Abraham and Newton in 1961.[33,42] The cephalosporins inhibit bacterial growth by binding to penicillin-binding proteins and preventing cell wall synthesis like other β-lactams. There have been over 50 semi-synthetic cephalosporins launched since the introduction of cefalotin in 1964.[13,42–44]

3 Late Stage NDAs and Clinical Candidates

At the end of 2008, there were five NP-derived compounds in the New Drug Application (NDA) development phase in the USA and/or Market Authorisation Application (MAA) in Europe and thirty-one NP-derived compounds undergoing Phase III clinical trials (Table 11.2). Details of late stage clinical trials, mechanism of action and derivation of each compound are described in this section.

The 36 late stage compounds are being developed predominantly in the oncology (twenty compounds) and antibacterial (nine compounds) therapeutic areas, as well as for the treatment of diabetes (three compounds), multiple sclerosis (MS) (two compounds), pain relief (one compound) and cardiovascular disease (one compound). There are 25 unique lead compounds of which 12 are derived from microorganisms (six actinomycetes, four fungi, one myxobacterium and one bacterium), nine from plants, two from marine sponges and two from mammals (one bovine and one lizard).

3.1 Antibacterial

An NDA for telavancin (TD-6424) **35** (Theravance and Astellas Pharma) was filed in December 2006 and an MAA in May 2007 for the treatment of cSSSIs, especially MRSA. In November 2008, the Anti-Infective Drugs Advisory Committee of the US Food and Drug Administration (FDA) gave telavancin **35** a favourable recommendation.[45] An additional NDA was filed in January 2009 for the use of telavancin **35** for the treatment of HAP.[46] Telavancin **35** is a semi-synthetic derivative of vancomycin **36** that was designed to have a dual mode of action by disrupting the bacterial plasma barrier membrane in addition to inhibition of cell wall synthesis.[47–51] Vancomycin **36** is a glycopeptide used to treat various bacterial infections that was originally reported in 1955 from *Amycolatopsis orientalis* (previously *Streptomyces orientalis*) and its structure was determined in 1978 by X-ray crystallography.[52] Vancomycin **36** and related glycopeptides inhibit bacterial growth by preventing transglycosylation and transpeptidation reactions essential for cell wall production by binding to the D-Ala–D-Ala termini of bacterial peptidoglycan precursors.

An NDA for oritavancin (Nuvocid®, LY-333328) **37** (Targanta Therapeutics) was filed in February 2008 for the treatment of cSSSIs. However, in November 2008, the FDA's Anti-Infective Drugs Advisory Committee recommended that oritavancin **37** should not be approved for the treatment of cSSSIs and, in December 2009, the FDA informed Targanta that no approval could be obtained unless further clinical trials were undertaken.[53] Shortly after the FDA decision, Targanta announced that it would be merging with the Medicines Company.[54] Oritavancin **37**[54–56] was also being evaluated in Phase II trials for the treatment of catheter related bacteraemia and nosocomial pneumonia. Oritavancin **37** is a semi-synthetic derivative of chloroorienticin A (LY264826, A82846B, chloroeremomycin) **38**,[57] a vancomycin **36** analogue originally isolated from *Nocardia orientalis* and synthesised by Eli Lilly.[58]

Table 11.2 Natural product derived drugs in late stage clinical development (NDA or equivalent and Phase III), lead source with year structure determined and disease area (current 31 December 2008).

Compound	*Classification*	*Lead (source, year structure determined[a])*	*Disease area*
NDA or equivalent			
telavancin **35**	semi-synthetic NP	vancomycin **36** (actinomycetes, 1978)[52]	antibacterial
oritavancin **37**	semi-synthetic NP	chloroorienticin A **38** (actinomycetes, 1988)[57]	antibacterial
cethromycin **39**	semi-synthetic NP	erythromycin **40** (actinomycetes, 1957)[71]	antibacterial
ruboxistaurin **41**	NP derived	staurosporine **42** (actinomycetes, 1978)[181]	diabetic retinopathy
vinflunine **43**	semi-synthetic NP	vinblastine **44** (plant, 1965)[108]	oncology
Phase III			
dalbavancin **45**	semi-synthetic NP	A40926 **46** (actinomycetes, 1987)[64]	antibacterial
tebipenem pivoxil **47**	NP derived	thienamycin **14** (actinomycetes, 1979)[23]	antibacterial
ceftaroline **48**	NP derived	cephalosporin **33** (fungus, 1961)[33]	antibacterial
omiganan **49**	NP derived	indolicidin **50** (bovine, 1992)[84]	antibacterial (catheter)
fidaxomicin **51**	NP	tiacumicin B **51** (actinomycetes, 1987)[89]	antibacterial
eritoran **52**	NP derived	RS-DPLA **53** (bacteria, 1991)[b,101]	antibacterial (sepsis)[c]
cositecan **54**	semi-synthetic NP	camptothecin **55** (plant, 1966)[115]	oncology
fosbretabulin **56**	semi-synthetic NP	combretastatin A-4 **57** (plant, 1989)[123]	oncology
ombrabulin **58**	NP derived	combretastatin A-4 **57** (plant, 1989)[123]	oncology
larotaxel **59**	semi-synthetic NP	paclitaxel **60** (plant, 1971)[131]	oncology
cabazitaxel **61**	semi-synthetic NP	paclitaxel **60** (plant, 1971)[131]	oncology

DHA-paclitaxel **62**	semi-synthetic NP	paclitaxel **60** (plant, 1971)[131]	oncology
omacetaxine mepesuccinate **63**	NP	homoharringtonine **63** (plant, 1970)[142]	oncology
phenoxodiol **64**	NP derived	genistein **65** (plant, 1926)[150]	oncology
alvocidib **66**	NP derived	rohitukine **67** (plant, 1979)[154]	oncology
tanespimycin **68**	semi-synthetic NP	geldanamycin **69** (actinomycetes, 1970)[163]	oncology
retaspimycin **70**	semi-synthetic NP	geldanamycin **69** (actinomycetes, 1970)[163]	oncology
deforolimus **71**	semi-synthetic NP	sirolimus **9** (actinomycetes, 1975)[19]	oncology
enzastaurin **72**	NP derived	staurosporine **42** (actinomycetes, 1978)[71]	oncology
lestaurtinib **73**	NP derived	K252a **74** (actinomycetes, 1986)[184]	oncology
midostaurin **75**	NP derived	staurosporine **42** (actinomycetes, 1978)[71]	oncology
patupilone **29**	NP	epothilone B **29** (myxobacteria, 1995)[31]	oncology
irofulven **76**	semi-synthetic NP	illudin S **77** (fungus, 1965)[202]	oncology
eribulin **78**	NP derived	halichondrin B **79** (sponge, 1985)[212]	oncology
panobinostat **80**	NP derived[d]	psammaplin A **81** (sponge, 1987)219–221	oncology
morphine-6-glucuronide **82**	semi-synthetic NP	morphine **31** (plant, 1923)[229]	pain
dapagliflozin **83**	NP derived	phlorizin **84** (plant, 1929)[241]	type 2 diabetes
lixisenatide **85**	NP-derived	extenatide-4 **21** (lizard, 1992)[27]	type 2 diabetes
SCH 530348 **86**	NP derived	himbacine **87** (plant, 1961)[247]	cardiovascular
voclosporin **88**	semi-synthetic NP	cyclosporin A **89** (fungus, 1976)[255]	multiple sclerosis
fingolimod **90**	NP derived	myriocin **91** (fungus, 1973)[266]	multiple sclerosis

[a]This is the year that the lead compound's correct structure was reported in a journal.
[b]Lipid A is the endotoxic principle of Gram-negative bacteria lipopolysaccharide. Its structure varies between bacteria.
[c]Sepsis could equally also be classed as immunomodulatory.
[d]Structure of panobinostat **80** also based upon the fungal metabolite trapoxin A **92** and vorinostat **93**.

An NDA for dalbavancin (Zeven®, BI-397) **45** (Pfizer) was filed in the USA in February 2005 for the treatment of cSSSIs but, in September 2008, Pfizer announced the worldwide withdrawal of all marketing applications.[59] Pfizer will undertake further Phase III trials of dalbavancin **45** for the treatment of cSSSIs and start a paediatric programme.[59] Dalbavancin **45**[60–63] is a semi-synthetic derivative of the teicoplanin analogue A40926 **46**, which was discovered by Biosearch Italia in 1987 from *Nonomuraea* sp. and licensed to Vicuron, which was subsequently bought by Pfizer.[64–66]

Cethromycin (ABT-773) **39** (Advanced Life Sciences) had an NDA filed in October 2008 for the treatment of CAP.[67] Advanced Life Sciences is also evaluating cethromycin **39** against other respiratory tract infections and in pre-clinical studies as a prophylactic treatment of anthrax post-exposure. Cethromycin **39**[68–70] is a semi-synthetic ketolide derivative of erythromycin **40**[71] originally synthesised by Abbott Laboratories,[72] which like erythromycin **40**, inhibits bacterial protein synthesis through binding to the peptidyl-transferase site of the bacterial 50*S* ribosomal subunit. Important macrolide antibiotics in clinical use today include erythromycin **40** itself, clarithromycin, azithromycin and, most recently, telithromycin (launched in 2001).

Tebipenem pivoxil (ME-1211, L-084) **47** (Meiji Seika Kaisha), for which an NDA was submitted in Japan for use as a broad-spectrum antibiotic,[73] is an orally active, pivaloyloxymethyl ester prodrug of the carbapenem tebipenem. Tebipenem pivoxil **47**[74–76] inhibits bacterial growth through binding to penicillin-binding proteins and preventing cell wall synthesis. The lead compound for carbapenems is the actinomycete-derived thienamycin **14**, which was first reported in 1979 from *Streptomyces cattleya*.[23] There are currently six clinically used carbapenems: imipenem, panipenem, meropenem, ertapenem, biapenem and doripenem.

Ceftaroline (PPI-0903, TAK-599) **48** (Forest Laboratories) has completed Phase III trials for the treatment of cSSSIs and is undergoing Phase III trials for the treatment of CAP.[77] Ceftaroline **48** is a semi-synthetic cephalosporin derivative that was originally discovered by Takeda.[78]

Omiganan (Omigard®, CPI-226, MBI-226) **49** (Cadence Pharmaceuticals) is being evaluated as a treatment of catheter-related infections in Phase III trials using a gel-based formulation and, if successful, an NDA will be filed in the second quarter of 2009.[78] Cadence had obtained promising results in a previous Phase III trial but its primary endpoint of a reduced rate of infection was not attained. Cutanea Life Sciences are also evaluating omiganan **49** (coded as CLS001) in late stage clinical trials for the treatment of acne and rosacea.[79] Omiganan **49**[80–83] is a cationic peptide discovered by MIGENIX whose structure was based on the antibacterial and antiviral peptide indolicidin **50**, which was originally purified from the cytoplasmic granules of bovine neutrophils.[84] As with other cationic peptides, omiganan **49** exerts its antibacterial activity through cytoplasmic membrane interaction.

Fidaxomicin (tiacumicin B, difimicin, OPT-80, PAR-101) **51** (Optimer Pharmaceuticals) is being evaluated in several Phase III trials for the treatment of patients infected with *Clostridium difficile*. *C. difficile* produce toxins that cause inflammation of the colon and severe diarrhoea, which can lead to death in serious infections.[84–88] People who contract *C. difficile*-associated disease

(CDAD) have usually undergone broad spectrum antibiotic treatments that have disrupted the normal gut bacteria flora. In November 2008, Optimer announced positive Phase II results that suggested fidaxomicin **51** was superior to vancomycin **36** in both clinical cure and reoccurrence of *C. difficile* infection.[84] Fidaxomicin **51** (then called tiacumicin B) is an actinomycete-derived macrolactone isolated by Abbott Laboratories[89] that is identical to lipiarmycin A and clostomicin B1,[90–92] which blocks bacterial growth by inhibiting RNA synthesis.[93]

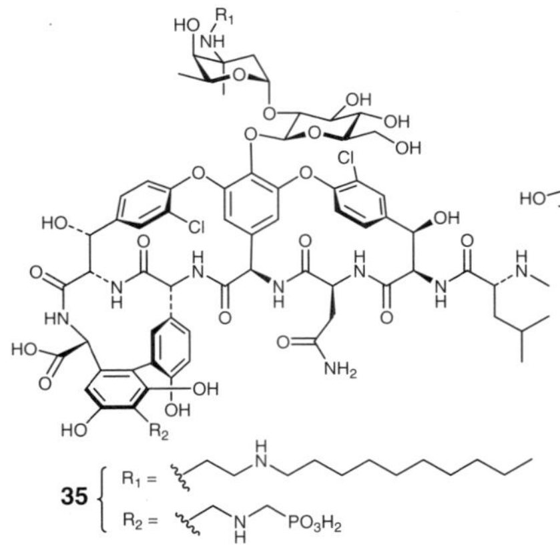

35 R_1 = ; R_2 =

36 R_1 = H, R_2 = H

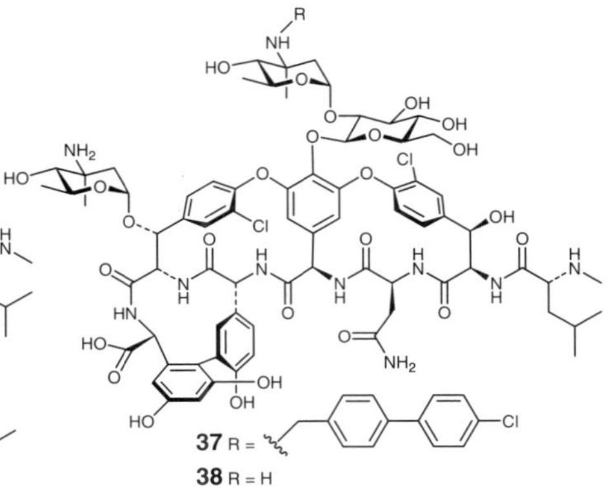

37 R =

38 R = H

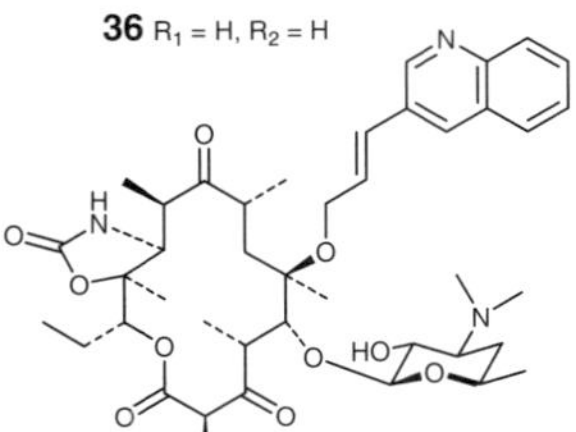

39

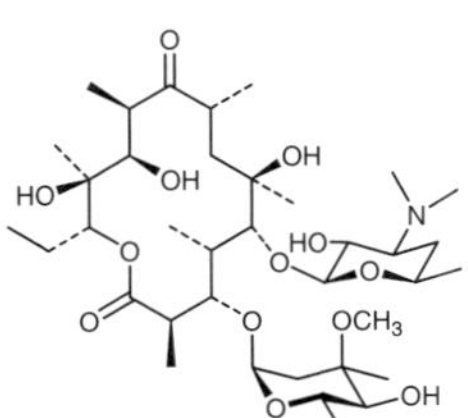

40

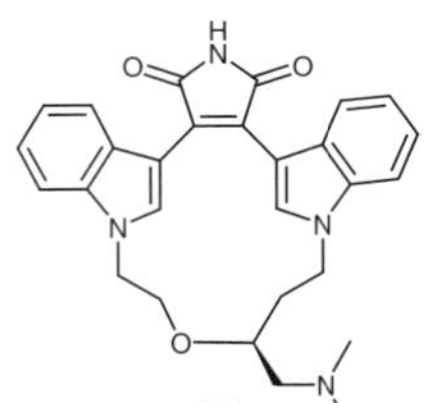

41

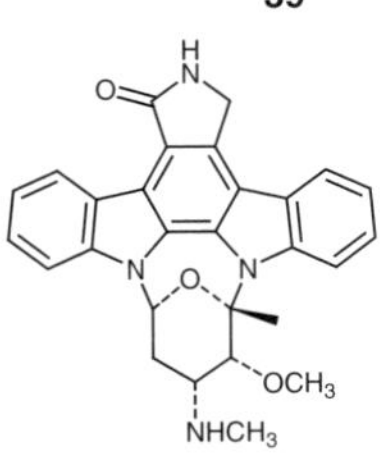

42

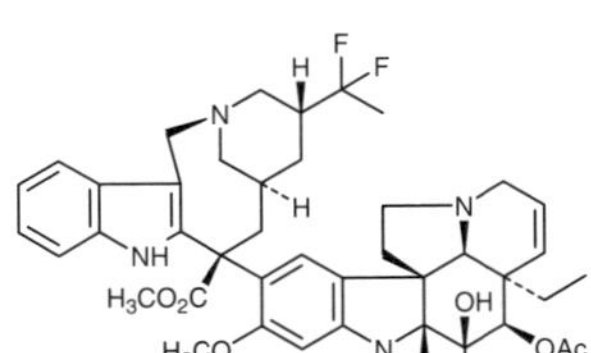

43

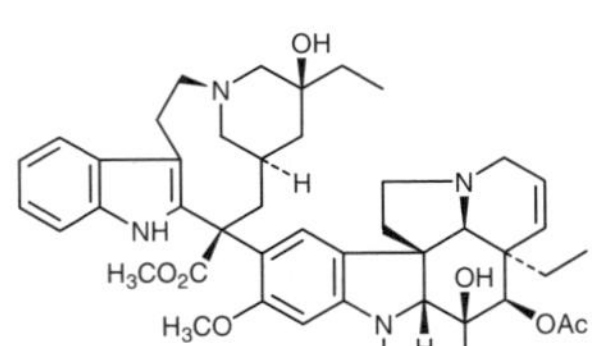

44

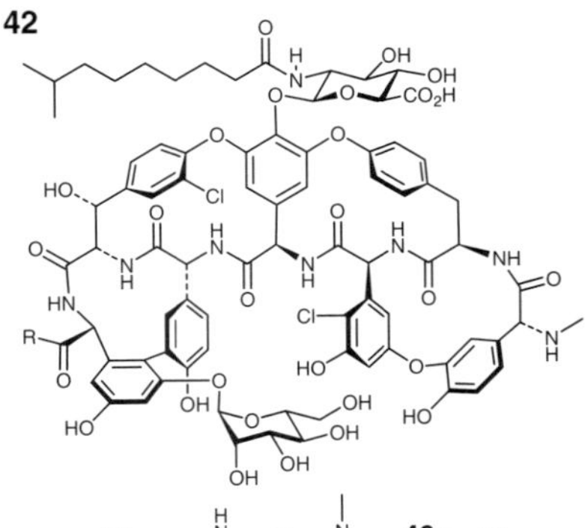

45 R = ; **46** R = OH

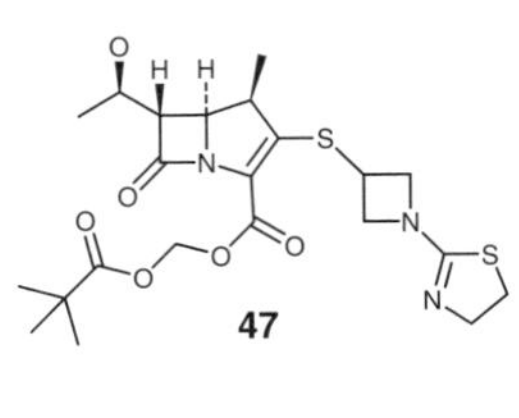

47

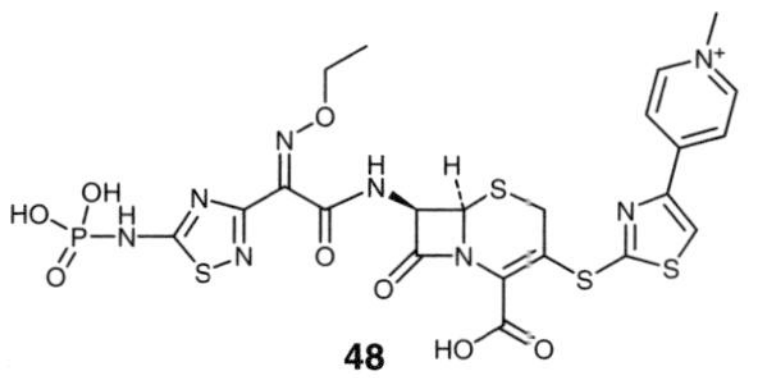

48

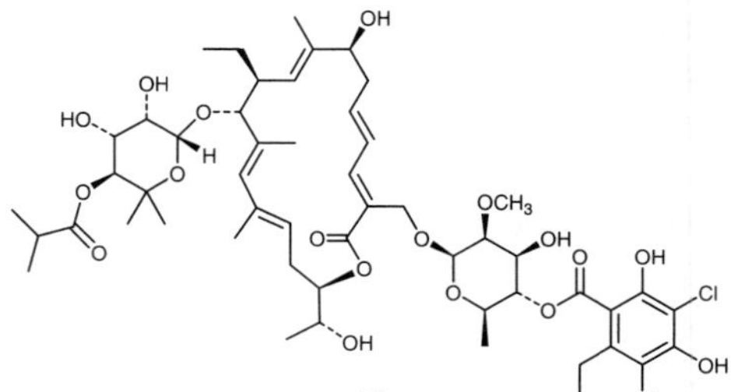

51

ILRWPWWPWRRK-NH$_2$ ILPWKWPWWPWRR-NH$_2$

49 50

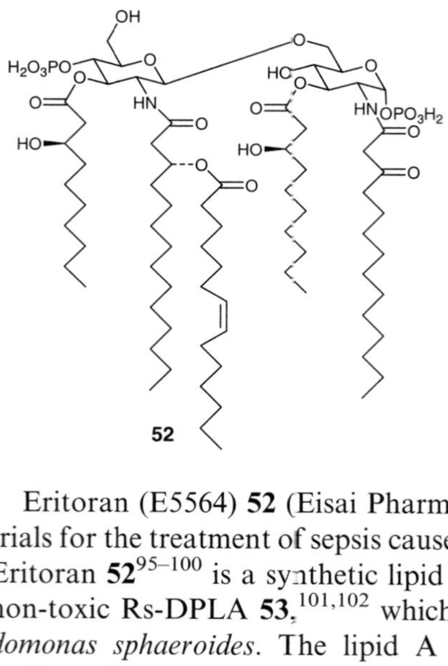

52

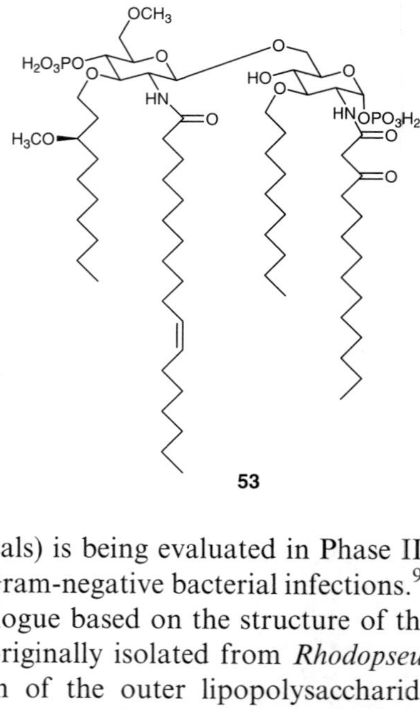

53

Eritoran (E5564) **52** (Eisai Pharmaceuticals) is being evaluated in Phase III trials for the treatment of sepsis caused by Gram-negative bacterial infections.[94] Eritoran **52**[95–100] is a synthetic lipid A analogue based on the structure of the non-toxic Rs-DPLA **53**,[101,102] which was originally isolated from *Rhodopseudomonas sphaeroides*. The lipid A portion of the outer lipopolysaccharide surface of Gram-negative bacteria activates a strong immune response in host cells that protects humans against further infection. However, increased and uncontrolled inflammation responses have been implicated in the induction of septic shock, which may lead to widespread organ damage and failure. Eritoran **52** was found to antagonise the human Toll-like receptor 4 (TLR4), which inhibits endotoxin response.

3.2 Oncology

An MAA for vinflunine (Javlor®) **43** (Pierre Fabre) has been submitted to the European Medicines Agency (EMEA) for the treatment of various cancers.[103] Pierre Fabre had been developing vinflunine **43**[104–106] in the USA in partnership with Bristol-Myers Squibb for the treatment of breast, bladder and lung cancers but development was halted in late 2007.[107] Four Vinca-type alkaloids have been approved for cancer treatment: vinblastine **44**[108] and vincristine isolated from the Madagascar periwinkle, *Catharanthus roseus*, and the semi-synthetic vinblastine derivatives, vindesine and vinorelbine. Vinflunine **43** is a semi-synthetic derivative of vinblastine **44** which is synthesised from vinorelbine by fluorination in superacid.[109] The Vinca alkaloids have been shown to inhibit tumour growth by blocking mitosis by binding to tubulin and blocking the assembly of microtubules.

Cositecan (Karenitecin®, BNP1350) **54** (BioNumerik and ASKA Pharmaceutical) is currently being evaluated in a Phase III trial for the treatment of patients with advanced ovarian cancer who have become resistant to platinum and taxane drugs.[110] Cositecan **54**,[111–114] which is also being evaluated against solid tumours in a Phase I trial, is an orally bioavailable, lipophilic 7-[2-(trimethylsilyl)ethyl] derivative of camptothecin **55** which is less sensitive to both common and camptothecin-specific resistance mechanisms. Camptothecin **55** was first isolated in 1958 from *Camptotheca acuminata* (Nyssaceae) and its structure was reported in 1966.[115–117] Camptothecin **55** was later shown to be a topoisomerase I inhibitor; two camptothecin derivatives, topotecan and irinotecan, are approved for chemotherapy use.

Fosbretabulin (combretastatin A-4 phosphate, CA4P, Zybrestat™) **56** (OXiGENE) is being evaluated in Phase II/III trials in combination with paclitaxel **60** and carboplatin as a treatment of anaplastic thyroid cancer, which is a highly lethal tumour that currently lacks an effective treatment.[118] Fosbretabulin **56** is also being evaluated against other solid tumours such as head and neck cancers (Phase I), non-small cell lung carcinoma (Phase II) and platinum-resistant ovarian cancer (Phase II), as well as for ophthalmological diseases (preclinical). Fosbretabulin **56**[119–121] is the phosphate ester of combretastatin A-4 **57**,[122] which was first reported in 1989 from the African medicinal plant *Combretum caffrum* by Pettit and co-workers.[123] Combretastatin A-4 **57** and fosbretabulin **56** are called vascular disrupting agents as they cause microtubule depolymerisation, which in turn disrupts the tumour blood vessels resulting in cell necrosis. Combretastatin A-4 **57** could not be developed as a drug due to poor water solubility.

Ombrabulin (AVE8062, AC-7700) **58** (Sanofi-Aventis) is being evaluated in Phase II/III trials as a treatment for advanced stage soft tissue sarcoma for patients who have failed previous anthracycline and ifosfamide treatments.[124] Ombrabulin **58**[125–128] is a synthetic combretastatin analogue that was licensed by Sanofi-Aventis from Ajinomoto and is also a vascular disrupting agent.

Larotaxel (XRP-9881, RPR 109881A) **59** (Sanofi-Aventis) is undergoing Phase III trials in patients with advanced pancreatic cancer who had been previously treated with gemcitabine, as well as in combination with cisplatin to treat locally advanced/metastatic urothelial tract or bladder cancer.[124] A Phase III trial for the treatment of advanced breast cancer has been completed. Larotaxel **59**[129,130] is a semi-synthetic derivative of 10-deacetyl baccatin III with a docetaxel-like side chain that has a low affinity for the P-glycoprotein drug efflux pump, an efflux mechanism that diminishes the effectiveness of the marketed drugs paclitaxel **60** and docetaxel. Importantly, this low affinity should enable larotaxel **59** to be effective in tumours resistant to paclitaxel **60** and docetaxel. Paclitaxel **60** (then called taxol) was first reported in 1971 from the Pacific yew tree *Taxus brevifolia* by Wall and Wani[131] and was later shown to stabilise microtubules.[132] 10-Deacetyl baccatin III is used as a synthetic template for most semi-synthetic taxanes as it is present in the leaves of the European yew tree, *Taxus baccata*, in a considerably larger quantity than paclitaxel **60**.[133]

XRP6258 (cabazitaxel, TXD-258, RPR-116258A) **61** (Sanofi-Aventis) is undergoing Phase III trials in combination with prednisone as a treatment of hormone resistant prostate cancer in patients who had previously been treated with a paclitaxel **60**.[124] XRP6258 **61**[134,135] is a semi-synthetic derivative of the paclitaxel **60** with a docetaxel-like side chain which, like larotaxel **59**, has a low affinity for the P-glycoprotein drug efflux pump.

DHA-paclitaxel (Taxoprexin®) **62** (Luitpold) is being evaluated in a Phase III trial for the treatment of advanced non-small cell lung cancer in combination with carboplatin,[136] as well against a variety of cancers in Phase II trials. DHA-paclitaxel **62**,[137–139] which is the C-2′ docosahexaenoic acid (DHA) ester of paclitaxel **60**, was licensed by Luitpold from Protarga in 2003. Polyunsaturated fatty acids have been reported to be taken up by tumour cells at a higher rate than normal cells and, as a consequence, a polyunsaturated fatty ester such as DHA-paclitaxel **62** may be absorbed more specifically in tumour cells. In preclinical studies, DHA-paclitaxel **62** exhibited increased activity in mice xenograft models and was more stable *in vivo* than paclitaxel **60**. However, DHA-paclitaxel **62** may still suffer from efflux problems associated with paclitaxel **60** and docetaxel resistant tumours.

Omacetaxine mepesuccinate (homoharringtonine, Ceflatonin®) **63** (ChemGenex) is being evaluated in Phase II/III trials as a treatment for patients with chronic myeloid leukaemia (CML) who did not responded to imatinib therapy and have the T315I bcr-abl point mutation.[140] Omacetaxine mepesuccinate **63** is also being evaluated in a Phase II trial for CML patients who have failed multiple protein tyrosine kinase (PTK) inhibitor therapies and for acute myelogenous leukaemia (AML). Omacetaxine mepesuccinate **63**[141,142] (first called homoharringtonine) is an alkaloid from the Chinese evergreen tree *Cephalotaxus harringtonia* that has been used in China since 1970s for the treatment of AML. Omacetaxine mepesuccinate **63** is found along with a series of closely related ester analogues in the leaves of *Cephalotaxus harringtonia*. Ester hydrolysis to form the inactive alcohol cephalotaxine and re-esterification produces semi-synthetic omacetaxine mepesuccinate **63**, which is being used for clinical studies.[143] It was recently reported that omacetaxine mepesuccinate **63** may act as a broad-spectrum PTK inhibitor by inhibition of the signal protein phosphorylation of key oncogenic proteins including JAK2V617F and BCR/BL.[144]

Phenoxodiol **64** (Marshall Edwards) is being evaluated in a Phase III trial for the treatment of ovarian cancer in combination with carboplatin, as well as in earlier stage trials as a monotherapy and in combination with docetaxel for the treatment of prostate and cervical cancer.[145] Phenoxodiol **64**[146–149] is a synthetic analogue of genistein **65**,[150] an isoflavone present in many plants including soybeans and red clover, which causes major downstream disturbances in cancer cell protein expression through binding to a cell membrane oxidase.

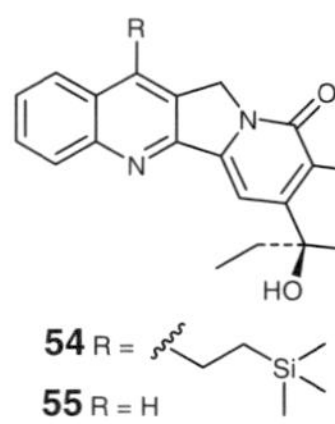

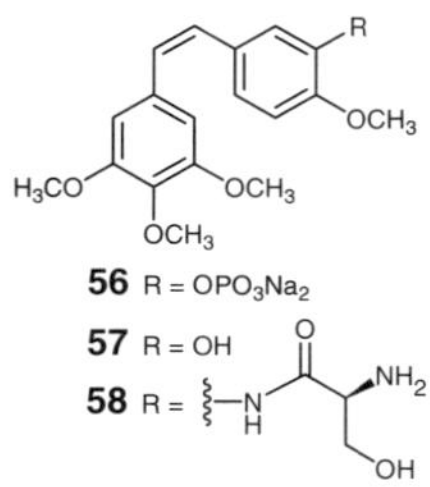

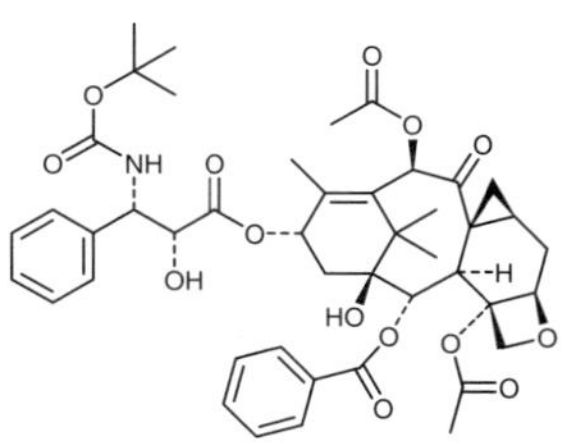

59

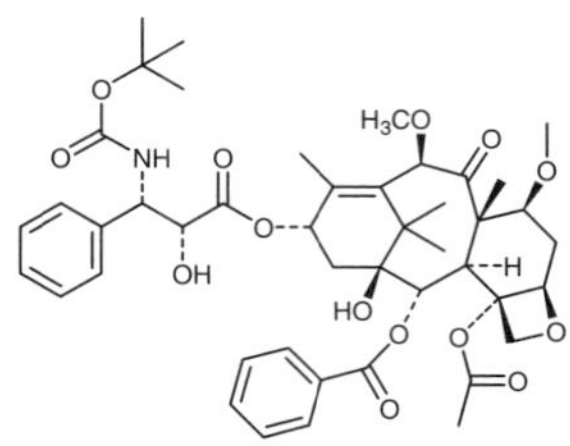

61

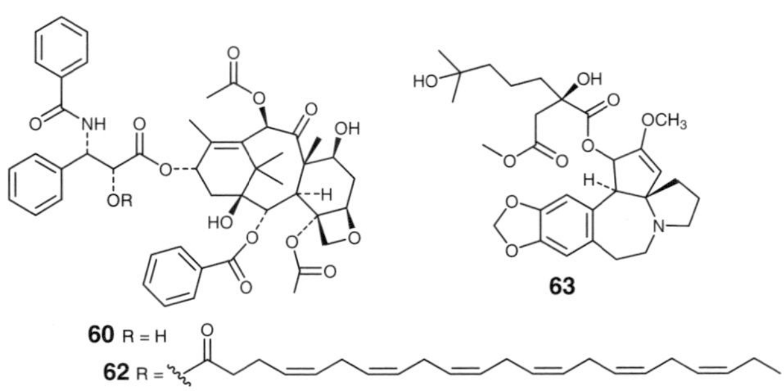

63

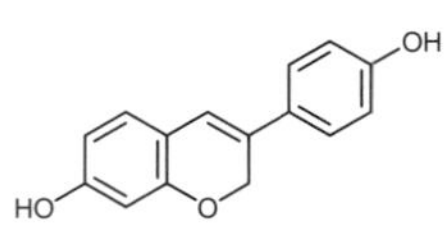

64

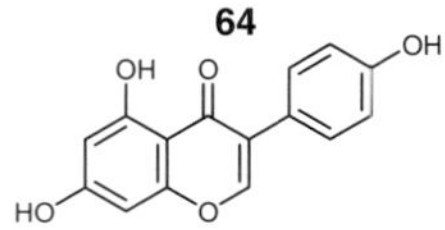

65

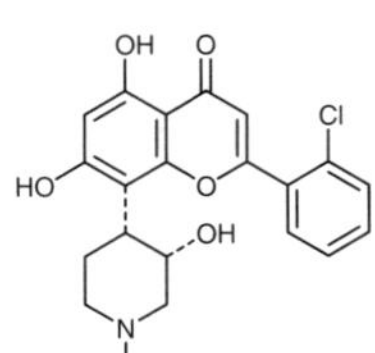

66

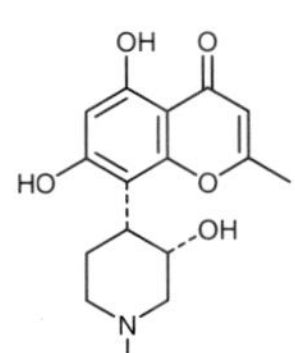

67

Alvocidib (flavopiridol, HMR 1275) **66** (Sanofi-Aventis) is being evaluated in Phase II/III trials for the treatment of chronic lymphocytic leukaemia in collaboration with Ohio State University and the US National Cancer Institute (NCI).[124] Alvocidib **66** is also being evaluated in Phase I and II trials against various haematological malignancies. Alvocidib **66**[151–153] exerts it activity though inhibition of cyclin-dependent kinase and is a synthetic derivative of rohitukine **67**, an alkaloid first isolated from *Amoora rohituka*[154] and subsequently from *Dysoxylum binectariferum*.[155]

Tanespimycin (17-AAG, KOS-953, NSC-330507) **68** (Bristol-Myers Squibb, acquired the previous developers Kosan in June 2008) is being evaluated in a Phase III trial in combination with bortezomib as treatment for patients with multiple myeloma in first relapse and in Phase II/III trials as a standalone agent.[156] Tanespimycin **68** is the 17-allylamino-17-demethoxy semi-synthetic derivative of geldanamycin **69** developed by the NCI.[157–161] Geldanamycin **69** was first reported from *Streptomyces hygroscopicus* in 1970[162,163] and was later shown to bind strongly to heat shock protein 90 (Hsp90), a component of a multichaperone complex with important roles in the development and progression of pathogenic cellular transformation. Geldanamycin **69** was too toxic for commercial development and tanespimycin **68** was found to have a more favourable toxicity profile while retaining potent Hsp90 affinity.

Retaspimycin (IPI-504) **70** (Infinity Pharmaceuticals) is being evaluated in a Phase III trial to treat patients with gastrointestinal stromal tumours that have failed previous treatments with imatinib and/or sunitinib.[164] Retaspimycin **70** is a hydroquinone derivative of tanespimycin **68** that has significantly improved solubility while retaining potent Hsp90 binding activity.[165–167]

Deforolimus (MK-8669, AP-23573) **71** (ARIAD Pharmaceuticals and Merck & Co) is being evaluated in a Phase III clinical trial to treat patients with metastatic soft-tissue or bone sarcomas, as well as in other Phase I and II trials against various tumours.[168] Deforolimus **71** is a semi-synthetic dimethylphosphinate derivative of sirolimus (rapamycin) **9** discovered by ARIAD, which has been co-developed with Merck since July 2007.[169–172] Like other sirolimus **9** derivatives, deforolimus **71** potently inhibits the mammalian target of rapamycin (mTOR), which is a central regulator of various signalling pathways. The parent NP sirolimus **9** (launched 1999) and the semi-synthetic derivative everolimus **8** (2003) are used clinically as immunosuppressants, while the semi-synthetic derivatives zotarolimus **18** (2005) and temsirolimus **26** (2007) are approved for use in a drug-eluting coronary stent and as a treatment for advanced renal cell carcinoma respectively.

Enzastaurin (LY317615) **72** (Eli Lilly) is being evaluated in a Phase III trial for the treatment of relapsed glioblastoma multiform, which is an aggressive and malignant form of brain cancer and in another Phase III trial to investigate whether enzastaurin **72** can prevent the relapse of patients with diffuse large B cell lymphoma.[173] Enzastaurin **72** is also being evaluated against a variety of other cancers in earlier stage clinical trials. Enzastaurin **72**[174–180] is a synthetically derived staurosporine **42** analogue that is a potent inhibitor of serine/threonine protein kinase Cβ (PKCβ). PKCβ is an important target as activated PKCβ phosphorylates GSK3β, which has been shown to lead to apoptosis suppression. Staurosporine **42** was first isolated from the actinomycetes *Saccharothrix aerocolonigenes* (originally *Streptomyces staurosporeus*) and its structure was determined by X-ray crystallography in 1978.[181,182]

Lestaurtinib (CEP-701, KT-5555) **73** (Cephalon) is being evaluated in a Phase III trial for the treatment of AML and is in Phase II trials against a variety of cancers and for the treatment of psoriasis.[183] Lestaurtinib **73** is a synthetic derivative of K252a **74**,[184] a indolocarbazole originally isolated from actinomycetes *Nonomuraea longicatena* that is related to staurosporine **42**, in which the methyl ester has been reduced to a primary alcohol. Lestaurtinib **73**[185–190] has been reported to be a potent inhibitor of FMS-like tyrosine kinase (FLT3), which is present in a mutant form in approximately a third of AML patients, and Janus kinas 2 (JAK2), which is tyrosine kinase that signals between cytokine receptors and other important downstream targets.

Midostaurin (*N*-benzoylstaurosporine, PKC412, GCP41251) **75** (Novartis) is being evaluated in a Phase III trial with or without daunorubicin and cytarabine, followed by treatment with cytarabine and midostaurin **75** combination, for patients with newly diagnosed FLT3 mutated acute myeloid leukaemia.[191] Midostaurin **75**[190,192–195] is the *N*-benzoyl derivative of staurosporine **42** and has a similar mechanism of action to lestaurtinib **73** though inhibition of various kinases such PKC, FLT3, VEGFR-2 (vascular endothelial cell growth factor receptor), c-KIT and PDGFR (platelet derived growth factor receptor).

Patupilone (epothilone B, EPO906) **29** (Novartis) is being evaluated in a Phase III trial for the treatment of patients with ovarian, primary fallopian or peritoneal cancer.[191,196,197] Patupilone (then called epothilone B) **29** was first reported by Höfle and co-workers[198–200,31] from the myxobacterium *Sorangium cellulosum* in a 1991 patent application and epothilones were shown by workers at Merck in 1995 to have tubulin-stabilising activity similar to that of paclitaxel **60**.[31] Ixabepilone **28**, which is the semi-synthetic lactam derivative of patupilone **29**, was the first epothilone derivative approved (October 2007) for the treatment of breast cancer.

Irofulven (MGI-114, E7850, HMAF) **76** (Eisai Pharmaceuticals, which acquired the previous developers MGI Pharma in February 2008) is being evaluated in a Phase III trial to treat patients with pancreatic cancer, as well as in other Phase II cancer studies.[201] Irofulven **76**[202–204] is a semi-synthetic derivative of illudin S **77**, which was originally isolated from the fungus *Omphalotus illudens*.[205] Irofulven **76**, which is made by treatment of illudin S **77** with dilute sulfuric acid, has two orders of magnitude increased DNA damaging activity compared with illudin S **77**.

Eribulin (E7389, NSC-707389) **78** (Eisai Pharmaceuticals) is being evaluated in a Phase III trial for the treatment of breast cancer and Phase II trials against non-small cell lung cancer, prostate cancer and sarcomas.[201] Eribulin **78**[206–211] is synthetically inspired from the sponge metabolite halichondrin B **79**, which was originally isolated from the *Halichondria okadai* in 1985.[212] Eribulin **78** is the right hand part of halichondrin B **79** in which the macrocyclic ester has been replaced with a ketone and the polyether ring system replaced with a tetrahydrofuran. The total synthesis of such a complex molecule was achieved by modifying Kishi's total synthetic approach to halichondrins.[213] It has been proposed recently that eribulin **78** blocks mitosis through suppression of spindle microtubule dynamics.[211]

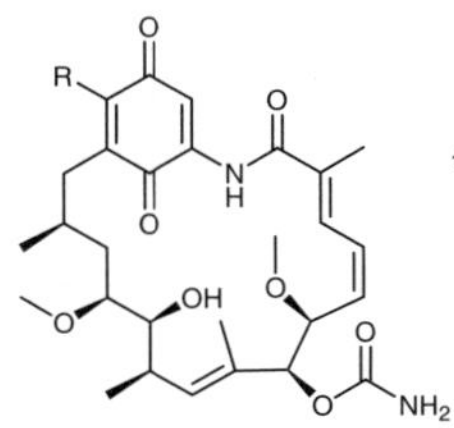

68 R = 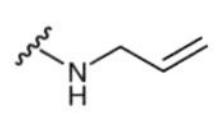

69 R = OCH_3

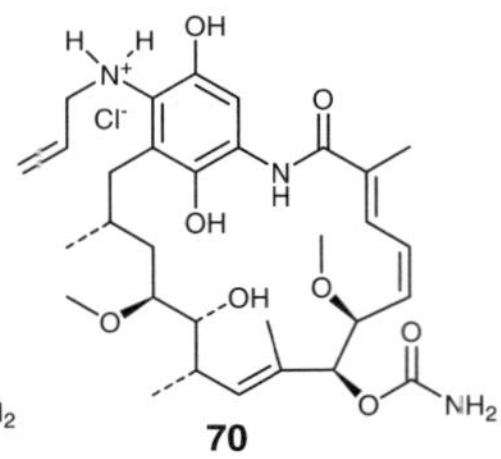

70

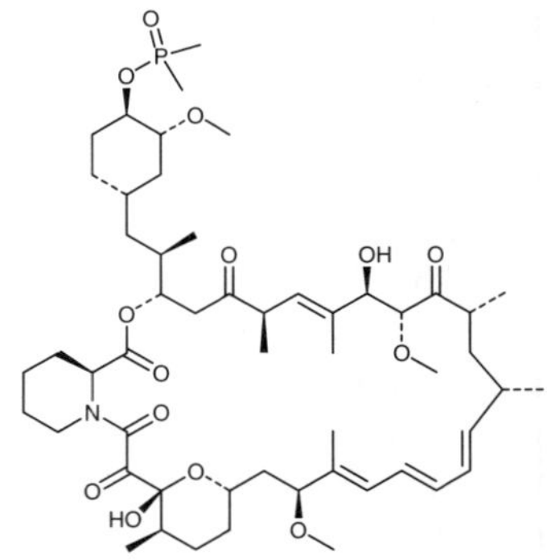

71

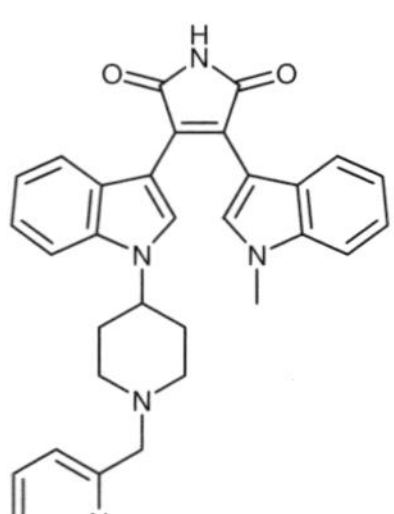

72

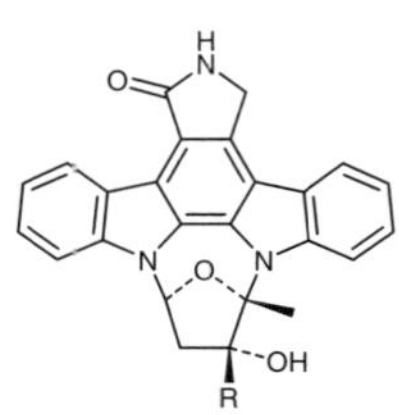

73 R = CH_2OH

74 R = CO_2CH_3

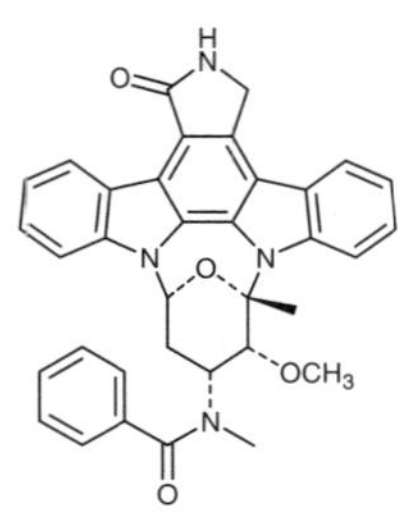

75

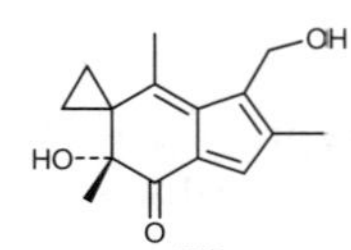

76

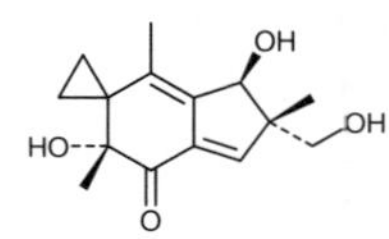

77

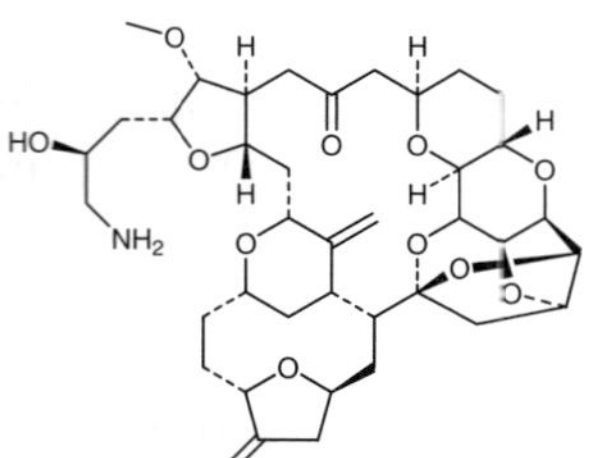

78

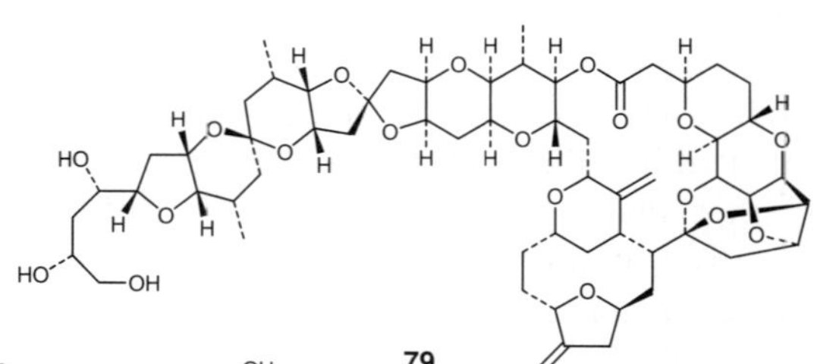

79

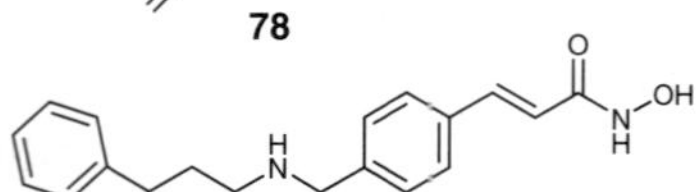

80

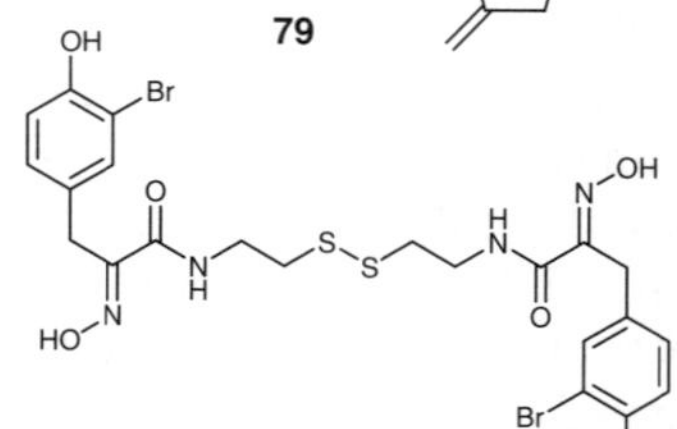

81

Panobinostat (LBH-589) **79** (Novartis) is being evaluated in Phase II/III trials for the treatment of refractory cutaneous T-cell lymphoma, while Phase II/III trials against refractory chronic myeloid leukaemia have been completed.[191] Panobinostat **79** is also being evaluated against a variety of other cancers in Phase I and II trials. Panobinostat **79**[214–218] is a synthetically derived compound whose structure is based upon the histone deacetylase (HDAC) inhibitors psammaplin A **80**, trapoxin B **92** and vorinostat **93**. Psammaplin A **79** was first reported in 1987 independently by the Crews,[219] Schmitz[220] and Scheuer[221] groups and has been subsequently isolated from a number of sponges, while trapoxin B **92** is a fungal-derived tetrapeptide.[222] Vorinostat **93** is a hydroxamic acid containing HDAC inhibitor that is approved for the treatment of cutaneous T cell lymphoma.[223] HDACs are a family of enzymes that remove acetyl groups from an *N*-acetyl lysine amino acid on histone tails, leading to chromatin compaction and transcriptional repression. In addition, HDACs can also influence DNA repair and deacetylate non-histone proteins involved in cell proliferation and death.

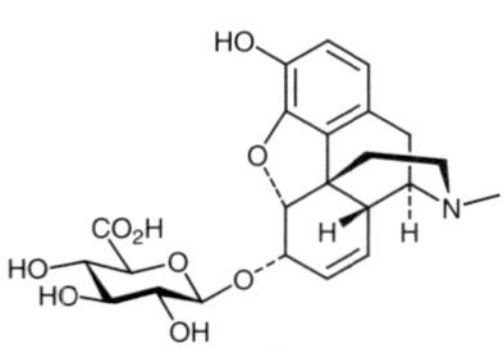

82

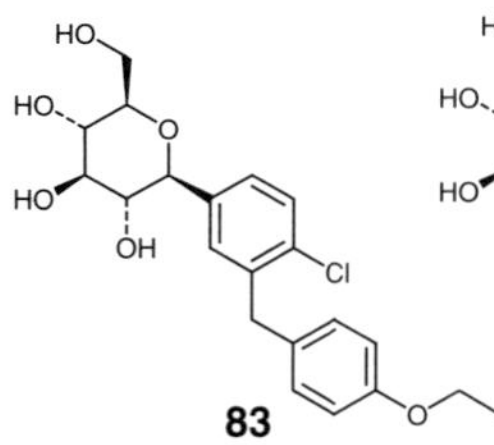

83

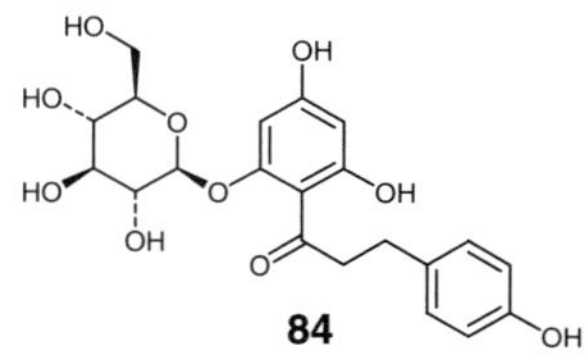

84

HGGGTFTSDLSKQMEEEAVRLFIGWLKNGGPSSGAPPSKKKKKK-NH_2

85

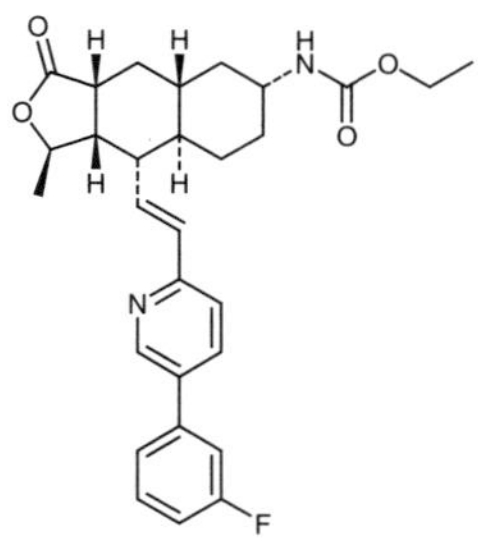

86

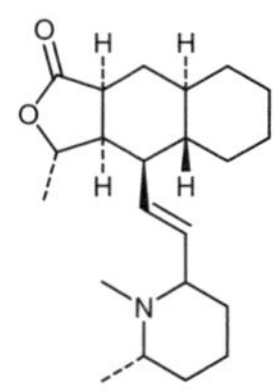

87

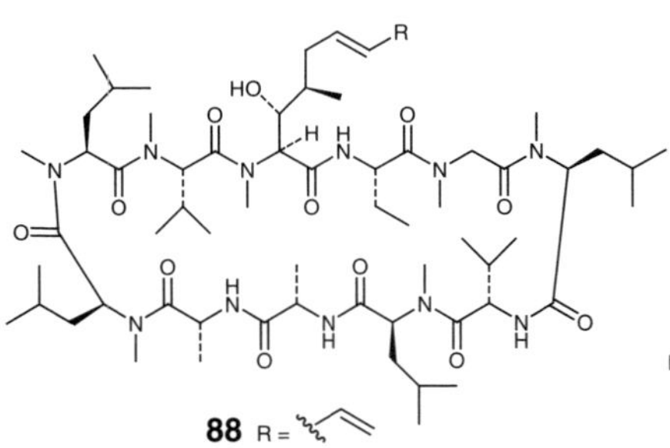

88 R =

89 R = H

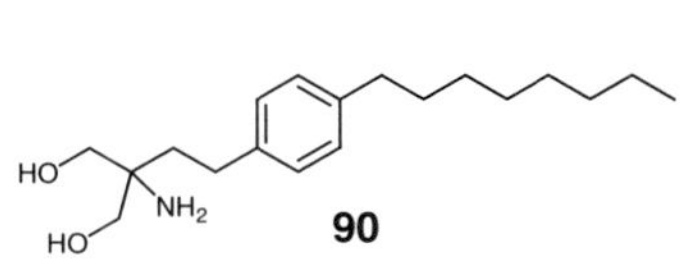

90

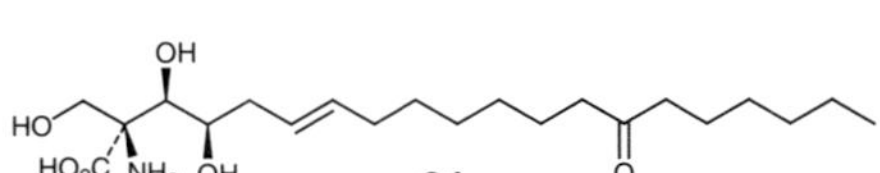

91

3.3 Other Therapeutic Areas

Ruboxistaurin (LY333531) **41** (Eli Lilly) is being evaluated in a Phase III trial for the treatment of diabetic macular oedema.[224] Lilly had submitted an NDA in February 2006 to the FDA for the treatment of diabetic retinopathy and received an Approvable Letter in September 2006 that requested another Phase III trial for additional efficacy data. The EMEA also required further clinical data and, as a consequence, Lilly withdrew its European MAA. Ruboxistaurin **41**[225–228] competitively inhibits adenosine triphosphate (ATP) binding to PKCβ and is a synthetic analogue of staurosporine **42**.

Morphine-6-glucuronide (M6G) **82** (PAION AG, which acquired the previous developers CeNeS Pharmaceuticals in June 2008), which is one of two morphine **31**[32,229] glucuronide metabolites formed in the human body,[230] has been evaluated in two Phase III trials for the treatment of patients suffering from post-operative pain following knee surgery and major abdominal surgery.[231] The knee surgery trial confirmed that M6G **82**[232–234] was effective at a single dose and demonstrated nausea levels similar to placebo. However, although the abdominal surgery trial showed that M6G **82** controlled analgesia as well as morphine **31**, M6G **82** narrowly failed to demonstrate a statistically significant reduction of nausea ($p=0.052$ against a target of <0.050).[230] There also appeared to be a reduction in sedation with M6G **82** during the early hours following surgery. Further analysis of Phase II and III data and modelling studies of dose–response relationships and pharmacodynamic effects suggested that M6G **82** has a wider therapeutic margin than morphine **31**, with lower incidence of post-operative nausea and vomiting at equivalent doses. Therefore, PAION will continue with existing partnerships to evaluate M6G **82** in another Phase III trial at higher doses.

Dapagliflozin (BMS-512148) **83** (Bristol-Myers Squibb and AstraZeneca) is being evaluated in various Phase II and III trials for the treatment of type 2 diabetes.[235] Dapagliflozin **83**[236–238] is a selective inhibitor of sodium glucose co-transporter-2 (SGLT2) whose structure is based upon phlorizin **84**, which was first isolated from the root bark of the apple tree in 1835.[239,240] Although the position of the sugar moiety of phlorizin **84** had been determined previously, Müller and Robertson unequivocally showed in 1933 that the sugar present was β-glucose.[241] Phlorizin **84** has been shown to lower glucose plasma levels and improve insulin resistance levels through inhibition of SGLTs, but could not be developed as a drug because of poor intestinal absorption and inactivation by lactase-phlorizin hydrolase.

Lixisenatide (AVE0010, ZP10) **85** (Sanofi-Aventis) is undergoing various Phase III trials as a potential treatment for type 2 diabetes.[242] Lixisenatide **85**[243,244] is a synthetic analogue of exenatide 4 **21**, a 39 amino acid peptide originally isolated from the oral secretions of the Gila monster lizard (*Heloderma suspectum*),[27] which was synthesised by Zealand Pharmaceuticals and licensed to Sanofi Aventis in 2003. Exenatide 4, now known as exenatide (Byetta®) **21** (Eli Lilly and Amylin), was launched in 2005 to help improve blood sugar control in patients with type 2 diabetes and has a structure similar to glucagon-like peptide-1 (GLP-1), a human hormone that helps the pancreas to regulate glucose-induced insulin secretion when the blood glucose levels are elevated. Although not strictly within the NP-derived definition used in this chapter, a long-acting human GLP-1 analogue, liraglutide (NN2211) (Novo Nordisk), had an NDA and a MAA filed in May 2008.[245]

SCH 530348 **86** (Schering Plough) is in Phase III trials for the secondary prevention of cardiovascular diseases such as atherosclerosis, ischemia, myocardial infarction and stroke.[246] The structure of SCH 530348 **86** is based upon the plant alkaloid himbacine **87**, which was first isolated in 1961 from the Australian plant, *Galbulimima baccata*.[247] There is an element of serendipity to the discovery of SCH 530348 **86** as scientists at Bristol-Myers Squibb were originally evaluating himbacine **87** derivatives in an Alzheimer's programme when it was discovered that they had thrombin receptor antagonist activity.[248] This led to the synthesis of over 2000 analogues, culminating in the discovery of the orally bioavailable derivative SCH 530348 **86**.[249]

Voclosporin (ISA-247, R1524) **88** (Isotechnika) is being evaluated in a Phase III trial for the treatment of psoriasis,[250] as well as a Phase III trial by Lux Biosciences as for the treatment of uveitis (coded as LX211, Luveniq™).[251,252] In addition, voclosporin **88** has completed a Phase IIb trial for the prevention of kidney graft rejection. Voclosporin **88**[253,254] is a slightly more potent but less toxic semi-synthetic derivative of the fungal-derived immunosuppressant cyclosporin A **89**, which has the same mechanism of calcineurin inhibition. Cyclosporin A **89** was first isolated from *Tolypocladium inflatum* by workers at Sandoz and its structure was published in 1976.[255,256]

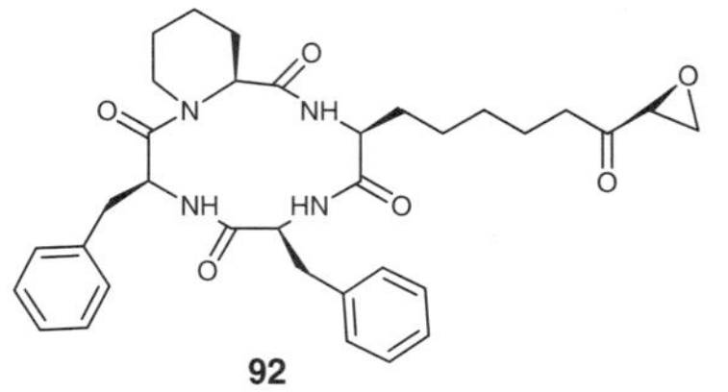

92

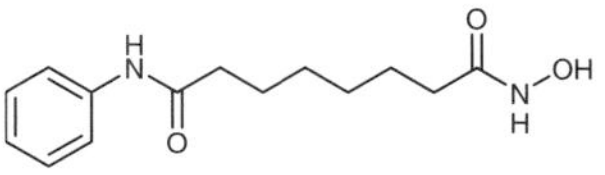

93

Fingolimod (FTY720) **90** (Novartis and Mitsubishi Tanabe) is being evaluated in Phase III trials for the treatment of multiple sclerosis (MS). Novartis reported in December 2008 that fingolimod **90** had shown superior efficacy compared with interferon beta-1a in the TRANSFORMS Phase III study for the treatment of patients with relapse remitting MS.[257] Fingolimod **90**[258–261] is a prodrug that is phosphorylated *in vivo* by sphingosine kinase to form a potent agonist of sphingosine-1-phosphate (S1P) receptors 1, 3, 4 and 5. S1Ps are critical regulators of cell growth and death that regulate the cellular balance of S1P and ceramide. In addition to being a potential MS drug, fingolimod **90** has been an important tool in the elucidation of S1P biological pathway. Fingolimod **90** is a synthetic compound whose structure was inspired by fungal metabolite myriocin **91**,[262] that was originally isolated as an antifungal agent and later identified as an immunosuppressant.[263–265]

4 Conclusions and Outlook

That NPs are still making an impact in drug development is undeniable with 21 NP-derived drugs launched since 2003 (Table 11.1) and 36 compounds (five NDAs and 31 in Phase III) in late stage development (Table 11.2). The 21 marketed NP-derived drugs are spread through a variety of therapeutic areas with anti-infectives (five antibacterial, one antifungal and one antiparasitic) and oncology (three compounds) well represented. The focus on anticancer and antibacterial agents is even more pronounced with late stage development compounds, 21 being evaluated in oncology and nine as antibacterials (including one for sepsis). The number of antibacterials probably reflects the resurgence of interest around 10–15 years ago in the development of new treatments for multidrug resistant Gram-positive bacterial infections such as MRSA. However, the FDA's move towards non-inferiority Phase III trials has considerably slowed the development of many late stage antibacterials in the USA.[266] There are a very large number of NP-derived oncology compounds in late stage development of which several have new templates that are not present in any approved cancer drugs (combretastatin, homoharringtonine, daidzein, rohitukine, geldanamycin, staurosporine, illudin, halichondrin and psammaplin). In addition, ixabepilone **28**, which was launched in 2007, is the first member of myxobacterial-derived epothilone class of anticancer agents. However, there is a high attrition rate in oncology programmes with Tufts Center for the Study of Drug Development (CSDD) showing that the FDA approves only 9% of oncology compounds that start Phase I clinical trials.[267] The remaining seven late stage clinical candidates are being developed for the treatment of diabetes (three compounds), multiple sclerosis (two compounds), pain relief (one compound) and cardiovascular disease (one compound). Of the 25 unique lead compounds (Table 11.2), 12 are derived from microorganisms (six actinomycetes, four fungi, one myxobacterium and one bacterium), nine from plants, two from marine sponges and two from mammals (one each from bovine and lizard). Although it is of no surprise that microorganisms and plants predominate, it is pleasing to see that other compound sources are also providing late stage drug candidates.

The late stage NP-derived compounds discussed in this chapter originate from research undertaken at least ten years ago. This is demonstrated by Tufts CSDD data, which shows there is an average time of 8.5 years from the start of clinical testing to FDA approval with a 21.5% success rate.[267] As a consequence, late stage clinical compounds represent a window into the past and certainly do not reflect the current state of NP-derived drug development.

As detailed in previous reviews,[1,2] there were 13 new NP compounds with new templates discovered between 1990 and 2008 which have entered clinical trials, but only fingolimod **90** (Phase III, MS), bevirimat (Phase II, HIV) (see Chapter 13), romidepsin (Phase II, oncology), salinosporamide A (Phase I, oncology) (see Chapter 12) and the semi-synthetic pladienolide derivative E7107 (Phase I, oncology) remain in clinical development at the end of 2008. This continues the worrying trend of the lack of truly novel NP templates entering clinical trials.

There has been a noticeable decline in NP research in the academic, industrial and governmental sectors over the last 20 years, which is reflected, in part, by the failure of many major universities to support NP chemistry programmes as well as the closure of NP drug discovery programmes in the industrial sector. This trend seems paradoxical given that NPs provide a chemical space not easily accessible with synthetic compounds and have a privileged place within certain therapeutic areas such as anti-infectives, oncology and immunosuppression. This concern is further heightened by the increasing interest in protecting biodiversity, which coincides with the rapid disappearance of ecological niches due to agricultural, industrial and urban development and global climate change. The question then remains as to the effects of this decline on the advancement of human medicines. At what point do we decide that these financial and ecological resources are needed for our own medical use? Can their use be made sustainable and how? At what level do we need to protect our precious NP resources, or can biological and synthetic drugs cover our future medical needs? The paucity of novel drug candidates and falling product approval rates would suggest that it would be foolhardy, if not irresponsible, to totally ignore NPs as sources of new drug leads.

References

1. M. S. Butler, *Nat. Prod. Rep.*, 2005, **22**, 162.
2. M. S. Butler, *Nat. Prod. Rep.*, 2008, **25**, 475.
3. D. J. Newman and G. M. Cragg, *J. Nat. Prod.*, 2007, **70**, 461.
4. A. Ganesan, *Curr. Opin. Chem. Biol.*, 2008, **12**, 306.
5. A. Harvey, *Drug Discov. Today*, 2008, **13**, 894.
6. D. D. Baker, M. Chu, U. Oza and V. Rajgarhia, *Nat. Prod. Rep.*, 2007, **24**, 1225.
7. K. S. Lam, *Trends Microbiol.*, 2007, **15**, 279.
8. W. P. Jones, Y. W. Chin and A. D. Kinghorn, *Curr. Drug Targets*, 2006, **7**, 247.
9. R. M. Wilson and S. J. Danishefsky, *J. Org. Chem.*, 2006, **71**, 8329.
10. V. P. Gullo, J. McAlpine, K. S. Lam, D. Baker and F. Petersen, *J. Ind. Microbiol. Biotechnol.*, 2006, **33**, 523.
11. F. E. Koehn and G. T. Carter, *Nat. Rev. Drug Discov.*, 2005, **4**, 206.
12. M. S. Butler, *J. Nat. Prod.*, 2004, **67**, 2141.
13. W. S. Sneader, *Drug Discovery: A History*, Wiley, Chichester, 2005.
14. D. J. Newman, G. M. Cragg and K. M. Snader, *Nat. Prod. Rep.*, 2000, **17**, 215.
15. S. Inouye, T. Tsuruoka and T. Nida, *J. Antibiot. (Tokyo)*, 1966, **19**, 288.
16. (a) J. H. Birkinshaw, H. Raistrick and D. J. Ross, *Biochem. J.*, 1952, **50**, 630; (b) R. Bentley, *Chem. Rev.*, 2000, **100**, 3801.
17. (a) A. Endo, M. Kuroda and Y. Tsujita, *J. Antibiot. (Tokyo)*, 1976, **29**, 1346; (b) A. G. Brown, T. C. Smale, T. J. King, R. Hasenkamp and R. H.

Thompson, *J. Chem. Soc. (Perkin Trans 1)*, 1976, 1165; (c) A. Endo, *Nat. Med.*, 2008, **14**, 1050.

18. (a) M. Debono, M. Barnhart, C. B. Carrell, J. A. Hoffmann, J. L. Occolowitz, B. J. Abbott, D. S. Fukuda, R. L. Hamill, K. Biemann and W. C. Herlihy, *J. Antibiot. (Tokyo)*, 1987, **40**, 761; (b) R. H. Baltz, V. Miao and S. K. Wrigley, *Nat. Prod. Rep.*, 2005, **22**, 717.
19. (a) C. Vézina, A. Kudelski and S. N. Sehgal, *J. Antibiot. (Tokyo)*, 1975, **28**, 721; (b) S. N. Sehgal, H. Baker and C. Vézina, *J. Antibiot. (Tokyo)*, 1975, **28**, 727.
20. (a) D. S. Tarbell, R. M. Carman, D. D. Chapman, K. R. Huffman and N. J. McCorkindale, *J. Am. Chem. Soc.*, 1960, **82**, 1005; (b) N. J. McCorkindale and J. G. Sime, *Proc. Chem. Soc.*, 1961, **331**.
21. Y. Gaoni and R. Mechoulam, *J. Am. Chem. Soc.*, 1964, **86**, 1646.
22. R. Michoulam and Y. Shvo, *Tetrahedron*, 1963, **19**, 2073.
23. J. S. Kahan, F. M. Kahan, R. Goegelman, S. A. Currie, M. Jackson, E. O. Stapley, T. W. Miller, A. K. Miller, D. Hendlin, S. Mochales, S. Hernandez, H. B. Woodruff and J. Birnbaum, *J. Antibiot. (Tokyo)*, 1979, **32**, 1.
24. (a) F. A. Hochstein, C. R. Stephens, L. H. Conover, P. P. Regna, R. Pasternack, K. J. Brunings and R. B. Woodward, *J. Am. Chem. Soc.*, 1952, **74**, 3708; (b) R. Stephens, L. H. Conover, F. A. Hochstein, P. P. Regna, F. J. Pilgrim and K. J. Brunings, *J. Am. Chem. Soc.*, 1952, **74**, 4976.
25. B. M. Olivera, L. J. Cruz, V. de Santos, G. W. LeCheminant, D. Griffin, R. Zeikus, J. M. McIntosh, R. Galyean, J. Varga, W. R. Gray and J. Rivier, *Biochemistry*, 1987, **26**, 2086.
26. R. Nyfeler and W. Keller-Schierlein, *Helv. Chim. Acta*, 1974, **57**, 2459.
27. J. Eng, W. A. Kleinman, L. Singh, G. Singh and J. P. Raufman, *J. Biol. Chem.*, 1992, **267**, 7402.
28. (a) E. Späth and R. Göhring, *Monatsh. Chem.*, 1920, **41**, 319; (b) K. K. Chen and C. H. Kao, *J. Am. Pharm. Assoc.*, 1926, **15**, 625.
29. D. Arigoni, *Gazz. Chim. Ital.*, 1962, **92**, 884.
30. R. Sakai, K. L. Rinehart, Y. Guan and A. H. Wang, *Proc. Natl. Acad. Sci. USA*, 1992, **89**, 11456.
31. (a) D. M. Bollag, P. A. McQueney, J. Zhu, O. Hensens, L. Koupal, J. Liesch, M. Goetz, E. Lazarides and C. M. Woods, *Cancer Res.*, 1995, **55**, 2325; (b) G. Höfle, N. Bedorf, H. Steinmetz, D. Schomburg, K. Gerth and H. Reichenbach, *Angew. Chem., Int. Ed. Eng.*, 1996, **35**, 1567.
32. (a) J. M. Gulland and R. Robinson, *J. Chem. Soc.*, 1923, 980; (b) M. MacKay and D. C. Hodgkin, *J. Chem. Soc.*, 1955, 3261; (c) K. W. Bentley and H. M. E. Cardwell, *J. Chem. Soc.*, 1955, 3245.
33. (a) E. P. Abraham and G. G. Newton, *Biochem. J.*, 1961, **79**, 377; (b) D. C. Hodgkin and E. N. Maslen, *Biochem. J.*, 1961, **79**, 393.
34. Progenics Pharmaceuticals: Press releases 1 April 2008, 24 April 2008 and 3 July 2008.
35. F. M. Reichle and P. F. Conzen, *Curr. Opin. Investig. Drugs*, 2008, **9**, 90.

36. J. Thomas, S. Karver, G. A. Cooney, B. H. Chamberlain, C. K. Watt, N. E. Slatkin, N. Stambler, A. B. Kremer and R. J. Israel, *N. Engl. J. Med.*, 2008, **358**, 2332.
37. Basilea Pharmaceutica: Press releases 30 June 2008 and 13 November 2008.
38. P. Hebeisen, I. Heinze-Krauss, P. Angehrn, P. Hohl, M. G. P. Page and R. L. Then, *Antimicrob. Agents Chemother.*, 2001, **45**, 825.
39. A. F. Widmer, *Clin. Infect. Dis.*, 2008, **46**, 656.
40. G. J. Noel, K. Bush, P. Bagchi, J. Ianus and R. Strauss, *Clin. Infect. Dis.*, 2008, **46**, 647.
41. K. Bush, M. Heep, M. J. Macielag and G. J. Noel, *Expert Opin. Investig. Drugs*, 2007, **16**, 419.
42. J. M. T. Hamilton-Miller, *Int. J. Antimicrob. Ag.*, 2000, **15**, 179.
43. M. G. P. Page, *Expert Opin. Invest. Drugs*, 2004, **13**, 973.
44. M. S. Barber, U. Giesecke, A. Reichert and W. Minas, *Adv. Biochem. Engin./ Biotechnol.*, 2004, **88**, 179.
45. Theravance: Press release 19 November 2008.
46. Theravance: Press release 26 January 2009.
47. M. R. Leadbetter, S. M. Adams, B. Bazzini, P. R. Fatheree, D. E. Karr, K. M. Krause, B. M. Lam, M. S. Linsell, M. B. Nodwell, J. L. Pace, K. Quast, J. P. Shaw, E. Soriano, S. G. Trapp, J. D. Villena, T. X. Wu, B. G. Christensen and J. K. Judice, *J. Antibiot. (Tokyo)*, 2004, **57**, 326.
48. J. K. Judice and J. L. Pace, *Bioorg. Med. Chem. Lett.*, 2003, **13**, 4165.
49. L. M. Dunbar, D. M. Tang and R. M. Manausa, *Ther. Clin. Risk Manag.*, 2008, **4**, 235.
50. E. C. Nannini and M. E. Stryjewski, *Expert Opin. Pharmacother.*, 2008, **9**, 2197.
51. M. E. Stryjewski, D. R. Graham, S. E. Wilson, W. O'Riordan, D. Young, A. Lentnek, D. P. Ross, V. G. Fowler, A. Hopkins, H. D. Friedland, S. L. Barriere, M. M. Kitt and G. R. Corey, *Clin. Infect. Dis.*, 2008, **46**, 1683.
52. G. M. Sheldrick, P. G. Jones, O. Kennard, D. H. Williams and G. A. Smith, *Nature*, 1978, **271**, 223.
53. Targanta Therapeutics: Press release 8 December 2009.
54. Targanta Therapeutics: Press release 12 January 2009.
55. G. Poulakou and H. Giamarellou, *Expert Opin. Investig. Drugs*, 2008, **17**, 225.
56. D. L. Anderson, *Drugs Today (Barc.)*, 2008, **44**, 563.
57. N. Tsuji, T. Kamigauchi, M. Kobayashi and Y. Terui, *J. Antibiot. (Tokyo)*, 1988, **41**, 1506.
58. R. D. G. Cooper, N. J. Snyder, M. J. Zweifel, M. A. Staszak, S. C. Wilkie, T. I. Nicas, D. L. Mullen, T. F. Butler, M. J. Rodriguez, B. E. Huff and R. C. Thompson, *J. Antibiot. (Tokyo)*, 1996, **49**, 575.
59. Pfizer: Press release 9 September 2008.
60. M. Billeter, M. J. Zervos, A. Y. Chen, J. R. Dalovisio and C. Kurukularatne, *Clin. Infect. Dis.*, 2008, **46**, 577.

61. A. Kim, J. L. Kuti and D. P. Nicolau, *Expert Opin. Investig. Drugs*, 2007, **16**, 717.
62. J. W. Bennett, J. S. Lewis II and M. W. Ellis, *Ther. Clin. Risk Manag.*, 2008, **4**, 31.
63. A. M. Roecker and S. D. Pope, *Expert Opin. Pharmacother.*, 2008, **9**, 1745.
64. B. P. Goldstein, E. Selva, L. Gastaldo, M. Berti, R. Pallanza, F. Ripamonti, P. Ferrari, M. Denaro, V. Arioli and G. Cassani, *Antimicrob. Agents Chemother.*, 1987, **31**, 1961.
65. A. Malabarba, R. Ciabatti, J. Kettenring, P. Ferrari, R. Scotti, B. P. Goldstein and M. Denaro, *J. Antibiot. (Tokyo)*, 1994, **47**, 1493.
66. M. Sosio and S. Donadio, *J. Ind. Microbiol. Biotechnol.*, 2006, **33**, 569.
67. Advanced Life Sciences: Press releases 1 October 2008 and 3 December 2008.
68. F. Van Bambeke, J. M. Harms, Y. Van Laethem and P. M. Tulkens, *Expert Opin. Pharmacother.*, 2008, **9**, 267.
69. E. Azoulay-Dupuis, J. Mohler, J. P. Bédos, C. Barau and B. Fantin, *Antimicrob. Agents Chemother.*, 2006, **50**, 3033.
70. M. R. Hammerschlag and R. Sharma, *Expert Opin. Investig. Drugs*, 2008, **17**, 387.
71. P. F. Wiley, K. Gerzon, E. H. Flynn, M. V. Sigal Jr, O. Weaver, U. C. Quarck, R. R. Chauvette and R. Monahan, *J. Am. Chem. Soc.*, 1957, **79**, 6062.
72. (a) Y. S. Or, R. F. Clark, S. Wang, D. T. Chu, A. M. Nilius, R. K. Flamm, M. Mitten, P. Ewing, J. Alder and Z. Ma, *J. Med. Chem.*, 2000, **43**, 1045; (b) Z. Ma, R. F. Clark, A. Brazzale, S. Wang, M. J. Rupp, L. Li, G. Griesgraber, S. Zhang, H. Yong, L. T. Phan, P. A. Nemoto, D. T. Chu, J. J. Plattner, X. Zhang, P. Zhong, Z. Cao, A. M. Nilius, V. D. Shortridge, R. Flamm, M. Mitten, J. Meulbroek, P. Ewing, J. Alder and Y. S. Or, *J. Med. Chem.*, 2001, **44**, 4137.
73. Meiji Seika Kaisha: Annual Report 2008.
74. M. Hikida, K. Itahashi, A. Igarashi, T. Shiba and M. Kitamura, *Antimicrob. Agents Chemother.*, 1999, **43**, 2010.
75. Y. Wang, J. Bolos and N. Serradell, *Drugs Future*, 2006, **31**, 676.
76. M. Yamada, T. Watanabe, N. Baba, Y. Takeuchi, F. Ohsawa and S. Gomi, *Antimicrob. Agents Chemother.*, 2008, **52**, 2053.
77. Forest Laboratories: Press release 27 October 2008.
78. T. Ishikawa, N. Matsunaga, H. Tawada, N. Kuroda, Y. Nakayama, Y. Ishibashi, M. Tomimoto, Y. Ikeda, Y. Tagawa, Y. Iizawa, K. Okonogi, S. Hashiguchi and A. Miyake, *Bioorg. Med. Chem.*, 2003, **11**, 2427.
79. Cadence Pharmaceuticals: Press release 6 November 2008.
80. Cutanea Life Sciences: Press release 17 October 2007.
81. M. N. Melo, D. Dugourd and M. A. R. B. Castanho, *Recent Pat. Anti-infect. Drug Discov.*, 2006, **1**, 201.
82. T. R. Fritsche, P. R. Rhomberg, H. S. Sader and R. N. Jones, *Antimicrob. Agents Chemother.*, 2008, **52**, 1187.

83. H. S. Sader, K. A. Fedler, R. P. Rennie, S. Stevens and R. N. Jones, *Antimicrob. Agents Chemother.*, 2008, **52**, 3112.
84. M. E. Selsted, M. J. Novotny, W. L. Morris, Y. Q. Tang, W. Smith and J. S. Cullor, *J. Biol. Chem.*, 1992, **267**, 4292.
85. Optimer Pharmaceuticals: Press release 10 November 2008.
86. A. P. Johnson, *Curr. Opin. Investig. Drugs*, 2007, **8**, 168.
87. M. Gerber and G. Ackermann, *Expert Opin. Investig. Drugs*, 2008, **17**, 547.
88. T. Louie, M. Miller, C. Donskey, K. Mullane and E. J. Goldstein, *Antimicrob. Agents Chemother.*, 2009, **53**, 223.
89. J. E. Hochlowski, S. J. Swanson, L. M. Ranfranz, D. N. Whittern, A. M. Buko and J. B. McAlpine, *J. Antibiot. (Tokyo)*, 1987, **40**, 575.
90. C. Coronelli, R. J. White, G. C. Lancini and F. Parenti, *J Antibiot. (Tokyo)*, 1975, **28**, 253.
91. S. Sergio, G. Pirali, R. White and F. Parenti, *J. Antibiot. (Tokyo)*, 1975, **28**, 543.
92. S. Ōmura, N. Imamura, R. Oiwa, H. Kuga, R. Iwata and R. Masuma, *J. Antibiot. (Tokyo)*, 1986, **39**, 1407.
93. M. Gualtieri, P. Villain-Guillot, J. Latouche, J.-P. Leonetti and L. Bastide, *Antimicrob. Agents Chemother.*, 2006, **50**, 401.
94. Eisai: Annual Report 2008.
95. D. P. Rossignol and M. Lynn, *Curr. Opin. Investig. Drugs*, 2005, **6**, 496.
96. L. D. Hawkins, W. J. Christ and D. P. Rossignol, *Curr. Top. Med. Chem.*, 2004, **4**, 1147.
97. M. Mullarkey, J. R. Rose, J. Bristol, T. Kawata, A. Kimura, S. Kobayashi, M. Przetak, J. Chow, F. Gusovsky, W. J. Christ and D. P. Rossignol, *J. Pharmacol. Exp. Ther.*, 2003, **304**, 1093.
98. D. P. Rossignol, K. M. Wasan, E. Choo, E. Yau, N. Wong, J. Rose, J. Moran and M. Lynn, *Antimicrob. Agents Chemother.*, 2004, **48**, 3233.
99. E. Bennett-Guerrero, H. P. Grocott, J. H. Levy, K. A. Stierer, C. W. Hogue, A. T. Cheung, M. F. Newman, A. A. Carter, D. P. Rossignol and C. D. Collard, *Anesth. Analg.*, 2007, **104**, 378.
100. D. P. Rossignol, N. Wong, R. Noveck and M. Lynn, *Innate Immun.*, 2008, **14**, 383.
101. N. Qureshi, K. Takayama and R. Kurtz, *Infect. Immun.*, 1991, **59**, 441.
102. I. A. Kaltashov, V. Doroshenko, R. J. Cotter, K. Takayama and N. Qureshi, *Anal. Chem.*, 1997, **69**, 2317.
103. Pierre Fabre: Press release June 2008.
104. Y. A. Yun-San, O. E. Yuen-Yuen and L. W. Chow, *Expert Opin. Investig. Drugs*, 2008, **17**, 583.
105. J. Bennouna, J. P. Delord, M. Campone and L. Nguyen, *Clin. Cancer Res.*, 2008, **14**, 1625.
106. A. A. Makarov, P. O. Tsvetkov, C. Villard, D. Esquieu, B. Pourroy, J. Fahy, D. Braguer, V. Peyrot and D. Lafitte, *Biochemistry*, 2007, **46**, 14899.
107. Bristol-Myers Squibb: Press release 23 November 2007.
108. J. W. Moncrief and W. N. Lipscomb, *J. Am. Chem. Soc.*, 1965, **89**, 4963.

109. (a) J. Fahy, A. Duflos, J.-P. Ribet, J.-C. Jacquesy, C. Berrier, M. P. Jouannetaud and F. Zunino, J. *Am. Chem. Soc.*, 1997, **119**, 8576; (b) J.-C. Jacquesy, C. Berrier, M.-P. Jouannetaud, F. Zunino, J. Fahy, A. Duflos and J.-P. Ribet, *J. Fluor. Chem.*, 2002, **114**, 139.
110. BioNumerik Pharmaceuticals: Press release 7 February 2008.
111. J. J. Kavanagh, M. W. Sill, P. T. Ramirez, D. Warshal, M. L. Pearl and M. A. Morgan, *Int. J. Gynecol. Cancer*, 2008, **18**, 460.
112. P. A. Thompson, S. L. Berg, A. Aleksic, J. Z. Kerr, L. McGuffey, R. Dauser, J. G. Nuchtern, F. Hausheer and S. M. Blaney, *Cancer Chemother. Pharmacol.*, 2004, **53**, 527.
113. A. H. Van Hattum, H. M. Schluper, F. H. Hausheer, H. M. Pinedo and E. Boven, *Int. J. Cancer.*, 2002, **100**, 22.
114. M.-B. Yin, G. Hapke, J. Wu, R. G. Azrak, C. Frank, C. Wrzosek and Y. M. Rustum, *Biochem. Biophys. Res. Commun.*, 2002, **295**, 435.
115. M. E. Wall, M. C. Wani, C. E. Cook, K. H. Palmer, A. T. McPhail and G. A. Sim, *J. Am. Chem. Soc.*, 1966, **88**, 3888.
116. N. H. Oberlies and D. J. Kroll, *J. Nat. Prod.*, 2004, **67**, 129.
117. J. Kluza, P. B. Arimondo, M.-H. David-Cordonnier and C. Bailly, in *Cancer Drug Design and Discovery*, ed. S. Neidle, Elsevier, Amsterdam, 2008, pp. 173–197.
118. OXiGENE: Press release 5 November 2008.
119. S. L. Young and D. J. Chaplin, *Expert Opin. Investig. Drugs*, 2004, **13**, 1171.
120. N. Mahindroo, J.-P. Liou, J.-Y. Chang and H.-P. Hsieh, *Expert Opin. Ther. Pat.*, 2006, **16**, 647.
121. P. Hinnen and F. A. Eskens, *Br. J. Cancer*, 2007, **96**, 1159.
122. G. R. Pettit, C. Temple Jr, V. L. Narayanan, R. Varma, M. J. Simpson, M. R. Boyd, G. A. Rener and N. Bansal, *Anticancer Drug Des.*, 1995, **10**, 299.
123. G. R. Pettit, S. B. Singh, E. Hamel, C. M. Lin, D. S. Alberts and D. Garcia-Kendall, *Experientia*, 1989, **45**, 209.
124. Sanofi-Aventis: Annual Report on Form 20-F 2007.
125. T. J. Kim, M. Ravcori, C. N. Landen, A. A. Kamat, L. Y. Han, C. Lu, Y. G. Lin, W. M. Merritt, N. Jennings, W. A. Spannuth, R. Langley, D. M. Gershenson, R. L. Coleman, V. Kundra and A. K. Sood, *Cancer Res.*, 2007, **67**, 9337.
126. Y. Nihei, Y. Suga, Y. Morinaga, K. Ohishi, A. Okano, K. Ohsumi, T. Hatanaka, R. Nakagawa, T. Tsuji, Y. Akiyama, S. Saito, K. Hori, Y. Sato and T. Tsuruo, *Jpn. J. Cancer Res.*, 1999, **90**, 1016.
127. K. Hori and S. Saito, *Br. J. Cancer*, 2003, **89**, 1334.
128. L.-H. Zhang, L. Wu, H. K. Raymon, R. S. Chen, L. Corral, M. A. Shirley, R. K. Narla, J. Gamez, G. W. Muller, D. I. Stirling, J. B. Bartlett, P. H. Schafer and F. Payvandi, *Cancer Res.*, 2006, **66**, 951.
129. P. Zatloukal, R. Gervais, J. Vansteenkiste, L. Bosquee, C. Sessa, E. Brain, E. Dansin, T. Urban, N. Dohollou, M. Besenval and E. Quoix, *J. Thorac. Oncol.*, 2008, **3**, 894.

130. V. Diéras, S. Limentani, G. Romieu, M. Tubiana-Hulin, A. Lortholary, P. Kaufman, V. Girre, M. Besenval and V. Valero, *Ann. Oncol.*, 2008, **19**, 1255.
131. M. C. Wani, H. L. Taylor, M. E. Wall, P. Coggon and A. T. McPhail, *J. Am. Chem. Soc.*, 1971, **93**, 2325.
132. P. B. Schiff, J. Fant and S. B. Horwitz, *Nature*, 1979, **277**, 665.
133. A. Wahl, F. Guéritte-Voegelein, D. Guénard, M.-T. Le Goff and P. Potier, *Tetrahedron*, 1992, **48**, 6965.
134. S. Cisternino, F. Bourasset, Y. Archimbaud, D. Sémiond, G. Sanderink and J. M. Scherrmann, *Br. J. Pharmacol.*, 2003, **138**, 1367.
135. X. Pivot, P. Koralewski, J. L. Hidalgo, A. Chan, A. Gonçalves, G. Schwartsmann, S. Assadourian and J. P. Lotz, *Ann. Oncol.*, 2008, **19**, 1547.
136. Luitpold: Information on trial NCT00243867 available at www.clinicaltrials.gov.
137. M. O. Bradley, C. S. Swindell, F. H. Anthony, P. A. Witman, P. Devanesan, N. L. Webb, S. D. Baker, A. C. Wolff and R. C. Donehower, *J. Control Release*, 2001, **74**, 233.
138. M. O. Bradley, N. L. Webb, F. H. Anthony, P. Devanesan, P. A. Witman, S. Hemamalini, M. C. Chander, S. D. Baker, L. He, S. B. Horwitz and C. S. Swindell, *Clin. Cancer Res.*, 2001, **7**, 3229.
139. R. J. Jones, R. E. Hawkins, M. M. Eatock, D. R. Ferry, F. A. Eskens, H. Wilke and T. R. Evans, *Cancer Chemother. Pharmacol.*, 2008, **61**, 435.
140. ChemGenex: Press release 9 December 2008.
141. A. Quintás-Cardama and J. Cortes, *IDrugs*, 2008, **11**, 356.
142. R. G. Powell, D. Weisleder, C. R. Smith Jr and W. K. Rohwedder, *Tetrahedron Lett.*, 1970, **11**, 815.
143. J.-P. Robin, J. Blanchard, S. Cavoleau, L. Chauviat, S. Charbonnel, R. Dhal, G. Dujardin, F. Fournier, C. Gilet, L. Girodier, L. Mevelec, S. Poutot and S. Rouaud, US Patent 6,613,900, 2003.
144. H. Tong, Y. Ren, F. Zhang and J. Jin, *Eur. J. Haematol.*, 2008, **81**, 259.
145. Marshall Edwards: Press release 12 November 2008.
146. A. B. Alvero, M. Kelly, P. Rossi, A. Leiser, D. Brown, T. Rutherford and G. Mor, *Future Oncol.*, 2008, **4**, 475.
147. P. M. Herst, T. Petersen, P. Jerram, J. Baty and M. V. Berridge, *Biochem. Pharmacol.*, 2007, **74**, 1587.
148. D. J. Morré, P. J. Chueh, K. Yagiz, A. Balicki, C. Kim and D. M. Morré, *Oncol. Res.*, 2007, **16**, 299.
149. (a) A. I. Constantinou and A. Husband, *Anticancer Res.*, 2002, **22**, 2581; (b) A. I. Constantinou, R. Mehta and A. Husband, *Eur. J. Cancer*, 2003, **39**, 1012.
150. W. Baker and R. Robinson, *J. Chem. Soc.*, 1926, 2713.
151. T. S. Lin, *Clin. Lymphoma Myeloma*, 2008, **8**(Suppl 4), S137.
152. B. A. Christian, M. R. Grever, J. C. Byrd and T. S. Lin, *Curr. Opin. Oncol.*, 2007, **19**, 573.

153. G. Kaur, M. Stetler-Stevenson, S. Sebers, P. Worland, H. Sedlacek, C. Myers, J. Czech, R. Naik and E. Sausville, *J. Natl. Cancer Inst.*, 1992, **84**, 1736.
154. A. D. Harmon, U. Weiss and J. V. Silverton, *Tetrahedron Lett.*, 1979, **20**, 721.
155. R. G. Naik, S. L. Kattige, S. B. Bhat, B. Alreja, N. J. de Souza and R. H. Rupp, *Tetrahedron*, 1988, **44**, 2081.
156. Bristol-Myers Squibb: Press release 26 June 2008.
157. R. C. Schnur, M. L. Corman, R. J. Gallaschun, B. A. Cooper, M. F. Dee, J. L. Doty, M. L. Muzzi, J. D. Moyer, C. I. DiOrio, E. G. Barbacci, P. E. Miller, V. A. Pollalk, D. M. Savage, D. E. Sloan, L. R. Pustilnik and M. P. Moyer, *J. Med. Chem.*, 1995, **38**, 3806.
158. S. Modi, A. T. Stopeck, M. S. Gordon, D. Mendelson, D. B. Solit, R. Bagatell, W. Ma, J. Wheler, N. Rosen, L. Norton, G. F. Cropp, R. G. Johnson, A. L. Hannah and C. A. Hudis, *J. Clin. Oncol.*, 2007, **25**, 5410.
159. L. Neckers, *Curr. Top. Med. Chem.*, 2006, **6**, 1163.
160. J. G. Supko, R. L. Hickman, M. R. Grever and L. Malspeis, *Cancer Chemother. Pharmacol.*, 1995, **36**, 305.
161. T. Taldone, A. Gozman, R. Maharaj and G. Chiosis, *Curr. Opin. Pharmacol.*, 2008, **8**, 370.
162. C. DeBoer, P. A. Meulman, R. J. Wnuk and D. H. Peterson, *J. Antibiot. (Tokyo)*, 1970, **23**, 442.
163. K. Sasaki, K. L. Rinehart Jr, G. Slomp, M. F. Grostic and E. C. Olson, *J. Am. Chem. Soc.*, 1970, **92**, 7591.
164. Infinity Pharmaceuticals: Press release 11 December 2008.
165. J. R. Sydor, E. Normant, C. S. Pien, J. R. Porter, J. Ge, L. Grenier, R. H. Pak, J. A. Ali, M. S. Dembski, J. Hudak, J. Patterson, C. Penders, M. Pink, M. A. Read, J. Sang, C. Woodward, Y. Zhang, D. S. Grayzel, J. Wright, J. A. Barrett, V. J. Palombella, J. Adams and J. K. Tong, *Proc. Natl. Acad. Sci. USA*, 2006, **103**, 17408.
166. J. Ge, E. Normant, J. R. Porter, J. A. Ali, M. S. Dembski, Y. Gao, A. T. Georges, L. Grenier, R. H. Pak, J. Patterson, J. R. Sydor, T. T. Tibbitts, J. K. Tong, J. Adams and V. J. Palombella, *J. Med. Chem.*, 2006, **49**, 4606.
167. J. Patterson, V. J. Palombella, C. Fritz and E. Normant, *Cancer Chemother. Pharmacol.*, 2008, **61**, 923.
168. ARIAD Pharmaceuticals: Press release 13 January 2009.
169. M. M. Mita, A. C. Mita, Q. S. Chu, E. K. Rowinsky, G. J. Fetterly, M. Goldston, A. Patnaik, L. Mathews, A. D. Ricart, T. Mays, H. Knowles, V. M. Rivera, J. Kreisberg, C. L. Bedrosian and A. W. Tolcher, *J. Clin. Oncol.*, 2008, **26**, 361.
170. D. A. Rizzieri, E. Feldman, J. F. Dipersio, N. Gabrail, W. Stock, R. Strair, V. M. Rivera, M. Albitar, C. L. Bedrosian and F. J. Giles, *Clin. Cancer Res.*, 2008, **14**, 2756.
171. M. Mita, K. Sankhala, I. Abdel-Karim, A. Mita and F. Giles, *Expert Opin. Investig. Drugs*, 2008, **17**, 1947.
172. A. Fasolo and C. Sessa, *Expert Opin. Investig. Drugs*, 2008, **17**, 1717.

173. Eli Lilly: Press release 11 December 2008.
174. M. M. Faul, J. L. Grutsch, M. E. Kobierski, M. E. Kopach, C. A. Krumrich, M. A. Staszak, U. Udodong, J. T. Vicenzi and K. A. Sullivan, *Tetrahedron*, 2003, **59**, 7215.
175. H. L. Pearce and M. Alice Miller, *Adv. Enzyme Regul.*, 2005, **45**, 229.
176. S. Ma and S. T. Rosen, *Curr. Opin. Oncol.*, 2007, **19**, 590.
177. Y. B. Chen and A. S. LaCasce, *Expert Opin. Investig. Drugs*, 2008, **17**, 939.
178. J. P. Leonard, P. Martin, J. Barrientos and R. Elstrom, *Semin. Hematol.*, 2008, **45**(Suppl 2), S11.
179. M. A. Rizvi, K. Ghias, K. M. Davies, C. Ma, F. Weinberg, H. G. Munshi, N. L. Krett and S. T. Rosen, *Mol. Cancer Ther.*, 2006, **5**, 1783.
180. A. C. Spalding, B. D. Zeitlin, K. Wilder-Romans, M. E. Davis, J. E. Nor, T. S. Lawrence and E. Ben-Josef, *Transl. Oncol.*, 2008, **1**, 195.
181. A. Furusaki, N. Hashiba, T. Matsumoto, A. Hirano, Y. Iwai and S. Ōmura, *J. Chem. Soc. Chem. Commun.*, 1978, 800.
182. N. Funato, H. Takayanagi, Y. Konda, Y. Toda, Y. Harigaya, Y. Iwai and S. Ōmura, *Tetrahedron Lett.*, 1994, **35**, 1251.
183. Cephalon: Press release 3 November 2008.
184. H. Kase, K. Iwahashi and Y. Matsuda, *J. Antibiot. (Tokyo)*, 1986, **39**, 1059.
185. S. J. Miknyoczki, H. Chang, A. Klein-Szanto, C. A. Dionne and B. A. Ruggeri, *Clin. Cancer Res.*, 1999, **5**, 2205.
186. B. D. Smith, M. Levis, M. Beran, F. Giles, H. Kantarjian, K. Berg, K. M. Murphy, T. Dauses, J. Allebach and D. Small, *Blood*, 2004, **103**, 3669.
187. S. Knapper, A. K. Burnett, T. Littlewood, W. J. Kell, S. Agrawal, R. Chopra, R. Clark, M. J. Levis and D. Small, *Blood*, 2006, **108**, 3262.
188. T. Illmer and G. Ehninger, *Clin. Lymphoma Myeloma*, 2007, **8**(Suppl 1), S24.
189. E. O. Hexner, C. Serdikoff, M. Jan, C. R. Swider, C. Robinson, S. Yang, T. Angeles, S. G. Emerson, M. Carroll, B. Ruggeri and P. Dobrzanski, *Blood*, 2008, **111**, 5663.
190. A. Kuendgen and U. Germing, *Cancer Treat. Rev.*, 2009, **35**, 97.
191. Novartis: Annual Report on Form 20-F 2008.
192. R. M. Stone, D. J. DeAngelo, V. Klimek, I. Galinsky, E. Estey, S. D. Nimer, W. Grandin, D. Lebwohl, Y. Wang, P. Cohen, E. A. Fox, D. Neuberg, J. Clark, D. G. Gilliland and J. D. Griffin, *Blood*, 2005, **105**, 54.
193. L. Möllgård, S. Deneberg, H. Nahi, S. Bengtzen, K. Jonsson-Videsäter, T. Fioretos, A. Andersson, C. Paul and S. Lehmann, *Cancer Chemother. Pharmacol.*, 2008, **62**, 439.
194. K. Zaugg, S. Rocha, H. Resch, I. Hegyi, C. Oehler, C. Glanzmann, D. Fabbro, S. Bodis and M. Pruschy, *Cancer Res.*, 2001, **61**, 732.
195. D. Fabbro, E. Buchdunger, J. Wood, J. Mestan, F. Hofmann, S. Ferrari, H. Mett, T. O'Reilly and T. Meyer, *Pharmacol. Ther.*, 1999, **82**, 293.
196. M. Trivedi, I. Budihardjo, K. Loureiro, T. R. Reid and J. D. Ma, *Future Oncol.*, 2008, **4**, 483.

197. K. H. Altmann and K. Memmert, *Prog. Drug Res.*, 2008, **66**, 275.
198. G. Hoefle, N. Bedorf, K. Gerth and H. Reichenbach, *German Patent* DE4,138,042, 1993.
199. K. Gerth, N. Bedorf, G. Höfle, H. Irschik and H. Reichenbach, *J. Antibiot. (Tokyo)*, 1996, **49**, 560.
200. H. Reichenbach and G. Höfle, *Drugs R. D.*, 2008, **9**, 1.
201. Eisai: Presentation for 2008 Merrill Lynch Japan Conference Value to Growth, 16 September 2008.
202. T. C. McMorris, M. J. Kelner, W. Wang, J. Yu, L. A. Estes and R. Taetle, *J. Nat. Prod.*, 1996, **59**, 896.
203. T. C. McMorris, *Bioorg. Med. Chem.*, 1999, **7**, 881.
204. Y. Wang, T. Wiltshire, J. Senft, E. Reed and W. Wang, *Biochem. Pharmacol.*, 2007, **73**, 469.
205. T. C. McMorris and M. Anchel, *J. Am. Chem. Soc.*, 1965, **87**, 1594.
206. M. J. Towle, K. A. Salvato, J. Budrow, B. F. Wels, G. Kuznetsov, K. K. Aalfs, S. Welsh, W. Zheng, B. M. Seletsky, M. H. Palme, G. J. Habgood, L. A. Singer, L. V. DiPietro, Y. Wang, J. J. Chen, D. A. Quincy, A. Davis, K. Yoshimatsu, Y. Kishi, M. J. Yu and B. A. Littlefield, *Cancer Res.*, 2001, **61**, 1013.
207. M. Seletsky, Y. Wang, L. D. Hawkins, M. H. Palme, G. J. Habgood, L. V. DiPietro, M. J. Towle, K. A. Salvato, B. F. Wels, K. K. Aalfs, Y. Kishi, B. A. Littlefield and M. J. Yu, *Bioorg. Med. Chem. Lett.*, 2004, **14**, 5547.
208. W. Zheng, B. M. Seletsky, M. H. Palme, P. J. Lydon, L. A. Singer, C. E. Chase, C. A. Lemelin, Y. Shen, H. Davis, L. Tremblay, M. J. Towle, K. A. Salvato, B. F. Wels, K. K. Aalfs, Y. Kishi, B. A. Littlefield and M. J. Yu, *Bioorg. Med. Chem. Lett.*, 2004, **14**, 5551.
209. G. Kuznetsov, M. J. Towle, H. Cheng, T. Kawamura, K. TenDyke, D. Liu, Y. Kishi, M. J. Yu and B. A. Littlefield, *Cancer Res.*, 2004, **64**, 5760.
210. S. Newman, *Curr. Opin. Investig. Drugs*, 2007, **8**, 1057.
211. T. Okouneva, O. Azarenko, L. Wilson, B. A. Littlefield and M. A. Jordan, *Mol. Cancer Ther.*, 2008, **7**, 2003.
212. D. Uemura, K. Takahashi, T. Yamamoto, C. Katayama, J. Tanaka, Y. Okumura and Y. Hirata, *J. Am. Chem. Soc.*, 1985, **107**, 4796; Y. Hirata and D. Uemura, *Pure Appl. Chem.*, 1986, **58**, 701.
213. T. D. Aicher, K. R. Buszek, F. G. Fang, C. J. Forsyth, S. H. Jung, Y. Kishi, M. C. Matelich, P. M. Scola, D. M. Spero and S. K. Yoon, *J. Am. Chem. Soc.*, 1992, **114**, 3162.
214. P. Jones and C. Steinkühler, *Curr. Pharm. Design*, 2008, **14**, 545.
215. O. Martínez-Iglesias, L. Ruiz-Llorente, R. Sánchez-Martínez, L. García, A. Zambrano and A. Aranda, *Clin. Transl. Oncol.*, 2008, **10**, 395.
216. A. V. Bieliauskas and M. K. H. Pflum, *Chem. Soc. Rev.*, 2008, **37**, 1402.
217. I. C. Piña, J. T. Gautschi, G.-Y.-S. Wang, M. L. Sanders, F. J. Schmitz, D. France, S. Cornell-Kennon, L. C. Sambucetti, S. W. Remiszewski, L. B. Perez, K. W. Bair and P. Crews, *J. Org. Chem.*, 2003, **68**, 3866.
218. A. Scuto, M. Kirschbaum, C. Kowolik, L. Kretzner, A. Juhasz, P. Atadja, V. Pullarkat, R. Bhatia, S. Forman, Y. Yen and R. Jove, *Blood*, 2008, **111**, 5093.

219. E. Quinoa and P. Crews, *Tetrahedron Lett.*, 1987, **28**, 3229.
220. L. Arabshahi and F. Schmitz, *J. Org. Chem.*, 1987, **52**, 3584.
221. A. Rodriguez, R. Akee and P. Scheuer, *Tetrahedron Lett.*, 1987, **28**, 4989.
222. H. Itazaki, K. Nagashima, K. Sugita, H. Yoshida, Y. Kawamura, Y. Yasuda, K. Matsumoto, K. Ishii, N. Uotani, H. Nakai, A. Terui, S. Yoshimatsu, Y. Ikenishi and Y. Nakagawa, *J. Antibiot. (Tokyo)*, 1990, **43**, 1524.
223. P. A. Marks and R. Breslow, *Nat. Biotechnol.*, 2007, **25**, 84.
224. Eli Lilly: Information on trial NCT00133952 available at www.clinicaltrials.gov.
225. M. R. Jirousek, J. R. Gillig, C. M. Gonzalez, W. F. Heath, J. H. McDonald III, D. A. Neel, C. J. Rito, U. Singh, L. E. Stramm, A. Melikian-Badalian, M. Baevsky, L. M. Ballas, S. E. Hall, L. L. Winneroski and M. M. Faul, *J. Med. Chem.*, 1996, **39**, 2664.
226. G. L. Engel, N. A. Farid, M. M. Faul, L. A. Richardson and L. L. Winneroski, *Int. J. Pharm.*, 2000, **198**, 239.
227. M. Clarke and P. M. Dodson, *Best Pract. Res. Clin. Endocrinol. Metab.*, 2007, **21**, 573.
228. T. W. Gardner and D. A. Antonetti, *Ophthalmology*, 2006, **113**, 2135.
229. R. J. Huxtable and S. K. W. Schwarz, *Mol. Interv.*, 2001, **1**, 189.
230. G. Andersen, L. Christrup and P. Sjøgren, *J. Pain Symptom. Manage.*, 2003, **25**, 74.
231. PAION AG: Press release 3 November 2008.
232. G. J. Kilpatrick and T. W. Smith, *Med. Res. Rev.*, 2005, **25**, 521.
233. E. L. van Dorp, A. Morariu and A. Dahan, *Expert Opin. Pharmacother.*, 2008, **9**, 1955.
234. G. P. Joshi, *Curr. Opin. Investig. Drugs*, 2008, **9**, 786.
235. Bristol-Myers Squibb: Press release 8 December 2008.
236. (a) W. Meng, B. A. Ellsworth, A. A. Nirschl, P. J. McCann, M. Patel, R. N. Girotra, G. Wu, P. M. Sher, E. P. Morrison, S. A. Biller, R. Zahler, P. P. Deshpande, A. Pullockaran, D. L. Hagan, N. Morgan, J. R. Taylor, M. T. Obermeier, W. G. Humphreys, A. Khanna, L. Discenza, J. G. Robertson, A. Wang, S. Han, J. R. Wetterau, E. B. Janovitz, O. P. Flint, J. M. Whaley and W. N. Washburn, *J. Med. Chem.*, 2008, **51**, 1145; (b) B. A. Ellsworth, W. Meng, M. Patel, R. N. Girotra, G. Wu, P. M. Sher, D. L. Hagan, M. T. Obermeier, W. G. Humphreys, J. G. Robertson, A. Wang, S. Han, T. L. Waldron, N. N. Morgan, J. M. Whaley and W. N. Washburn, *Bioorg. Med. Chem. Lett.*, 2008, **18**, 4770.
237. S. Han, D. L. Hagan, J. R. Taylor, L. Xin, W. Meng, S. A. Biller, J. R. Wetterau, W. N. Washburn and J. M. Whaley, *Diabetes*, 2008, **57**, 1723.
238. S. A. Jabbour and B. J. Goldstein, *Int. J. Clin. Pract.*, 2008, **62**, 1279.
239. F. W. McKee and W. B. Hawkins, *Physiol. Rev.*, 1945, **25**, 255.
240. J. R. Ehrenkranz, N. G. Lewis, C. R. Kahn and J. Roth, *Diabetes Metab. Res. Rev.*, 2005, **21**, 31.
241. (a) F. Wessely and K. Sturm, *Monatsh. Chem.*, 1929, **53**, 554; (b) A. Müller and A. Robertson, *J. Chem. Soc.*, 1933, 1170.
242. Zealand Pharma: Press release 12 February 2008.

243. B. D. Larsen, J. D. Mikkelsen and S. Neve, *PCT Int. Appl.*, WO 2001004156 A1, 2001.
244. H. B. Vaidya and R. K. Goyal, *Curr. Drug Targets*, 2008, **9**, 911.
245. T. Vilsbøll, *Expert Opin. Investig. Drugs*, 2007, **16**, 231.
246. Schering Plough: Information on trials NCT00526474, NCT00527943 and NCT00617123 available at www.clinicaltrials.gov.
247. J. T. Pinhey, E. Ritchie and W. C. Taylor, *Aust. J. Chem.*, 1961, **14**, 106.
248. S. Borman, *Chem. Eng. News*, 2005, **83**, 40.
249. S. Chackalamannil, Y. Wang, W. J. Greenlee, Z. Hu, Y. Xia, H. S. Ahn, G. Boykow, Y. Hsieh, J. Palamanda, J. Agans-Fantuzzi, S. Kurowski, M. Graziano and M. Chintala, *J. Med. Chem.*, 2008, **51**, 3061.
250. Isotechnika: Press release 22 January 2009.
251. Lux Biosciences: Press release 26 November 2008.
252. M. A. Cunningham, B. A. Austin, Z. Li, B. Liu, S. Yeh, C. C. Chan, E. Anglade, P. Velagaleti and R. B. Nussenblatt, *Invest. Ophthalmol. Vis. Sci.*, 2009, **50**, 249.
253. K. Papp, R. Bissonnette, L. Rosoph, N. Wasel, C. W. Lynde, G. Searles, N. H. Shear, R. B. Huizinga and W. P. Maksymowych, *Lancet*, 2008, **371**, 1337.
254. E. Anglade, R. Yatscoff, R. Foster and U. Grau, *Expert Opin. Investig. Drugs*, 2007, **16**, 1525.
255. (a) A. Rüegger, M. Kuhn, H. Lichti, H. R. Loosli, R. Huguenin, C. Quiquerez and A. von Wartburg, *Helv. Chim. Acta*, 1976, **59**, 1075; (b) J. F. Borel, C. Feurer and A. Rüegger, *Helv. Chim. Acta*, 1976, **59**, 1480.
256. H. F. Stähelin, *Experientia*, 1996, **52**, 13.
257. Novartis: Press release 12 December 2008.
258. H. Rosen and E. J. Goetzl, *Nat. Rev. Immunol.*, 2005, **5**, 560.
259. M. Mehling, V. Brinkmann, J. Antel, A. Bar-Or, N. Goebels, C. Vedrine, C. Kristofic, J. Kuhle, R. L. Lindberg and L. Kappos, *Neurology*, 2008, **71**, 1261.
260. K. R. Lynch and T. L. Macdonald, *Biochim. Biophys. Acta*, 2008, **1781**, 508.
261. A. Kihara and Y. Igarashi, *Biochim. Biophys. Acta*, 2008, **1781**, 496.
262. J. F. Bagli, D. Kluepfel and M. St.-Jacques, *J. Org. Chem.*, 1973, **38**, 1253.
263. T. Fujita, K. Inoue, S. Yamamoto, T. Ikumoto, S. Sasaki, R. Toyama, K. Chiba, Y. Hoshino and T. Okumoto, *J. Antibiot. (Tokyo)*, 1994, **47**, 208.
264. M. Kiuchi, K. Adachi, T. Kohara, M. Minoguchi, T. Hanano, Y. Aoki, T. Mishina, M. Arita, N. Nakao, M. Ohtsuki, Y. Hoshino, K. Teshima, K. Chiba, S. Sasaki and T. Fujita, *J. Med. Chem.*, 2000, **43**, 2946.
265. P. C. Hiestand, M. Rausch, D. P. Meier and C. A. Foster, *Prog. Drug Res.*, 2008, **66**, 361.
266. Anon., *Nat. Rev. Drug Discov.*, 2009, **8**, 10.
267. I. Kola, *Clin. Pharmacol. Ther.*, 2008, **83**, 227.

CHAPTER 12

From Natural Product to Clinical Trials: NPI-0052 (Salinosporamide A), a Marine Actinomycete-Derived Anticancer Agent

KIN S. LAM, G. KENNETH LLOYD, SASKIA T. C. NEUTEBOOM, MICHAEL A. PALLADINO, KOBI M. SETHNA, MATTHEW A. SPEAR AND BARBARA C. POTTS*

Nereus Pharmaceuticals, Inc., 10480 Wateridge Circle, San Diego CA 92121, USA

1 Introduction

1.1 Bioprospecting Marine Actinomycetes and the Discovery of *Salinispora* and NPI-0052

The exploitation of marine actinomycetes as a source for new drug leads is in its infancy. Even with the limited screening efforts that have been dedicated to date, the discovery rate of novel bioactive metabolites from marine actinomycetes has recently surpassed that of their terrestrial counterparts.[1–5] Culture-dependent methods have demonstrated that actinomycetes with tremendous

RSC Biomolecular Sciences No. 18
Natural Product Chemistry for Drug Discovery
Edited by Antony D. Buss and Mark S. Butler

Published by the Royal Society of Chemistry, www.rsc.org

diversity and novelty are widely distributed in different marine ecosystems.[6,7] These marine actinomycetes produce different types of new bioactive metabolites that have the potential to be developed as therapeutic agents.[4,5]

However, these findings may reflect only a tiny fraction of the true therapeutic potential of marine actinomycetes. Recent culture-independent studies have shown that marine environments contain a high diversity of actinomycetes that are rarely, if ever, recovered by culture-dependent methods.[8] Most of these unculturable actinomycetes have very different 16S rRNA sequences when compared with their terrestrial counterparts,[6,9] and the ability to cultivate these novel actinomycetes will provide a new source for the discovery of novel drug leads. The future success of bioprospecting marine actinomycetes relies on our ability to isolate and grow novel actinomycetes from the marine environments.

One successful example of bioprospecting of marine actinomycetes is the isolation of salinosporamide A from the novel marine genus *Salinispora* by Fenical's research group at the Scripps Institution of Oceanography;[10–12] the first account of its discovery and development has recently been reported.[13] Salinosporamide A (designated as NPI-0052 by Nereus Pharmaceuticals at the commencement of preclinical development) was first isolated from the fermentation broth of *Salinispora tropica* strain CNB392.[12] Structurally, it comprises a fused γ-lactam-β-lactone ring system that is decorated with a cyclohexenyl carbinol at the C4 ring junction, a chloroethyl substituent at C2 and a methyl group at the C3 ring junction (Figure 12.1). This highly potent 20*S* proteasome inhibitor[12,14,15] was licensed to Nereus Pharmaceuticals by the University of California, San Diego, in 2001 and is currently undergoing Phase I clinical studies for the treatment of various haematological and solid tumour malignancies.

In this chapter, we highlight the unique aspects of developing a marine actinomycete-derived therapeutic agent, NPI-0052, as we chronicle events from the early mechanism of action and preclinical studies that supported its entry into clinical trials in cancer patients, to the current strategy for its continued development as an anticancer agent.

1.2 The Ubiquitin–Proteasome System as a Target for Drug Development

Concurrent with the discovery of *S. tropica* and salinosporamide A,[10,12] the ubiquitin–proteasome system was receiving considerable attention for its role in the degradation of intracellular proteins and as a target for the treatment of cancer and other diseases.[16–19] In 2003, approval by the US Food and Drug Administration (FDA) of the 20*S* proteasome inhibitor bortezomib (PS-341, Velcade®; Figure 12.1) for the treatment of relapsed and relapsed/refractory multiple myeloma[20,21] effectively validated the proteasome as a target in cancer chemotherapy; in 2004, the Nobel Prize for Chemistry was awarded to Ciechanover, Hershko and Rose in recognition of their roles in elucidating the importance of proteolytic degradation inside cells and the role of ubiquitin in proteolytic pathways, *i.e.* the ubiquitin proteasome system.[22–24]

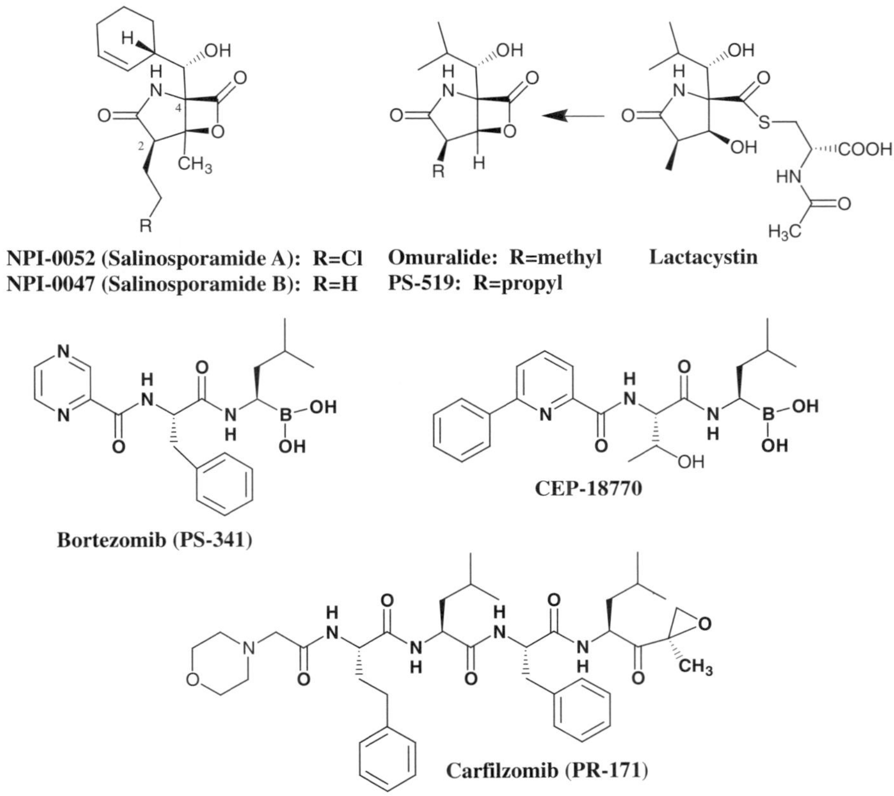

Figure 12.1 Structures of proteasome inhibitors discussed in this chapter.

Ubiquitinylation of proteins is achieved by three sequential enzymatic steps involving ubiquitin activating enzyme (E1), ubiquitin carrier protein (E2) and a ubiquitin ligase (E3), resulting in polyubiquitinated proteins that are targeted for degradation by the proteasome. Protein degradation can be blocked by inhibiting the proteasome, with numerous potential downstream consequences including:

- inhibition of the activation of the NF-κB pathway;
- interference with the degradation of cyclins and other proteins that regulate the cell cycle;
- stabilisation of pro-apoptotic proteins.

These mechanistic findings supported the exploration of the therapeutic potential of proteasome inhibitors in oncology, with the first clinical trials involving a proteasome inhibitor (bortezomib) commencing about ten years ago. "Lessons learned" from bench to bedside during this time period have recently been reviewed.[19]

In terms of structure and function, the 26*S* proteasome is an ATP-dependent multicatalytic enzyme complex comprising one or two 19*S* regulatory caps and a proteolytic 20*S* core particle within which protein degradation occurs.[18,25,26] The 20*S* proteasome contains three pairs of proteolytic subunits, β5, β2 and β1, for which chymotrypsin-like (CT-L), trypsin-like (T-L) and caspase-like (C-L) activities have been ascribed, respectively, based on their substrate preferences.[27]

Polyubiquitinated proteins tagged for degradation by the proteasome are recognised by the 19*S* regulatory caps and unfolded for entry into the "barrel" of the 20*S* core particle, where they are hydrolysed into small peptides by the proteolytic subunits. Binding of protein substrate involves amino acid side chain recognition by proteolytic subunit binding pockets (*e.g.* S1, S3, S4) near the active site. Once bound, hydrolysis of the peptide bond adjacent to the S1-binding pocket is catalysed by the *N*-terminal threonine residue; this mechanism is unique from many other proteases and thus, the proteolytic subunits of the 20*S* proteasome are classed among the *N*-terminal threonine hydrolases.[26,28]

Elucidation of the role of the 20*S* proteasome in protein degradation and as a target for cancer chemotherapy could not have been achieved without small molecule inhibitors, some of which have served strictly as research tools, while others have progressed through preclinical development and clinical trials.[18,28,29] These molecules represent a variety of structural classes, including peptide boronic acids such as bortezomib and CEP-18770,[30] epoxyketones (*e.g.* carfilzomib)[31] and the γ-lactam-β-lactone family of inhibitors (Figure 12.1).

The first identified member of this latter family was omuralide, the β-lactone product of the *Streptomyces*-derived secondary metabolite lactacystin.[32–34] Although omuralide was never developed by the pharmaceutical industry, it continues to be used as an important biochemical tool. Moreover, it served as a model for the synthetic analogue PS-519 (Figure 12.1), which was evaluated in Phase I clinical trials based on preclinical data that indicated a protective effect in models of cerebral ischaemia.[35]

The structure of NPI-0052 offered unique substitutions about the γ-lactam-β-lactone ring system with functional groups that significantly enhanced its potency and potential for drug development, as highlighted in this account.

2 Mechanism of Action

Some of our earliest studies on NPI-0052 were focused on establishing its mechanism of action. The structural relationship of NPI-0052 to omuralide immediately suggested that the two molecules may share a common molecular target. This hypothesis was confirmed by screening the two compounds against purified rabbit 20*S* proteasomes. NPI-0052 inhibited all three proteolytic functions, *i.e.* the CT-L, T-L and C-L activities, with IC_{50} values in the low to mid nanomolar range and was found to be significantly more potent than omuralide.[12] A similar trend was later established in human 20*S* proteasomes.

Structural biology provided further confirmation of the molecular target. The crystal structure of NPI-0052 in complex with the yeast 20*S* proteasome showed that the molecule occupies the active sites of all three pairs of proteolytic subunits (*vide infra*).[15]

While the collective data on isolated 20*S* proteasomes from different species provided strong support for the mechanism of action of NPI-0052, it was critical to establish the target *in vivo*. Initial studies in mice demonstrated that treatment with NPI-0052 induced a prolonged duration of inhibition of all three proteolytic functions of the 20*S* proteasome in packed whole blood (Figure 12.2A),[14] thereby confirming the molecular target *in vivo*. These and other studies provided a strong basis for continued preclinical evaluation of NPI-0052.

3 Microbiology of *Salinispora tropica*, Fermentation and Scale-up

Any successful preclinical development programme requires a reliable source of the drug. The original fermentation conditions and the production strain (*S. tropica* CNB476) transferred from Fenical's research group to Nereus afforded the production of a few milligrams per litre of NPI-0052 in shake flask cultures. The original seed and production media contained animal-derived media components and natural seawater that cannot be used for cGMP (current Good Manufacturing Practice) production of NPI-0052. Extensive fermentation development in improving the production of NPI-0052 was carried out at Nereus Pharmaceuticals and the improvement processes are summarised in Table 12.1.

The key to the initial success of yield improvement was the addition of solid resins to the production culture (step 1, Table 12.1). The inherent instability of the β-lactone ring of NPI-0052 in aqueous solution,[36] such as in the submerged saline fermentation, was overcome by addition of solid resin to the fermentation in order to bind and capture NPI-0052. The addition of resin to the production culture led to an 18-fold increase in yield in a preliminary study (Table 12.1). Further investigation of the resin stabilisation effect on NPI-0052 using the production strain NPS21184 established the conditions for the large-scale resin addition process.[37]

The second key yield improvement was the isolation of *S. tropica* strain NPS21184, a single colony isolate directly derived from strain CNB476 without mutation and genetic manipulation (step 4, Table 12.1). Besides supporting higher production of NPI-0052, strain NPS21184 produces a significantly lower amount (three-fold less) of the interfering deschloro analogue NPI-0047 (salinosporamide B),[38] which must be removed during purification, than the parent strain CNB476.

Media formulation studies (steps 3 and 5, Table 12.1) were successfully carried out to replace natural seawater and animal-derived media components

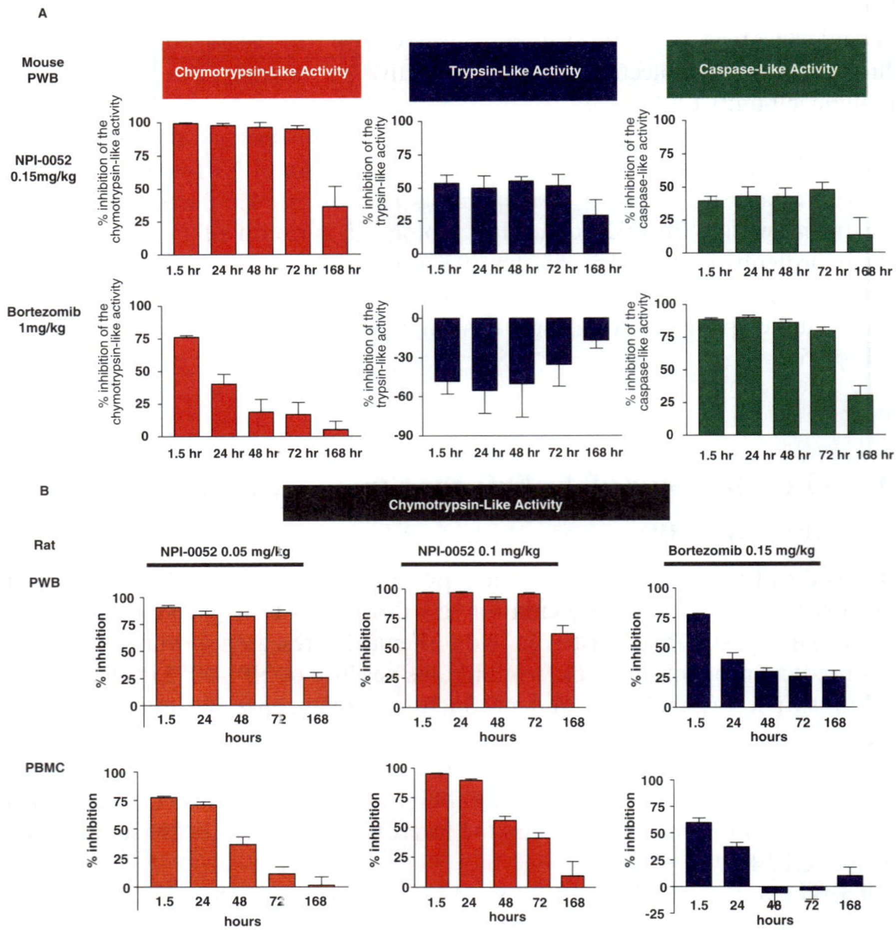

Figure 12.2 Pharmacodynamic profiles of NPI-0052 and bortezomib after a single IV administration in mice or rats. (**A**) Inhibition of CT-L, T-L and C-L 20*S* proteasome activities in packed whole blood (PWB) lysates after a single IV administration of NPI-0052 (0.15 mg/kg) or bortezomib (1 mg/kg) in mice. NPI-0052 exhibits a broader and longer 20*S* proteasome inhibition profile than bortezomib.[14] (**B**) CT-L 20*S* proteasome activity recovers more quickly in peripheral blood mononuclear cell (PBMC) lysates compared with PWB lysates after NPI-0052 administration to rats at 0.05 mg/kg or 0.1 mg/kg.

with media components that are acceptable for cGMP. Furthermore, additional yield improvement was achieved *via* media formulation studies. A greater than 100-fold increase in the production of NPI-0052 in shake flask culture was obtained after the above yield improvement processes with a production titre of 450 mg/L.

Table 12.1 Fermentation yield improvement of NPI-0052 in shake flask and laboratory fermenter.

Improvement step	*Improvement process*	*Shake flask (mg/L)*	*Fermenter (mg/L)*
–	Original condition	4	–
1	Resin addition	70	25
2	Optimisation of seed, production and resin addition cycles	120	120
3	Initial media improvement	220	220
4	Single colony isolation	330	330
5	Further media optimisation (statistical design)	450	360

The production of NPI-0052 by marine actinomycete strain NPS21184 was carried out *via* a saline fermentation process. Saline fermentation poses a major challenge in scale-up since published literature suggested that the 316-type stainless steel fermenters commonly found in manufacturing facilities are not resistant to the corrosive effect of the saline media.[39] Using a 316 stainless steel B. Braun Biostat-C fermenter (42 L total volume), we developed a process to overcome the corrosive effect of saline culture media based on this stainless steel fermenter. The foaming, aeration and agitation issues associated with the scale-up production of NPI-0052 in fermenters were also addressed using the B. Braun Biostat-C fermenter.

We successfully transferred the yield improvement conditions developed in shake flasks to lab fermenter as shown in Table 12.1. A NPI-0052 titre of 360 mg/L was achieved in the 42L lab fermenter, which is lower than the maximum titre of 450 mg/L detected in shake flasks. The discrepancy in titres is due to the foaming problem that occurred in the fermenter (but not in shake flasks) when rich media containing high concentrations of starch and soy type products were used. Based on the potency of NPI-0052, a titre of 360 mg/L in fermenter scale is sufficient to support the clinical development of NPI-0052.

This intensive optimisation of production from shake flask to laboratory-scale fermenter provided the foundation for a smooth transition to large-scale manufacturing of the drug substance, as discussed in Section 7.

4 Structural Biology and Structure–Activity Relationship Studies

NPI-0052 represents the case of a natural product that entered the clinic without structural modification through synthesis or semi-synthesis. A growing body of structure–activity relationship (SAR) studies and structural biology data offer clear insights into its naturally optimised molecular design.

The crystal structure of NPI-0052 in complex with the yeast 20*S* proteasome core particle (CP)[15] demonstrated that the inhibitor occupies all three proteolytic subunits (β5, β2 and β1), consistent with the broad inhibition profile against the CT-L, T-L and C-L activities, respectively. The cyclohexene ring makes sufficiently favourable contacts with the S1 binding/recognition pockets of all three sites, despite their individual preferences (albeit non-exclusivity) for hydrophobic, positively charged and negatively charged residues, respectively. Enzyme inhibition kinetics confirmed that the relative binding affinity for the three sites follows the order CT-L > T-L > > C-L,[40] consistent with the trend in IC_{50} values.

Comparison of the crystal structure of the 20*S* proteasome CP:NPI-0052 complex with that of salinosporamide B (NPI-0047)[15] and lactacystin[25] allowed the chemical mechanism of inhibition by NPI-0052 to be established (Figure 12.3).[15] Once bound to the active site, the catalytic *N*-terminal Thr1O$^{\gamma}$ forms an ester linkage with the carbonyl derived from the β-lactone ring. The remaining γ-lactam circumvents free rotation of the C3–C4 bond, a configuration that positions Thr1NH_2 within hydrogen bonding distance of the C3 OH. This prearrangement enables abstraction of the C3 OH proton by

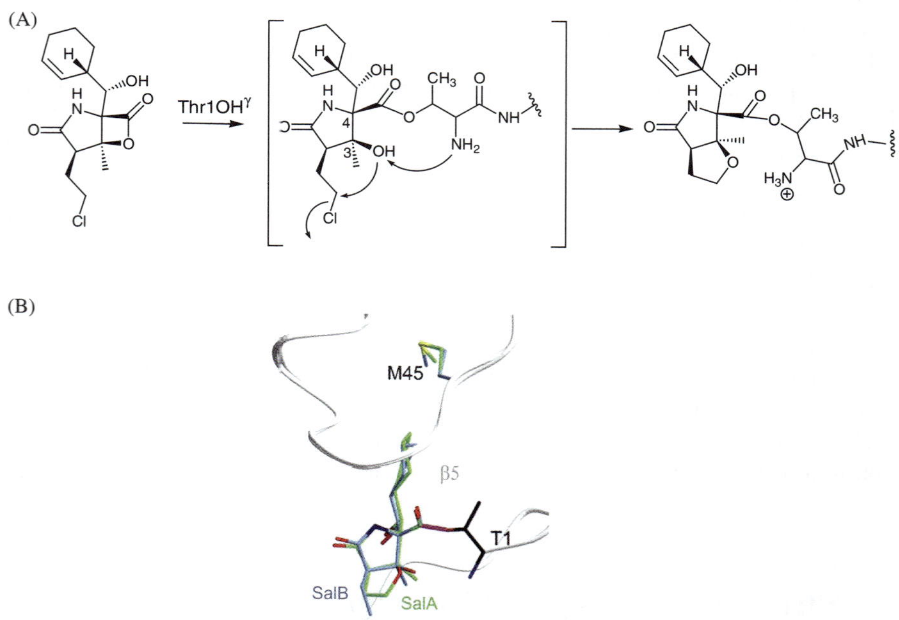

Figure 12.3 (**A**) Mechanism of inactivation of the 20*S* proteasome by NPI-0052. (**B**) Superposition of NPI-0052 (SalA) and salinosporamide B (SalB) in the β5 subunit of the yeast 20*S* proteasome. Each inhibitor is covalently bound to the N-terminal threonine (T1) *via* an ester linkage between Thr1O$^{\gamma}$ and the carbonyl derived from the β-lactone ring. In the case of NPI-0052, the chlorine atom is displaced to form a five-membered cyclic ether ring.[15]

Thr1NH_2, such that C3 O displaces chloride, giving rise to a five-membered cyclic ether ring and a fully protonated *N*-terminus (Thr1NH_3^+), which cannot catalyse ester hydrolysis as long as it remains in the protonated state. Moreover, the position of C3 O sterically intervenes with the approach of water, further obstructing hydrolysis of the ester linkage between the inhibitor and the enzyme, while cleavage of the ester linkage *via* reformation of the β-lactone is not possible because C3 O is confined to the cyclic ether ring. The result is a highly stabilised end product.[15,40] As a result of this irreversible binding mechanism, *de novo* synthesis of proteasomes is required to regain enzyme activity.

These mechanistic studies reveal NPI-0052 as a minimalist among proteasome inhibitors, in that each atom is used to maximum efficiency for inhibitor binding to the active site followed by the two-step reaction that renders it irreversibly bound. In fact, NPI-0052 clearly exceeds atom-for-atom efficiency of any currently known proteasome inhibitor.

SAR studies of analogues with a broad range of potential leaving groups (in place of chlorine) further support the above mechanism and were utilised to demonstrate that the presence of a leaving group results in prolonged duration of proteasome inhibition *in vitro* (no recovery after 24 h dialysis).[40] These and other SAR studies were made possible through natural product chemistry and semi-synthesis,[38,40–42] enantioselective total synthesis,[43–46] directed biosynthesis[47] and bioengineering of *S. tropica*,[48,49] which allowed access to new structural space. This included replacement of the cyclohexene ring with the isopropyl group found in omuralide,[49–51] along with other congeners,[49] albeit without enhancements in potency over the parent natural product. However, these enabling technologies suggest that SAR is still in its early days and has, at a minimum, provided critical retro-evaluation of the structural features of NPI-0052 that reveal it to be a remarkably well-designed proteasome inhibitor gifted by nature.

5 Translational Biology

Our initial findings that NPI-0052 inhibited all three proteolytic functions of the 20*S* proteasome (*vide supra*) led us to compare its profile with other proteasome inhibitors (Figure 12.1) such as bortezomib,[14] carfilzomib (PR-171)[31] and CEP-18770.[30] These agents inhibit the CT-L activity to a similar degree as NPI-0052 but exhibit different inhibition profiles for the T-L and C-L activities. In addition, NPI-0052 exhibits a different recovery profile of proteasome functions in whole blood, normal organs, tumour and peripheral blood mononuclear cell preparations compared with other agents.[52]

The ability of NPI-0052 to inhibit all three proteolytic sites is considered an important and potentially marked advantage, as elegant studies by Kisselev *et al.*[53] have demonstrated that the C-L and T-L sites also mediate a significant role in protein breakdown and their relative importance varies with the target protein, particularly when the CT-L site is markedly inhibited. Moreover, inhibiting only CT-L activity may not be sufficient to block protein degradation. Thus, measuring only CT-L activity may not accurately reflect the degree of proteasome inhibition and the level of protein degradation.[53] Additional studies have defined the specificity of NPI-0052 for the 20*S* proteolytic activities of the proteasome *vs.* other proteases such as chymotrypsin, cathepsin A, cathepsin B and trypsin. The data demonstrates that NPI-0052 exhibits at least a three-log selectivity for the proteasome proteolytic activities *vs.* the other proteases.

The identification of NPI-0052 as a novel chemical entity with an established mechanism of action against a validated target provided a strong foundation for commencement of an accelerated preclinical development programme. In view of the clinical activity of bortezomib in multiple myeloma, NPI-0052 was initially evaluated in a human multiple myeloma xenograft model in immunodeficient mice, where it was efficacious at low doses after twice weekly treatment either by the IV (0.15 mg/kg) or oral (0.25 mg/kg) routes.[14] In mice, treatment with NPI-0052 induced a prolonged duration of inhibition of all three proteolytic functions of the 20*S* proteasome in packed whole blood compared with bortezomib (Figure 12.2A).

NPI-0052 has been evaluated in a wide range of haematologic and solid tumour malignancies studies including models for multiple myeloma,[14] colon,[54] pancreatic,[55] and non-small cell lung carcinomas,[56] melanoma,[57] non-Hodgkin's lymphoma,[58] mantle cell lymphoma,[59] Waldenstrom's macroglobulinemia (WM),[60] acute lymphocytic leukaemia (ALL) and acute myeloid leukaemia (AML)[61] and chronic lymphocytic leukaemia (CLL).[62] Further evaluations indicated that NPI-0052 inhibited the activation of NF-κB and multiple genes regulated by NF-κB, a major downstream event of proteasome inhibition that triggers apoptosis in multiple myeloma and other cells.[63]

The data suggests that NPI-0052 may be efficacious against haematological and solid tumours either as a single agent and/or in combination with biologics, chemotherapeutics and targeted therapeutic agents such as the histone deacetylase inhibitor, vorinostat (suberoylanilide hydroxamic acid (SAHA), Zolinza®).

Our studies clearly differentiate NPI-0052 from other proteasome inhibitors in the speed and duration of action and the inhibition profile of the 20*S* proteasome. These differences and the possible off-target activities of bortezomib indicate that NPI-0052 may provide a greater therapeutic index and greater activity in diseases where bortezomib shows minimal activity.[64]

6 IND-Enabling Studies of NPI-0052

With the established mechanism of action and early indications of efficacy in multiple preclinical models, an intensive programme was undertaken to support the filing of an Investigational New Drug Application (IND) with the FDA. This encompassed production of the active pharmaceutical ingredient (API), a formulated drug product for IND-enabling safety/toxicology studies in two animal species and clinical trials, the establishment of drug pharmacodynamics (*i.e.* inhibition of CT-L, T-L and C-L activities in packed whole blood) and pharmacokinetics, as described below. The IND was filed in December 2005.

7 API Manufacturing

NPI-0052 is currently manufactured under cGMP through a robust saline fermentation process by *S. tropica* strain NPS21184. It was quite an effort to find the proper contract manufacturing organisations (CMOs) that would accept and adapt our saline fermentation process, developed in laboratory fermenters, to their industrial-scale production fermenters and also have the proper containment facility to handle the downstream processing (DSP) of the highly potent NPI-0052.

Despite our development of protocols to minimise the corrosive effect of saline media on laboratory stainless steel fermenters and processing equipment on a relatively small scale, saline fermentation was not a common, acceptable practice in a large scale production process and the corrosiveness of saline media on seed and production fermenters and processing equipment remained the major concern of many CMOs. Nonetheless, we identified several CMOs that accepted our manufacturing proposal and selected two for the large-scale production of NPI-0052.

The final process fermentation development standardised parameters such as temperature exposure, operating parameters, cleaning and passivation to overcome the corrosive effect of saline fermentation and was performed in 500–1500 L industrial-scale stainless steel fermenters. This, together with careful design of the timing and method for introducing the resin to the production fermenter, resulted in production titres of 250–300 mg/L in 500–1500 L industrial fermenters.

During the peak of the production cycle, the resin-bound drug is collected, filtered, extracted with organic solvent and concentrated for DSP in an environment with appropriate containment for a high biological potency substance. To maintain optimal stability of the API, all DSP steps are executed in non-aqueous solvent systems. The crude extract from the resin undergoes purification involving a highly effective silica gel flash chromatography step, which removes all unrelated substances as well as most congeners, such as the des-chloro analogue. In fact, the purity of the product increases from ~55% to ~95% (UV area) after this single step.

This highly purified substance may contain up to ~3% of a diastereomeric impurity. By using an evaporative crystallisation process that exploits subtle solubility differences between the parent compound and the diastereomer, this impurity level is reduced to <1% and the API is isolated as a white crystalline solid. The final pharmaceutical grade cGMP drug substance is obtained in >98% purity with overall ~50% recovery from the crude extract.

Based on the potency of NPI-0052, the production titre is adequate for both clinical development and commercial production. To our knowledge, this represents the first manufacture of clinical trial materials by saline fermentation.

8 Formulation Development and Drug Product Manufacturing

The development of a suitable parenteral formulation for NPI-0052 required careful consideration of several factors including its solubility, high potency and β-lactone ring stability. The Phase I clinical trials formulation for β-lactone proteasome inhibitor PS-519 (40% propylene glycol, 10% ethanol and 50% saline)[35] suggested that a co-solvent system might be suitable for NPI-0052. However, the pH stability profile of NPI-0052[36] did not support the use of a neutral pH aqueous component such as saline, but instead suggested that a more stable dosing solution could be achieved at lower pH. The API was ultimately formulated as a propylene glycol/ethanol drug product concentrate, which is stable for at least two years when stored at low temperature. The dosing solution (0.1 mg/mL NPI-0052 in 40% propylene glycol, 10% ethanol, 50% citrate buffer pH 5) is obtained by diluting the drug product concentrate with citrate buffer pH 5/ethanol at 2–8 °C prior to intravenous (IV) injection.

While the co-solvent formulation is sufficiently robust to support clinical trials, a second generation formulation was developed to enhance product shelf life and user convenience. A lyophilised drug product was targeted to take advantage of the excellent stability of the API in the solid form. Lyophilisation is typically performed from aqueous bulk formulation solutions, but non-aqueous co-solvent systems are also amenable to this process and offer potential advantages, including increased drug solubility or stability in the bulk formulation solution, as well as increased stability in the lyophilised product.[65] However, the process requires proper safe handling of flammable solvents and control of residual solvent levels.

Several characteristics of *tert*-butyl alcohol (TBA), including its high vapour pressure, low freezing point, favourable sublimation profile and low toxicity (albeit currently unclassified in the US Pharmacopeia), have made TBA–water bulk formulation solutions the most widely applied co-solvent systems in the manufacture of lyophilised pharmaceuticals[65] and this approach was adopted for NPI-0052. The manufacturing process for NPI-0052 lyophilised drug product for injection involves freeze-drying of a bulk solution of API and sucrose excipient from ~85% TBA. This bulk solution composition was selected to provide sufficient sucrose solubility while minimising the potential for aqueous hydrolysis of the API during the compounding and fill/finish processes. Lyophilisation from ~85% TBA far exceeds that of other products[65] and may represent the current upper limit in lyophilised drug product manufacturing.

The TBA and water are removed during the lyophilisation cycle and the resulting lyophilised powder or cake is reconstituted with propylene glycol, ethanol and citrate buffer pH 5 prior to administration. Thus, the final dosing solution composition is similar to that of the original co-solvent formulation.

9 Pharmacodynamics

Monitoring the pharmacodynamics (activity at its biological target, *i.e.* inhibition of the proteasome in the case of NPI-0052) can be very helpful in guiding dosing and scheduling in IND-enabling safety studies and clinical trials. Results of pharmacodynamic assays may be illustrative and useful in defining an optimal biological dose, especially for molecularly targeted agents that are rapidly removed from the vascular compartment and distributed widely throughout the body, but with sustained biological effects.

The pharmacodynamic profile of NPI-0052 is different from other proteasome inhibitors (bortezomib and carfilzomib) in that upon a single IV administration to mice, a sustained inhibition ($\geq$ 72 hours) of the main three 20*S* proteolytic activities is observed in packed whole blood lysates (Figure 12.2A). Bortezomib has been reported to either have no effect on or to enhance the T-L activity, while carfilzomib specifically inhibits the CT-L 20*S* proteasome activity.[14,31] Repeated NPI-0052 administration to rodents and monkeys leads to sustained dose-dependent inhibition of whole blood 20*S* proteasome activity, with higher inhibition observed after subsequent administrations and with partial recovery between consecutive NPI-0052 treatments.

One important issue to bear in mind when assessing the biological effect of an irreversible inhibitor over time is the turnover rate of its biological target, in this case the proteasome, which will depend upon the cell type. This was nicely demonstrated by the fact that after a single NPI-0052 administration, recovery of 20*S* proteasome activity in packed whole blood lysates ($\geq$ 72 h; $>$99% red blood cells with a half-life of 15–17 weeks) significantly differed from the recovery of 20*S* proteasome activity in isolated peripheral blood mononuclear cells (48–72 h; nucleated cells with a half life of a few days; Figure 12.2B), as well as normal tissues such as liver, lung, spleen and kidney.[52] This is in contrast with the findings for bortezomib, where recovery of proteasome activity was readily observed in both packed whole blood and peripheral blood mononuclear cell lysates (Figure 12.2B).

In the ongoing Phase I clinical trials, the effects of NPI-0052 on proteasome activities in both packed whole blood and peripheral blood mononuclear cells are being monitored. In approximately 80 patients treated to date with NPI-0052, a dose-dependent inhibition of the whole blood 20*S* proteasome activity is observed, with increasing inhibition upon multiple administrations and partial recovery between consecutive doses (Figure 12.4). Nicely paralleling the preclinical studies, a more pronounced recovery of 20*S* proteasome CT-L activity is observed between consecutive NPI-0052 administrations in the peripheral mononuclear cell population of patients compared to packed whole blood.

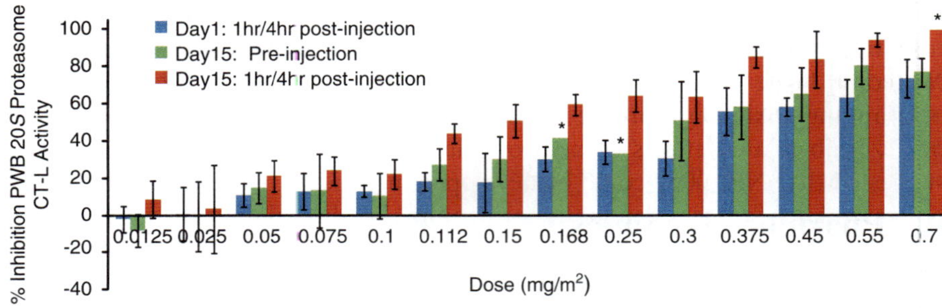

Figure 12.4 Inhibition of the packed whole blood (PWB) CT-L 20*S* proteasome activity in patient samples increases with dose and is more pronounced after the third NPI-0052 administration. NPI-0052 is administered IV on Days 1, 8 and 15 at the doses indicated. Proteasome activity does not restore to baseline levels as indicated by the inhibition observed on Day 15 before the third NPI-0052 administration. Results are the average of three or more patients per cohort, except where indicated (*), the average results of two patients is shown.

Monitoring changes in proteasome activity for evaluating drug effects is one aspect, but can proteasome activity be used as a diagnostic biomarker for predicting clinical response? Initial research has shown that circulating proteasome levels are an independent prognostic factor for survival in multiple myeloma.[66] Studies to monitor plasma and cell proteasome activities before and after NPI-0052 treatment are ongoing.[67] Moreover, recent work has shown that the activity patterns of the various proteasome subunits reflect bortezomib sensitivity of haematologic malignancies.[68] Clearly, much more research is needed, but the initial work suggests that, eventually, proteasome activity profiling of patients may lead to patient stratification and help tailor individualised cancer therapy.

10 Pharmacokinetics

NPI-0052 clears rapidly from the vascular compartment and is distributed to major organ systems, but appears not to enter the central nervous system. The whole blood half-life is up to 30 minutes. The area under the whole blood total concentration-time curve (AUC_{TOTAL}) increases with dose, as does the inhibition of CT-L activity in packed whole blood lysates. There is a large volume of distribution that is consistent across doses and species.

11 Clinical Trials

A clinical development programme for NPI-0052 was initiated based on the preclinical data indicating that with a novel structure, NPI-0052 has different signal transduction, safety and efficacy profiles than other proteasome inhibitors. In particular, the ability of NPI-0052 to synergise with bortezomib[69] and overcome bortezomib resistance with a greater therapeutic index and non-overlapping toxicology profile (NPI-0052 did not demonstrate the common bortezomib related effects of neutropenia, thrombocytopenia or neuropathy) suggested that NPI-0052 could be developed in patients that had failed or were not candidates for treatment with bortezomib. Preclinical data showing efficacy in cancers such as chronic lymphocytic CLL and solid tumour malignancies, where bortezomib has not shown efficacy in clinical trials,[62] also suggested different development strategies.

Clinical development of NPI-0052 began with a typical Phase 1 first-in-human (FIH) dose escalation study in patients with solid tumours or lymphomas.[59] Secondary to preclinical and clinical data, the programme was expanded to include clinical trials in patients with other diagnoses such as multiple myeloma and leukaemia. In order to take advantage of the extended duration of proteasome inhibition with NPI-0052, the dosing regimen utilised in these trials consisted of once a week administration, as opposed to twice per week with other proteasome inhibitors. Each of these trials incorporated pharmacodynamic assessment through proteasome inhibition in blood, as well as standard pharmacokinetic assessments. Dose escalation was carried out through a dose of 0.7 mg/m^2, with adverse events thought likely related to drug being quite tolerable; the most common being fatigue, nausea/vomiting, transient peri-infusion site discomfort, with increasing platelet counts and decreasing haemoglobin and lymphocyte counts, both within and outside of normal ranges, also often being seen.[70]

More importantly, proof-of-mechanism with proteasome inhibition levels in packed whole blood (inhibition of CT-L activity increasing with time and dose up to 100%; Figure 12.4) reaching and exceeding those reported with therapeutic doses of bortezomib was attained at lower doses, suggesting potential for a significantly improved therapeutic ratio. Of particular interest, the profile of adverse events associated with NPI-0052 was quite different from those generally reported with bortezomib.

Given these results and the very favourable preclinical data in combination with other standard of care oncology agents such as cytotoxic agents, immunomodulatory drugs and histone deacetylase inhibitors, combination clinical trials have been initiated with NPI-0052. Ongoing clinical trials include patients that have failed bortezomib treatment, as well as patients with diagnoses where other proteasome inhibitors have not demonstrated efficacy. It is expected the development of NPI-0052 to a New Drug Application (NDA) would focus initially in demonstrating efficacy in patients where bortezomib therapy is indicated (multiple myeloma and mantle cell lymphoma) but has failed, followed by proof of efficacy in randomised trials compared with standard of care agents alone in diagnoses where proteasome inhibitor therapy is not yet indicated.

12 Concluding Remarks

The successful entry of NPI-0052 into several concurrent Phase I clinical trials marks another milestone in the continuing validation of marine bioprospecting for drug discovery and development. NPI-0052 expands the dimensions of the existing body of marine natural products under clinical evaluation in several unique ways. Its discovery from a marine actinomycete, *S. tropica*, offers proof that marine microbiology is a fertile discovery resource for new chemistry with clinical pharmaceutical applications. Moreover, large-scale API manufacturing from *S. tropica* demonstrates that saline fermentation is a viable pharmaceutical manufacturing process and that marine natural products of microbial origin need not be limited by the "supply issue".

Natural products offer structures with unique functionalities and the successful API manufacturing and formulation of NPI-0052 show that the challenge of developing a compound with a potentially labile β-lactone ring and chloroethyl trigger can be overcome. NPI-0052 represents the first marine-derived proteasome inhibitor, effectively broadening the growing list of molecular targets for marine natural products. The establishment of the mechanism of action of NPI-0052 provided a strong basis for launching a successful preclinical development programme, which in turn added value to the knowledge base for proteasome inhibitor biology while providing the foundation for the ongoing clinical trials. The entry of a marine microbial natural product into clinical trials in oncology should stimulate both academia and industry to further explore and exploit marine microbiology for drug discovery and development.

Acknowledgements

The authors are indebted to the many Nereus employees and collaborators who contributed to the development of NPI-0052. We thank Michael Groll for contributing Figure 12.3B and Marcia Katz for editorial assistance during the preparation of this manuscript.

References

1. E. W. Schmidt, *Trends Biotechnol.*, 2005, **23**, 437.
2. J. W. Blunt, B. R. Copp, W. P. Hu, M. H. Munro, P. T. Northcote and M. R. Prinsep, *Nat. Prod. Rep.*, 2008, **25**, 35.
3. J. W. Blunt, B. R. Copp, W. P. Hu, M. H. Munro, P. T. Northcote and M. R. Prinsep, *Nat. Prod. Rep.*, 2007, **24**, 31.
4. W. Fenical and P. R. Jensen, *Nat. Chem. Biol.*, 2006, **2**, 666.
5. A. T. Bull and E. M. Stach, *Trends Microbiol.*, 2007, **15**, 491.
6. L. A. Maldonado, W. Fenical, P. R. Jensen, C. A. Kauffman, T. J. Mincer, A. C. Ward, A. T. Bull and M. Goodfellow, *Int. J. Syst. Evol. Microbiol.*, 2005, **55**, 1759.
7. P. R. Jensen, E. Gontang, C. Mafnas, T. J. Mincer and W. Fenical, *Environ. Microbiol.*, 2005, **7**, 1039.

8. P. R. Jensen and F. M. Lauro, *Antonie Van Leeuwenhoek*, 2008, **94**, 51.
9. R. T. Hill, in *Microbial Diversity and Bioprospecting*, ed. A. T. Bull, ASM Press, 2004, pp. 177–190.
10. T. J. Mincer, P. R. Jensen, C. A. Kauffman and W. Fenical, *Appl. Environ. Microbiol.*, 2002, **68**, 5005.
11. T. J. Mincer, W. Fenical and P. R. Jensen, *Appl. Environ. Microbiol.*, 2005, **71**, 7019.
12. R. H. Feling, G. O. Buchanan, T. J. Mincer, C. A. Kauffman, P. R. Jensen and W. Fenical, *Angew. Chem., Int. Ed. Engl.*, 2003, **42**, 355.
13. W. Fenical, P. R. Jensen, M. A. Palladino, K. S. Lam, G. K. Lloyd and B. C. Potts, *Bioorg. Med. Chem.*, 2008, **17**, 2175.
14. D. Chauhan, L. Catley, G. Li, K. Podar, T. Hideshima, M. Velankar, C. Mitsiades, N. Mitsiades, H. Yasui, A. Letai, H. Ovaa, C. Berkers, B. Nicholson, T. H. Chao, S. T. Neuteboom, P. Richardson, M. Palladino and K. C. Anderson, *Cancer Cell*, 2005, **8**, 407.
15. M. Groll, R. Huber and B. C. M. Potts, *J. Am. Chem. Soc.*, 2006, **128**, 5136.
16. A. Hershko and A. Ciechanover, *Ann. Rev. Biochem.*, 1998, **67**, 425.
17. J. Adams, *Proteasome Inhibitors in Cancer Therapy*, Humana Press, Totowa, NJ, 2004.
18. A. F. Kisselev and A. L. Goldberg, *Chem. Biol.*, 2001, **8**, 739.
19. R. Z. Orlowski and D. J. Kuhn, *Clin. Cancer Res.*, 2008, **14**, 1649.
20. P. G. Richarson, B. Barlogie, J. Berenson, S. Singhal, S. Jagannath, D. Irwin, S. V. Rajkumar, G. Srkalovic, M. Alsina, R. Alexanian, D. Seigel, R. Z. Orlowski, D. Kuter, S. A. Limentani, S. Lee, T. Hideshima, D. L. Esseltine, M. Kauffman, J. Adams, D. P. Schenkein and K. C. Anderson, *N. Engl. J. Med.*, 2003, **348**, 2609.
21. P. F. Bross, R. Kane, A. T. Farrell, S. Abraham, K. Benson, M. E. Brower, S. Bradley, J. V. Gobburu, A. Goheer, S.-L. Lee, J. Leighton, C. Y. Liang, R. T. Lostritto, W. D. McGuinn, D. E. Morse, A. Rahman, L. A. Rosario, S. L. Verbois, G. Williams, Y.-C. Wang and R. Pazdur, *Clin. Cancer Res.*, 2004, **10**, 3954.
22. A. Hershko, *Ang. Chem., Int. Ed. Engl.*, 2005, **44**, 5932.
23. A. Ciechanover, *Angew. Chem., Int. Ed. Engl.*, 2005, **44**, 5944.
24. I. Rose, *Angew. Chem., Int. Ed. Engl.*, 2005, **44**, 5926.
25. M. Groll, L. Ditzel, J. Löwe, D. Stock, M. Bochtler, H. D. Bartunik and R. Huber, *Nature*, 1997, **386**, 463.
26. J. Löwe, D. Stock, B. Jap, P. Zwickl, W. Baumeister and R. Huber, *Science*, 1995, **268**, 533.
27. S. Wilk, M. Pereira and B. Yu, *Biomed. Biochim. Acta*, 1991, **50**, 471.
28. L. Borissenko and M. Groll, *Chem. Rev.*, 2007, **107**, 687.
29. K. B. Kim and C. M. Crews, *J. Med. Chem.*, 2008, **51**, 2600.
30. R. Piva, B. Ruggeri, M. Williams, G. Costa, I. Tamagno, D. Ferrero, V. Giai, M. Coscia, S. Peola, M. Massaia, G. Pezzoni, C. Allievi, N. Pescalli, M. Cassin, S. di Giovine, P. Nicoli, P. de Feudis, I. Strepponi, I. Roato, R. Ferracini, B. Bussolati, G. Camussi, S. Jones-Bolin, K. Hunter, H. Zhao,

A. Neri, A. Palumbo, C. Berkers, H. Ovaa, A. Bernareggi and G. Inghirami, *Blood*, 2008, **111**, 2765.
31. S. D. Demo, C. J. Kirk, M. A. Aujay, T. J. Buchholz, M. Dajee, M. N. Ho, J. Jiang, G. J. Laidig, E. R. Lewis, F. Parlati, K. D. Shenk, M. S. Smyth, C. M. Sun, M. K. Vallone, T. M. Woo, C. J. Molineaux and M. K. Bennett, *Cancer Res.*, 2007, **67**, 6383.
32. S. Ōmura, K. Matsuzaki, T. Fujimoto, K. Kosuge, T. Furuya, S. Fujita and A. Nakagawa, *J. Antibiot. (Tokyo)*, 1991, **44**, 117.
33. G. Fenteany, R. F. Standaert, W. S. Lane, S. Choi, E. J. Corey and S. L. Schreiber, *Science*, 1995, **268**, 726.
34. L. R. Dick, A. A. Cruikshank, L. Grenier, F. D. Melandri, S. L. Nunes and R. L. Stein, *J. Biol. Chem.*, 1996, **271**, 7273.
35. I. M. Shah, K. R. Lees, C. P. Pien and P. J. Elliot, *J. Clin. Pharmacol.*, 2002, **54**, 269.
36. N. Denora, B. C. M. Potts and V. J. Stella, *J. Pharm. Sci.*, 2007, **96**, 2037.
37. G. Tsueng and K. S. Lam, *J. Antibiot. (Tokyo)*, 2007, **60**, 469.
38. P. G. Williams, G. O. Buchanan, R. H. Feling, C. A. Kauffman, P. R. Jensen and W. Fenical, *J. Org. Chem.*, 2005, **70**, 6196.
39. J. A. Sedriks, *Corrosions of Stainless Steels*, Wiley, NY, 2nd edn. 1996.
40. R. R. Manam, K. A. McArthur, T.-H. Chao, J. Weiss, J. A. Ali, V. J. Palombella, M. Groll, G. K. Lloyd, M. A. Palladino, S. T. C. Neuteboom, V. R. Macherla and B. C. M. Potts, *J. Med. Chem.*, 2008, **51**, 6711.
41. V. R. Macherla, S. S. Mitchell, R. R. Manam, K. Reed, T.-H. Chao, B. Nicholson, G. Deyanat-Yazdi, B. Mai, P. R. Jensen, W. Fenical, S. T. C. Neuteboom, K. S. Lam, M. A. Palladino and B. C. M. Potts, *J. Med. Chem.*, 2005, **48**, 3684.
42. K. A. Reed, R. R. Manam, S. S. Mitchell, J. Xu, S. Teisan, T.-H. Chao, G. Deyanat-Yazdi, S. T. C. Neuteboom, K. S. Lam and B. C. M. Potts, *J. Nat. Prod.*, 2007, **70**, 269.
43. L. R. Reddy, P. Saravanan and E. J. Corey, *J. Am. Chem. Soc.*, 2004, **126**, 6230.
44. A. Endo and S. J. Danishefsky, *J. Am. Chem. Soc.*, 2005, **127**, 8298.
45. T. Ling, V. R. Macherla, R. R. Manam, K. A. McArthur and B. C. Potts, *Org. Lett.*, 2007, **9**, 2289.
46. M. Shibasaki, M. Kanai and N. Fukuda, *Chem. Asian J.*, 2007, **2**, 20.
47. K. S. Lam, G. Tsueng, K. A. McArthur, S. S. Mitchell, B. C. M. Potts and J. Xu, *J. Antibiot. (Tokyo)*, 2007, **60**, 13.
48. A. S. Eustaquio and B. S. Moore, *Angew. Chem., Int. Ed. Engl.*, 2008, **47**, 3936.
49. R. P. McGlinchey, M. Nett, A. S. Eustaquio, R. N. Asolkar, W. Fenical and B. S. Moore, *J. Am. Chem. Soc.*, 2008, **130**, 7822.
50. L. R. Reddy, J.-F. Fournier, B. V. S. Reddy and E. J. Corey, *J. Am. Chem. Soc.*, 2005, **127**, 8974.
51. R.R. Manam, V.R. Macherla, G. Tsueng, C. Dring, K.S. Lam and B.C. Potts, *J. Nat. Prod.*, 2009, **72**, 295.

52. A. Singh, G. K. Lloyd, M. A. Palladino, D. Chauhan and K. C. Anderson, *Blood*, 2008, **112**, Abstract 3665.
53. A. F. Kisselev, A. Callard and A. L. Goldberg, *J. Biol. Chem.*, 2006, **281**, 8582.
54. J. Cusack, R. Liu, L. Xia, C. Pien, W. Niu, V. Palombella, T. Chao, S. Neuteboom and M. Palladino, *Clin. Cancer Res.*, 2006, **12**, 6758.
55. C. M. Sloss, F. Wang, R. Liu, M. Houston, D. Ljungman, M. A. Palladino and J. C. Cusack, *Clin. Cancer Res.*, 2008, **14**, 5116.
56. H. Drabkin, personal communication.
57. D. McConkey, personal communication.
58. S. Baritaki, E. Suzuki, K. Umezawa, D. Spandidos, J. Berenson, T. R. Daniels, M. L. Penichet, M. A. Palladino and B. Bonavida, *J. Immunol.*, 2008, **180**, 6199.
59. C. A. Aghajanian, P. Hamlin, M. S. Gordon, D. S. Hong, A. Naing, A. Younes, A. Hannah, M. A. Palladino, M. A. Spear and R. Kurzrock, *2008 American Society of Clinical Oncology Annual Meeting Proceedings, Supplement to J. Clin. Oncol.*, 2008, 26, No. 15S Part 1 of 2, p. 171S, Abstract 3574.
60. A. Roccaro, X. Jia, A. Sacco, M. Melhem, A. Moreau, X. Leleu, H. Ngo, J. Runels, A. Azab, F. Azab, N. Burwick, M. Farag, S. Treon, M. Palladino, T. Hideshima, D. Chauhan, K. Anderson and I. Ghobrial, *Blood*, 2008, **3**, 4752.
61. C. Miller, K. Ban, M. Dujka, D. McConkey, M. Palladino and J. Chandra, *Blood*, 2007, **110**, 267.
62. S. Ruiz, Y. Krupnik, M. Keating, J. Chandra, M. Palladino and D. McConkey, *Mol. Cancer Ther.*, 2006, **5**, 1836.
63. K. S. Ahn, G. Sethi, T. H. Chao, S. Neuteboom, M. Chaturvedi, M. Palladino, A. Younes and B. Aggarwal, *Blood*, 2007, **110**, 2286.
64. I. Nickeleit, S. Zender, F. Sasse, R. Geffers, G. Brandes, I. Sörensen, H. Steinmetz, S. Kubicka, T. Carlomagno, D. Menche, I. Gütgemann, J. Buer, A. Gossler, M. P. Manns, M. Kalesse, R. Frank and N. P. Malek, *Cancer Cell*, 2008, **14**, 23.
65. D. L. Teargarden and D. S. Baker, *Eur. J. Pharm. Sci.*, 2002, **15**, 115.
66. C. Jakob, K. Egerer, P. Liebisch, S. Türkmen, I. Zavrski, U. Kuckelkorn, U. Heider, M. Kaiser, C. Fleissner, J. Sterz, L. Kleeberg, E. Feist, G. R. Burmester, P. M. Kloetzel and O. Sezer, *Blood*, 2007, **109**, 2100.
67. M. A. Palladino, T. Chao, S. T. Neuteboom, M. Spear, W. Ma and M. Albitar, *Eur. J. Cancer, EJC Supplements,* 2008, **6**, 74 (Abstract 235).
68. M. Kraus, T. Rückrich, M. Reich, J. Gogel, A. Beck, W. Kammer, C. R. Berkers, D. Burg, H. Overkleeft, H. Ovaa and C. Driessen, *Leukemia*, 2007, **21**, 84.
69. D. Chauhan, A. Singh, M. Brahmandam, K. Podar, T. Hideshima, P. Richardson, N. Munshi, M. Palladino and K. Anderson, *Blood*, 2008, **111**, 1654.
70. P. Padrik, T. J. Price, M. A. Spear, A. Townsend, A. Longenecker, M. A. Palladino, G. K. Lloyd, G. F. Cropp and M. Millward, *Ann. Oncol.*, 2008, **19**(Suppl 8), viii162(489P).

CHAPTER 13

From Natural Product to Clinical Trials: Bevirimat, a Plant-Derived Anti-AIDS Drug†

KEDUO QIAN,[a] THEODORE J. NITZ,[b] DONGLEI YU,[a] GRAHAM P. ALLAWAY,[b] SUSAN L. MORRIS-NATSCHKE[a] AND KUO-HSIUNG LEE*[a]

[a] Natural Products Research Laboratories, Eshelman School of Pharmacy, University of North Carolina, Chapel Hill, NC 27599, USA; [b] Panacos Pharmaceuticals Inc., 209 Perry Parkway, Gaithersburg, MD 20878, USA

1 Introduction

Theoretically, any of the multiple steps in the life-cycle of human immunodeficiency virus (HIV) such as viral attachment, co-receptor binding, fusion, reverse transcription, integration, translation, proteolytic cleavage, glycosylation, assembly, budding and release can be attacked chemotherapeutically by anti-HIV agents.[1–4] However, among the 29 approved drugs (23 single compounds and six combination drugs) marketed to treat HIV-1 infection, 23 target one of two viral enzymes, reverse transcriptase (RT) or protease (PR). Two HIV entry inhibitors are currently available:

- Fuzeon (enfuvirtide, ENF, T-20) which inhibits viral entry by blocking fusion of the viral particle into the host target cell;[5,6]

† Anti-AIDS Agents 76. For paper 75, see K. Qian, S.L. Morris-Natschke and K.H. Lee, *Med. Res. Rev.*, 2009, **29**, 369.

RSC Biomolecular Sciences No. 18
Natural Product Chemistry for Drug Discovery
Edited by Antony D. Buss and Mark S. Butler

Published by the Royal Society of Chemistry, www.rsc.org

- Selzentry (maraviroc, MVC, UK-427,857) which functions as a CCR5 co-receptor antagonist.[7]

Most recently, Isentress (raltegravir), an integrase strand transfer inhibitor, was approved by the US Food and Drug Administration (FDA).[8]

Introduction of highly active antiretroviral therapy (HAART), which employs a combination of drugs with different targets, was a major beneficial advance that improved both the quality and length of patient survival.[9,10] However, virus isolates resistant to the approved drugs eventually appear and reduce the choice and effectiveness of treatment options.[11] It is estimated that up to 78% of HIV-1-infected individuals harbour drug-resistant virus with a rapidly growing subgroup (5–10%) exhibiting resistance to all classes of RT and PR inhibitors.[12,13] Therefore, novel potent antiretroviral agents with unique modes of action that can retain potency against resistant viral strains still remain a major unmet need.

2 Bioactivity-directed Fractionation and Isolation

Natural products are a reservoir of biologically active compounds. At the Natural Products Research Laboratories (NPRL) (Eshelman School of Pharmacy, University of North Carolina), our research focuses on the discovery of novel hits from natural products, followed by an extensive period of investigation whereby leads are identified and optimised, ultimately identifying new clinical trial candidates.

Bioactivity-directed fractionation and isolation (BDFI) is a major approach for new lead generation (Figure 13.1). In our anti-HIV drug discovery programme, much effort has been expended on finding anti-HIV natural products with unique structural characteristics or mechanisms of action.[11–14] Lead optimisation is based on rational drug design, synthesis and bio-evaluation, as well as structure–activity relationship (SAR) and quantitative-SAR (QSAR) studies (Figure 13.2). Selection of a pre-clinical candidate involves rapid, efficient assays to gather absorption, distribution, metabolism, excretion and toxicity (ADMET) data, permitting an effective evaluation of drug candidates in the early stage of the drug discovery process.

Bevirimat (also referred to as DSB, YK-FH312 and PA-457 in other studies[14,15]), identified by our group at NPRL, provides an excellent example of BDFI-based anti-HIV drug discovery and development.

3 Lead Identification

With nearly 20 years of experience on plant-derived natural products, NPRL has identified numerous active anti-HIV compounds including polycyclic diones, saponins, alkaloids, triterpenes, polyphenols, flavonoids and coumarins. Triterpenes have diverse structures and pharmacological activities.

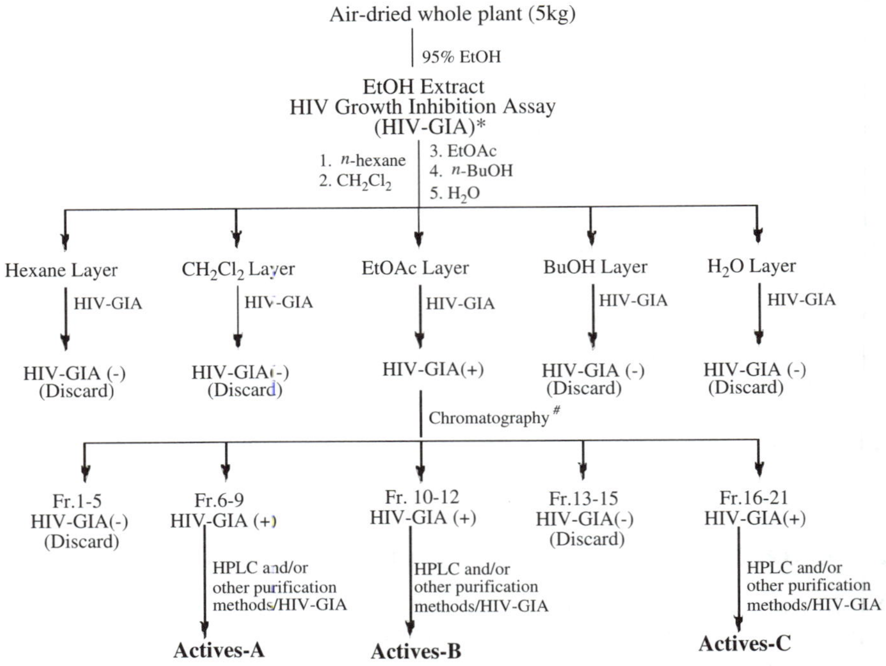

Figure 13.1 Example of the bioassay-derived fractionation and isolation of anti-HIV compounds from a typical high priority plant of interest.
*See M. Nishizawa, T. Yamagishi, G. E. Durtschman, W. B. Parker, A. J. Bodner, R. E. Kilkuskie, Y. C. Cheng, and K. H. Lee, *J. Nat. Prod.*, 1989, **52**, 762, for detailed bioassay procedure.
(+) means fraction which inhibits HIV replicated (by $>30\%$) at a therapeutic index of >5 [a concentration which is at least 5-fold less than the concentration which exhibits cytotoxicity (CC_{50}) ($>30\%$ cell growth inhibition)].

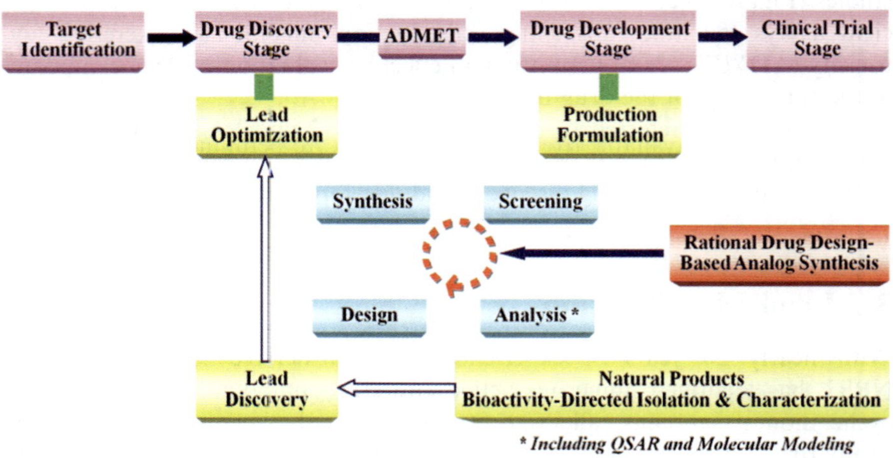

Figure 13.2 Flowchart for discovery and development of plant-derived anti-HIV agents.

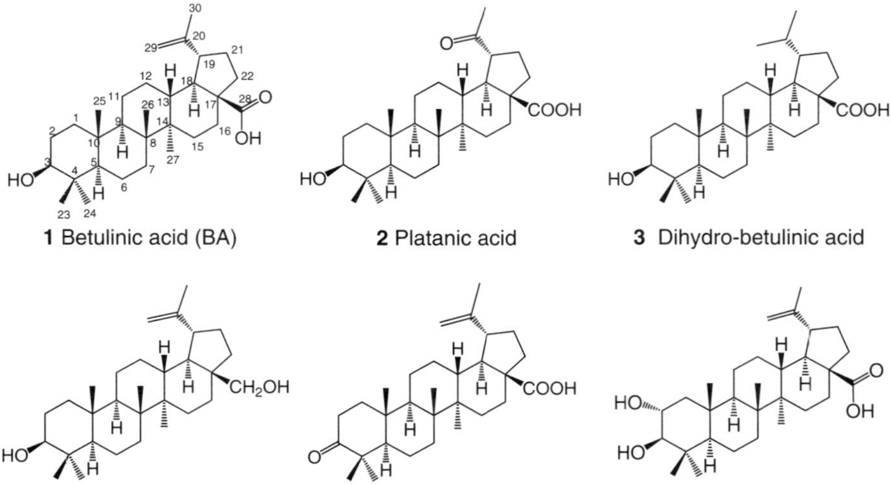

1 Betulinic acid (BA) **2** Platanic acid **3** Dihydro-betulinic acid

4 Betulin **5** Betulonic acid **6** Alphitolic acid

Figure 13.3 Structures of BA (**1**) and related skeleton-modified compounds **2–6**.

Several naturally occurring triterpenes have been reported to show anti-HIV activity, *e.g.* betulinic acid, platanic acid, glycyrrhizin, mimusopic acid, ganoderiol and geumonoid.[16–20] These active compounds hold the potential to serve as hits or leads for anti-HIV drug development.[21,22] The focus of this chapter is bevirimat, the first in a new class of compounds termed HIV maturation inhibitors (MIs). Our discussion covers modification, structure–activity relationship (SAR), mechanism of action studies and clinical trials of bevirimat.

In 1994, betulinic acid **1** (BA, 3β-hydroxylup-20(29)-en-28-oic acid) and platanic acid **2** (Figure 13.3) were discovered at NPRL to show inhibitory effects against $HIV\text{-}1_{IIIB}$ replication, with EC_{50} values of 1.4 and 6.5 μM, respectively.[16] They belong to the lupane family, a group of pentacyclic triterpenes with a five-membered E ring. BA is present in many plant species, *e.g. Tryphyllum peltatum*, *Ancistrocladus heyneanus*, *Ziziphi fructus*, *Diospyros leucomelas*, *Tetracera boliviana* and *Syzygium formosanum*,[23–25] and can be obtained in quantity from the bark of the London plane tree, *Platanus acerifolia*.[26,27] In addition, several methods are available for the conversion of betulin **4**, readily available from white birch bark, to betulinic acid **1**.[28] Due to its promising anti-HIV activity and ready availability from natural and semisynthetic sources, BA was selected as the template for structural modification with the primary goal of improving antiviral activity.

4 Lead Optimisation and SAR Study

4.1 Modification of the BA Triterpene Skeleton

Although not without some synthetic challenges due to the sterically demanding and rigid framework of the pentacyclic triterpene, betulinic acid **1** has three sites that readily lend themselves to chemical modification and analogue syntheses.[29] The C-3 alcohol can be readily acylated, the process that led to the synthesis of bevirimat. In addition, the C-3 alcohol can be eliminated or oxidised for further modification of the A-ring. The isopropenyl group at C-19 undergoes chemistry typical of the allylic group. Lastly, the C28 carboxyl group can be derivatised providing esters and amides.

Catalytic hydrogenation of the BA C-19 isopropenyl group using palladium on carbon (Pd-C) provides dihydrobetulinic acid **3**,[16] which exhibited similar anti-HIV activity with an EC_{50} value of 0.9 μM and therapeutic index (TI) value of 14. Betulin **4** showed weaker anti-HIV activity than BA with an EC_{50} of 23 μM and a TI of 1.9, indicating that the C-28 carboxyl group is required for optimal potency. Oxidation of the C-3 hydroxyl group provides betulonic acid **5**, which is more active than BA (0.22 μM), but is also significantly more cytotoxic with an CC_{50} value of 1.8 μM.[30] Alphitolic acid **6**, isolated from *Rosa woodsii* (Rosaceae) showed a reduced anti-HIV activity with an EC_{50} of 42.3 μM (Figure 13.3).[31] Additional examples of BA skeleton modifications are discussed in the following sections.

4.2 Modification on C-3 Position of BA

As mentioned previously, the C-3 hydroxyl group can be readily acylated with a wide variety of anhydrides and acid chlorides.[30] Modification on this site of BA **1** and dihydro-BA **3** in our laboratory afforded a group of C-3 ester derivatives **7–22** (Table 13.1).

Among these compounds, seven analogues showed significantly improved antiviral activities, demonstrating that the C-3 position is a pharmacophore for anti-HIV potency. BA derivatives with 3′,3′-dimethylsuccinyl **8** (bevirimat) and 3′,3′-dimethylglutaryl **9** substituents exhibited extremely potent anti-HIV activity, with EC_{50} values in the nanomolar range (0.00035 and 0.0023 μM, respectively).[32] The corresponding substituted dihydro-BA derivatives **15** and **16** showed similar potencies when compared with bevirimat **8** and **9**, respectively.[33] Moving the dimethyl substitution on the C-3 side chain to the 2′ position (as seen in compounds **10** and **17**) resulted in a significant reduction in antiviral activity. Compounds **13** and **20–22**, which lack the terminal carboxylic acid, were much less potent or exhibited no activity at the highest concentration tested. The C-3 esters **7**, **11**, **12**, **14** and **18**, which incorporate the terminal acid functionality but lack the dimethyl substitution, retain partial activity compared with the geminal disubstituted analogues. Analogues with a single methyl substitution at the C-3′ position (Figure 13.4) were synthesised in order to determine if there was a stereochemical preference for either of the geminal dimethyl groups.[34] It was discovered that (3′*S*)-monomethylsuccinyl BA **23** [(3′*S*)-MSB)] and bevirimat **8** showed similar activity (EC_{50} 0.0087 and 0.0013 μM, respectively) in this assay format, while the (3′*R*)-MSB isomer **24** exhibited only moderate anti-HIV activity (EC_{50} 0.12 μM). This result indicates that the two C-3′ methyl groups on the bevirimat side chain contribute differently to the extremely potent anti-HIV activity of this compound class.[34]

Table 13.1 Anti-HIV activities of C-3 modified BA and dihydro-BA derivatives **7–22** in acutely infected H9 lymphocytes.

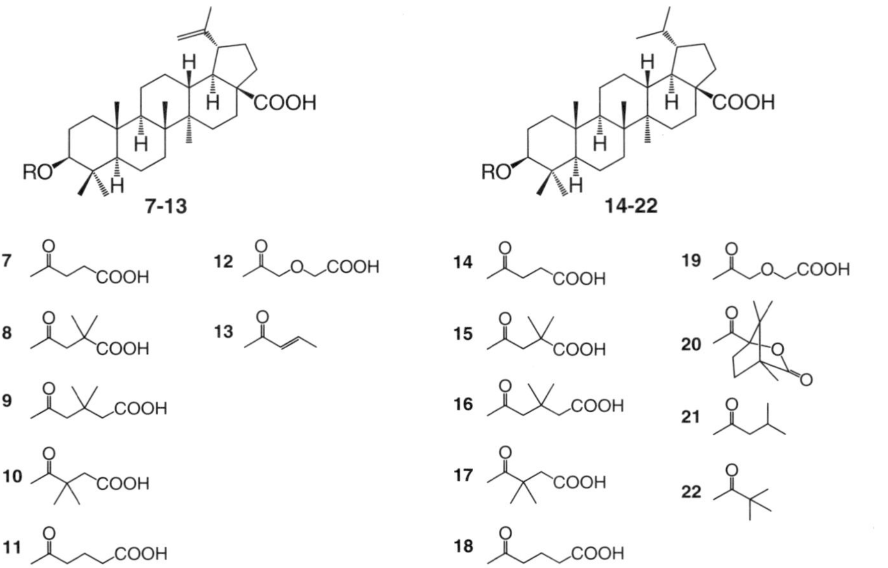

Comp'd	*CC_{50} (μM)*	*EC_{50} (μM)*	*TI*	*Comp'd*	*CC_{50} (μM)*	*EC_{50} (μM)*	*TI*
1 (BA)	13.0	1.4	9.3	**3**	12.6	0.9	14
7	16	4.0	4	**14**	13.4	1.8	7.5
8 (DSB)	7	0.00035	20000	**15** (DSD)	4.9	0.00035	14000
9	4.5	0.0023	2000	**16**	13.1	0.0056	2344
10	15.9	2.7	6.7	**17**	7.7	0.56	13.8
11	12.8	0.044	292	**18**	7.9	0.9	9
12	11.7	0.01	1170	**19**	5.8	0.0057	1000
13	48	19	2.5	**20**	1	0.5	2
				21	83	1.5	56
				22		NS	

CC_{50} = 50% cytotoxic concentration, the drug concentration that inhibits cell growth by 50% *in vitro*.
EC_{50} = 50% effective concentration, the drug concentration that inhibits viral replication by 50% *in vitro*.
NS = no suppression.
TI = CC_{50}/EC_{50}.

We postulate that an interaction of the (3′*S*)-methyl group with the viral target might be essential to the anti-HIV activity of **8**.

To summarise the SAR of the C-3 pharmacophore, we discovered that within the C-3 side chain, a terminal carboxylic acid and a dimethyl substitution at the C-3′ position contribute significantly to the anti-HIV activity. Regarding the dimethyl pattern, the (3′*S*)-methyl group contributes more to increased potency.

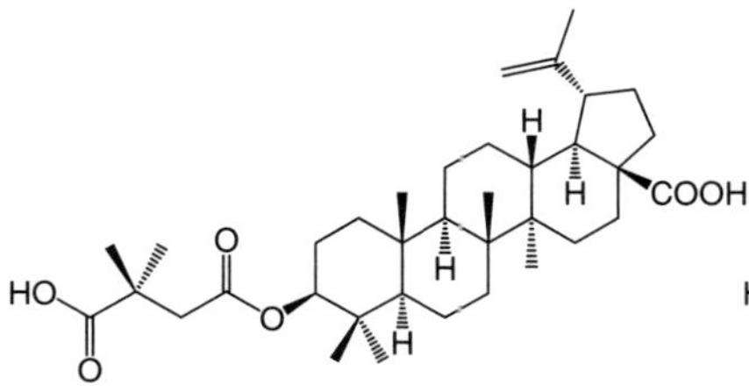

8 DSB (EC_{50} = 0.0013 uM)

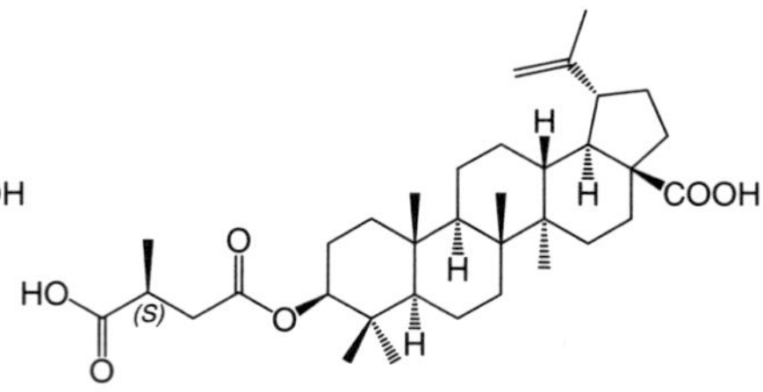

23 3′*S*-MSB (EC_{50} = 0.0087 uM)

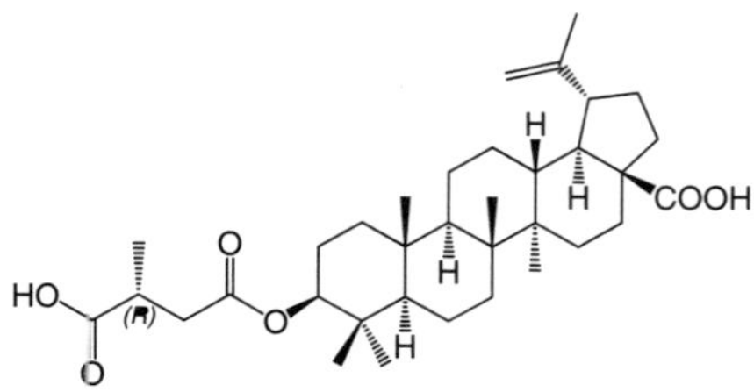

24 3′R-MSB (EC_{50} = 0.12 uM)

Figure 13.4 Antiviral activities of monomethylsuccinyl BA (MSB) analogues **23** and **24**.

As we mentioned earlier, betulin **4** differs from BA **1** only by a C-28 hydroxymethyl group rather than a carboxylic acid. To further verify our SAR information, diverse acid anhydrides were reacted with **4** and dihydrobetulin **25** to yield a group of 3,28-disubstituted compounds **26–37** (Table 13.2).[35]

Compound 3,28-di-(3′,3′-dimethylglutaryl) betulin **29** showed the most potent activity in this group with an EC_{50} value of 0.00066 μM.[35] Without the geminal dimethyl moiety in their side chains, succinyl compound **26** and glutaryl compound **27** exhibited weak activity, as was also observed for corresponding BA analogues. A significant gain in potency is achieved by introducing a single methyl group, with the 3′-methyl **28** being almost 3 $\log_{10}$ more potent then the unsubstituted analogue **27** and nearly as active as the geminally disubstituted analogue **29**. The 3′-methyl-3′-ethyl **30** substitution showed approximately eight-fold decreased activity compared with **29**. It should be noted that both **28** and **30** are diastereomeric mixtures at the 3′-position. Similar activity trends were also observed in the dihydrobetulin series. These data confirmed that, at the C-3 position of BA and betulin, geminal dimethyl substitution of the side chain C-3′ position coupled with a terminal carboxylic acid are important to the enhanced antiviral activity for both of these triterpene templates. In addition, some A-ring modified 28-DMG betulin analogues were also synthesised. However, both the 3-α-**29** analogue **38** and 3-epi-**29** analogue **39**, showed dramatically reduced antiviral activity compared with **29**, indicating the importance of the 3-position stereocentre (Figure 13.5).

Bioisosteric substitutions are a commonly used strategy in medicinal chemistry drug design as an approach to enhance the desired biological or physical

Table 13.2 Structures and activities of disubstituted betulins and dihydrobetulins **25–37**.

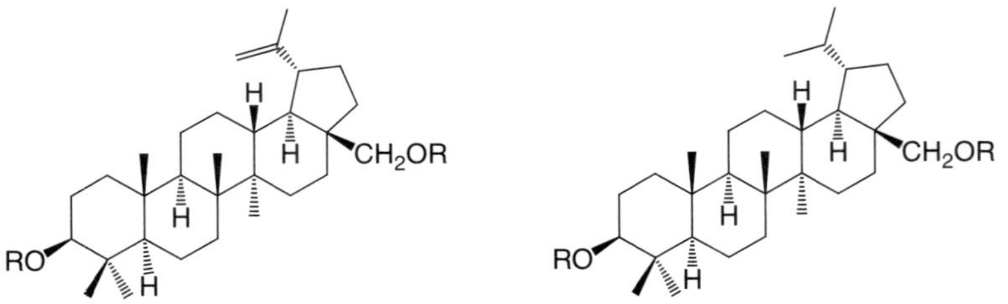

R		*CC₅₀ (μM)*	*EC₅₀ (μM)*	*TI*		*CC₅₀ (μM)*	*EC₅₀ (μM)*	*TI*
H	**4**	43.7	23	1.9	**25**		–	
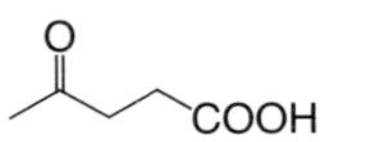	**26**	35.3	3.8	9.3	**32**	28.8	4.7	6.2
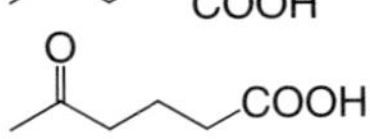	**27**	25.8	3.8	6.8	**33**	26.3	1.6	17
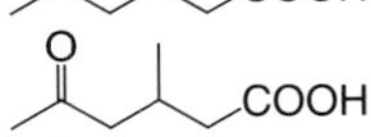	**28**	20.7	0.0039	5308	**34**	19.2	0.059	325
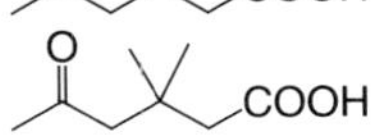	**29**	14.2	0.00066	21515	**35**	10.6	0.0047	2253
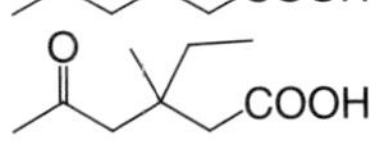	**30**	18.4	0.0053	3476	**36**	18.7	0.075	248
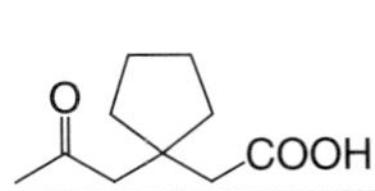	**31**	20.5	0.077	267	**37**	21.6	0.58	37

– Not tested due to limited solubility.

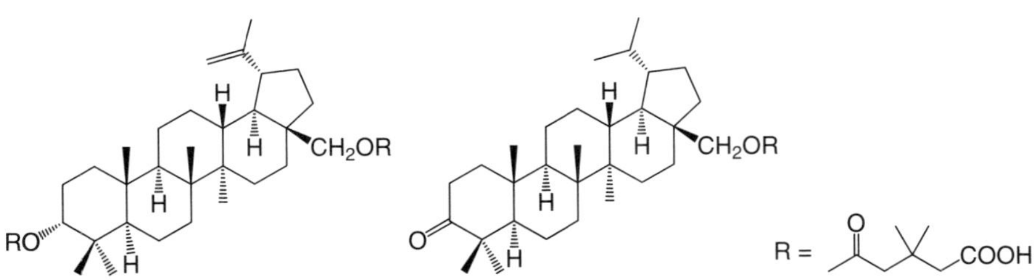

38 (3-α-**29**, EC_{50} > 13.8 uM) **39** (3-epi-**29**, EC_{50} = 10.0 uM)

Figure 13.5 Structures and antiviral activities of 3-α-29 **38** and 3-epi-29 **39**.

properties of a compound without making significant changes in the overall chemical structure. In our study, the possible bioisosteric replacement of the C-3 ester O with an NH provided C-3 amide derivatives of BA and betulin. The anti-HIV activity of the 3*β*- and 3*α*-alkylamido-deoxy-BA derivatives **40–45**

Table 13.3 Anti-HIV activities of 3-alkylamido-deoxy-BA derivatives **40–45**.

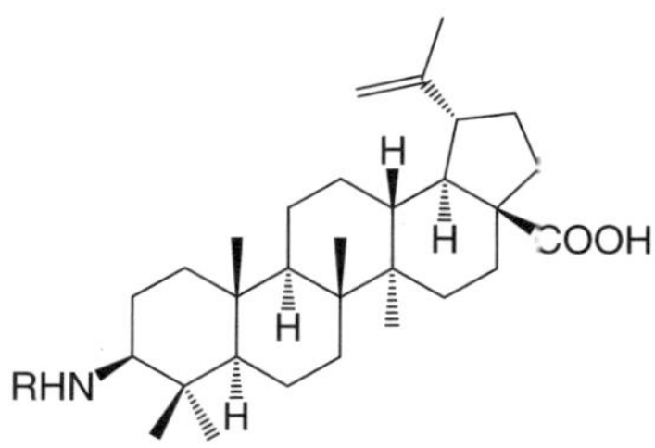

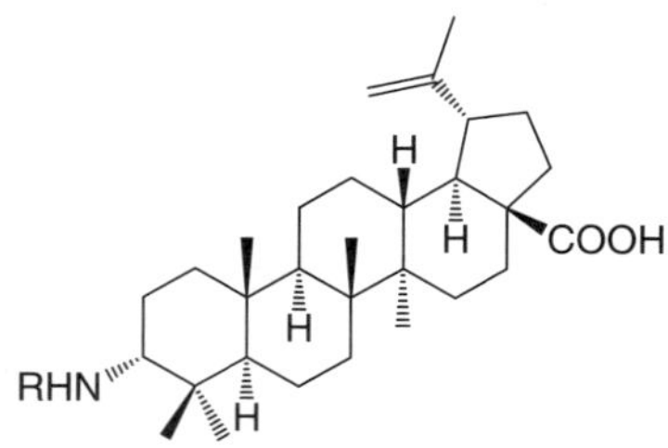

R		*CC₅₀* (*μM*)	*EC₅₀* (*μM*)	*TI*		*CC₅₀* (*μM*)	*EC₅₀* (*μM*)	*TI*
	**40**	32.4	NS		**43**	–		
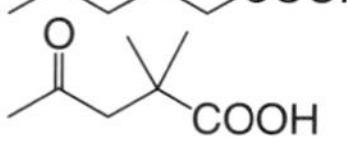	**41**	35.8	NS		**44**	35.8	NS	
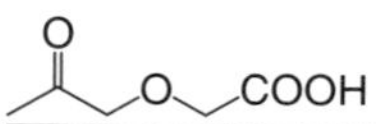	**42**	38.7	7.9	4.9	**45**	174.9	0.24	728

(Table 13.3) decreased significantly.[36] Only **42** and **45** showed weak activity with EC_{50} values of 7.9 and 0.24 μM, respectively.[36] A similar approach was applied in the betulin series (Figure 13.6). It was found that the intermediate oxime **46** (EC_{50}: 1.07 μM, CC_{50}: 5.47 μM) showed better anti-HIV activity than betulin **4**, but was cytotoxic, as was the amine **47** (EC_{50}: 1.07 μM, CC_{50}: 0.52 μM).[35] The 3,28-disubstituted alkylamido products **48–50**, exhibited similar antiviral activity (EC_{50}: 0.57–4.57 μM) and decreased cytotoxicity (CC_{50} $>$ 15.0 μM) compared with **46** and **47**.[35] These results strongly indicated that the C-3 ester group is essential for the potent anti-HIV activity that is observed in bevirimat **8** and its analogues and that an alkylamido moiety at this position is not favourable.

Based on our extensive *in vitro* and *in vivo* antiviral and ADMET preclinical studies (see Sections 5 and 6), bevirimat **8** was chosen as a clinical candidate for the treatment of HIV infection.

4.3 Introduction of C-28 Side Chain into BA

Extensive research has been conducted to explore the C-28 side chain modifications of BA.[30,37,38] A series of C-28 ω-aminoalkanoic acid derivatives of BA, which were first synthesised by Soler *et al.*,[39] functions by blocking the viral entry into the host cells. Incremental chain lengthening significantly influenced the anti-HIV potency of the derivatives.[4] Compounds with amide side chains between aminooctanoic acid and aminododecanoic acid (7–11 carbons between the C-28 amide moiety and the terminal carboxylic acid

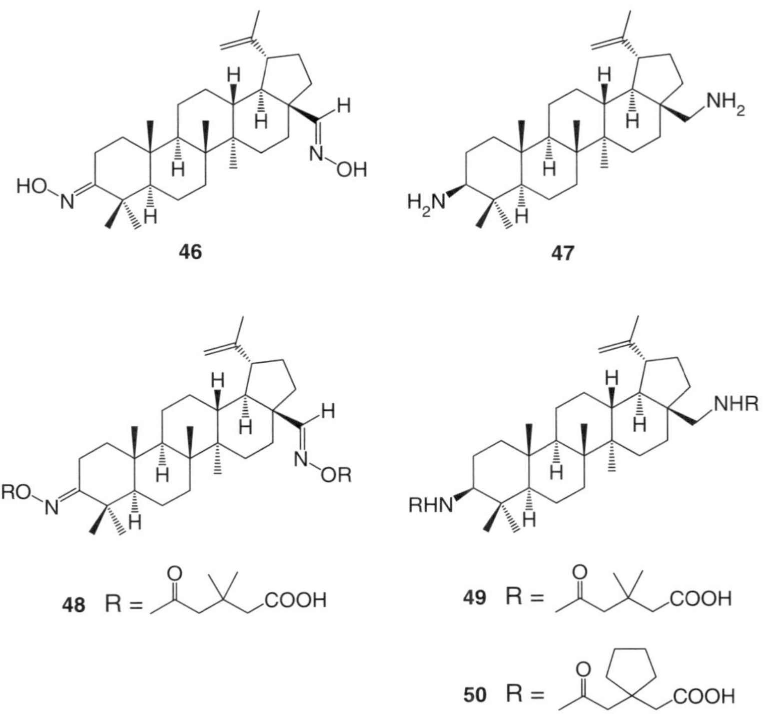

Figure 13.6 Structures of alkylamido betulin derivatives **46–50**.

group) showed much increased antiviral potency.[40] Introduction of a second aminoalkanoic acid at the end of these C-28 ω-aminoalkanoic acid derivatives could further modulate the antiviral potency.[4,39]

This effort led to the discovery of RPR103611 **51**, a BA derivative, which inhibits the infectivity of several HIV-1 strains in EC_{50} values in the range of 0.05–2 μM.[39,41] IC9564 **52**, discovered in our laboratories, is a stereoisomer of RPR103611 and is equipotent in antiviral activity.[37] Dihydro-IC9564 **53** was also found to be equipotent against HIV-1 virus when compared with RPR103611 and IC9564.[37] Although RPR103611 showed potent antiviral activity *in vitro*, its clinical development by Rhone-Poulenc (now Sanofi-Aventis) was less successful and was stopped due to poor "pharmacodynamic properties",[42] suggesting that further modification of this compound class as HIV entry inhibitors is still necessary (Figure 13.7).

4.4 Bifunctional BA Analogues—Potential for Maturation Inhibitor Development

Interestingly, the anti-HIV-1 targets of triterpene analogues can vary depending on the side chain modification positions.[21] Some C-28 modified BA analogues are potent HIV entry inhibitors, while some C-3 modified BA derivatives

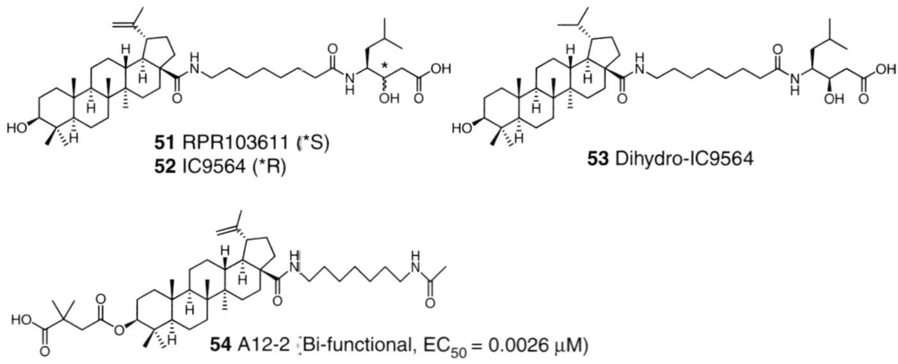

Figure 13.7 Structures of C-28 modified BA analogues as HIV entry inhibitors **51–53** and bifunctional BA derivatives **54**.

function by blocking virus maturation.[33] Because the two functionalities are on the opposite position of the BA skeleton, a design combining both modifications led to the development of bi-functional BA analogues.[43] A12-2 **54** (Figure 13.7), a representative in this class, showed an EC_{50} value of 0.0026 μM and was more potent than both the C-3 and C-28 parent compounds.[44] This category preserves both the anti-HIV-1 entry and anti-HIV-1 maturation activities and shows a synergistic effect on antiviral potency, which may serve as a trend for the future development of maturation inhibitors (MIs).

5 Mechanism of Action Studies of Bevirimat

Results from a series of *in vitro* assays first demonstrated that bevirimat does not affect the function of viral RT.[15,32,45] This observation was supported by the fact that bevirimat retained its nanomolar antiviral potency against HIV-1 isolates resistant to RT inhibitors.

To investigate whether bevirimat functions as a PR inhibitor, several bioassays were conducted, including a cell-free fluorometric assay using a synthetic peptide substrate of PR and experiments using a recombinant form of the Gag precursor protein $Pr55^{Gag}$. Both assays showed no effect of bevirimat on PR function compared with the PR inhibitor indinavir. Indeed, bevirimat at nanomolar concentrations could inhibit the replication of HIV-1 isolates resistant to PR inhibitors.

A multinuclear-activation galactosidase indicator (MAGI) infectivity assay was then used to determine the inhibitory stage of bevirimat. It was realised that bevirimat blocks virus replication at a time point after the completion of viral DNA integration and Tat expression.[15] Furthermore, several assays, including quantitative radioimmunoprecipitation analysis, demonstrated that bevirimat does not affect virus particle release.

To define the target of bevirimat, scientists from Panacos Pharmaceuticals Inc. and the National Institutes of Health (NIH) characterised the virus produced by cells treated with the compound. It was then observed that bevirimat inhibited the processing of the viral *Gag* polyprotein at a specific step, the conversion of the capsid precursor p25 to mature capsid p24, in a dose-dependent manner.[15] Normally, the viral PR cleaves $Pr55^{Gag}$ and generates mature *Gag* proteins: matrix (MA), capsid (CA), nucleocapsid (NC) and p6, as well as two small Gag spacer peptides (SP1 and SP2).[46] This cleavage triggers a structural rearrangement termed maturation, during which the nascent particle transitions to a mature virion characterised by an electron-dense, conical core. The efficiencies with which protease cleaves its target sequences vary widely, resulting in a highly ordered *Gag* and *GagPol* processing cascade[47–49] and even partial inhibition of *Gag* processing profoundly impairs virus maturation and infectivity.[50]

Both radioimmunoprecipitation analyses and Western blot analyses revealed that bevirimat specifically blocks processing of CA-SP1 (p25) to CA (p24), which is necessary for final capsid condensation and formation of infectious viral particles. Electron microscopy studies confirmed that virion particles treated with bevirimat exhibit abnormal morphology, including spherical rather than the normal conical cores and a thin electron dense layer immediately under the viral membrane.[15] Identical morphological defects were observed in viruses containing mutations of the CA-SP1 cleavage site in *Gag*.[51] Significantly, bevirimat has no effect on processing at other sites of *Gag*, unlike protease inhibitors (PIs) that interact with the protease enzyme and globally inhibit *Gag* processing.[15] All these data suggested that bevirimat inhibits virus replication by disrupting the release of functional CA protein, which results in a blockage of virus maturation and leads to the production of non-infectious viral particles. This novel mechanism is illustrated in Figure 13.8.

In subsequent studies, bevirimat-resistant isolates were selected by serial passage of HIV-1 in the presence of sub-inhibitory concentrations of the compound. Sequence analyses of the resistant virus found several point mutations proximal to the cleavage site of CA-SP1, including CA H226Y, L231M, L231F, SP1 A1V and A3V (Figure 13.9). Significantly, no mutations have been observed in other regions of *Gag* or in the PR coding region.[52] When these mutations were introduced into the backbone of virus isolate NL4-3, respectively, the wild-type virus became resistant to bevirimat. This may also explain why HIV-2 and simian immunodeficiency virus (SIV) are normally insensitive to bevirimat, since both viruses carry a different amino acid sequence at the CA-SP1 cleavage site. Results from the *in vitro* resistance selection experiments further supported that bevirimat functions by disrupting the processing of CA-SP1, which is distinct from the mechanism of any of the currently approved classes of anti-retroviral drugs.

6 Preclinical Studies of Bevirimat

Bevirimat exhibited a mean EC_{50} value of 10.3 nM against a large panel of primary subtype B virus isolates, which was similar to that observed with the approved antiretroviral drugs AZT, nevirapine and indinavir.[15] Most

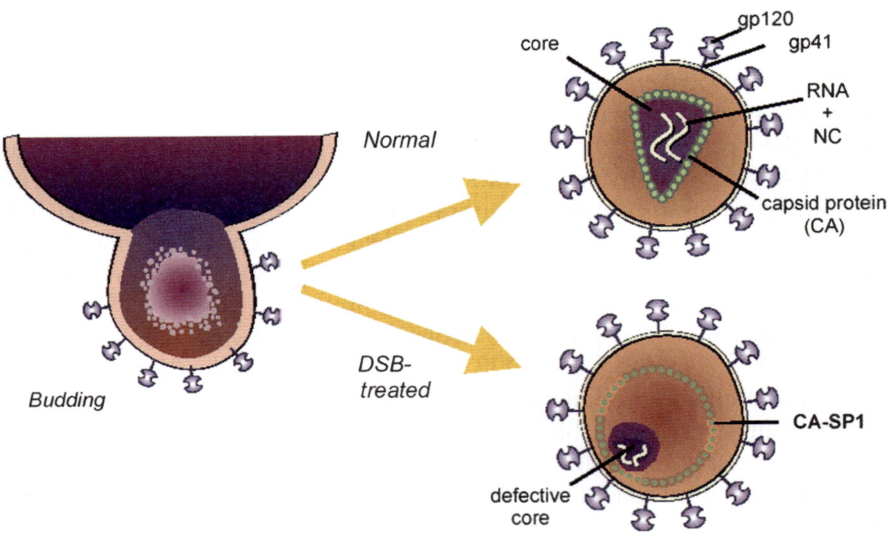

Figure 13.8 Bevirimat inhibits HIV-1 maturation by blocking the processing of CA-SP1 to CA, resulting in the release of aberrant, non-infectious viral particles.[15] Reprinted with permission.

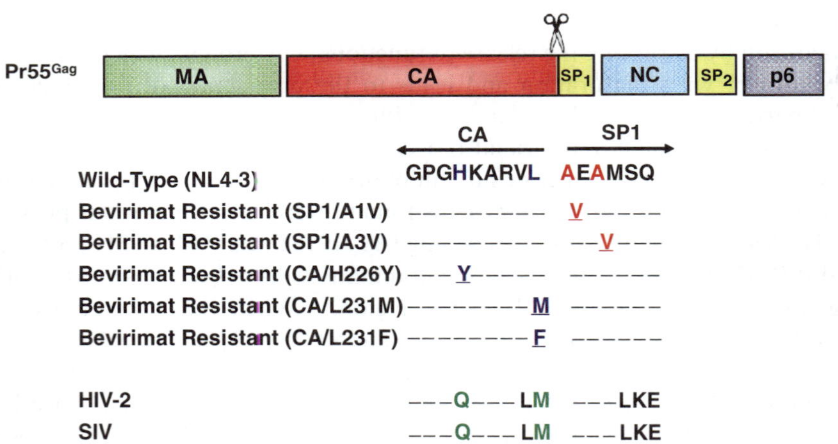

Figure 13.9 Sequence analyses of the bevirimat resistant virus as well as HIV-2 and SIV.

importantly, bevirimat retains its nanomolar inhibitory activity (average 7.8 nM) against drug resistant HIV strains, including isolates resistant to the major classes of approved nucleoside reverse transcriptase inhibitors (NRTIs), non-nucleoside reverse transcriptase inhibitors (NNRTIs) and protease inhibitors (PIs).[15,32] Furthermore, this antiviral activity is HIV-1 specific, with no activity against HIV-2 and SIV. Extended research showed that bevirimat has no effect against a range of enveloped viruses including influenza, herpes simplex type 1 and HTLV-1.[41]

The efficacy of bevirimat was evaluated *in vivo* using the SCID-hu Thy/Liv mouse model, a well accepted animal model for testing HIV drugs.[53,54] Twice-daily oral administration of bevirimat to HIV-1-infected SCID-hu Thy/Liv mice reduced implant viral loads in a dose-dependent manner, causing reductions of $>2\ \log_{10}$ in HIV-1 RNA and $\geq 90\%$ both in implant p24 concentration and in percentage of *Gag*-p24^{+} thymocytes at 100 mg/kg per day, while preserving immature and mature T-cell populations.[55] Antiviral activity was observed in the mice at plasma concentrations that are achievable in humans by oral dosing.[55]

Additional preclinical studies showed that bevirimat has good oral bioavailability in animals and is metabolised primarily by glucuronidation, suggesting it would not be subject to drug–drug interactions when used in combination with the majority of HIV drugs that are metabolised by the cytochrome P450 enzyme system. A panel of preclinical safety studies was completed, including toxicology studies in two species, with results supporting the filing of an IND and initiation of clinical studies in humans.

7 Clinical Trials and Current Status of Bevirimat

In 2004, Panacos Pharmaceuticals completed single and multiple dose Phase I clinical trials of bevirimat in healthy volunteers.[56,57] The drug candidate was administered as an oral solution. It was well-tolerated in these studies, with a half-life of ~2.5–3 days, supporting a once daily dosing regimen. The plasma concentrations of the drug reached levels significantly greater than those predicted to provide a therapeutic benefit in HIV-infected patients.

In August 2005, a Phase IIa clinical study of bevirimat was completed successfully. In this randomised double-blind Phase IIa study, bevirimat monotherapy for ten days resulted in statistically significant reductions in viral load compared with placebo, with individual decreases of up to 1.7 $\log_{10}$, at the 100 and 200 mg doses. Genetic analysis of HIV in patients pre- and post-treatment showed no evidence of the development of resistance to the drug.

In 2008, a Phase IIb study of bevirimat in patients failing HIV therapy due to drug resistance was completed successfully. The results of this study demonstrated that patients who have virus with the most commonly occurring amino acids at positions 369, 370, or 371 on *Gag* are much more likely to respond to bevirimat treatment. In contrast, those patients whose virus has polymorphisms (variants) at these positions are less likely to respond to the drug. Furthermore, pharmacokinetic/pharmacodynamic modelling demonstrated that a trough plasma concentration of greater than 20 μg/mL bevirimat is required for a robust response.

In the Phase IIb study, the mean viral load reduction was $-1.18\ \log_{10}$ copies/mL after 14 days of bevirimat treatment in the patients who were free of key baseline *Gag* polymorphisms and who had bevirimat trough levels above the minimum target of 20 μg/mL. In addition, 91% of patients with these two response predictors had at least a 0.5 $\log_{10}$ viral load reduction by week 2 with a maximum treatment response of 2.03 $\log_{10}$. Consistent with earlier studies, bevirimat was well-tolerated, with a safety profile comparable to placebo through the 14 days of treatment.

Most recently, Panacos announced that bevirimat tablet formulation dosed twice daily achieved target plasma levels. After 14 days of bevirimat treatment given twice daily at doses of 200 mg or 300 mg (using the 50 mg tablet), 100% of 32 treatment-naive and treatment-experienced patients in this study had bevirimat plasma concentrations well above the previously identified minimum target of 20 μg/mL. Phase III clinical studies of bevirimat are currently being planned.

8 Conclusions

Although the application of HAART has led to a significant improvement in the health and life span of HIV-1-infected patients, several key issues have negatively impacted the efficacy of this treatment approach. A major problem is the increasing prevalence of virus strains that are resistant to approved drugs, which can have a significant adverse impact on treatment effectiveness and disease outcome, highlighting the need for new HIV-1 treatment options.

One strategy to address these problems is to develop antiretroviral drugs with targets other than reverse transcriptase and protease. In the collaboration of our laboratories at the University of North Carolina and Panacos Pharmaceuticals, we have taken advantage of the huge molecular diversity found in natural products. Plants are a major source of biologically active compounds and can provide good leads that are structurally unique and/or have new mechanisms of action.

In a decade of extensive research, we have made great progress in the identification of novel anti-HIV drugs, which has led to the discovery of bevirimat, a compound representing a completely new class of antiretroviral agents that block HIV-1 replication by disrupting virus maturation. Bevirimat is currently in Phase IIb clinical development and has a great potential to offer a valuable new option for the treatment of HIV/AIDS.

Acknowledgements

This investigation was supported in part by grants AI-33066 and AI-077417 (awarded to K. H. Lee) and R44 AI051047 (awarded to G. P. Allaway) from the National Institute of Allergies and Infectious Diseases.

References

1. A. Pani, A. G. Loi, M. Mura, T. Marceddu, P. La Colla and M. E. Marongiu, *Curr. Drug Targets Infect. Disord.*, 2002, **2**, 17.

2. M. D. Miller and D. J. Hazuda, *Curr. Opin. Microbiol.*, 2001, **4**, 535.
3. J. P. Moore and M. Stevenson, *Nat. Rev. Mol. Cell Biol.*, 2000, **1**, 40.
4. K. Qian, S. L. Morris-Natschke and K. H. Lee, *Med. Res. Rev.*, 2009, **29**, 369.
5. H. B. Fung and Y. Guo, *Clin. Ther.*, 2004, **26**, 352.
6. J. S. Cervia and M. A. Smith, *Clin. Infect. Dis.*, 2003, **37**, 1102.
7. N. J. Carter and G. M. Keating, *Drugs*, 2007, **67**, 2277.
8. J. D. Croxtall, K. A. Lyseng-Williamson and C. M. Perry, *Drugs*, 2008, **68**, 131.
9. J. Stephenson, *JAMA*, 1997, **277**, 614.
10. M. B. Klein, P. Willemot, T. Murphy and R. G. Lalonde, *AIDS*, 2004, **18**, 1895.
11. G. M. Lucas, R. E. Chaisson and R. D. Moore, *Ann. Intern. Med.*, 1999, **131**, 81.
12. D. Richman, S. Bozzette, S. Morton, S. Chien, T. Wrin, K. Dawson and N. Hellmann, in *Proceedings of the 41st Interscience Conference on Antimicrobial Agents and Chemotherapy, 16–19 December 2001, Chicago, IL*, American Society for Microbiology, 2001.
13. J. LaBonte, J. Lebbos and P. Kirkpatrick, *Nat. Rev. Drug Discov.*, 2003, **2**, 345.
14. T. Kanamoto, Y. Kashiwada, K. Kanbara, K. Gotoh, M. Yoshimori, T. Goto, K. Sano and H. Nakashima, *Antimicrob. Agents Chemother.*, 2001, **45**, 1225.
15. F. Li, R. Goila-Gaur, K. Salzwedel, N. R. Kilgore, M. Reddick, C. Matallana, A. Castillo, D. Zoumplis, D. E. Martin, J. M. Orenstein, G. P. Allaway, E. O. Freed and C. T. Wild, *Proc. Natl. Acad. Sci. USA*, 2003, **100**, 13555.
16. T. Fujioka, Y. Kashiwada, R. E. Kilkuskie, L. M. Cosentino, L. M. Ballas, J. B. Jiang, W. P. Janzen, I. S. Chen and K. H. Lee, *J. Nat. Prod.*, 1994, **57**, 243.
17. K. Hirabayashi, S. Iwata, H. Matsumoto, T. Mori, S. Shibata, M. Baba, M. Ito, S. Shigeta, H. Nakashima and N. Yamamoto, *Chem. Pharm. Bull.*, 1991, **39**, 112.
18. N. P. Sahu, *Phytochemistry*, 1996, **41**, 883.
19. S. el-Mekkawy, M. R. Meselhy, N. Nakamura, Y. Tezuka, M. Hattori, N. Kakiuchi, K. Shimotohno, T. Kawahata and T. Otake, *Phytochemistry*, 1998, **49**, 1651.
20. H. -X. Xu, D. -S. Ming, H. Dong and P. P. -H. But, *Chem. Pharm. Bull.*, 2000, **48**, 1367.
21. L. Huang and C. H. Chen, *Curr. Drug Targets Infect. Disord.*, 2002, **2**, 33.
22. E. De Clercq, *Med. Res. Rev.*, 2000, **20**, 323.
23. C. -W. Chang, T. -S. Wu, Y. -S. Hsieh, S. -C. Kuo and P. -D. L. Chao, *J. Nat. Prod.*, 1999, **62**, 327.
24. G. Bringmann, W. Saeb, L. A. Assi, G. Francois, A. S. S. Narayanan, K. Peters and E. M. Peters, *Planta Med.*, 1997, **63**, 255.
25. K. H. Bae, S. M. Lee, E. S. Lee, J. S. Lee and J. S. Kang, *Yakhak Hoeji*, 1996, **40**, 558.

26. P. A. Krasutsky, R. M. Carlson, V. V. Nesterenko, I. V. Kolomitsyn and C. F. Edwardson, 2001, *Application: WO Patent* 2001010885.
27. D. G. Birgit, T. Neubert and R. Wohlrab, 2001, *US Patent 6,175,035*.
28. A. S. Barthel and S. Csuk, *Tetrahedron*, 2008, **64**, 9225.
29. R. H. Cichewicz and S. A. Kouzi, *Med. Res. Rev.*, 2004, **24**, 90.
30. F. Hashimoto, Y. Kashiwada, L. M. Cosentino, C. -H. Chen, P. E. Garrett and K. -H. Lee, *Bioorg. Med. Chem.*, 1997, **5**, 2133.
31. Y. Kashiwada, H. K. Wang, T. Nagao, S. Kitanaka, I. Yasuda, T. Fujioka, T. Yamagishi, L. M. Cosentino, M. Kozuka, H. Okabe, Y. Ikeshiro, C. Q. Hu, E. Yeh and K. H. Lee, *J. Nat. Prod.*, 1998, **61**, 1090.
32. Y. Kashiwada, F. Hashimoto, L. M. Cosentino, C. H. Chen, P. E. Garrett and K. H. Lee, *J. Med. Chem.*, 1996, **39**, 1016.
33. D. Yu, C. T. Wild, D. E. Martin, S. L. Morris-Natschke, C. H. Chen, G. P. Allaway and K. H. Lee, *Expert Opin. Investig. Drugs*, 2005, **14**, 681.
34. K. Qian, K. Nakagawa-Goto, D. Yu, S. L. Morris-Natschke, T. J. Nitz, N. Kilgore, G. P. Allaway and K. H. Lee, *Bioorg. Med. Chem. Lett.*, 2007, **17**, 6553.
35. I. C. Sun, H. K. Wang, Y. Kashiwada, J. K. Shen, L. M. Cosentino, C. H. Chen, L. M. Yang and K. H. Lee, *J. Med. Chem.*, 1998, **41**, 4648.
36. Y. Kashiwada, J. Chiyo, Y. Ikeshiro, T. Nagao, H. Okabe, L. M. Cosentino, K. Fowke, S. L. Morris-Natschke and K. -H. Lee, *Chem. Pharm. Bull.*, 2000, **48**, 1387.
37. I. C. Sun, C. H. Chen, Y. Kashiwada, J. H. Wu, H. K. Wang and K. H. Lee, *J. Med. Chem.*, 2002, **45**, 4271.
38. G. N. W. Robinson, C.T.; Ashton, M.; Thomas, R.; Montalbetti, C.; Coulter, T.S.; Magaraci, F.; Townsend, R.J.; Nitz, T.J., 2004, *WO Patent* 2006053255.
39. F. Soler, C. Poujade, M. Evers, J. C. Carry, Y. Henin, A. Bousseau, T. Huet, R. Pauwels, E. De Clercq, J. F. Mayaux, J. B. Le Pecq and N. Dereu, *J. Med. Chem.*, 1996, **39**, 1069.
40. M. Evers, C. Poujade, F. Soler, Y. Ribeill, C. James, Y. Lelievre, J. C. Gueguen, D. Reisdorf, I. Morize, R. Pauwels, E. De Clercq, Y. Henin, A. Bousseau, J. F. Mayaux, J. B. Le Pecq and N. Dereu, *J. Med. Chem.*, 1996, **39**, 1056.
41. J. F. Mayaux, A. Bousseau, R. Pauwels, T. Huet, Y. Henin, N. Dereu, M. Evers, F. Soler, C. Poujade and E. De Clercq, *Proc. Natl. Acad. Sci. USA*, 1994, **91**, 3564.
42. L. Huang, L. Zhang and C. H. Chen, *Curr. Pharm. Des.*, 2003, **9**, 1453.
43. K. Qian, D. Yu, C. H. Chen, L. Huang, S. L. Morris-Natschke, T. J. Nitz, K. Salzwedel, M. Reddick, G. P. Allaway and K. H. Lee, *J. Med. Chem.*, 2009, **52**, 3248.
44. L. Huang, P. Ho, K. H. Lee and C. H. Chen, *Bioorg. Med. Chem.*, 2006, **14**, 2279.
45. Anon., *MMW Fortschr. Med.*, 2003, **145**, 51.
46. V. M. Vogt, *Curr. Top. Microbiol. Immunol.*, 1996, **214**, 95.

47. H. G. Krausslich, H. Schneider, G. Zybarth, C. A. Carter and E. Wimmer, *J. Virol.*, 1988, **62**, 4393.
48. R. J. Mervis, N. Ahmad, E. P. Lillehoj, M. G. Raum, F. H. Salazar, H. W. Chan and S. Venkatesan, *J. Virol.*, 1988, **62**, 3993.
49. S. Erickson-Viitanen, J. Manfredi, P. Viitanen, D. E. Tribe, R. Tritch, C. A. Hutchison 3rd, D. D. Loeb and R. Swanstrom, *AIDS Res. Hum. Retroviruses*, 1989, **5**, 577.
50. A. H. Kaplan, J. A. Zack, M. Knigge, D. A. Paul, D. J. Kempf, D. W. Norbeck and R. Swanstrom, *J. Virol.*, 1993, **67**, 4050.
51. J. Zhou, X. Yuan, D. Dismuke, B. M. Forshey, C. Lundquist, K.-H. Lee, C. Aiken and C. H. Chen, *J. Virol.*, 2004, **78**, 922.
52. K. Salzwedel, R. Goila-Gaur, F. Li, A. Castillo, N. Kilgore, M. Reddick, C. Matallana, D. Zoumplis, D. Martin, G. P. Allaway, E. Freed and C. Wild, in *Proceedings of the 15th International AIDS Conference, 11–16 July 2004, Bangkok, Thailand*, International Aids Society, 2004.
53. R. Namikawa, K. N. Weilbaecher, H. Kaneshima, E. J. Yee and J. M. McCune, *J. Exp. Med.*, 1990, **172**, 1055.
54. L. Rabin, M. Hincenbergs, M. B. Moreno, S. Warren, V. Linquist, R. Datema, B. Charpiot, J. Seifert, H. Kaneshima and J. M. McCune, *Antimicrob. Agents Chemother.*, 1996, **40**, 755.
55. C. A. Stoddart, P. Joshi, B. Sloan, J. C. Bare, P. C. Smith, G. P. Allaway, C. T. Wild and D. E. Martin, *PLoS ONE*, 2007, **2**, e1251.
56. D. Martin, P. Smith, C. Wild and G. Allaway, in *Proceedings of the 15th International AIDS Conference, 11–16 July 2004, Bangkok, Thailand*, International Aids Society, 2004.
57. D. Martin, C. Ballow, J. Doto, R. Blum, C. Wild and G. Allaway, in *Proceedings of the 12th Conference on Retroviruses and Opportunistic Infections, 22–25 February 2005, Boston, MA*, Foundation for Retrovirology and Human Health, 2005.

Section 5: Case Studies of Marketed Natural Product-derived Drugs

CHAPTER 14

Daptomycin

RICHARD H. BALTZ

Cubist Pharmaceuticals, Inc., Lexington, MA 02421, USA

1 Introduction

Daptomycin is an important antibiotic approved for the treatment of complicated skin and skin structure infections caused by Gram-positive pathogens[1] and for treatment of bacteraemia, including right-sided endocarditis caused by *Staphylococcus aureus* strains, including those resistant to methicillin (MRSA).[2]

Daptomycin is a cyclic lipopeptide produced by *Streptomyces roseosporus* and has a novel mechanism of action. Thus, it can be used to treat Gram-positive infections by organisms resistant to other antibiotic classes. The core structure of daptomycin is amenable to limited chemical structure–activity relationship (SAR) modification, mainly by changing the lipid tail,[3] but recent studies indicate that the peptide portion is amenable to biosynthetic engineering to produce novel derivatives.[4–6] A21978C factors, the natural lipopeptides produced by *S. roseosporus*, were discovered by Eli Lilly and Company[7] and daptomycin[8] was developed through Phase II clinical trials before abandonment.[9]

Owing to a chance encounter and the tenacity of the late Dr Frank Tally, daptomycin was licensed from Lilly to Cubist Pharmaceuticals, where it was developed successfully for the indications mentioned above. In this chapter, I review the history of the discovery and development of daptomycin and the passing of the baton from Lilly to Cubist.

RSC Biomolecular Sciences No. 18
Natural Product Chemistry for Drug Discovery
Edited by Antony D. Buss and Mark S. Butler

Published by the Royal Society of Chemistry, www.rsc.org

2 Discovery of A21987C and Daptomycin

The lipopeptide antibiotic A21978C factors (Figure 14.1) are produced by *S. roseosporus*, an actinomycete isolated from a soil sample from Mount Ararat. A21978C factors are composed of a thirteen-member peptide that is cyclised to form a ten-member ring with a three-member exocyclic tail. The main three factors (A21978C_{1-3}) have different long chain fatty acids (*anteiso*-undecanoate, *iso*-dodecanoate and *anteiso*-tridecanoate) attached to the *N*-terminus of L-Trp_1. Daptomycin[8] has a decanoate side chain attached to the *N*-terminus of L-Trp (as discussed below). Daptomycin and A21978C_{1-3} factors contain three non-proteinogenic amino acids [L-ornithine (Orn), L-*threo*-3-methyl-glutamic acid (3mGlu) and L-kynurinine (Kyn)] and three D-amino acids (Figure 14.1).

2.1 Enzymatic Cleavage of the Fatty Acid Side Chain

None of the A21978C_{1-3} factors was sufficiently promising to develop as a clinical candidate, so Lilly explored methods to chemically modify the

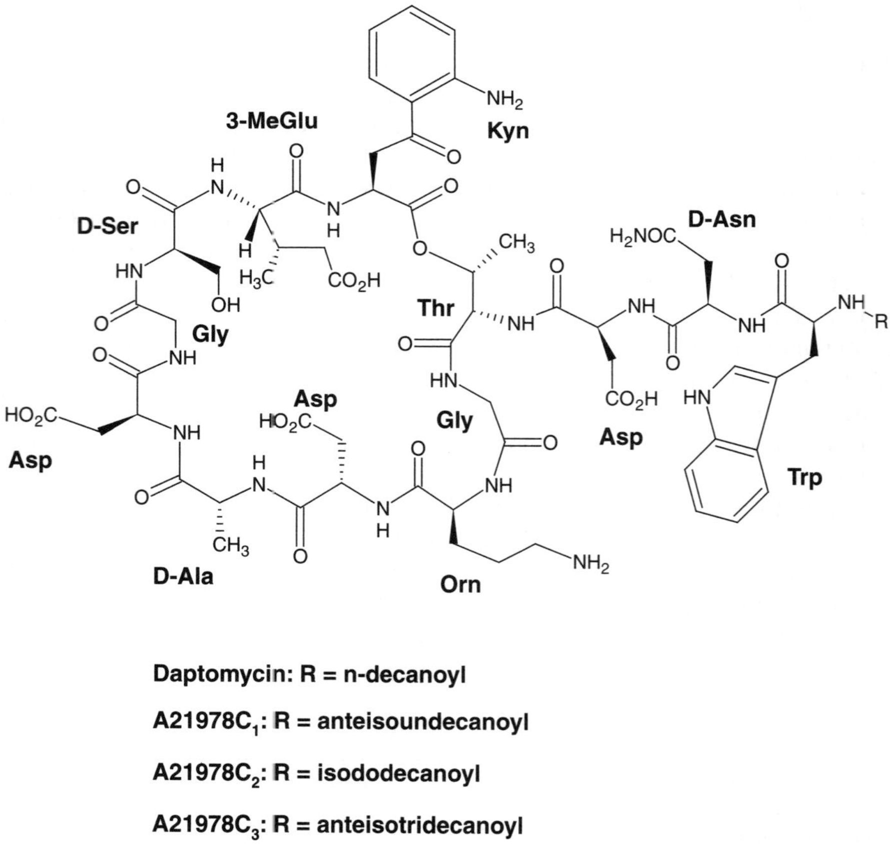

Figure 14.1 Structures of A21978C factors and daptomycin (Reprinted with permission from Baltz *et al.*[3]).

A21978C core peptide. Lilly scientists reasoned that, if they could remove the lipid side chain, it would open up the molecule to facile SAR studies with other lipid side chains. They discovered that *Actinoplanes utahensis* produced a deacylase enzyme that cleaved the lipid side chains from all A21978C factors.[10] This provided a robust route to prepare the core peptide for chemical modification.

Lilly scientists cloned the deacylase gene in a high copy number vector in *Streptomyces lividans*; the recombinant produced substantially higher levels of the enzyme than the original *A. utahensis* culture.[11] The deacylase was also used by Lilly scientists to remove the linoleoyl side chain from echinocandin B,[12] thus providing a route to discover and develop LY303366 (anidulafungin),[13] an antifungal agent approved in 2006 and marketed by Pfizer.

In retrospect, neither of these important antimicrobial agents could have been developed if Lilly did not have a fully integrated natural products discovery group that included scientists dedicated to bioconversions of natural products and medicinal chemists dedicated to modifying complex cyclic peptides and other secondary metabolites. This may be a lesson learned to help guide antibiotic discovery and development in the 21st century.

2.2 Chemical Modifications of the A21978C Core Peptide

Having developed an enzymatic route to remove the fatty acid side chains of the natural A21978C factors, Lilly scientists were able to reacylate the core peptide with different fatty acids to optimise antibacterial activity while minimising toxicity. This led to the discovery of daptomycin, which has a decanoic acid side chain.[8]

3 Biosynthesis

Like many other peptide antibiotics—including the lipopeptides A54145, CDA, amphomycin, laspartomycin and friulimicin—daptomycin is produced by a nonribosomal peptide synthetase (NRPS) mechanism.[3,6,14] It is produced in fermentation by *S. roseosporus* by feeding decanoic acid, which is incorporated as the fatty acid starter unit. Much has been learned about the biosynthetic process by fermentation feeding studies and by the analysis of the daptomycin biosynthetic genes.[3,6,14,15]

3.1 Analysis of the Daptomycin Biosynthetic Gene Cluster

The cloning of the daptomycin biosynthetic gene cluster was initiated at Eli Lilly and Company in the early 1990s. During that timeframe, a number of molecular genetic tools were developed to initiate the genetic engineering of *S. roseosporus*.[16–18] This enabled the localisation of the daptomycin biosynthetic genes to one end of the linear chromosome, the cloning of the genes in cosmids[19] and initiation of sequencing the gene cluster by Cubist. Cubist also cloned the daptomycin genes in a bacterial artificial chromosome (BAC) vector,[14] which facilitated the completion of the sequencing. One BAC clone contained the complete set of genes, as verified by the production of A21978C lipopeptides in a recombinant strain of *S. lividans*.[14,20]

Daptomycin is assembled by a NRPS that contains three giant multi-modular multi-enzyme subunits[3,6,14] that appear to be translated from a single giant mRNA.[21] The daptomycin biosynthetic cluster also contains genes involved in the biosynthesis of 3mGlu and Kyn.[14,22] Perhaps as an integral part of the NRPS, there are two smaller proteins, an acyl-CoA ligase and acyl carrier protein (ACP), that are required for the activation and coupling of the long chain fatty acids to the *N*-terminal Trp_1 to initiate biosynthesis.[14]

If we consider the requirements for the coupling of the lipid and 13 amino acids, followed by ring closure, the multi-subunit NRPS contains 45 enzymatic functions (13 condensation (C) domains, 13 adenylation (A) domains, 13 thiolation (T) domains, three epimerase (E) domains, one thioesterase (Te) domain, one acyl-CoA ligase and one ACP; Figure 14.2).

Since the daptomycin cluster also contains genes involved in the production of 3mGlu and Kyn, gene regulation, resistance and transport and the daptomycin core peptide has three non-proteinogenic amino acids (Orn_6, $3mGlu_{12}$ and Kyn_{13}), three D-isomers of proteinogenic amino acids (D-Asn_2, D-Ala_8 and D-Ser_{11}) and seven typical L-amino acids, the biosynthesis of daptomycin can be appreciated as a truly complex, highly coordinated, biochemical process.

3.2 Daptomycin Structure

One interesting outcome of sequencing the daptomycin gene cluster was the prediction of the stereochemical structure. The original structure proposed by

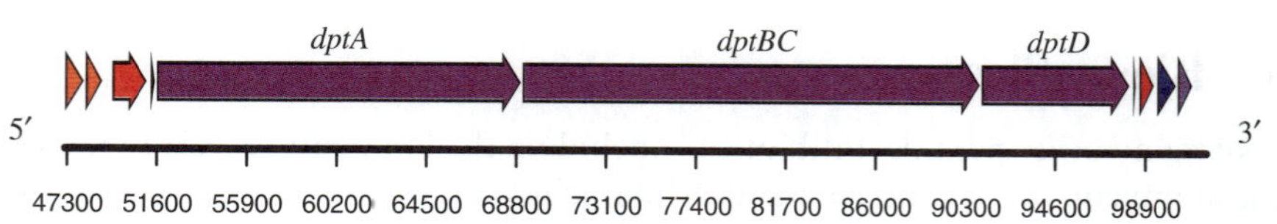

dptA **(5) C*A$_{\text{Trp1}}$T •CA$_{\text{Asn2}}$TE •CA$_{\text{Asp3}}$T •CA$_{\text{Thr4}}$T •CA$_{\text{Gly5}}$T**

dptBC **(6) CA$_{\text{Orn6}}$T •CA$_{\text{Asp7}}$T •CA$_{\text{Ala8}}$TE •CA$_{\text{Asp9}}$T •CA$_{\text{Gly10}}$T •CA$_{\text{Ser11}}$TE**

dptD **(2) CA$_{\text{3mGlu12}}$T •CA$_{\text{Kyn13}}$TTe**

Figure 14.2 A segment of the daptomycin gene cluster containing the NRPS genes. The location (in base pairs) of the giant *dptA*, *dptBC* and *dptD* genes cloned on BAC pVC1 which contains a 128 000 base pair insert.[14] The *dptA*, *dptBC* and *dptD* genes encode five, six and two modules, respectively. The specificity of A domains is shown with amino acid subscripts. The C* condensation domain differs from the others in that it couples the long chain fatty acids to the *N*-terminus of Trp_1 to initiate daptomycin biosynthesis. Note that modules 2, 8 and 11 have CATE modules to incorporate D-amino acids. The Kyn_{13} module has a terminal Te (as CATTe) to cyclise and release the completed lipopeptide.

Lilly, based upon optical rotation of individual amino acids from the core peptide, assigned only two D-amino acids, D-Ala_8 and D-Ser_{11}.[7,8] A typical NRPS module that processes an L-amino acid has three enzymatic domains: condensation (C), adenylation (A) and thiolation (T) or peptidyl carrier protein (PCP) organised as CAT (Figure 14.2). NRPS modules that incorporate D-amino acids typically bind L-amino acids and convert them to D-isomers by incorporating an epimerase (E) function in the module as CATE.[6] The DNA sequence of the daptomycin NRPS genes had three CATE modules, including one that suggested the presence of D-Asn_2, which was confirmed by chemical analysis.[14]

The Cubist DNA sequencing work is an example of the powerful predictive nature of the DNA sequences encoding secondary metabolites, which should be useful in the discovery of novel antibiotics from fully sequenced actinomycete genomes.[23]

4 Mechanism of Action Studies

There have been a number of studies addressing the mechanism of action of daptomycin. Early studies at Lilly provided evidence that daptomycin inhibited peptidoglycan biosynthesis.[24] Subsequent studies demonstrated that daptomycin depolarised membranes in *S. aureus* and *Bacillus megaterium*,[25–27] perhaps facilitated by micelle formation.[28,29] Recent transcriptome analysis in *S. aureus* supports the notion that daptomycin has a dual mechanism of action, which includes inhibition of peptidoglycan biosynthesis and membrane depolarisation.[30] Like vancomycin and oxacillin, daptomycin induces the cell wall stress stimulon member genes in *S. aureus*. Of particular note, it strongly induces transcription of the *vraSR* genes which encode a two-component system that positively regulates cell-wall biosynthesis.[31–33] The *vraSR* genes are strongly induced by treatment with cell-wall active antibiotics including vancomycin, oxacillin, bacitracin and D-cycloserine,[30,31] but not by membrane active compounds (CCCP or nisin).[30] Daptomycin also induced a set of genes which were also induced by CCCP but not by vancomycin or oxacillin,[30] consistent with a dual mechanism of peptidoglycan and membrane targets.

In *Bacillus subtilis*, daptomycin strongly induced the transcription of the *liaRS* two-component regulatory system which is orthologous to the *vraRS* system in *S. aureus*.[34] Analysis of the genes induced by daptomycin treatment (referred to as the daptomycin stimulon) identified a cluster of genes induced by inhibitors of peptidoglycan biosynthesis and a cluster induced by membrane perturbation. They also showed that daptomycin preferentially inserts into dividing *B. subtilis* cells in the region of cell division septum formation. This region is enriched for phosphatidylglycerol (PG) and depletion of PG in a conditional *pgsA* mutant led to significantly reduced concentration of daptomycin at the nascent septum and an eight-fold increase in minimum inhibitory concentration (MIC).

This dual mechanism model of peptidoglycan and membrane targets is compatible with the recent finding that daptomycin is bactericidal against stationary phase *S. aureus*, which might be attributed to membrane depolarisation.[35] The target of daptomycin in peptidoglycan biosynthesis remains to be determined, but a candidate protein has emerged from daptomycin resistance studies (discussed below).

4.1 Daptomycin Resistant Mutants

For many antibiotics, the mechanism of antibiotic resistance can help establish the mechanism of action. Therefore, it may be instructive to determine the mechanism(s) of daptomycin resistance to see if it yields insights into the mechanism of action. The incidence of reduced susceptibility to daptomycin in clinical isolates is very low and resistant strains are usually associated with deep-seated infections in compromised patients.[2,9,36,37]

Daptomycin resistant (DapR) clinical isolates show small increases in MICs, unlike many other antibiotics (*e.g.* streptomycin, rifampin or fluoroquinolones) which show large increases in resistance associated with target site mutations. Friedman *et al.*[38] explored mutations that accumulated in *S. aureus* during serial passage in media containing increasing levels of daptomycin. A key finding was that no single mutation gave high level resistance. Mutations in *mprF*, *rpoB*, *rpoC* and *yycG* individually gave about two-fold increases in MICs. Combinations of three or four mutations gave rise to 5–10 fold increases in MIC. Interestingly, daptomycin-nonsusceptible *S. aureus* strains isolated post therapy had mutations in *mprF*[38,39] or *yycG*.[38]

So how might these mutations relate to the mechanism(s) of action (MOA) of daptomycin? The MprF protein catalyses the coupling of lysine residues to phosphatidylglycerol (PG) to give lysyl-PG (LPG). The *mprF* gene was named for "multiple peptide resistance factor" because a transposon insertion mutation caused *S. aureus* to become highly susceptible to antimicrobial cationic peptides, including human neutrophil defensin HNP-1, porcine leukocyte protegrins 3 and 5, and others.[40] They also showed that the *mprF* mutant was killed more rapidly by neutrophils than the parent strain and was less virulent in a mouse model; hence, MprF is considered to be a virulence factor.

The level of LPG also influences the susceptibility of an MRSA strain to other antibiotics. A transposition mutant defective in *mprF* was less susceptible to moenomycin, but more susceptible to oxacillin, methicillin and gentamicin.[41] Recent studies have shown that DapR strains have increased ratios of LPG/PG,[42] or increased amounts of LPG in the outer leaflet of the membrane, thus increasing the surface positive charge.[43] This indicates that the mutations have enhanced ability to couple lysine to PG or to flip LPG to the outer leaflet. Deletion of *mprF* increases susceptibility to daptomycin by about four-fold in *S. aureus*.[43,44] Disruption of *mprF* in *B. subtilis* caused a two-fold reduction in daptomycin MIC.[34]

These combined results support the idea that Ca^{++}-bound daptomycin functions as a cationic peptide;[28,29] thus the increased positive charge imparted by LPG in the membrane outer leaflet in DapR mutants likely repulses or retards daptomycin penetration into the membrane.

The mutations in *rpoB* and *rpoC* in DapR *S. aureus* strains may alter the patterns of transcription, thus influencing membrane composition or key target protein levels associated with the dual mechanism of action.

Mutations in *yycG* were observed in the laboratory experiment and among clinical isolates of *S. aureus*. The YycG protein is the membrane spanning sensor of a two component response regulator system that acts as a master regulator for cell wall metabolism and biofilm formation.[45,46] The *yycG* and its partner *yycF* genes are present in many low G+C Gram-positive bacteria, where they are required for viability,[46] but are not found in Gram-negative bacteria. Enhanced expression of YycGF in *S. aureus* caused elevated peptidoglycan biosynthesis and turnover and increased biofilm formation, whereas depletion of YycGF caused cell death without lysis.[45] YycG has been localised to the cell division septum in *B. subtilis* where it regulates cell division and wall restructuring.[47] YycG is co-localised with FtsZ, a key protein in cell division and was demonstrated to be present in an immune complex precipitated by anti-FtsZ antibody.

Daptomycin localises to the cell division septum in *B. subtilis*;[34] it causes rapid cell death without lysis,[48] causes the formation of aberrant cell wall septa[48] and treats biofilms effectively in *S. aureus*.[49] Daptomycin induces the cell wall stress stimulon in *S. aureus* and *B. subtilis*,[30,34] but has very poor activity (MIC of 128) against an *E. coli imp* mutant defective in outer membrane assembly,[5] whereas vancomycin, another bulky peptide that normally does not penetrate the outer membrane, has an MIC of 0.8 against *E. coli imp*.[50]

These data are consistent with a second possible mechanism of action of daptomycin: inhibition of YycG function in *S. aureus* and perhaps in some other low G + C Gram-positive bacteria, but not in *E. coli* or other Gram-negative bacteria which lack this target. This model is also consistent with the early studies showing that daptomycin inhibits cell wall biosynthesis at an unspecified early step(s).[24] This partial mechanism of action is consistent with daptomycin synergy with gentamicin and certain β-lactam antibiotics (see below).

5 Antibacterial Activities

5.1 *In vitro* Activities

Daptomycin has potent bactericidal activity against a wide spectrum of Gram-positive pathogens including strains resistant to methicillin, vancomycin, erythromycin and linezolid.[9,51–55] Importantly, this includes both hospital-acquired and community-acquired MRSA strains.[55] The daptomycin MICs for pathogenic Gram-positive bacteria have not changed from the 1980s to the present time. Daptomycin does not have significant antibacterial activity against Gram-negative bacteria, including the *E. coli imp* mutant[5] that is susceptible to vancomycin.[50]

Daptomycin displays synergistic activity with rifampin against >75% of clinical isolates of *Enterococcus faecium* resistant to linezolid and vancomycin at the MICs for both antibiotics.[56] Strikingly, about 65% of these strains were resistant to rifampin (RifR). At sub-inhibitory concentrations, daptomycin causes substantial reductions of MICs for rifampin in some RifR VanR *E. faecium* strains, but no synergism in others.[57] In more recent studies, it was shown that RifR, VanR *E. faecium* strains with typical *rpoB* mutations do not show synergism between rifampin and daptomycin, whereas RifR strains of unknown mechanism show synergy.[58] It will be interesting to identify this novel mechanism of RifR in *E. faecium* and to see how it interfaces with the MOA of daptomycin. To date, there is no compelling evidence for synergy between daptomycin and rifampin in *S. aureus*.[59]

In 34 of 50 *S. aureus* strains with a variety of antibiotic resistance profiles, daptomycin was synergistic with gentamicin.[59] Combinations of gentamicin and daptomycin kill *S. aureus* faster than daptomycin alone *in vitro*.[60] These observations are inconsistent with a single mechanism of action involving dissipation of membrane potential because gentamicin requires membrane potential for uptake and bactericidal activity in *S. aureus*.[61,62] One possible mechanism to explain daptomycin synergy with gentamicin is inhibition of translation of *vraRS* mRNA by gentamicin, thus blocking the expression of the cell wall stimulon normally induced by daptomycin. Daptomycin has also been shown to be synergistic with certain β-lactam antibiotics,[63] an observation that might be exploited clinically.

5.2 *In vivo* Activities in Animal Models

Daptomycin has proven efficacy in a number of *in vivo* animal models, including soft tissue infections by MRSA, bacteraemia caused by *S. aureus* or vancomycin-resistant enterococci (VRE), *Enterococcus faecalis* pyelonephritis, MRSA osteomyelitis, MRSA and *Bacillus anthracis* pulmonary infections, Gram-positive endocarditis, *Clostridium difficile* colitis and *S. pneumoniae* and *S. aureus* meningitis.[9,64–66]

6 Clinical Studies

6.1 Eli Lilly and Company

Lilly conducted Phase I and Phase II clinical studies on daptomycin in the 1980s and early 1990s for skin and soft tissue infections, bacteraemia and endocarditis.[9] In a Phase I safety study to evaluate increasing the dose to 4 mg/kg every 12 hours to treat endocarditis, two of five volunteers exhibited symptoms of muscle toxicity, including elevated creatine phosphokinase (CPK). After observing muscle toxicity with this dosing regimen, Lilly discontinued clinical development of daptomycin in 1991.

6.2 The Passing of the Baton

An interesting aspect of the daptomycin development story is the passing of the baton from Lilly to Cubist. Without this, daptomycin may have ended up on the trash heap of undeveloped antibiotics and other pharmaceutically active substances discovered in big pharma.

In the early 1990s, my laboratory at Lilly had initiated a programme to clone the genes involved in daptomycin biosynthesis and to develop molecular genetic tools to engineer the daptomycin biosynthetic pathway to produce novel lipopeptide antibiotics. The goal was to produce derivatives of daptomycin with improved therapeutic index by making amino acid substitutions in the core peptide. Unfortunately, daptomycin had become "*chimica non grata*" around 1993 and no-one was permitted to work on producing derivatives of daptomycin, with the exception of my postdoctoral student. Our work established the basic framework for the genetic engineering of the daptomycin gene cluster to produce novel derivatives, but we were not supported to complete the project.

In 1996, Lilly downsized its Infectious Disease Discovery Division and began a slow dismantling of its Natural Products Discovery Division despite unprecedented past successes and the recent discovery and development of daptomycin, oritavancin and anidulafungin.[67] In the same year I met Dr Frank Tally of Cubist Pharmaceuticals at a scientific meeting. Frank knew that Lilly was downsizing its Infectious Disease Discovery Division and was eager to recruit certain Lilly scientists to Cubist. Frank invited me to interview for a position at Cubist and I accepted the invitation. During the visit, I presented the work on cloning the daptomycin biosynthetic genes as part of a seminar and discussed the merits of improving the therapeutic index of daptomycin by engineering the biosynthesis of the peptide core of the molecule. Frank became interested in daptomycin, initiated discussions with Lilly and licensed the compound for Cubist in 1997.

6.3 Cubist Pharmaceuticals

Cubist was initially interested in developing daptomycin as an oral agent to purge vancomycin-resistant enterococci from the gastrointestinal tract and as a topical agent to treat Gram-positive skin infections.[68] Subsequently, Frank Tally and Rick Oleson designed a dog toxicity experiment directed at determining what pharmacokinetic or pharmacodynamic parameter(s) might be associated with muscle toxicity. They found that neither the peak antibiotic concentration (C_{max}) nor the total amount of drug exposure over a 24-hour period (area under the curve, AUC) primarily accounted for the toxicity. Instead, toxicity was primarily associated with dosing interval. Administration of the full dose once every 24 hours was substantially less toxic that splitting the dose into thirds and administering every eight hours.[69] This breakthrough information was incorporated in the clinical trial design by Cubist.

With the once-a-day dosing regimen, daptomycin has been approved for the treatment of complicated skin and skin structure infections (cSSSIs) caused by Gram-positive pathogens[1] and for treatment of bacteraemia, including right-sided endocarditis caused by *S. aureus* strains.[2] Analysis of a subset of patients that were treated for diabetic foot ulcers indicated that daptomycin treatment outcomes were not statistically different from outcomes with comparators (vancomycin or a semi-synthetic penicillin).[70] Analysis of a relatively uniform subset of patients from South Africa in the cSSSI trials indicated that daptomycin had comparable clinical success rates relative to comparator treatments, but that daptomycin-treated patients improved more quickly and required shorter durations of treatment.[71] They suggested that this might be a direct consequence of the rapid bactericidal activity without cell lysis, thus minimising inflammation at the site of infection.

Using a *post hoc* analysis of the patients treated in the *S. aureus* bacteraemia clinical trial, Lalani *et al.*[72] reviewed the outcomes of patients with osteo-articular infections (OAI) and found that daptomycin may be effective at treating OAI associated with staphylococcal bacteraemia. Daptomycin has also been used successfully to treat meningitis caused by MRSA.[73]

Daptomycin failed to show non-inferiority to controls in a clinical trial for community acquired pneumonia (CAP).[74] The clinical failure was explained by subsequent experiments demonstrating that daptomycin is sequestered in lung surfactant.[64] This shortcoming of daptomycin is limited to pulmonary pneumonia involving alveoli and does not extend to haematogenous pneumonia caused by *S. aureus*.[74]

While Lilly observed muscle toxicity when administering daptomycin at 4 mg/kg twice a day, Cubist has recently shown that daptomycin is well tolerated up to 12 mg/kg administered once a day.[75] This may indicate that daptomycin can be administered safely at higher doses than those used in the Phase III clinical trials for bacteraemia and endocarditis to treat life-threatening infections. In a relatively small, prospective controlled trial examining once a day dosing of 10 mg/kg for four days, the dose regimen was well-tolerated and the clinical outcomes of patients treated with daptomycin were not statistically different than those treated with standard care, although the latter were higher.[76] The data suggests that a larger trial is warranted to further investigate and optimise the high-dose, short duration (HDSD) use of daptomycin.

7 Lessons Learned

The process of drug discovery and development is often not straightforward and daptomycin is a good case in point. A soil sample was isolated from Mount Ararat in the 1960s, when it was much easier to sample exotic locations. To my knowledge, A21978C producing cultures were only isolated by Lilly, so they may not be widely distributed around the globe. The development of daptomycin did not move quickly at Lilly. The 1960s and 1970s were dominated by the development of cephalosporin and aminoglycoside antibiotics.[67] Even vancomycin sales were relatively small in the early 1970s, so there was not much need for an antibiotic like daptomycin.

Daptomycin might not have been developed if Lilly did not have a group working on enzymatic bioconversions. Without this capability, the deacylase enzyme from *A. utahensis*[10] would not have been discovered and so the chemical SAR studies around the lipid tail leading to daptomycin might not have been approachable.

It may not be easy to duplicate the series of events that led to the discovery and development of daptomycin today, but there is reason to believe that other important molecules are yet to be discovered and there are approaches that integrate new biology and chemistry not available at the time of daptomycin discovery.[23,77]

8 Epilogue

There are undoubtedly many stories that accompany the discovery and development of important drugs. The daptomycin story points out how fragile the process is, but success ultimately was achieved through a dedicated champion, Dr Francis P. (Frank) Tally. I had the privilege to work with Frank to help initiate the daptomycin project at Cubist[68] and I dedicate this chapter to his memory.

References

1. R. D. Arbeit, D. Maki, F. P Tally, E. Campanaro and B. E. Eisenstein, *Clin. Infect. Dis.*, 2004, **38**, 1673.
2. V. G. Fowler, H. W. Boucher, G. R. Corey, E. Abrutyn, A. W. Karchmer, M. E. Rupp, D. P. Levine, H. F. Chambers, F. P. Tally, G. A. Vigliani, C. H. Cabell, A. S. Link, I. DeMeyer, S. G. Filler, M. Zervos, P. Cook, J. Parsonnet, J. M. Bernstein, C. S. Price, G. N. Forest, G. Fatkenheuer, M. Gareca, S. J. Rehm, H. R. Brodt, A. Tice, G. R. Corey and S. E. Cosgrove, *N. Engl. J. Med.*, 2006, **355**, 653.
3. R. H. Baltz, V. Miao and S. K. Wrigley, *Nat. Prod. Rep.*, 2005, **22**, 717.
4. V. Miao, M.-F. Coëffet-LeGal, K. Nguyen, P. Brian, A. Whiting, J. Steele, X. D. Kau, S. Martin, R. Ford, T. Gibson, M. Bouchard, S. K. Wrigley and R. H. Baltz, *Chem. Biol.*, 2006, **13**, 269.
5. K. Nguyen, D. Ritz, J.-Q. Gu, D. Alexander, M. Chu, V. Miao, P. Brian and R. H. Baltz, *Proc. Nat. Acad. Sci. USA*, 2006, **103**, 17462.
6. R. H. Baltz, *Curr. Top. Med. Chem.*, 2008, **8**, 618.
7. M. Debono, M. Barnhart, C. B. Carrell, J. A. Hoffman, J. L. Occolowitz, B. J. Abbott, D. S. Fukuda and R. L. Hamill, *J. Antibiot. (Tokyo)*, 1987, **40**, 761.
8. M. Debono, B. J. Abbott, R. M. Molloy, D. S. Fukuda, A. H. Hunt, V. M. Daupert, F. T. Counter, J. L. Ott, C. B. Carrell, L. C. Howard, L. D. Boeck and R. L. Hamill, *J. Antibiot. (Tokyo)*, 1988, **41**, 1093.

9. F. P. Tally, M. Zeckel, M. M. Wasilewski, C. Carini, C. L. Berman, G. L. Drusano and F. B. Oleson, *Expert Opin. Investig. Drugs*, 1999, **8**, 1223.
10. L. D. Boeck, D. S. Fukuda, B. J. Abbott and M. Debono, *J. Antibiot. (Tokyo)*, 1988, **41**, 1085.
11. A. J. Kreuzman, R. L. Hodges, J. R. Swartling, T. E. Pohl, S. K Ghag, P. J. Baker, D. McGilvray and W. K. Yeh, *J. Ind. Microbiol. Biotechnol.*, 2000, **24**, 173.
12. L. D. Boeck, D. S. Fukuda, B. J. Abbott and M. Debono, *J. Antibiot. (Tokyo)*, 1989, **42**, 382.
13. M. Debono, W. W. Turner, L. LaGrandeur, F. J. Burkhardt, J. S. Nissen, K. K. Nichols, M. J. Rodriguez, M. J. Zweifel, D. J. Zeckner, R. S. Gordee, J. Tang and T. R. Parr Jr, *J. Med. Chem.*, 1995, **38**, 3271.
14. V. Miao, M.-F. Coëffet-Le Gal, P. Brian, R. Brost, J. Penn, A. Whiting, S. Martin, R. Ford, I. Parr, M. Bouchard, C. J. Silva, S. K. Wrigley and R. H. Baltz, *Microbiology*, 2005, **151**, 1507.
15. F. M. Huber, R. L. Pieper and A. J. Tietz, *J. Biotechnol.*, 1988, **7**, 283.
16. M. A. McHenney and R. H. Baltz, *Microbiology*, 1996, **142**, 2363.
17. T. J. Hosted and R. H. Baltz, *Microbiology*, 1996, **142**, 2803.
18. T. J. Hosted and R. H. Baltz, *J. Bacteriol.*, 1997, **179**, 180.
19. M. A. McHenney, T. J. Hosted, B. S. DeHoff, P. R. Rosteck Jr and R. H. Baltz, *J. Bacteriol.*, 1998, **180**, 143.
20. J. Penn, X. Li, A. Whiting, M. Latif, T. Gibson, C. J. Silva, P. Brian, J. Davies, V. Miao, S. K. Wrigley and R. H. Baltz, *J. Indust. Microbiol. Biotechnol.*, 2006, **33**, 121.
21. M.-F. Coëffet-Le Gal, L. Thurson, P. Rich, V. Miao and R. H. Baltz, *Microbiology*, 2006, **152**, 2993.
22. K. T. Nguyen, D. Kau, J.-Q. Gu, P. Brian, S. K. Wrigley and R. H. Baltz, *Mol. Microbiol.*, 2006, **61**, 1294.
23. R. H. Baltz, *Curr. Opin. Pharm.*, 2008, **8**, 557.
24. N. E. Allen, J. N. Hobbs and W. E. Alborn Jr, *Antimicrob. Agents Chemother.*, 1987, **31**, 1093.
25. W. E. Alborn Jr, N. E. Allen and D. A. Preston, *Antimicrob. Agents Chemother.*, 1991, **35**, 2282.
26. N. E. Allen, W. E. Alborn Jr and J. N. Hobbs, *Antimicrob. Agents Chemother.*, 1991, **35**, 2639.
27. J. A. Silverman, N. G. Perlmutter and H. M. Shapiro, *Antimicrob. Agents Chemother.*, 2003, **47**, 2538.
28. S. K. Straus and R. E. W. Hancock, *Biochim. Biophys. Acta*, 2006, **1758**, 1215.
29. S. W. Ho, D. Jung, J. R. Calhoun, J. D. Lear, M. Okon, W. R. P. Scott, R. E. W. Hancock and S. K. Straus, *Eur. Biophys. J.*, 2008, **37**, 421.
30. A. Muthaiyan, J. A. Silverman, R. K. Jayaswal and B. J. Wilkerson, *Antimicrob. Agents Chemother.*, 2008, **52**, 980.
31. M. Kuroda, H. Kuroda, T. Oshima, F. Takeuchi, H. Mori and K. Hiramatsu, *Mol. Microbiol.*, 2003, **49**, 807.

32. S. Gardete, S. W. Wu, S. Gill and A. Tomasz, *Antimicrob. Agents Chemother.*, 2006, **50**, 3424.
33. A. Belcheva and D. Golemi-Kotra, *J. Biol. Chem.*, 2008, **283**, 12354.
34. A.-B. Hachmann, E. R. Angert and J. D. Helmann, *Antimicrob. Agents Chemother.*, 2009, **53**, 1598.
35. C. T. M. Mascio, J. D. Alder and J. A. Silverman, *Antimicrob. Agents Chemother.*, 2007, **51**, 4255.
36. S. Y. Lee, H. W. Fan, J. L. Kuti and D. P. Nicolau, *Expert Opin. Pharmacother.*, 2006, **7**, 1381.
37. R. L. Holmes and J. H. Jorgensen, *Antimicrob. Agents Chemother.*, 2008, **52**, 757.
38. L. Friedman, J. D. Alder and J. A. Silverman, *Antimicrob. Agents Chemother.*, 2006, **50**, 2137.
39. K. Julian, K. Kosowska-Shick, C. Whitener, M. Roos, H. Labishinski, A. Rubio, L. Parent, L. Ednie, L. Koeth, T. Bogdanovich and P. C. Applebaum, *Antimicrob. Agents Chemother.*, 2007, **51**, 3445.
40. A. Peschel, R. W. Jack, M. Otto, L. V. Collins, P. Staubitz, G. Nicholson, H. Kalbacher, W. F. Nieuwenhuizen, G. Jung, A. Tarkowski, K. P. M. van Kessel and J. A. G. van Strijp, *J. Exp. Med.*, 2001, **193**, 1067.
41. H. Nishi, H. Kamatsuzawa, T. Fujiwara, N. McCallum and M. Sugai, *Antimicrob. Agents Chemother.*, 2004, **48**, 4800.
42. A. Rubio, J. Moore, W. Shaw, M. Conrad and J. A. Silverman, *Abstracts of the 48th Annual ICAAC Meeting, 25–28 October 2008, Washington, DC*, American Society for Microbiology, 2008.
43. T. Jones, M. R. Yeaman, G. Sakoulas, S.-J. Yang, R. A. Procter, H.-G. Sahl, J. Schrenzel, Y. Q. Xiong and A. S. Bayer, *Antimicrob. Agents Chemother.*, 2008, **52**, 269.
44. A. Rubio, M. Conrad, R. Haselbeck, G. C. Kedar, V. Driver, J. Finn and J. Silverman, *Abstracts of the 48th Annual ICAAC Meeting, 25–28 October 2008, Washington, DC*, American Society for Microbiology, 2008.
45. S. Dubrac, I. G. Boneca, O. Poupel and T. Msadek, *J. Bacteriol.*, 2007, **189**, 8257.
46. M. E. Winkler and J. A. Hock, *J. Bacteriol.*, 2008, **190**, 2645.
47. T. Fukushima, H. Szurmant, E.-J. Kim, M. Perego and J. A. Hoch, *Mol. Microbiol.*, 2008, **69**, 621.
48. N. Cotroneo, R. Harris, N. Perlmutter, T. Beveridge and J. A. Silverman, *Antimicrob. Agents Chemother.*, 2008, **52**, 2223.
49. I. Raad, H. Hanna, Y. Jiang, T. Dvorak, R. Reitzel, G. Chaiban, R. Sherertz and R. Hachem, *Antimicrob. Agents Chemother.*, 2007, **51**, 1656.
50. U. S. Eggert, N. Ruiz, B. V. Falcone, A. A. Branstrom, R. C. Goldman, T. J. Silhavy and D. Kahne, *Science*, 2001, **294**, 361.
51. R. H. Baltz, in *Biotechnology of Antibiotics*, ed. W. R. Strohl, Marcel Dekker, New York, 1997, pp. 415–435.
52. J. N. Steenbergen, J. Alder, G. M. Thorne and F. P. Tally, *J. Antimicrob. Chemother.*, 2005, **55**, 283.

53. J. M. Streit, J. N. Steenbergen, G. M. Thorne, J. Alder and R. N. Jones, *J. Antimicrob. Chemother.*, 2005, **55**, 574.
54. D. M. Anastasiou, G. M. Thorne, S. A. Luperchio and J. D. Alder, *Int. J. Antimicrob. Agents*, 2006, **28**, 385.
55. D. M. Anastasiou, M. Morgan, P. J. Ruane, J. N. Steenbergen, B. D. Katz, J. D. Alder and G. M. Thorne, *Diagn. Microbiol. Infect. Dis.*, 2008, **61**, 339.
56. G. Pankey, D. Ashcraft and N. Patel, *Antimicrob. Agents Chemother.*, 2005, **48**, 5166.
57. K. H. Rand and H. Houck, *J. Antimicrob. Chemother.*, 2004, **53**, 530.
58. K. H. Rand, H. J. Houck and J. A. Silverman, *J. Antimicrob. Chemother.*, 2007, **59**, 1017.
59. K. Credito, G. Lin and P. C. Appelbaum, *Antimicrob. Agents Chemother.*, 2007, **51**, 1504.
60. B. T. Tsuji and M. J. Rybak, *Antimicrob. Agents Chemother.*, 2005, **49**, 2735.
61. S. M. Mates, E. S. Eisenberg, L. J. Mandel, L. Patel, H. R. Kaback and M. H. Miller, *Proc. Nat. Acad. Sci. USA*, 1982, **79**, 6693.
62. E. S. Eisenberg, L. J. Mandel, H. R. Kaback and M. H. Miller, *J. Bacteriol.*, 1984, **157**, 863.
63. K. H. Rand and H. J. Houck, *Antimicrob. Agents Chemother.*, 2004, **48**, 2871.
64. J. A. Silverman, L. I. Morton, A. D. Vanpraagh, T. Li and J. Alder, *J. Infect. Dis.*, 2005, **191**, 2149.
65. P. Cottagnoud, M. Pfister, F. Acousta, M. Cottagnoud, L. Flatz, F. Kühn, H. P. Müller and A. Stucki, *Antimicrob. Agents Chemother.*, 2004, **48**, 3928.
66. P. Gerber, A. Stucki, F. Acousa, M. Cottagnoud and P. Coutagnoud, *J. Antimicrob. Chemother.*, 2006, **57**, 720.
67. R. H. Baltz, *SIM News*, 2005, **55**, 5.
68. B. I. Eisenstein, F. B. Oleson Jr and R. H. Baltz, *Clin. Infect. Dis.*, 2009, in press.
69. F. B. Oleson, C. L. Berman, J. B. Kirkpatrick, K. S. Regan, J.-J. Lai and F. P. Tally, *Antimicrob. Agents Chemother.*, 2000, **44**, 2948.
70. B. A. Lipsky and U. Stoutenburgh, *J. Antimicrob. Chemother.*, 2005, **55**, 240.
71. J. E. Krige, K. Lindfield, L. Friedrich, C. Otradovec, W. J. Martone, D. E. Katz and F. Tally, *Curr. Med. Res. Opin.*, 2007, **23**, 2147.
72. T. Lalani, H. W. Boucher, S. E. Cosgrove, V. G. Fowler, Z. A. Kanafani, G. A. Vigliani, M. Campion, E. Abrutyn, D. P Levine, C. S. Price, S. J. Rehm, G. R. Corey and A. W. Karchmer, *J. Antimicrob. Chemother.*, 2008, **61**, 177.
73. D. H. Lee, B. Palermo and M. Chowdhury, *Clin. Inf. Dis.*, 2008, **47**, 589.
74. P. E. Pertel, P. Bernardo, C. Fogerty, P. Matthews, R. Northland, M. Benvenuto, G. M. Thorne, S. A. Luperchio, R. D. Arbeit and J. Alder, *Clin. Infect. Dis.*, 2008, **46**, 1142.

75. M. Benvenuto, D. P. Benzinger, S. Yankelev and G. Vigliani, *Antimicrob. Agents Chemother.*, 2006, **50**, 3245.
76. D. E. Katz, K. C. Lindfield, J. N. Steenbergen, D. P. Benzinger, K. J. Blakerby, A. G. Knapp and W. J. Martone, *Int. J. Clin. Pract.*, 2008, **62**, 1183.
77. R. H. Baltz, *SIM News*, 2005, **55**, 186.

CHAPTER 15
Micafungin

AKIHIKO FUJIE,[a] SHUICHI TAWARA[a] AND SEIJI HASHIMOTO[b]

[a] Fermentation Research Labs, Astellas Pharma Inc., 5-2-3, Tokodai, Tsukuba, Ibaraki 300-2598, Japan; [b] Toyama Prefectural University, 5180 Kurokawa, Imizu, Toyama 939-0398, Japan

1 Introduction

Fungal infections are known to cause not only superficial diseases, such as athlete's foot and onychomycoses, but also those that become disseminated and life-threatening; serious invasive fungal infections caused by *Candida* spp., *Cryptococcus neoformans*, *Aspergillus* spp., *Pneumocystis carinii* and *Histoplasma capsulatum* pose an increasing threat to human health. The prevalence of these systemic fungal infections has increased significantly in recent years. Major factors responsible for this dramatic increase include widespread use of broad-spectrum antibiotics, growing numbers of immunocompromised transplant patients as well as those suffering from AIDS and cancer, the use of central venous catheters and the increase in the number of aged patients.

Before the 1970s, only a few compounds, including the polyenes (nystatin and amphotericin B) and flucytosine, were available for antifungal chemotherapy. Although the development of azole drugs began in the early 1970s, the number of antifungal agents available for the treatment of life-threatening fungal infections in the late 20th century was still limited. Moreover, these antifungal agents had some drawbacks, such as the significant nephrotoxicity of amphotericin B and emerging resistance to the azoles. To overcome these problems, lipid polyene formulations with reduced toxicity and new triazoles (voriconazole, ravuconazole and posaconazole) with an improved antifungal

RSC Biomolecular Sciences No. 18
Natural Product Chemistry for Drug Discovery
Edited by Antony D. Buss and Mark S. Butler

Published by the Royal Society of Chemistry, www.rsc.org

spectrum and efficacy against azole-resistant isolates were introduced onto the market. Despite these advances, new antifungal agents with new mechanisms of action are eagerly awaited.

Unlike the antibacterial arena, where many bacterium-specific drug target categories are recognised, antifungal research has been hampered because the targets are less selective. Fungi, being eukaryotic, have a metabolism similar to that of mammals. This makes finding treatments that will selectively affect the fungal pathogen, but not the patient, much more difficult. Attempts to find fungus-specific novel antifungals and the successful discovery of micafungin are discussed below.

1.1 New Antifungal Compounds Discovered at Fujisawa (a Predecessor of Astellas Pharma Inc.)

Fujisawa had discovered many antifungal compounds of microbial origin (Figure 15.1). The first of these was pyrrolnitrin, which has been widely used as a drug for dermatophytosis.[1] Two pyrrolnitrin-producing strains belonging to *Pseudomonas* were isolated and identified as new species. The production of pyrrolnitrin was strictly phosphate buffer-dependent. Culture conditions for its production required that the medium be maintained at a pH of 6.2 throughout fermentation by adding 0.2 M phosphate buffer to the nutrient broth. The substitution of buffers other than phosphate resulted in no antibiotic production, as did the use of inorganic acids and alkali to maintain pH. Given these outcomes, we speculated that the role of phosphate buffer extended beyond pH control to direct involvement in pyrrolnitrin biosynthesis. The effects of pH and phosphate concentration on the production of pyrrolnitrin were tested within the ranges of 5.4 to 7.2 and 0.05 to 0.4M, respectively. The optimal condition was found to be pH 6.2 in 0.2 M phosphate-buffered medium. This concentration of phosphate was equivalent to nearly 3% by weight, which was about ten times that in conventional media. This fact led us to consider the effect of osmolarity on the production of pyrrolnitrin. However, adding NaCl to increase the osmolarity of the nutrient broth had no effect. The effects of organic phosphate compounds such as phytin, adenosine monophosphate (AMP) and inosine monophosphate (IMP) were also investigated, but no effect on antibiotic production was observed.

In addition to the experiments described above, numerous studies were also carried out to elucidate the role of phosphate in pyrrolnitrin biosynthesis. Despite our efforts, the precise role of phosphate buffer has not been completely elucidated. However, this phenomenon prompted us to use phosphate-buffered medium to search for other novel compounds, which led to the discovery of the new antifungal compounds described in this chapter.

No *in vitro* antimicrobial assay systems that correctly predict the *in vivo* effect have yet been established for antifungal research. As an alternative approach, an *in vivo* assay using a mouse *Candida albicans* infection model was adopted for use in finding new antifungal compounds. FR109615 was isolated from the

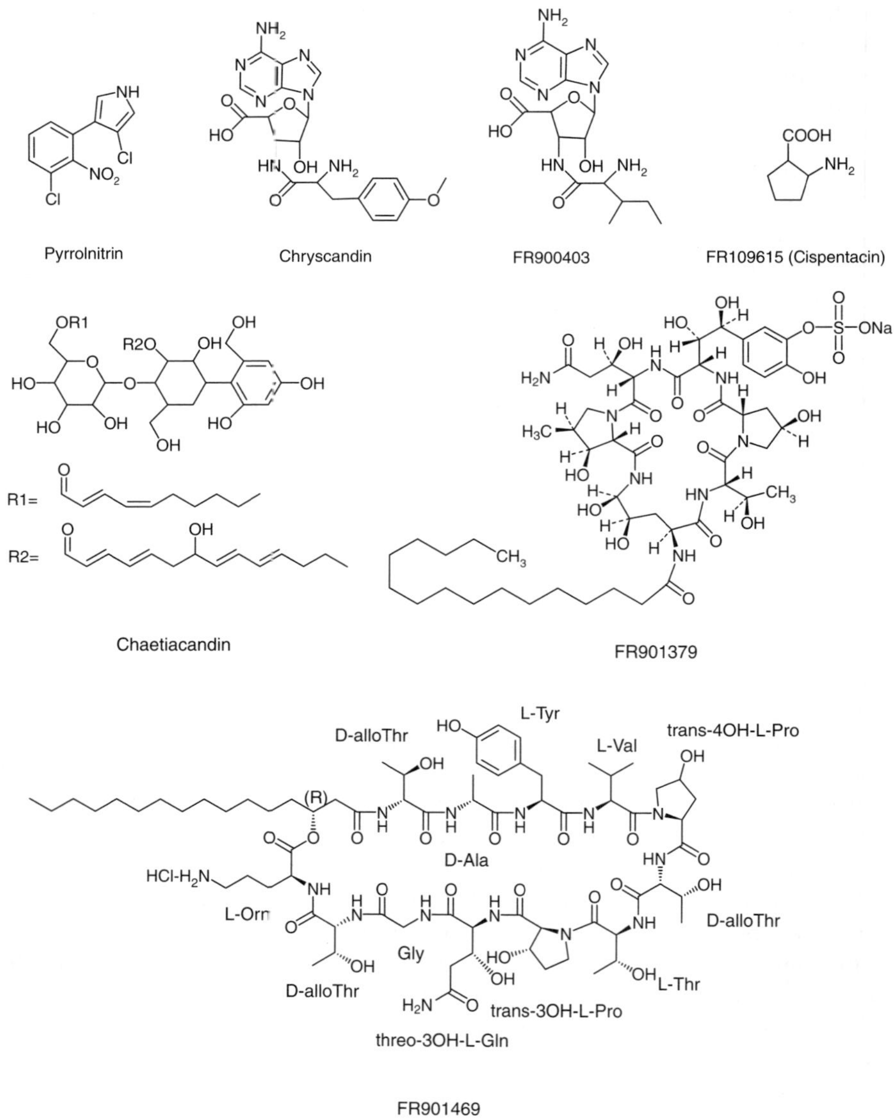

Figure 15.1 Antifungals discovered at Fujisawa.

broth of a *Streptomyces* species based on its potent *in vivo* efficacy in this model.[2] This compound was also discovered by Bristol-Myers' group as a metabolite of a *Bacillus* species and named cispentacin.

Another strategy employed was the rather classical concept of selective toxicity. Although both fungal and mammalian cells are eukaryotes, there are differences in structure and metabolism. We focused our interest on the biosynthesis of the cell wall, which is absent in mammalian cells.

Rapid screening for novel inhibitors of fungal cell wall synthesis was accomplished by using the fungal protoplast regeneration assay. This involved examining morphological changes in protoplasts of *C. albicans* under a microscope to selectively screen for cell wall inhibition. During the course of this screening programme, chryscandin, chaetiacandin and FR900403 were discovered in cultured broths of *Crysosporium pannorum* F-4629, *Monochaetia dimorphospora* F-5187 and *Kernia* species F-19849, respectively.[3–5] The precise modes of action of chryscandin and FR900403 are not clear; however, when *C. albicans* cells were treated with a lethal concentration of the compounds in hypertonic media, cell swelling and lysis occurred. Chaetiacandin is related to the papulacandin family, which are known 1,3-β-glucan synthase inhibitors.

Several compounds, including lipopeptides such as echinocandin B and peptidyl nucleosides such as polyoxin, have been reported to inhibit fungal cell wall biosynthesis. Moreover, pioneering research showed that they inhibit the key enzymes involved in the biosynthesis of 1,3-β-glucan and chitin, respectively. Novel lipopeptides such as FR901379 and related compounds were also discovered using a unique screening programme along with both protoplast and *in vivo* assays.[6]

Although *in vitro* enzyme assay systems for measuring the activity of cell wall synthesis had been developed and were used to find new inhibitors, these assay systems were not sensitive enough to detect low concentrations of inhibitory compounds in fermentation broths. Therefore, we tried to improve the cell-free assay by changing both the method of enzyme preparation and the addition of co-factors. As a result, our revised assay system became ten times more sensitive to known inhibitors than the original. This improved assay enabled us to find the new 1,3-β-glucan synthase inhibitor, FR901469, which is a metabolite of a rare fungus.[7]

FR901469 is a water-soluble, 40-membered macrocyclic lipopeptidolactone consisting of 12 amino acids and a 3-hydroxypalmitoyl moiety. The compound inhibits 1,3-β-glucan synthase prepared from *C. albicans* 6406 with an IC_{50} value of 0.05 μg/mL. As shown in Table 15.1, FR901469 is the most potent inhibitor of 1,3-β-glucan synthase among the compounds isolated so far.

1.2 1,3-β-Glucan Synthase Inhibition and Echinocandins

The inhibition of fungal cell wall biosynthesis has long been an attractive target in the search for novel antifungals. Inhibitors of 1,3-β-glucan synthase in particular have been investigated extensively. These inhibitors are grouped by structure into three classes: papulacandins; echinocandins; and others. Chaetiacandin, FR901379 and FR901469, which belong to these classes, respectively, were isolated by our group.

The echinocandins were the first new antifungal drug class to be introduced in more than 15 years (Figure 15.2). The first echinocandin product to be marketed was caspofungin acetate (Cancidas, Merck), followed by micafungin [Funguard (Japan) and Mycamine (other countries), Fujisawa] and

Table 15.1 Inhibitory activity of FR901379 and related compounds on 1,3-β-glucan synthase.

Compound		*IC_{50} (μg/ml)*
FR901469		0.05
FR901379	(WF11899A)	0.7
	WF11899B	0.7
	WF11899C	1.8
Aculeacin A		1.3
Echinocandin B		2.6

anidulafungin (Eraxis, Pfizer). The marketed echinocandins are all synthetically modified lipopeptides originally derived from the fermentation broths of various fungi. Different structures have been elucidated, including aculeacin A (from *Aspergillus aculeatus*), echinocandin B [from *Aspergillus rugulovalvus* (formerly *Aspergillus rugulosus*, a close relative of *Aspergillus nidulans*)], pneumocandin B (from *Glarea losoyensis*) and FR901379 (from *Coleophoma empetri*) (Figure 15.3). Anidulafungin was originally identified by Eli Lilly (as LY303366) and subsequently licensed to Vicuron (formerly Versicor) as VER-002. Pfizer finally launched this compound successfully after the acquisition of Vicuron.

Although the natural echinocandins had potent antifungal activity *in vitro*, their ADME (absorption, distribution, metabolism and excretion) characteristics had to be improved through chemical modification to meet practical pharmacological requirements. Lilly was the first to initiate such an approach; its modification of echinocandin B yielded cilofungin, which reached Phase II clinical trials but was then dropped due to toxicity. Conversion of the phenolic hydroxy of this compound to a sodium phosphate ester led to the synthesis of the more soluble prodrug LY307853.[8] Subsequently, Merck used pneumocandin B0 as the starting material for production of MK-0991,[9] a compound that demonstrates good activity against *C. albicans* and other pathogenic fungi.

2 From the Discovery of FR901379 to Clinical Studies of FK463 (Micafungin)

2.1 Discovery of FR901379

The seed compounds of micafungin, FR901379 and related analogues were discovered after about 6000 microbial broth samples had been screened in our original programme (described above). These new compounds were found to be members of the echinocandin-like class of lipopeptides, which include echinocandin B, pneumocandin B0, *etc.* These lipopeptides are characterised structurally by a cyclic hexapeptide acylated with a long side chain and have excellent anti-*Candida* activity, which is attributable to the selective inhibition of 1,3-β-glucan synthesis—although their intrinsic insolubility in water was a major problem during drug development.

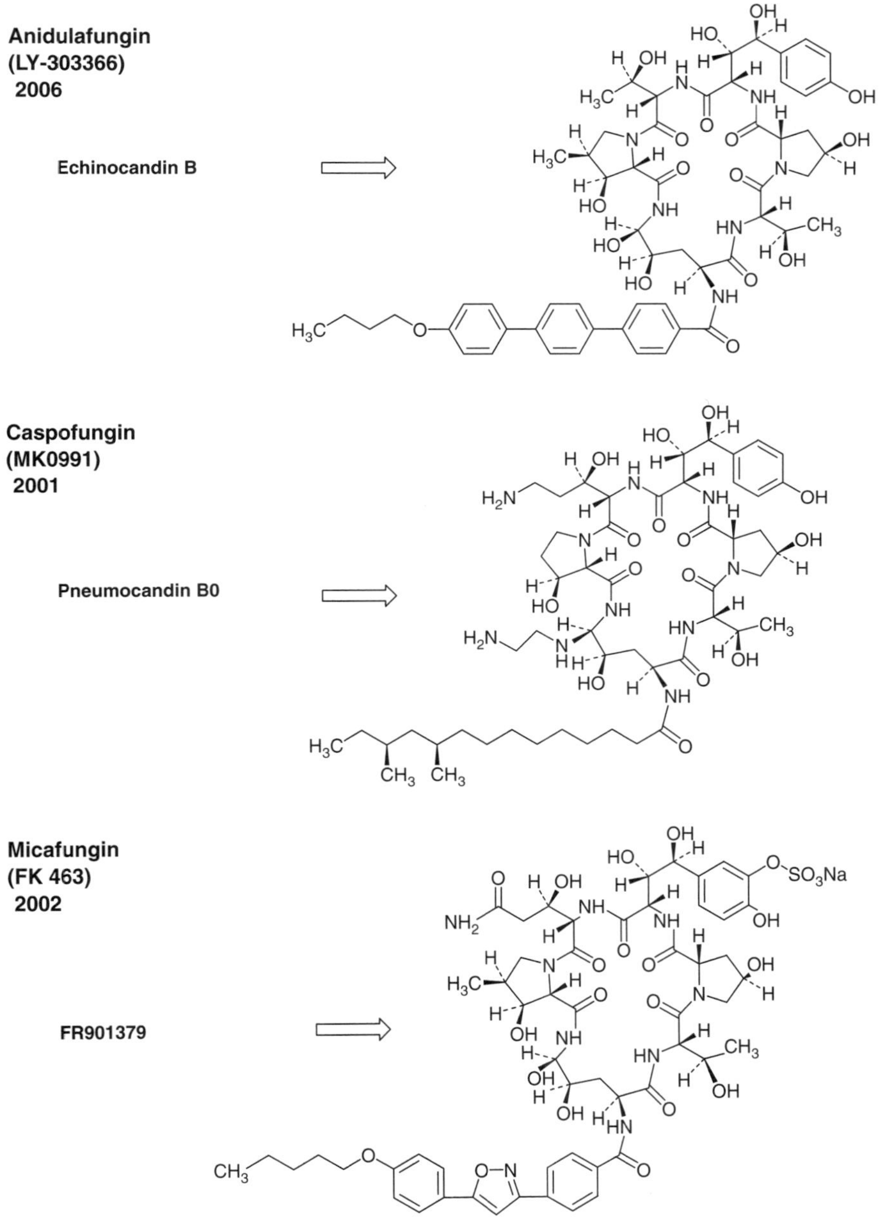

Figure 15.2 Launched semi-synthetic echinocandins.

In contrast, FR901379 and related compounds were highly soluble in water and also demonstrated a strong antifungal effect against *Candida* species. The structural difference between FR901379 and the other echinocandins is FR901379's sulfate moiety (Figure 15.3). We speculated that this portion of the

	R_1	R_2
FR901379	OH	OH
FR901381	OH	H
FR901382	H	OH

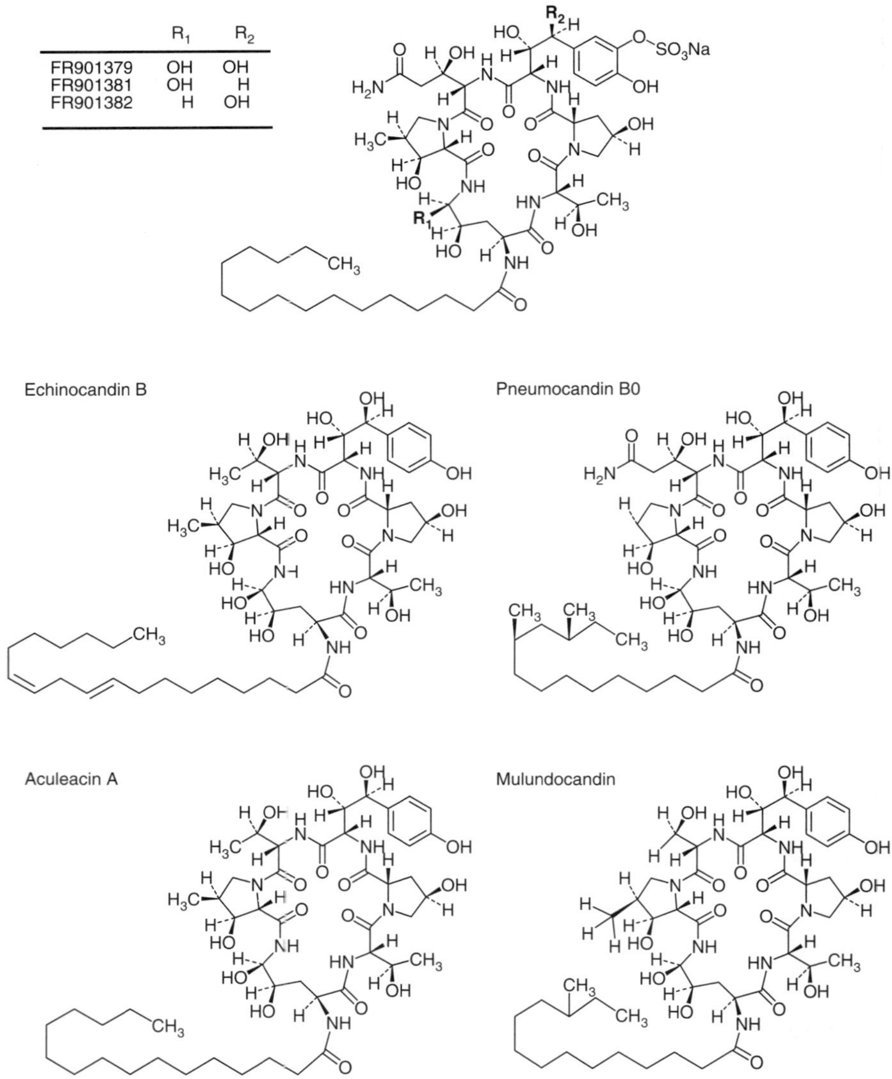

Figure 15.3 Natural echinocandins.

molecule might be responsible for the high water solubility because, while the other compounds are almost insoluble, FR901379 is readily soluble in water, even at a concentration of 50 mg/mL (Table 15.2). To prove this hypothesis, we digested FR901379 with arylsulfatase from *Aerobacter aerogenes*. The water solubility of the desulfated molecule (FR133302) decreased to 1 mg/mL, even though the inhibitory activity of 1,3-β-glucan synthase did not drop dramatically. This result suggested that FR901379's excellent water solubility is attributable to its sulfate moiety.

Table 15.2 Water solubility and inhibitory effect of echinocandins on 1,3-β-glucan synthase.

Compound	*Solubility in water (mg/ml)*	*Inhibition of 1,3-β-glucan synthase (μg/ml)*
FR901379	> 50	0.7
FR133302	1	1.3
Echinocandin B	0.008	2.6
Cilofungin	0.1	nt

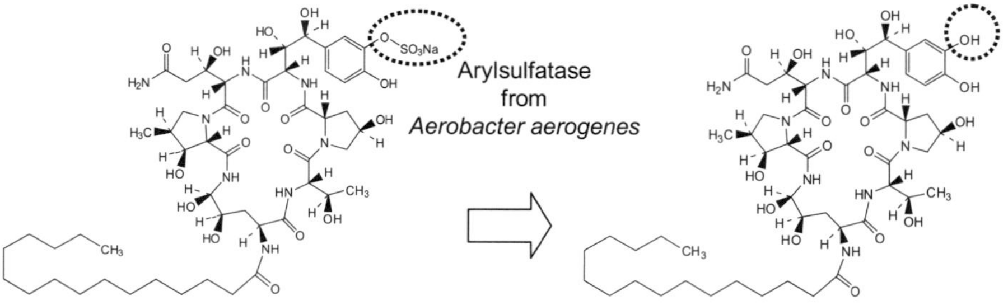

Table 15.3 *In vitro* antifungal activity of FR901379 and related compounds.

	IC_{50} (μg/ml)			
Test organism	*FR901379*	*FR901381*	*FR901382*	*Aculeacin A*
Candida albicans FR578	0.008	0.008	0.008	0.008
C. albicans FP582	0.025	0.015	0.03	0.06
C. albicans FP629	0.008	0.004	0.008	0.015
C. albicans FP633	0.025	0.025	0.03	0.06
C. tropicalis YC118	0.025	0.05	0.015	0.31
C. krusei YC109	0.16	0.16	0.16	0.62
C. utilis YC123	0.03	0.003	0.003	0.06
Aspergillus fumigatus FD050	1.9	1.6	0.62	2.5
A. niger ATCC9642	0.03	0.03	0.03	2.5
C. neoformans YC203	>2.5	>2.5	>2.5	>2.5

The IC_{50} values of FR901379 and related compounds against 1,3-β-glucan synthase are 0.7, 0.7 and 1.8 μg/mL, respectively, which is stronger inhibition than that of echinocandin B (Table 15.2). The *in vitro* antifungal activity of FR901379 and related compounds against both *C. albicans* and *Aspergillus fumigatus* is more potent than that of aculeacin A (Table 15.3). However, FR901379 is only weakly active against *A. fumigatus*. None of these compounds exert antifungal activity against *Cryptococcus neoformans*.

Table 15.4 shows the therapeutic effect of FR901379 in a mouse *C. albicans* infection model. The compounds were administered subcutaneously for four consecutive days. FR901379 and related compounds significantly prolonged

Table 15.4 *In vivo* efficacy in a neutropenic mouse model of disseminated candidiasis.[a]

Compound	ED_{50} *(mg/kg)*
FR901379	2.7
Aculeacin A	6.4
Fluconazole	4.5

[a]Infection: *Candida albicans* FP633

Table 15.5 Haemolytic activity.

Compound	*MLC*[a] *(μg/ml)*
FR901379	62
Aculeacin A	31
Echinocandin B	125
Amphotericin B	8

[a]Minimum lytic concentration

the survival of infected mice. FR901379 was the most potent compound with an ED_{50} value of 2.7 mg/kg on day 14 after challenge. This value was almost comparable to that of fluconazole. Despite its good water solubility and potent activity against fungi, low concentrations of FR901379 lysed red blood cells (Table 15.5). Although the lytic activity of FR901379 was weaker than that of amphotericin B, it was still too high for clinical use.

The strain producing FR901379 was isolated originally from a soil sample collected at Iwaki-City, Fukushima Prefecture, Japan. Because this strain only produced conidial structures when grown on a leaf, morphological characteristics were determined using cultures grown on a sterilised azalea leaf affixed to a Miura's LCA plate. It was identified as *Coleophoma empetri* F-11899 (Figure 15.4).

2.2 Generation of Lead Compound FR131535

FR901379 is a highly selective antifungal agent and an inhibitor of 1,3-β-glucan synthase. Despite its haemolytic activity, this compound offers some distinct advantages over other analogues, one of which is good water solubility. Therefore, focus was placed on transforming its acyl side chain to reduce the haemolytic activity, while keeping the sodium sulfate group responsible for its solubility intact. To that end, we attempted to replace the acyl side chain, just as Lilly's researchers had done previously.[10]

Initial modification of the acyl side chain yielded FR131535.[11] The synthesis of this novel echinocandin-like lipopeptide is outlined in Figure 15.5. The palmitoyl group was removed from FR901379 by treating it with acylase from *Actinoplanes utahensis*, which yielded FR179642. A new acyl side chain was prepared starting with 1-bromooctane and 4-hydroxybenzoic acid.

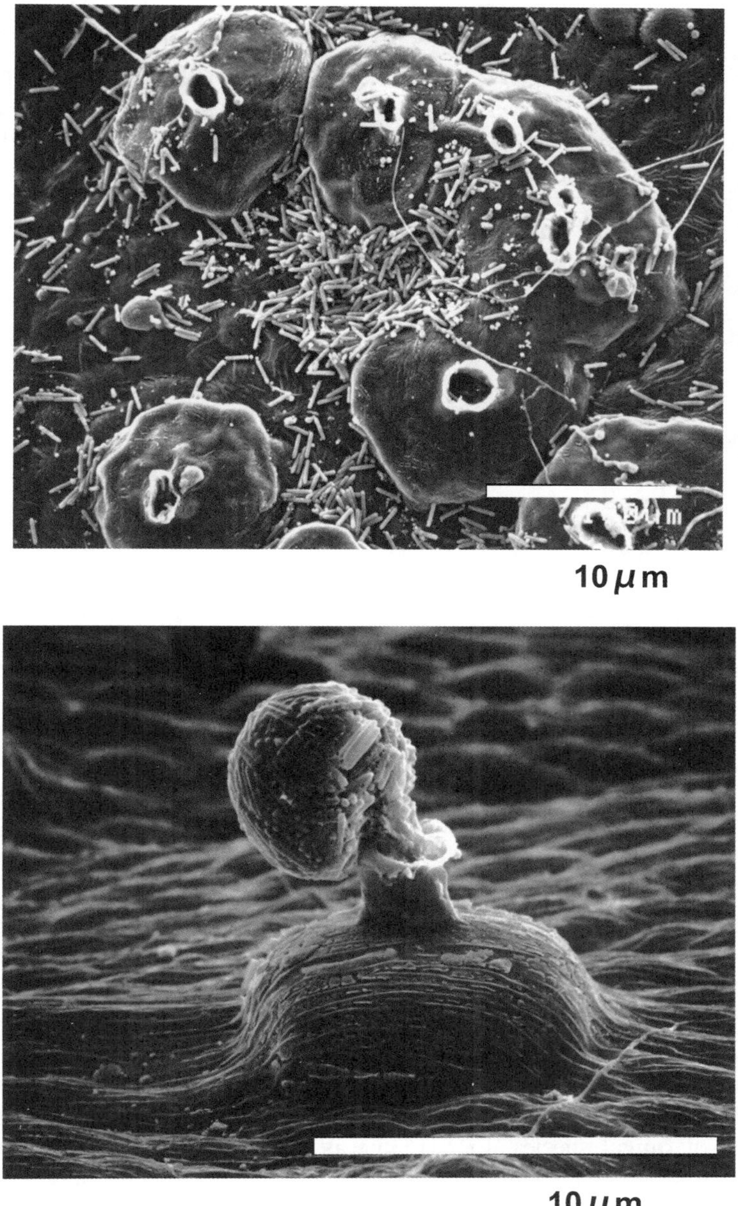

Figure 15.4 Scanning electron micrograph of *Coleophoma empetri* F-11899.

2,4,5-Trichlorophenyl 4-(*n*-octyloxy) benzoate was then obtained from 4-(*n*-octyloxy) benzoic acid and 2,4,5-trichlorophenol using *N*′-dicyclohexylcarbodiimide in ether. The reacylation of FR179642 was then carried out using the 2,4,5-trichlorophenoyl active ester to yield FR131535.

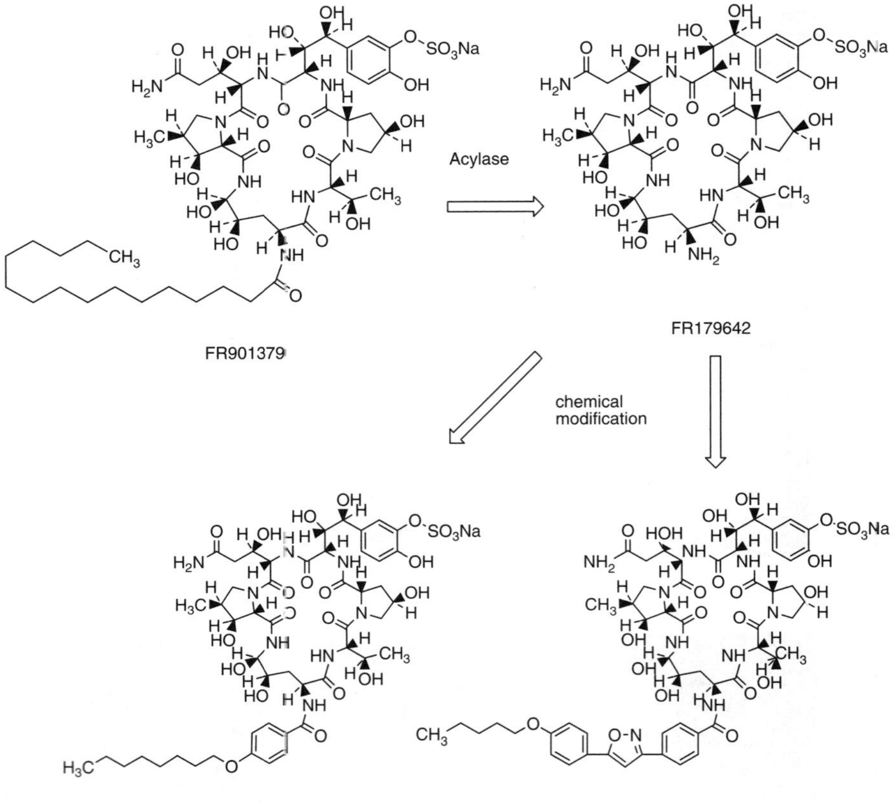

Figure 15.5 Semi-synthesis of FR131535 and micafungin (FK463).

The water solubility of FR131535 remained as high as that of FR901379, even after replacement of the acyl side chain. Echinocandin B and cilofungin did not dissolve in water under the same conditions. FR131535 non-competitively (Ki: 4.0 μM) inhibited 1,3-β-glucan synthase prepared from *C. albicans* 6406 with an IC_{50} value of 2.8 μg/mL. When tested using the microbroth dilution method, this compound displayed potent broad spectrum activity against a variety of fungal species. FR131535 was active against most *Candida* and *Aspergillus* species. The protective efficacy of FR131535 administered subcutaneously to mice systemically infected with *C. albicans* was examined. As shown in Table 15.6, the ED_{50} of FR131535 was 3.7 mg/kg. This compound was superior to echinocandin B and cilofungin in the above model. Furthermore, the *in vivo* efficacy of FR131535 was almost as potent as that of fluconazole, which is fungistatic against fungal pathogens. FR131535 is an inhibitor of cell wall biosynthesis, fungicidal against *Candida* species and shows potent *in vivo* activity against *A. fumigatus* (ED_{50} 4.3 mg/kg). Since there were no reports of echinocandins with good anti-*Aspergillus* activity at that time, this

Table 15.6 Influence of acyl side chain group.

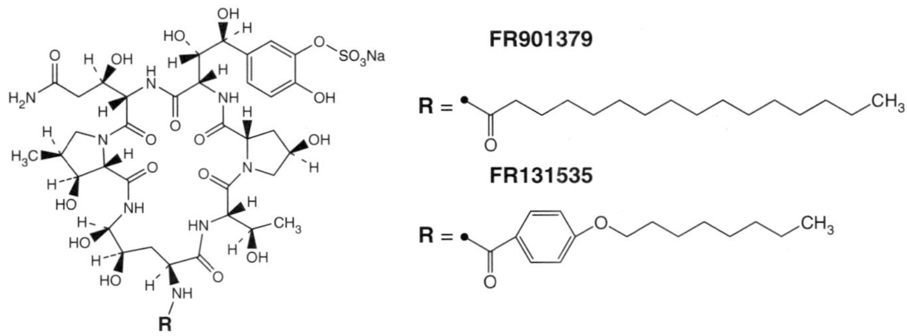

	C. albicans FP633		*A. fumigatus FP1305*	*Haemolysis*
Compound	*MIC (ug/ml)*	ED_{50} *(mg/kg)*	ED_{50} *(mg/kg)*	$LC_{30}{}^{a}$ *(mg/ml)*
FR901379	0.2	1.8	70	0.456
FR131535	0.78	3.7	4.3	>8

[a]Lytic concentration 30%

result encouraged us to expand our project. In addition, the haemolytic activity of FR131535 was significantly lower compared with that of FR901379 (Table 15.6). Our chemists were also especially encouraged by these results and, therefore, the synthesis and evaluation of derivatives with novel acyl side chains became the focus.

2.3 Lead Optimisation Leading to the Discovery of FK463[12,13]

Since replacing the acyl side chain caused the antifungal spectrum to expand to include *Aspergillus* spp., the relationship between the lipophilicity of the side chain and antifungal activity was examined next. Naphthalene side chain derivatives were chosen as the initial acyl side chains as they are compact and modify lipophilicity. Meanwhile, the relationship between antifungal activity and haemolysis was examined by varying the length of the alkyl chain to change the lipophilicity.

As shown in Figure 15.6, an increase in lipophilicity resulted in improved anti-*Candida* activity, which was most potent with an octyloxy group ($n = 7$). Furthermore, *in vivo* studies in mice reflected the *in vitro* antifungal activity. As a tool to aid analogue design, the ClogP value [octanol–water partition coefficient (calculated value)], which is a measure of the lipophilicity of the side chains, correlated well with anti-*Candida* activity. The strongest *in vivo* effect was obtained when the ClogP value was set at approximately 6. However, the longer alkyl chains resulted in greater haemolysis. This correlation allowed us to design novel side chains with enhanced activity; these chains were then synthesised to adjust the lipophilicity, measured by ClogP, to approximately 6. Conversion of the benzene ring in the aromatic side chain moiety of FR131535

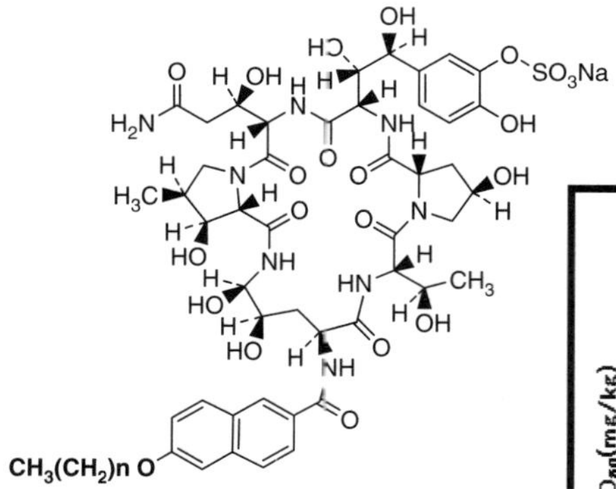

n	*C. albicans* FP633 MIC (mg/ml)	Haemolysis (% at 2mg/ml)
3	12.5	1
5	0.78	5
6	0.39	-
7	0.1	34
9	0.2	100
11	0.78	100

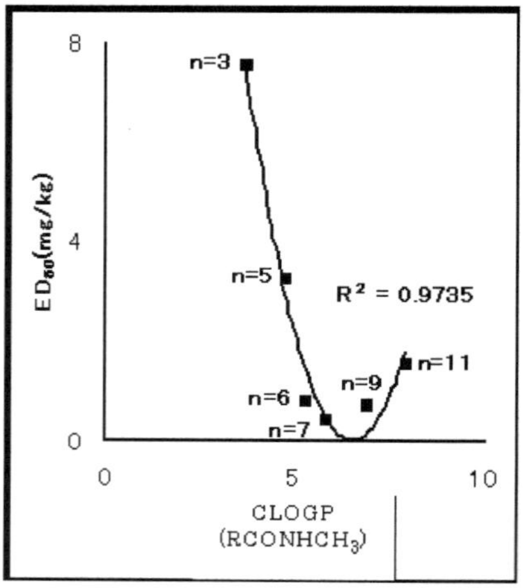

Figure 15.6 Effect of lipophilicity on MIC and haemolysis.

into a naphthalene ring improved anti-*Candida* activity; further introduction of aromatic rings into the side chain also resulted in increased activity (Table 15.7).

Anti-*Candida* activity tended to improve as the number of benzene rings increased, with compound **7** having the lowest minimum inhibitory concentration (MIC). The anti-*Aspergillus* activity of compound **4**, which contains a naphthalene ring, was much greater than that of FR131535. However, we encountered another problem: even though compound **7** had the lowest MIC, its *in vivo* effect [ED_{50} ratio (0.14)] was only slightly better than that of compound **5** [ED_{50} ratio (0.2)], which means that the MIC of compound **5** was five times lower than that of compound **7**.

To improve the *in vivo* effect (ED_{50}), we attempted to synthesise compounds with lower MIC values. Adding mouse serum to the medium when measuring the MIC allowed us to find a correlation between *in vivo* effect (ED_{50} ratio) and *in vitro* activity (serum MIC ratio). The addition of serum increased the MICs as a direct consequence of the reduced availability of the unbound form due to serum binding. After this finding, prediction of *in vivo* effects became feasible by measuring the serum MIC of synthetic derivatives, which allowed rapid establishment of structure-activity relationships. Consequently, this structure–activity correlation revealed that compound **7** type derivatives, which contain three linearly linked aromatic rings, have strong anti-*Candida* and anti-*Aspergillus* activity. However, these analogues were still haemolytic.

We tried to solve this problem by converting the central benzene ring of compound **7** into various heterocycles (Table 15.8). Initially, the reduction of

Table 15.7 Side chain modification part 1(introduction of aromatic rings).

No.	Compound: Acyl side-chain group	CLOGP	*C. albicans FP633*: MIC ($\mu g/ml$)[a]	*C. albicans FP633*: Serum MIC ($\mu g/ml$)[b]	*C. albicans FP633*: ED_{50} (mg/kg)[c]	*A. fumigatus FP1305*: ED_{50} (mg/kg)	Haemolysis: LC_{30} (mg/ml)
1		4.77	0.78(1)	25(1)	1.5–4.3(1)	4.31	>10
2	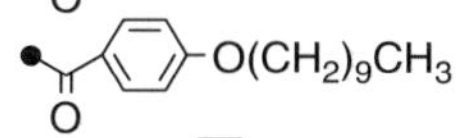	5.80	0.2(0.26)	–	1(0.7)	22.9	>10
3	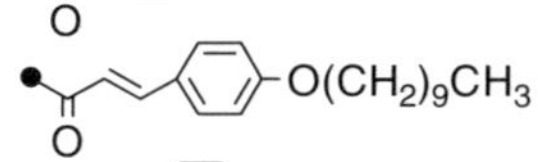	5.38	0.78(1)	–	4.3(1)	–	>10
4	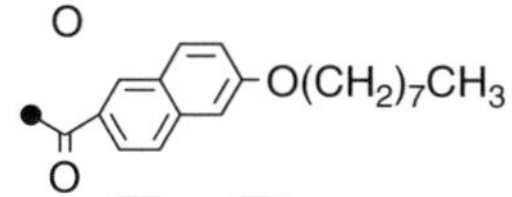	5.80	0.1(0.13)	6.25(0.25)	0.742(0.23)	0.788	10
5	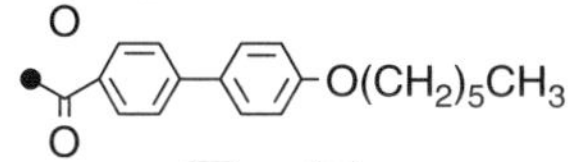	5.37	0.2(0.26)	6.25(0.25)	0.658(0.2)	–	1.74
6	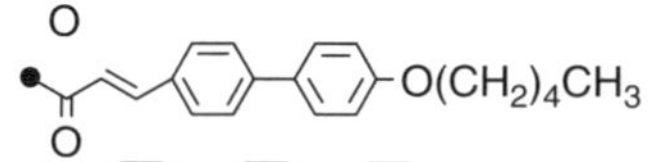	5.68	0.05(0.06)	3.13(0.13)	0.563(0.3)	0.894	3.95
7	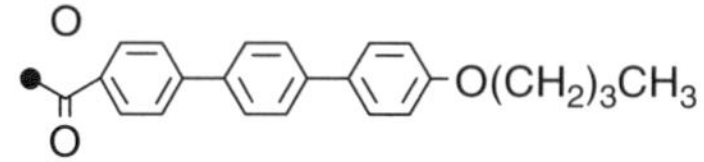	6.14	0.0125(0.02)	1.56(0.06)	0.447(0.14)	–	0.37

[a]Figures in parentheses indicates the ratio of MIC(ED_{50})(drug)/MIC(ED_{50})(FR131535)
[b]Represents the range of values of ED_{50} for FR131535 over a number of experiments
[c]Lytic Concentration 30%

Table 15.8 Side chain modification part 2 (introduction of heterocycles).

Compound			*C. albicans FP633*		*A. fumigatus FP1305*	
No.	*Acyl side-chain group*	*CLOGP*	*Serum MIC (μg/ml)*[a]	*ED_{50} (mg/kg)*[b]	*ED_{50} (mg/kg)*	*Haemolysis (%, 1mg/ml)*
1		4.77	25(1)	1.5–4.3(1)	4.31	<20
7	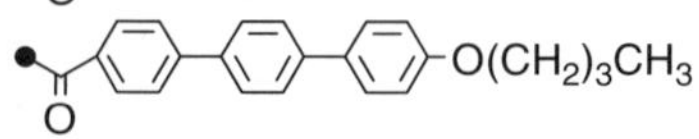	6.14	1.56(0.06)	0.447(0.14)	–	79
8	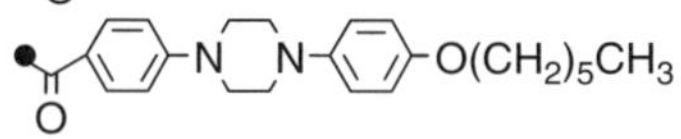	6.16	0.78(1)	4.3(1)	0.53(0.15)	<20
9	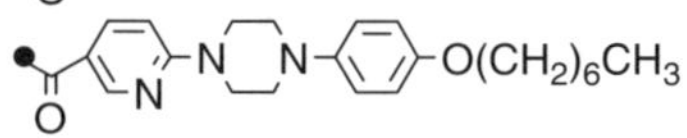	6.29	0.1(0.13)	0.742(0.23)	–	<20
FK463	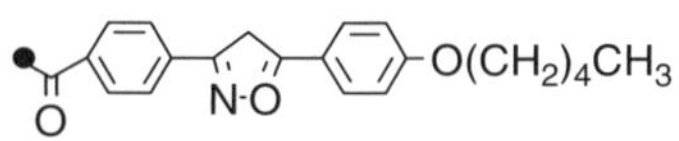	5.31	0.2(0.26)	0.658(0.2)	0.228(0.06)	<20
10	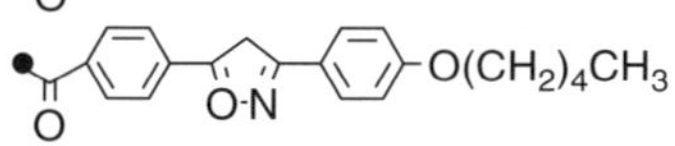	5.31	0.05(0.06)	0.563(0.3)	–	38
11	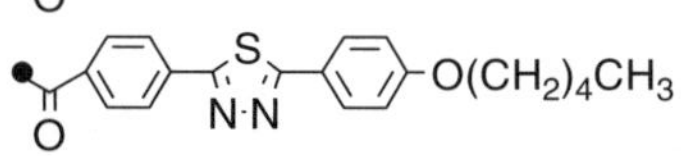	6.24	0.0125(0.02)	0.447(0.14)	–	82

[a]Figures in parentheses indicate the ratio of $MIC(ED_{50})$(drug)/$MIC(ED_{50})$(FR131535)
[b]Represents the range of values of ED_{50} for FR131535 over a number of experiments

haemolytic activity seemed difficult because FR901379 derivatives have an amphiphilic structure and surfactant-like activity. However, we found that the amount of branched fatty acids in the cell membranes of erythroid cells and eukaryotic cells are different. Therefore, we attempted to reduce the haemolytic potential by decreasing the linearity of the acyl side chains. As expected, the introduction of a heterocycle into the acyl side chain lessened the haemolytic activity of compound **7** without reducing its potent antifungal activity. Finally, the cyclic peptide nucleus FR179642, obtained through enzymatic cleavage of the natural product FR901379, was reacylated with a new side chain containing an isoxazole ring to yield FK463 (later named micafungin) (Figure 15.5).

2.4 Preclinical Studies of FK463

Of all the candidate compounds prepared, FK463 had the most potent *in vivo* effect against *Candida* and *Aspergillus*. The efficacy of FK463 was evaluated in neutropenic mouse models of disseminated candidiasis and aspergillosis, and was compared with those of amphotericin B and fluconazole.[14] Table 15.9 shows the ED_{50} calculated on the basis of survival rate 15 days after infection. The ED_{50} of FK463 against disseminated infections of *C. albicans*, *C. glabrata*, *C. tropicalis* and *C. krusei* ranged from 0.14 to 0.77 mg/kg. Although these efficacy values were 1.4–3.1 times weaker than those of amphotericin B (0.09–0.26 mg/kg), they were 9.6 to >77 times stronger than those of fluconazole. The ED_{50} of FK463 against disseminated *C. parapsilosis* infection was 1.0 mg/kg, which was 11 times more potent than that of fluconazole (10.9 mg/kg) and 18 times weaker than that of amphotericin B (0.06 mg/kg). FK463 showed good activity against disseminated *A. fumigatus* infection, with an ED_{50} in the range 0.25–0.50 mg/kg. The efficacy of FK463 was 1.7–2.3 times inferior to that of amphotericin B (0.11–0.29 mg/kg) and >80 times superior to that of fluconazole. These results indicate that micafungin is a potent parenteral therapeutic

Table 15.9 *In vivo* efficacy of micafungin (FK463) in neutropenic mouse model of disseminated candidiasis and aspergillosis.

	ED_{50} (mg/kg)[a]		
Organisms	*FK463*	*Fluconazole*	*Amphotericin B*
Candida albicans FP633	0.14	2.15	0.08
C. albicans 16010	0.21	4.51	0.12
C. albicans FP1839[b]	0.26	>20.0	0.18
C. glabrata 13002	0.30	6.27	0.11
C. tropicalis 16009	0.28	3.71	0.09
C. krusei Fp1866	0.77	9.52	0.26
C. parapsilosis FP1946	1.00	10.9	0.06
A. fumigatus TIMM0063	0.25	>20.0	0.11
A. fumigatus IFM41209	0.50	>20.0	0.29

[a]Once daily treatment for 4 days, starting at 1 h after infection
[b]Fluconazole resistant

agent for disseminated candidiasis and aspergillosis in the neutropenic mice model.

2.5 Industrial Manufacturing of Micafungin

In 1990, the Fermentation Development Laboratories at Fujisawa commenced the following developmental research steps to establish an industrial manufacturing method for micafungin:

(1) Strain improvement of *Coleophoma empetri* F-11899.
(2) Screening of new acylase (FR901379 acylase) producing microorganisms.
(3) Studies to scale up the fermentation process for FR901379 and "FR901379 acylase".
(4) Determination of effective purification procedures for FR901379, a key intermediate of FR179642 and FK463.
(5) Development of a high-performance liquid chromatography (HPLC) assay for measuring the amount of objective compounds and impurities.

At first, *Actinoplanes utahensis* was used as an acylase source, but the identification of a highly active enzyme-producing strain was necessary in order to supply the large amount of FR179642 needed. Fujisawa's original panel of acylase-producing microorganisms was, therefore, screened using a specially devised effective screening system. As a result, a wild strain, *Streptomyces* sp. No. 6907, was found which produced a new FR901379 acylase, which catalysed the deacylation ten times faster than original *A. utahensis* strain.[15]

2.6 Clinical Studies of FK463

Many of the clinical studies conducted so far have examined the efficacy of micafungin as a prophylaxis and treatment for mycoses. One of representative studies on the treatment of invasive aspergillosis (IA) is summarised below.[16]

A multinational, non-comparative study was conducted to examine proven or probable *Aspergillus* species infection in a wide variety of patients. The study employed an open-label design utilising micafungin alone or in combination with another systemic antifungal agent. Criteria for IA and therapeutic response were judged by an independent panel.

Of the 331 patients enrolled, only 225 met the diagnostic criteria for IA as determined by the independent panel. These participants received at least one dose of micafungin. Out of the 225 qualifying patients, 98 had undergone haematopoietic stem cell transplantation (HSCT) (88/98 allogeneic), 48 had undergone graft *versus* host disease (GVHD) and 83 received chemotherapy for haematological malignancy. A favourable response rate at the end of therapy was seen in 35.6% (80/225) of patients. Of those treated with micafungin alone, favourable responses were seen in 6/12 (50%) of the primary and 9/22 (40.9%) of the salvage therapy group, with corresponding numbers of 5/17 (29.4%) and 60/174 (34.5%) for the combination treatment groups. Of the 326 patients treated with micafungin, 183 (56.1%) died during therapy or during the six-week follow-up phase, 107 (58.5%) of which were attributable to IA.

Micafungin as primary or salvage therapy proved efficacious and safe in high-risk patients with IA.

3 Conclusions

This chapter describes the discovery of micafungin, which was the result of screening for novel antifungals from microbial products.[17] We believe that this discovery is not simply a fortuitous event, but rather the fruit of long-term efforts and enthusiasm fuelled by the initial discovery of pyrrolnitrin.

Micafungin is a semi-synthetic compound that is superior to the original product, FR901379 and exerts potent activity against not only *C. albicans*, but also *A. fumigatus*. Furthermore, micafungin is water-soluble and without the haemolytic activity seen with FR901379. Micafungin is marketed in Japan, North America and the EU as a "candin-class" parenteral antifungal agent for life-threatening mycoses.

Acknowledgements

We are honoured to have been involved in the discovery of micafungin and to be able to contribute this chapter. We express our sincere appreciation to the many colleagues at Fujisawa Pharmaceutical Co., Ltd who participated in the discovery and development of micafungin.

References

1. K. Arima, H. Imanaka, M. Kohsaka, A. Fukuda and G. Tamura, *Agr. Biol. Chem.*, 1964, **28**, 575.
2. T. Iwamoto, E. Tsujii, M. Ezaki, A. Fujie, S. Hashimoto, M. Okuhara, M. Kohsaka, H. Imanaka, K. Kawabata, Y. Inamoto and K. Sakane, *J. Antibiot. (Tokyo)*, 1990, **43**, 1.
3. M. Yamashita, Y. Tsurumi, J. Hosoda, T. Komori, M. Kohsaka and H. Imanaka, *J. Antibiot. (Tokyo)*, 1984, **37**, 1279.
4. T. Komori, M. Yamashita, Y. Tsurumi and M. Kohsaka, *J. Antibiot. (Tokyo)*, 1985, **38**, 455.
5. T. Iwamoto, A. Fujie, Y. Tsurumi, K. Nitta, S. Hashimoto and M. Okuhara, *J. Antibiot. (Tokyo)*, 1990, **43**, 1183.
6. T. Iwamoto, A. Fujie, K. Sakamoto, Y. Tsurumi, N. Shigematsu, M. Yamashita, S. Hashimoto, M. Okuhara and M. Kohsaka, *J. Antibiot. (Tokyo)*, 1994, **47**, 1084.
7. A. Fujie, T. Iwamoto, H. Muramatsu, T. Okudaira, K. Nitta, T. Nakanishi, K. Sakamoto, Y. Hori, M. Hino, S. Hashimoto and M. Okuhara, *J. Antibiot. (Tokyo)*, 2000, **53**, 912.

8. M. Debono, W. W. Turner, L. LaGrandeur, F. J. Burkhardt, J. S. Nissen, K. K. Nichols, M. J. Rodriguez, M. J. Zweifel, D. J. Zeckner, R. S. Gordee, J. Tang and R. P. Thomas, *J. Med. Chem.*, 1995, **38**, 3271.
9. K. Bartizal, C. J. Gill, G. K. Abruzzo, A. M. Flattery, L. Kong, P. M. Scott, J. G. Smith, C. E. Leighton, A. Bouffard, J. F. Dropinski and J. Balkovec, *Antimicrob. Agents Chemother.*, 1997, **41**, 2326.
10. M. Debono, B. J. Abbott, D. S. Fukuda, M. Barnhart, K. E. Willard, R. M. Molloy, K. H. Michel, J. R. Turner, T. F. Butler and A. H. Hunt, *J. Antibiot. (Tokyo)*, 1989, **42**, 389.
11. A. Fujie, T. Iwamoto, B. Sato, H. Muramatsu, C. Kasahara, T. Furuta, Y. Hori, M. Hino and S. Hashimoto, *Bioorg. Med. Chem. Lett.*, 2001, **11**, 399.
12. D. Barrett, *Biochim. Biophys. Acta*, 2002, **1587**, 224.
13. M. Tomishima, H. Ohki, A. Yamada, K. Maki and F. Ikeda, *Bioorg. Med. Chem. Lett.*, 2008, **18**, 2886.
14. F. Ikeda, Y. Wakai, S. Matsumoto, K. Maki, E. Watabe, S. Tawara, T. Goto, Y. Watanabe, F. Matsumoto and S. Kuwahara, *Antimicrob. Agents Chemother.*, 2000, **44**, 614.
15. S. Ueda, M. Tanaka, M. Ezaki, K. Sakamoto, S. Hashimoto, N. Oohata, M. Tsuboi and M. Yamashita, *US Patent* 6,537,789, 2003.
16. D. Denning, K. Marr, W. Lau, D. Facklam, V. Ratanatharathorn, C. Becker, A. Ullman, N. Seibel, P. Flynn and J. van Burik, *J. Infect.*, 2006, **53**, 337.
17. A. Fujie, *Pure Appl. Chem.*, 2007, **79**, 603.

Subject Index

Note: page numbers in *italics* refer to figures and tables